KING'S COLLEGE LONDON
INFORMATION SERVICES CENTRE
FRANKLIN WILKINS BUILDING
150 STAMFORD STREET
LONDON SE1 9NN

Thermodynamics of Biochemical Reactions

Thermodynamics of Biochemical Reactions

Robert A. Alberty

Massachusetts Institute of Technology
Cambridge, MA

WILEY-
INTERSCIENCE

A John Wiley and Sons Publication

Published by John Wiley & Sons, Inc., Hoboken, New Jersey.
Published simultaneously in Canada.

For general information on our other products and services please contact our Customer Care
Department within the U.S. at 877-762-2974, outside the U.S. at 317-572-3993 or fax 317-572-4002.

Wiley also publishes its books in a variety of electronic formats. Some content that appears in print,
however, may not be available in electronic format.

Library of Congress Cataloging-in-Publication Data:
   Alberty, Robert A.
      Thermodynamics of biochemical reactions / Robert A. Alberty.
         p. cm.
   Includes bibliographical references and index.
   ISBN 0-471-22851-6 (cloth : acid-free paper)
   1. Thermodynamics. 2. Physical biochemistry. 3. Bioenergetics. I.
   Title.
      QP517.T48 A42 2003
      547'.2--dc21   2002155481

Printed in the United States of America.

10 9 8 7 6 5 4 3 2 1

Contents

Part Two: *Mathematica* Solutions to Problems

BasicBiochemData2.nb

Solutions to Problems

Index of *Mathematica* Programs

Preface

This book is about the thermodynamics of enzyme-catalyzed reactions that make up the metabolism of living organisms. It is not an introductory text, but the fundamental principles of thermodynamics are reviewed. The reader does need some background in thermodynamics, such as that provided by a first course in physical chemistry. The book uses a generalized approach to thermodynamics that makes it possible to calculate the effects of changing pH, free concenrations of metal ions that are bound by reactants, and steady-state concentrations of coenzymes. This approach can be extended to other types of work that may be involved in a living organism.

The concepts involved in this approach are simple, but the equations become rather complicated. Biochemical reactions are written in terms of reactants like ATP that are made up of sums of species, and they are referred to as biochemical reactions to differentiate them from the underlying chemical reactions that are written in terms of species. The thermodynamics of biochemical reactions is independent of the properties of the enzymes that catalyze them. However, the fact that enzymes may couple reactions that might otherwise occur separately increases the number of constraints that have to be considered in thermodynamics.

Biochemical thermodynamics is complicated for several reasons: (1) Biochemical reactants consist of sums of species whenever a reactant has a pK within about two units of the pH of interest or binds metal ions reversibly. (2) Species of a biochemical reactant are often ions, and the activity coefficients of ions are functions of ionic strength. (3) Enzyme catalysis may introduce constraints in biochemical reactions in addition to balances of atoms of elements. (4) Metabolism is sufficiently complicated that it is important to find ways to obtain a more global view. (5) In biochemistry other kinds of work, such as electric work, elongation work, and surface work may be involved. It is remarkable that the same basic reactions are found in all living systems. The most important thing about these reactions is that they provide the means to carry out the oxidation of organic matter in a sequence of steps that store energy that is needed for the synthesis of organic molecules, mechanical work, and other functions required for life.

The theme of this book is that Legendre transforms make the application of thermodynamics more convenient for the users. The logic used here is a continuation of the process described by Gibbs that introduced the enthalpy H,

Helmholtz energy A, and the Gibbs energy G by use of Legendre transforms of the internal energy U. In Chapter 4 a Legendre transform is used to introduce pH and pMg as independent intensive variables. In Chapter 6 the steady-state concentrations of various coenzymes are introduced as independent intensive variables in discussing systems of enzyme-catalyzed reactions. In Chapter 8 a Legendre transform is used to introduce the electric potential of a phase as an independent intensive variable. These uses of Legendre transforms illustrate the comment by Callen (1985) that "The choice of variables in terms of which a given problem is formulated, while a seemingly innocuous step, is often the most crucial step in the solution." Choices of dependent and independent variables are not unique, and so choices can be made to suit the convenience of the experimenter. Gibbs has provided a mathematical structure for thermodynamics that is expandable in many directions and is rich in interrelationships between measurable properties because thermodynamic properties obey all the rules of calculus.

This book on thermodynamics differs from others in its emphasis on the fundamental equations of thermodynamics and the application of these equations to systems of biochemical reactions. The emphasis on fundamental equations leads to new thermodynamic potentials that provide criteria for spontaneous change and equilibrium under the conditions in a living cell. The equilibrium composition of a reaction system involving one or more enzyme-catalyzed reactions usually depends on the pH, and so the Gibbs energy G does not provide the criterion for spontaneous change and equilibrium. It is necessary to use a Legendre transform to define a transformed Gibbs energy G' that provides the criterion for spontaneous change and equilibrium at the specified pH. This process brings in a transformed entropy S' and a transformed enthalpy H', but this new world of thermodynamics is similar to the familiar world of G, S, and H, in spite of the fact that there are significant differences.

Since coenzymes, and perhaps other reactants, are in steady states in living cells, it is of interest to use a Legendre transform to define a further transformed Gibbs energy G'' that provides the criterion for spontaneous change and equilibrium at a specified pH and specified concentrations of coenzymes. This process brings in a further transformed entropy S'' and a further transformed enthalpy H'', but the relations between these properties have the familiar form.

Quantitative calculations on systems of biochemical reactions are sufficiently complicated that it is necessary to use a personal computer with a mathematical application. *Mathematica*® (Wolfram Research, Inc. 100 World Trade Center, Champaign, IL, 61820-7237) is well suited for these purposes and is used in this book to make calculations, construct tables and figures, and solve problems. The last third of the book provides a computer-readable database, programs, and worked-out solutions to computer problems. The database BasicBiochemData2 is available on the Web at http://www.mathsource.com/cgi-bin/msitem?0211-662.

Systems of biochemical reactions can be represented by stoichiometric number matrices and conservation matrices, which contain the same information and can be interconverted by use of linear algebra. Both are needed. The advantage of writing computer programs in terms of matrices is that they can then be used with larger systems without change.

This field owes a tremendous debt to the experimentalists who have measured apparent equilibrium constants and heats of enzyme-catalyzed reactions and to those who have made previous thermodynamic tables that contain information needed in biochemical thermodynamics.

Although I have been involved with the thermodynamics of biochemical reactions since 1950, I did not understand the usefulness of Legendre transforms until I had spent the decade of the 1980s working on the thermodynamics of petroleum processing. During this period I learned from Irwin Oppenheim (MIT) and Fred Krambeck (Mobil Research and Development) about Legendre transforms, calculations using matrices, and semigrand partition functions. In the 1990s I returned to biochemical thermodynamics and profited from many helpful discussions with Robert N. Goldberg (NIST). The new nomenclature that is used here was recommended by an IUPAC-IUBMB report by Alberty, Cornish-Bowden, Gibson, Goldberg, Hammes, Jencks, Tipton, and Veech in 1994. The use of Legendre transforms in chemical thermodynamics is the subject of an IUPAC Technical Report by Alberty, Barthel, Cohen, Ewing, Goldberg, and Wilhelm (2001). I am indebted to NIH for award 5-R01-GM48458 for support of my research on biochemical thermodynamics. My Associate Managing Editor, Kristin Cooke Fasano of John Wiley and Sons, was very helpful.

Robert A. Alberty
Cambridge, Massachusetts

Introduction to Apparent Equilibrium Constants

Two types of equilibrium constant expressions are needed in biochemistry. The thermodynamics of biochemical reactions can be discussed in terms of species like ATP^{4-}, $HATP^{3-}$, and $MgATP^{2-}$ or in terms of reactants (sums of species) like ATP. The use of species corresponds with writing chemical reactions that balance atoms of elements and electric charges; the corresponding equilibrium constants are represented by K. This approach is required when chemical details are being discussed, as in considering the mechanism of enzymatic catalysis. But discussion in terms of metabolism must involve, in great deal detail, acid dissociation constants and dissociation constants of complexes with metal ions. Therefore metabolism is discussed by writing biochemical reactions in terms of reactants—that is, sums of species, like ATP—at a specified pH and perhaps specified concentrations of free metal ions that are bound reversibly by reactants. Biochemical reactions do not balance hydrogen ions because the pH is held constant, and they do not balance metal ions for which free concentrations are held constant. When the pH is held constant, there is the implication that acid or alkali will be added to the system to hold the pH constant if the reaction produces or consumes hydrogen ions. In actual practice a buffer is used to hold the pH nearly constant, and the pH is measured at equilibrium. The corresponding equilibrium constants are represented by K', which are referred to as apparent equilibrium constants because they are functions of pH and perhaps the free concentrations of one or more free metal ions. Biochemical thermodynamics is more complicated than the chemical thermodynamics of reactions in aqueous solutions because there are more independent variables that have to be specified. This introductory chapter is primarily concerned with the hydrolysis of ATP at specified T, P, pH, pMg, and ionic strength. The thermodynamics of the hydrolysis of ATP and closely related reactions have received a good deal of attention because of the importance of these reactions in energy metabolism.

■ 1.1 BRIEF HISTORY OF THE THERMODYNAMICS OF BIOCHEMICAL REACTIONS

The first major publication on the thermodynamics of biochemical reactions was by Burton in Krebs and Kornberg, *Energy Transformations in Living Matter*, 1957. Before that time, apparent equilibrium constants had been measured for a number of enzyme-catalyzed reactions, but Burton recognized that these apparent equilibrium constants together with standard Gibbs energies of formation $\Delta_f G^0$ of species determined by chemical methods can yield $\Delta_f G^0$ for biochemical species to make a table that can be used to calculate equilibrium constants of biochemical reactions that have not been studied (Burton and Krebs, 1953). In retrospect it is easy to see that in 1953 to 1957 there were some problems that were apparently not clearly recognized or solved. Since Burton was the first, it is worth saying a little more about his 1957 thermodynamic tables. The first table gives $\Delta_f G^0$ values for about 100 species in biochemical reactions. A large number of these values were taken from chemical thermodynamic tables available in the 1950s, but a number were new values calculated from measured apparent equilibrium constants for enzyme-catalyzed reactions. $\Delta_f G^0$ values of species can be readily calculated when the reactants in the enzyme-catalyzed reaction are all single species and $\Delta_f G^0$ values are known for all of the reactants except one. It is noteworthy that Burton omitted the species of orthophosphate from his table and that he was not able to include species of ATP, ADP, NAD_{ox}, and NAD_{red}. His second table gives standard Gibbs energy changes at pH 7 for oxidation-reduction reactions that were calculated using the convention that $[H^+] = 1$ mol L^{-1} at pH 7; the symbol $\Delta G'$ was used for this quantity. This table also gives the corresponding standard cell potentials for these reactions. The third table gives $\Delta G'$ values at pH 7 for a number of reactions in glycolysis and alcoholic fermentation. The fourth table is on the citric acid cycle, and the fifth table is on Gibbs energies of hydrolysis. When a biochemical reaction is studied at a pH where there is a predominant chemical reaction, it is possible to discuss thermodynamics in terms of species. But when some reactants are represented by an equilibrium distribution of several species with different numbers of hydrogen atoms, this approach is not satisfactory. The quantitative treatment of reactions involving reactants with pKs in the neighborhood of pH 7 was not possible until acid dissociation constants of these reactants had been determined. Some measurements of acid dissociation constants of ATP and related substances (Alberty, Smith, and Bock, 1951) and dissociation constants of ionic complexes of these substances with divalent cations (Smith and Alberty, 1956) were made in this period.

In the 1960s there was a good deal of interest in the thermodynamics of the hydrolysis of ATP and of other organic phosphates (Alberty, 1968, 1969; Phillips, George, and Rutman, 1969), but standard Gibbs energies of species were not calculated. The measurement of apparent equilibrium constants for biochemical reactions was extended in the 1970s (Guynn and Veech, 1973; Veech et al., 1979) and 1980s (Tewari and Goldberg, 1988).

In 1969 Wilhoit picked up where Burton had left off and compiled the standard thermodynamic properties $\Delta_f G^0$ and $\Delta_f H^0$ of species involved in biochemical reactions. He recognized the problems involved in including species of ATP in such a table and made a suggestion as to how to handle it. In 1977 Thauer, Jungermann, and Decker published a table of standard Gibbs energies of formation of many species of biochemical interest, and showed how to adjust standard Gibbs energies of reaction to pH 7 by adding $m\Delta_f G^0(H^+)$, where m is the net number of protons in the reaction.

During the 1960s and 1970s, new nomenclature for treating the thermodynamics of biochemical reactions was developed, including the use of K' for the apparent equilibrium constant written in terms of sums of species, but omitting $[H^+]$. These changes led to the publication of *Recommendations for Measurement and Presentation of Biochemical Equilibrium Data* by an IUPAC-IUB Committee (Wadsö, Gutfreund, Privalov, Edsall, Jencks, Armstrong, and Biltonen, 1976).

Goldberg and Tewari published an evaluation of thermodynamic and transport properties of carbohydrates and their monophosphates in 1989 and of the ATP series in 1991. Miller and Smith-Magowan published on the thermodynamics of the Krebs cycle and related compounds in 1990.

Alberty (1992a, b) pointed out that when the pH or the free concentration of a metal ion is specified, the Gibbs energy G does not provide the criterion for spontaneous change and equilibrium. When intensive variables in addition to the temperature and pressure are held constant, it is necessary to define a transformed Gibbs energy G' by use of a Legendre transform, as discussed in Chapters 2 and 4. This leads to a complete set of transformed thermodynamic properties at specified pH, that is, a transformed entropy S', transformed enthalpy H', and a transformed heat capacity at constant pressure C'_{Pm}. These changes led to the publication of *Recommendations for Nomenclature and Tables in Biochemical Thermodynamics* by an IUPAC-IUBMB Committee (Alberty, Cornish-Bowden, Gibson, Goldberg, Hammes, Jencks, Tipton, Veech, Westerhoff, and Webb, 1994).

This introductory chapter describes the thermodynamics of biochemical reactions in terms of equilibrium constants and apparent equilibrium constants and avoids references to other thermodynamic properties, which are introduced later.

■ 1.2 ACID DISSOCIATION CONSTANTS AND DISSOCIATION CONSTANTS OF COMPLEX IONS

Strictly speaking, equilibrium constant expressions for chemical reactions involving ions in aqueous solutions should be written in terms of activities a_i of species, rather than concentrations. The **activity** of species i is given by $a_i = \gamma_i c_i$, where γ_i is the **activity coefficient**, which is a function of ionic strength. Activity coefficients of neutral molecules are close to unity in dilute aqueous solutions, but the activity coefficients of ions may deviate significantly from unity, depending on their electric charges and the ionic strength. The **ionic strength** of a solution is defined by $I = (\frac{1}{2})\Sigma z_i^2 c_i$, where z_i is the charge on ion i and c_i is its concentration on the molar scale. When dilute aqueous solutions are studied, the ionic strength is under the control of the investigator and is essentially constant when the composition changes during a reaction. Thus it is convenient to take equilibrium constants and other thermodynamic properties to be functions of the ionic strength so that equilibrium constant expressions can be written in terms of concentrations. The exception to this statement is H_2O. In dilute aqueous solutions the convention in thermodynamics is to omit $[H_2O]$ in the expression for the equilibrium constant because its activity remains essentially at unity.

In 1923 Debye and Hückel showed that the activity coefficient γ_i of an ion decreases with increasing ionic strength, according to

$$\log \gamma_i = -Az_i^2 I^{1/2} \qquad (1.2\text{-}1)$$

where $A = 0.510651 \text{ L}^{-1/2} \text{ mol}^{1/2}$ at 298.15 K in water at a pressure of 1 bar. This is referred to as a limiting law because it becomes more accurate as the ionic strength approaches zero. At ionic strengths in the physiological range, 0.05 to 0.25 M, there are significant deviations from equation 1.2-1. Of the several ways to extend this equation empirically to provide approximate activity coefficients in the physiological range, the most widely used equation is

$$\log \gamma_i = -\frac{A z_i^2 I^{1/2}}{1 + B I^{1/2}} \qquad (1.2\text{-}2)$$

This is referred to as the extended Debye-Hückel equation. It is an approximation that gives a good fit of data at low ionic strengths (Goldberg and Tewari, 1991) when $B = 1.6 \ \mathrm{L}^{1/2} \ \mathrm{mol}^{-1/2}$. Better fits can be obtained with more complicated equations with more parameters, but these parameters are not known for solutions involved in studying biochemical reactions. The way that thermodynamic properties vary with the ionic strength is discussed in more detail in Section 3.6.

Since hydrogen ions and metal ions, like Mg^{2+}, are often reactants, it is convenient to define the pH_c as $-\log[H^+]$, where c refers to concentrations, and pMg as $-\log[Mg^{2+}]$. However, a glass electrode measures $pH_a = -\log\{\gamma(H^+)[H^+]\}$ where a refers to activity. Thus

$$pH_a = -\log\{\gamma(H^+)\} + pH_c \qquad (1.2\text{-}3)$$

Substituting the extended Debye-Hckel equation in this equation yields (Alberty, 2001d)

$$pH_a - pH_c = \frac{A I^{1/2}}{1 + 1.6 I^{1/2}} \qquad (1.2\text{-}4)$$

The differences between the measured pH_a and the pH_c used in biochemical thermodynamics are given as a function of ionic strength and temperature in Table 1.1.

These are the adjustments to be subtracted from pH_a obtained with a pH meter to obtain pH_c, which is used in the equations in this book. pH_c is lower than pH_a because the ion atmosphere of H^+ reduces its activity. In the rest of the book, the subscript "c" on pH will be omitted so that $pH = -\log[H^+]$.

In considering reactions in biochemical systems it is convenient to move the activity coefficients into the equilibrium constants. For example, the equilibrium constant expression for the dissociation of a weak acid can be written as follows:

$$HA = H^+ + A^- \qquad (1.2\text{-}5)$$

$$K_a(HA) = \frac{a(H^+)a(A^-)}{a(HA)} = \frac{\gamma(H^+)[H^+]\gamma(A^-)[A^-]}{\gamma(HA)[HA]} \qquad (1.2\text{-}6)$$

$$K_c(HA) = \frac{K_a(HA)\gamma(HA)}{\gamma(H^+)\gamma(A^-)} = \frac{[H^+][A^-]}{[HA]} = \frac{10^{-pH}[A^-]}{[HA]} \qquad (1.2\text{-}7)$$

The acid dissociation constant K_a is independent of ionic strength, but the acid dissociation constant K_c depends on the ionic strength, as indicated by equation 1.2-7. The equilibrium constant expression in equation 1.2-7 will be used in the rest of the book, but the subscript "c" will be omitted. This will make it possible for us to deal with concentrations of species, rather than activities.

The same considerations apply to the dissociations of complex ions. For example, the equilibrium expression for the dissociation of a complex ion with a magnesium ion can be written as follows:

$$MgA^+ = Mg^{2+} + A^- \qquad (1.2\text{-}8)$$

$$K_{MgA} = \frac{[Mg^{2+}][A^-]}{[MgA^+]} = \frac{10^{-pMg}[A^-]}{[MgA^+]} \qquad (1.2\text{-}9)$$

Table 1.1 pH_a–pH_c as a Function of Ionic
Strength and Temperature

I/M	$10°C$	$25°C$	$40°C$
0	0	0	0
0.05	0.082	0.084	0.086
0.1	0.105	0.107	0.110
0.15	0.119	0.122	0.125
0.2	0.130	0.133	0.137
0.25	0.138	0.142	0.146

Source: R. A. Alberty, *J. Phys. Chem. B* 105, 7865 (2001).
Copyright 2001 American Chemical Society.

where $pMg = -\log[Mg^{2+}]$ and K_{MgA} is a function of the ionic strength, as well as temperature.

Strictly speaking, equations 1.2-7 and 1.2-9 should have $c°$ in the denominator, where $c° = 1\,M$ is the standard concentration, to make the equilibrium constant dimensionless (Mills et al., 1993). However, the $c°$ is omitted in this book in order to simplify expressions for equilibrium constants. Nevertheless, equilibrium constants are still considered to be dimensionless, so their logarithm can be taken.

In using acid dissociation constants and the dissociation constants of complex ions, it is convenient to take the base 10 logarithms of equations 1.2-7 and 1.2-9 to obtain

$$pH = pK_{HA} + \log\left(\frac{[A^-]}{[HA]}\right) \tag{1.2-10}$$

$$pMg = pM_{MgA} + \log\left(\frac{[A^-]}{[MgA^+]}\right) \tag{1.2-11}$$

where $pK_{HA} = -\log K_{HA}$ and $pK_{MgA} = -\log K_{MgA}$ are functions of ionic strength at constant temperature. Table 1.3 in the last section of this chapter gives the pKs of some weak acids of interest in biochemistry as a function of ionic strength. Note that the effect of ionic strength is larger for acids with larger charges. For polyprotic acids pK_1 applies to the weakest acid group, pK_2 to the second weakest, and so on, in the pH range considered (usually 5 to 9). The calculation of Table 1.3 is based on the extended Debye-Hückel equation.

■ 1.3 BINDING OF HYDROGEN IONS AND MAGNESIUM IONS BY ADENOSINE TRIPHOSPHATE

Acid dissociation constants and dissociation constants of complex ions determine the concentrations of species that are present in a solution at equilibrium under specified conditions. Ionic dissociation reactions occur rapidly and tend to remain at equilibrium during an enzyme-catalyzed reaction. Since ATP (see Fig. 1.1) is the primary carrier of energy in biochemical systems and since a good deal is known about its binding properties, these properties are considered here in some detail.

An ATP ion with four negative charges can bind five hydrogen ions in strongly acidic solutions, but biochemistry is primarily concerned with the neutral region. We will consider only the hydrogen ion bindings that affect equilibrium in this region, namely the terminal phosphate group with a pK about 7 and the

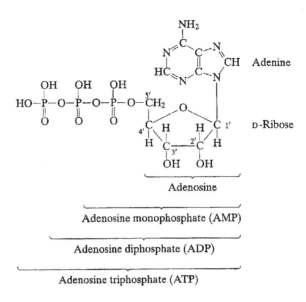

Figure 1.1   Structure of adenosine triphosphate.

adenine group with pK about 4. The other three pKs are in the neighborhood of 2 or below and can be ignored in treating biochemical reactions. The anions of ATP bind metal ions as well as hydrogen ions. The dissociation constants for the complex ions that are formed can be determined by use of acid titrations because the binding of a metal ion reduces the apparent pK for the phosphate group (Alberty, Smith, and Bock, 1951; Smith and Alberty, 1956; Silbey and Alberty, 2001). The apparent pK of the phosphate group is the midpoint of the titration of $H_2PO_4^{2-}$ in the presence of magnesium ions at the desired concentration of free metal ions. Because of the importance of ATP in energy metabolism, a great deal of data on the acid dissociation constants and the dissociation constants of complex ions of ATP, ADP, AMP, and P_i are available. Goldberg and Tewari (1991) and Larson, Tewari, and Goldberg (1993) critically evaluated these data including that on glucose 6-phosphate (G6P). The values for acid dissociation constants and magnesium complex ion dissociation constants involved in the ATP series are given in Table 1.2.

Since ATP is made up of three species in the physiological pH range in the absence of metal ions that are bound, its concentration is given by

$$[ATP] = [ATP^{4-}] + [HATP^{3-}] + [H_2ATP^{2-}] \qquad (1.3\text{-}1)$$

Substituting the expressions for the two acid dissociation constants yields

$$[ATP] = [ATP^{4-}]\left(1 + \frac{[H^+]}{K_{1ATP}} + \frac{[H^+]^2}{K_{1ATP}K_{2ATP}}\right) \qquad (1.3\text{-}2)$$

The mole fraction r of the ATP in the ATP^{4-} form at a specified concentration of hydrogen ions is given by

$$r(ATP^{-4}) = \cfrac{1}{1 + \cfrac{[H^+]}{K_{1ATP}} + \cfrac{[H^+]^2}{K_{1ATP}K_{2ATP}}} \qquad (1.3\text{-}3)$$

The mole fractions of ATP in the other two forms are readily derived:

$$r(HATP^{3-}) = \cfrac{\cfrac{[H^+]}{K_{1ATP}}}{1 + \cfrac{[H^+]}{K_{1ATP}} + \cfrac{[H^+]^2}{K_{1ATP}K_{2ATP}}} \qquad (1.3\text{-}4)$$

Table 1.2   Equilibrium Constants in the ATP Series at 298.15 K

Reaction		$pK(I=0)$	$K(I=0)$	$K(I=0.25\ M)$
$HAMP^- = H^+ + AMP^{2-}$		6.73	1.862×10^{-7}	6.877×10^{-7}
$H_2AMP = H^+ + HAMP^-$		3.99	1.023×10^{-4}	1.966×10^{-4}
$MgAMP = Mg^{2+} + AMP^{2-}$		2.79	1.622×10^{-3}	2.212×10^{-2}
$HADP^{2-} = H^+ + ADP^{-3}$	K_{1ADP}	7.18	6.607×10^{-8}	4.689×10^{-7}
$H_2ADP^- = H^+ + HADP^{2-}$	K_{2ADP}	4.36	4.365×10^{-5}	1.612×10^{-4}
$MgADP^- = Mg^{2+} + ADP^{3-}$	K_{3ADP}	4.65	2.239×10^{-5}	1.128×10^{-3}
$MgHADP = Mg^{2+} + HADP^{2-}$	K_{4ADP}	2.50	3.162×10^{-3}	4.313×10^{-2}
$HATP^{3-} = H^+ + ATP^{4-}$	K_{1ATP}	7.60	2.512×10^{-8}	3.426×10^{-7}
$H_2ATP^{2-} = H^+ + HATP^{3-}$	K_{2ATP}	4.68	2.089×10^{-5}	1.483×10^{-4}
$MgATP^{2-} = Mg^{2+} + ATP^{4-}$	K_{3ATP}	6.18	6.607×10^{-7}	1.229×10^{-4}
$MgHATP^- = Mg^{2+} + HATP^{3-}$	K_{4ATP}	3.63	2.344×10^{-4}	1.181×10^{-2}
$Mg_2ATP = Mg^{2+} + MgATP^{2-}$	K_{5ATP}	2.69	2.042×10^{-3}	2.785×10^{-2}
$H_2PO_4^- = H^+ + HPO_4^{2-}$		7.22	6.026×10^{-8}	2.225×10^{-7}
$MgHPO_4 = Mg^{2+} + HPO_4^{2-}$		2.71	1.950×10^{-3}	2.66×10^{-2}
$HG6P^- = H^+ + G6P^{2-}$		6.42	3.802×10^{-7}	1.404×10^{-6}
$MgG6P = Mg^{2+} + G6P^{2-}$		2.60	2.512×10^{-3}	3.462×10^{-2}
$Hadenosine^+ = H^+ + adenosine$		3.50	3.162×10^{-4}	3.162×10^{-4}
$ATP^{4-} + H_2O = ADP^{3-} + HPO_4^{2-} + H^+$			2.946×10^{-1}	
$ADP^{3-} + H_2O = AMP^{2-} + HPO_4^{2-} + H^+$			6.622×10^{-2}	
$AMP^{2-} + H_2O = adenosine + HPO_4^{2-}$			1.894×10^{2}	
$G6P^{2-} + H_2O = glucose + HPO_4^{2-}$			8.023×10^{1}	
$ATP^{4-} + glucose = ADP^{3-} + G6P^{2-} + H^+$			3.671×10^{-3}	
$2ADP^{3-} = ATP^{4-} + AMP^{2-}$			2.248×10^{-1}	

Source: R. A. Alberty and R. N. Goldberg, *Biochem.*, 31, 10612 (1992). Copyright 1992 American Chemical Society.

$$r(H_2ATP^{2-}) = \frac{\dfrac{[H^+]^2}{K_{1ATP}K_{2ATP}}}{1 + \dfrac{[H^+]}{K_{1ATP}} + \dfrac{[H^+]^2}{K_{1ATP}}K_{2ATP}} \tag{1.3-5}$$

These mole fractions are plotted versus pH at 298.15 K and $I = 0.25$ M in Fig. 1.2.

Since it is possible to calculate the mole fractions of the various species of ATP at a specified pH, the **average binding of hydrogen ions** $\bar{N}_H$ can be calculated by use of

$$\bar{N}_H = \frac{0[ATP^{4-}] + 1[HATP^{3-}] + 2[H_2ATP^{2-}]}{[ATP]} \tag{1.3-6}$$

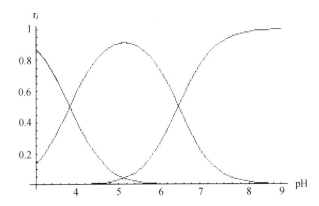

Figure 1.2   Mole fractions of three species of ATP plotted versus pH at 298.15 K and $I = 0.25$ M (see Problem 1.1).

The numbers of hydrogen ions bound that are calculated using this equation are based on the arbitrary convention of not counting the additional 12 hydrogen atoms in ATP. Thus, the average number $\bar{N}_H$ of hydrogen ions bound by ATP is given by

$$\bar{N}_H = \frac{\dfrac{[H^+]}{K_{1ATP}} + \dfrac{2[H^+]^2}{K_{1ATP}K_{2ATP}}}{1 + \dfrac{[H^+]}{K_{1ATP}} + \dfrac{[H^+]^2}{K_{1ATP}K_{2ATP}}} \tag{1.3-7}$$

At very high pH, the binding of H^+ approaches zero, and below pH 4 it approaches 2.

In dealing with binding, it is convenient to use the concept of a binding polynomial (Wyman 1948, 1964, 1965, 1975; Edsall and Wyman, 1958; Hermans and Scheraga, 1961; Schellman, 1975, 1976; Wyman and Gill, 1990). The polynomial in the denominator of equation 1.3-7 is referred to as the **binding polynomial** P. It is actually a kind of partition function because it gives the partition of a reactant between the various species that make it up. The binding polynomial for the binding of hydrogen ions by ATP is given by

$$P = 1 + \frac{[H^+]}{K_{1ATP}} + \frac{[H^+]^2}{K_{1ATP}K_{2ATP}} \tag{1.3-8}$$

The average binding of hydrogen ions is given by

$$\bar{N}_H = \frac{[H^+]}{P}\frac{dP}{d[H^+]} = \frac{d\ln P}{d\ln[H^+]} = \frac{-1}{\ln(10)}\frac{d\ln P}{dpH} \tag{1.3-9}$$

Equation 1.3-7 is readily obtained from equation 1.3-8 by use of equation 1.3-9.

Substituting the values of the two acid dissociation constants of ATP at 298.15 K, 1 bar, and $I = 0.25$ M from Table 1.2 into equation 1.3-7 or 1.3-9 yields the plot of $\bar{N}_H$ versus pH that is shown in Fig. 1.3.

Figure 1.3 shows that the acid titration curve for a weak acid can be calculated from its pKs, and this raises the question as to how the pKs can be calculated from the titration curve. This can be done by first integrating equation 1.3-9 to obtain the natural logarithm of the binding potential P:

$$\int \frac{\bar{N}_H}{[H^+]} d[H^+] = \int d\ln P = \ln P + \text{const.} \tag{1.3-10}$$

or

$$-\ln(10) \int \bar{N}_H \, pH = \int d\ln P = \ln P + \text{const.} \tag{1.3-11}$$

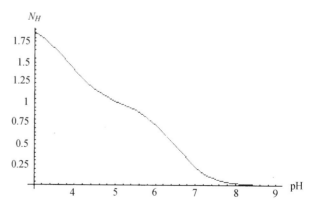

Figure 1.3 Binding of hydrogen ions by ATP at 298.15 K and $I = 0.25$ M (see Problem 1.2).

The acid dissociation constants can be calculated from $\ln P$ by fitting the plot of P versus $[H^+]$ with a power series in $[H^+]$.

ATP also binds magnesium ions as shown by the three complex ion dissociation constants in Table 1.2. Incorporating these species into equations 1.3-1 and 1.3-2 yields the following binding polynomial for ATP:

$$P = 1 + \frac{[H^+]}{K_{1ATP}} + \frac{[H^+]^2}{K_{1ATP}K_{2ATP}} + \frac{[Mg^{2+}]}{K_{3ATP}} + \frac{[Mg^{2+}][H^+]}{K_{1ATP}K_{4ATP}} + \frac{[Mg^{2+}]^2}{K_{3ATP}K_{5ATP}}$$

$$(1.3\text{-}12)$$

Now the binding of hydrogen ions is given by the following partial derivatives of the binding polynomial:

$$\bar{N}_H = \frac{[H^+]}{P}\left(\frac{\partial P}{\partial[H^+]}\right)_{pMg} = \frac{-1}{\ln(10)}\left(\frac{\partial \ln P}{\partial pH}\right)_{pMg} = [H^+]\left(\frac{\partial \ln P}{\partial[H^+]}\right)_{pMg}$$

$$(1.3\text{-}13)$$

The average binding of magnesium ions $\bar{N}_H$ is given by the following partial derivatives of the binding polynomial:

$$\bar{N}_{Mg} = \frac{[Mg^{2+}]}{P}\left(\frac{\partial P}{\partial[Mg^{2+}]}\right)_{pH} = \frac{-1}{\ln(10)}\left(\frac{\partial \ln P}{\partial pMg}\right)_{pH} = [Mg^{2+}]\left(\frac{\partial \ln P}{\partial[Mg^{2+}]}\right)_{pH}$$

$$(1.3\text{-}14)$$

These differentiations yield

$$\bar{N}_H = \frac{\dfrac{[H^+]}{K_{1ATP}} + \dfrac{2[H^+]^2}{K_{1ATP}K_{2ATP}} + \dfrac{[Mg^{2+}][H^+]}{K_{1ATP}K_{4ATP}}}{1 + \dfrac{[H^+]}{K_{1ATP}} + \dfrac{[H^+]^2}{K_{1ATP}K_{2ATP}} + \dfrac{[Mg^{2+}]}{K_{3ATP}} + \dfrac{[Mg^{2+}][H^+]}{K_{1ATP}K_{4ATP}} + \dfrac{[Mg^{2+}]^2}{K_{3ATP}K_{5ATP}}}$$

$$(1.3\text{-}15)$$

$$\bar{N}_{Mg} = \frac{\dfrac{[Mg^{2+}]}{K_{3ATP}} + \dfrac{[Mg^{2+}][H^+]}{K_{1ATP}K_{4ATP}} + \dfrac{2[Mg^{2+}]^2}{K_{3ATP}K_{5ATP}}}{1 + \dfrac{[H^+]}{K_{1ATP}} + \dfrac{[H^+]^2}{K_{1ATP}K_{2ATP}} + \dfrac{[Mg^{2+}]}{K_{3ATP}} + \dfrac{[Mg^{2+}][H^+]}{K_{1ATP}K_{4ATP}} + \dfrac{[Mg^{2+}]^2}{K_{3ATP}K_{5ATP}}}$$

$$(1.3\text{-}16)$$

Figure 1.4 shows a plot of $\bar{N}_H$ versus pH at several values of pMg. It is evident that the apparent pK of ATP in the neighborhood of 7 is reduced to about 5 in

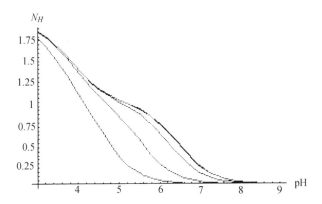

Figure 1.4 Binding of hydrogen ions by ATP at 298.15 K, $I = 0.25$ M, and pMg 2, 3, 4, 5, and 6 (see Problem 1.3).

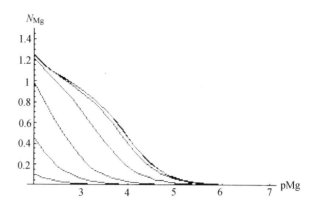

Figure 1.5 Binding of magnesium ions by ATP at 298.15 K, $I = 0.25$ M, and pH 3, 4, 5, 6, 7, 8, and 9 (see Problem 1.4).

a $MgCl_2$ solution with $[Mg^{2+}] = 10^{-2}$ M at ionic strength 0.25 M. Experimental plots of this type make it possible to calculate K_{4ATP} and K_{5ATP} (Smith and Alberty, 1956).

Figure 1.5 shows a plot of $\bar{N}_{Mg}$ versus pMg at several values of pH. As the hydrogen ion concentration is increased, the binding of magnesium ions is decreased because of the competition for the same sites.

Equations 1.3-15 and 1.3-16 can be used to make three-dimensional plots of $\bar{N}_{Mg}$ and as functions of pH and pMg. These plots are given in Figs. 1.6 and 1.7. The back plane of Fig. 1.6 gives the hydrogen ion binding of ATP in the essential absence of Mg (more accurately, pMg > 6). At pMg 2 the apparent second pK of ATP is less than 5. Figure 1.7 shows that below pMg 5 there is essentially no binding of magnesium ion and that binding increases to a number a little greater than 1 at pMg 2 and pH > 6 but is eliminated by further reduction of the pH. Figures 1.4 to 1.7 can also be obtained by plotting derivatives of the binding potential (see equations 1.3-13 and 1.3-14), rather than by use of equations 1.3-15 and 1.3-16 (see Problems 1.5 and 1.6).

A remarkable fact about Figs. 1.6 and 1.7 is that at any given pH and pMg, in Fig. 1.6, the slope in the pMg direction is the same as the slope in pH direction in Fig. 1.7 at that pH and pMg. This is because the mixed partial derivatives of

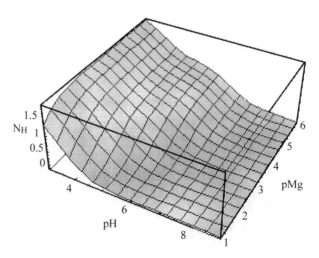

Figure 1.6 Plot of $\bar{N}_H$ versus pH and pMg for ATP at 298.15 K and $I = 0.25$ M (see Problem 1.5).

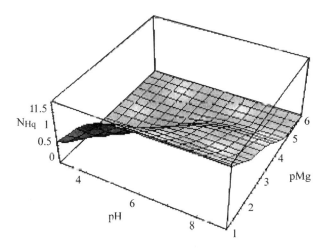

Figure 1.7   Plot of $\bar{N}_{Mg}$ versus pH and pMg at 298.15 K and $I = 0.25$ M (see Problem 1.6).

a function like P are equal.

$$\left(\frac{\partial \bar{N}_H}{\partial pMg}\right)_{T,P,pH} = \left(\frac{\partial \bar{N}_{Mg}}{\partial pH}\right)_{T,P,pMg} \tag{1.3-17}$$

In thermodynamics, this is referred to as a **Maxwell equation**. This equation is derived later in Section 4.8. Thus the effect of pMg on the binding of hydrogen ions is the same as the effect of pH on the binding of magnesium ions; in short, these are **reciprocal effects**. The bindings of these two ions are referred to as **linked functions**. Equation 1.3-17 can be confirmed by plotting these two derivatives, and the same plot is obtained in both cases. This would be a lot of work to do by hand, but since *Mathematica*[R] can take partial derivatives, this can be done readily with a computer. The two plots are identical and are given in Fig. 1.8.

■ 1.4   APPARENT EQUILIBRIUM CONSTANTS OF BIOCHEMICAL REACTIONS

In this section we consider the hydrolysis of adenosine triphosphate to adenosine diphosphate and inorganic phosphate, first at a specified pH in the absence of metal ions that are bound and then in the presence of magnesium ions. At

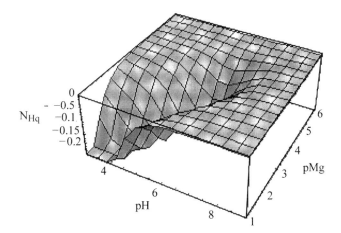

Figure 1.8   Plot of $(\partial \bar{N}_{Mg}/\partial pH)$ or $(\partial \bar{N}_H/\partial pMg)$ versus pH and pMg at 298.15 K and $I = 0.25$ M (see Problem 1.6).

specified pH (and pMg) the **biochemical reaction** is written in terms of sums of species:

$$ATP + H_2O = ADP + P_i \qquad (1.4\text{-}1)$$

Biochemical textbooks often add a H^+ on the right-hand side, but this is stoichiometrically incorrect when the pH is held constant, as we will see in the next section. It is also wrong, in principle, as we will see in Chapter 4, since hydrogen atoms are not balanced by biochemical reactions because the pH is held constant. The statement that the pH is constant means that in principle acid or alkali is added to the reaction system as the reaction occurs to hold the pH constant. In practice, a buffer is used to hold the pH nearly constant, and the pH is measured at equilibrium.

The expression for the apparent equilibrium constant K' for reaction 1.4-1 is

$$K' = \frac{[ADP][P_i]}{[ATP]} \qquad (1.4\text{-}2)$$

because the activity of water is taken as unity in dilute aqueous solutions at each temperature. The apparent equilibrium constant K' is a function of T, P, pH, pMg, and ionic strength. In the neutral region in the absence of magnesium ions, ATP, ADP, and P_i each consist of two species, and so

$$
\begin{aligned}
K' &= \frac{([ADP^{3-}] + [HADP^{2-}])([HPO_4^{2-}] + [H_2PO_4^-])}{[ATP^{4-}] + [HATP^{3-}]} \\
&= \frac{[ADP^{3-}][HPO_4^{2-}]}{[ATP^{4-}]} \frac{(1 + [H^+]/K_{1ADP})(1 + [H^+]/K_{1Pi})}{(1 + [H^+]/K_{1ATP})} \\
&= \frac{K_{ref}}{[H^+]} \frac{(1 + [H^+]/K_{1ADP})(1 + [H^+]/K_{1Pi})}{(1 + [H^+]/K_{1ATP})}
\end{aligned} \qquad (1.4\text{-}3)
$$

where K_{ref} is the chemical equilibrium constant for the chemical **reference reaction**

$$ATP^{4-} + H_2O = ADP^{3-} + HPO_4^{2-} + H^+ \qquad (1.4\text{-}4)$$

$$K_{ref} = \frac{[ADP^{3-}][HPO_4^{2-}][H^+]}{[ATP^{4-}]} \qquad (1.4\text{-}5)$$

Since the acid dissociation constants are known, the value of K_{ref} can be calculated from the value of K' at a pH in the neutral region in the absence of metal ions by using equation 1.4-3. Values of K_{ref} at zero ionic strength are given in Table 1.2 for six reference reactions.

When magnesium ions or other metal ions are bound reversibly and a wider range of pH is considered, equation 1.4-3 becomes more complicated. Therefore it is convenient to use the nomenclature of binding polynomials introduced in equation 1.3-8. The binding polynomial of ATP is given in equation 1.3-12, and the binding potentials for ADP and P_i are as follows:

$$P_{ADP} = 1 + \frac{[H^+]}{K_{1ADP}} + \frac{[H^+]^2}{K_{1ADP}K_{2ADP}} + \frac{[Mg^{2+}]}{K_{3ADP}} + \frac{[Mg^{2+}][H^+]}{K_{1ADP}K_{4ADP}} \qquad (1.4\text{-}6)$$

$$P_{P_i} = 1 + \frac{[H^+]}{K_{1Pi}} + \frac{[Mg^{2+}]}{K_{2Pi}} \qquad (1.4\text{-}7)$$

Thus the apparent equilibrium constant for the hydrolysis of ATP as a function of $[H^+]$ and $[Mg^{2+}]$ is given by

$$K' = \frac{K_{ref}P_{ADP}P_{P_i}}{[H^+]P_{ATP}} \qquad (1.4\text{-}8)$$

Since the chemical equilibrium constants in this equation are known at zero ionic strength at 298.15 K and are given in Table 1.2, K' can be calculated at any pH

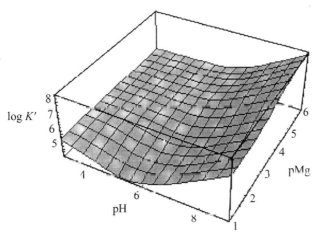

Figure 1.9 Plot of the base 10 logarithm of the apparent equilibrium constant for the hydrolysis of ATP to ADP and P_i at 298.15 K and 0.25 M ionic strength (see Problem 1.7).

in the range 3 to 9 and any pMg in the range 2 to 7 at a specified ionic strength. The dependence of K' on pH and pMg is shown in Fig. 1.9; log K' is plotted versus pH and pMg since K' varies over many powers of ten in these ranges of pH and pMg.

In Chapter 4 we will be interested in $-RT \ln K'$, where the gas constant R is 8.31451 J K^{-1} mol^{-1}, and so this quantity in kJ mol^{-1} is plotted versus pH and pMg in Fig. 1.10. The pH dependencies of the apparent equilibrium constants of biochemical reactions were discussed by Alberty and Cornish-Bowden in 1993.

■ 1.5 PRODUCTION OF HYDROGEN IONS AND MAGNESIUM IONS IN THE HYDROLYSIS OF ADENOSINE TRIPHOSPHATE

The calculation of the binding of hydrogen ions $\bar{N}_H$ for ATP discussed in Section 1.3 can be applied to ADP and P_i so that the change in binding of H^+ in the hydrolysis of ATP can be calculated using

$$\Delta_r N_H = \bar{N}_H(\text{ADP}) + \bar{N}_H(P_i) - \bar{N}_H(\text{ATP}) - 1 \qquad (1.5\text{-}1)$$

where the -1 is for the two protons in water minus the proton in HPO_4^-, which is treated as the base species of inorganic phosphate in the reference reaction. The

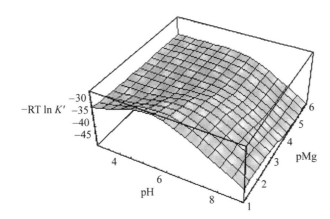

Figure 1.10 Plot of $-RT \ln K'$ in kJ mol^{-1} versus pH at 298.15 K and 0.25 M ionic strength (see Problem 1.7).

change in the binding of magnesium ions in the hydrolysis of ATP at specified pH is given by

$$\Delta_r N_{Mg} = \bar{N}_{Mg}(ADP) + \bar{N}_{Mg}(P_i) - \bar{N}_{Mg}(ATP) \qquad (1.5\text{-}2)$$

Since $\bar{N}_H$ and $\bar{N}_{Mg}$ can be calculated for these reactants, $\Delta_r N_H$ and $\Delta_r N_{Mg}$ can be calculated as a function of pH and pMg. However, when a computer is available there is an easier way to do this by using equation 1.3-13 for the binding of hydrogen ions and 1.3-14 for the binding of magnesium ions. For example, equation 1.5-1 can be written

$$\Delta_r N_H = -\frac{1}{\ln(10)}\left(\frac{\partial \ln P_{ADP}}{\partial pH}\right)_{pMg} - \frac{1}{\ln(10)}\left(\frac{\partial \ln P_{P_i}}{\partial pH}\right)_{pMg}$$
$$+ \frac{1}{\ln(10)}\left(\frac{\partial \ln P_{ATP}}{\partial pH}\right)_{pMg} - 1 \qquad (1.5\text{-}3)$$

where of course T and P are also held constant. Note that this same result is obtained by simply differentiating the expression for $\ln K'$ (equation 1.4-8) with respect to pH. Thus

$$\Delta_r N_H = -\frac{1}{\ln(10)}\left(\frac{\partial \ln K'}{\partial pH}\right)_{pMg} \qquad (1.5\text{-}4)$$

The change in binding of Mg^{2+} ions can be calculated using

$$\Delta_r N_{Mg} = -\frac{1}{\ln(10)}\left(\frac{\partial \ln K'}{\partial pMg}\right)_{pH} \qquad (1.5\text{-}5)$$

Since K' is a pretty complicated function of pH and pMg, it would be very difficult to carry these calculations out by hand. However, with *Mathematica* the calculations can be done quickly. Figure 1.11 shows the change in the binding of hydrogen ions in the hydrolysis of ATP as a function of pH and pMg. At high pH the change in binding is -1 mole of H^+ per mole of ATP hydrolyzed, as expected from the reference reaction, which predominates at high pH. The products bind fewer hydrogen ions, and so there is a net production of hydrogen ions in the biochemical reaction. In the presence of magnesium ions there are conditions where the change in binding is positive, which indicates that hydrogen ions are consumed in the hydrolysis of ATP under these conditions.

Figure 1.12 shows the change in binding of magnesium ions as a function of pH and pMg. Magnesium ions are always produced in the hydrolysis because

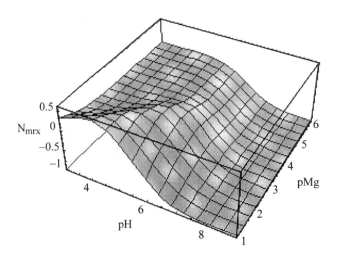

Figure 1.11   Change in the binding of hydrogen ions in the hydrolysis of ATP as a function of pH and pMg at 298.15 K and 0.25 M ionic strength (see Problem 1.8).

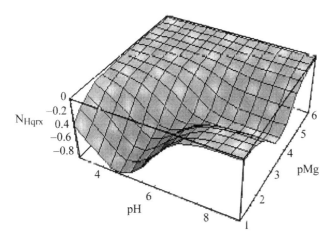

Figure 1.12 Change in the binding of magnesium ions in the hydrolysis of ATP at 298.15 K and 0.25 M ionic strength (see Problem 1.9).

they are more strongly bound by ATP than by ADP and P$_i$. The change in binding approaches zero as the concentration of free magnesium ions approaches zero, and it also approaches zero at high concentrations of magnesium ion and high pH, where the principal reaction is $Mg_2ATP + H_2O = MgADP^- + MgHPO_4 + H^+$.

Figures 1.11 and 1.12 are related in the same way as the binding curves for a single reactant (see equation 1.3-15); that is, the slope of the plot of $\Delta_r N_H$ in the pMg direction is the same as the slope of the plot of $\Delta_r N_{Mg}$ in the pH direction. This is a consequence of the reciprocity relation:

$$\left(\frac{\partial \Delta_r N_H}{\partial pMg}\right)_{T,P,pH} = \left(\frac{\partial \Delta_r N_{Mg}}{\partial pH}\right)_{T,P,pMg} \tag{1.5-6}$$

This equation is derived later in Section 4.8.

The change in the value of the apparent equilibrium constant with pH and pMg and the production or consumption of hydrogen ions and magnesium ions by the biochemical reaction are really two sides of the same coin. The effects of pH and pMg on K' are due to the fact that the biochemical reaction produces or consumes these ions. This is an example of **Le Chatelier's principle**, which states that when an independent variable of a system at equilibrium is changed, the equilibrium shifts in the direction that tends to reduce the effect of the change. If the reaction produces hydrogen ions, lowering the pH will cause K to decrease because the system is doing what it can to reduce the effect of the pH change.

■ 1.6 *p*Ks OF WEAK ACIDS

In this chapter we have seen that acid dissociation constants are needed to calculate the dependence of apparent equilibrium constants on pH. In Chapter 3 we will discuss the calculation of the effects of ionic strength and temperature on acid dissociation constants. The database described later can be used to calculate *p*Ks of reactants at 298.15 K at desired ionic strengths. Because of the importance of *p*Ks of weak acids, Table 1.3 is provided here. More experimental measurements of acid dissociation constants and dissociation constants of complex ions with metal ions are needed because they are essential for the interpretation of experimental equilibrium constants and heats of reactions. A major database of acid dissociation constants and dissociation constants of metal ion complexes is provided by Martell, Smith, and Motekaitis (2001).

Table 1.3   pKs of Weak Acids at 298.15 K in Dilute Aqueous Solutions as a Function of Ionic Strength (See Problem 1.10)

	$I = 0$ M	$I = 0.10$ M	$I = 0.25$ M
acetate	4.75	4.54	4.47
acetylphosphate K_1	8.69	8.26	8.12
acetylphosphate K_2	5.11	4.90	4.83
adenine	4.20	4.20	4.20
ammonia	9.25	9.25	9.25
ATP K_1	7.60	6.74	6.47
ATP K_2	4.68	4.04	3.83
ADP K_1	7.18	6.53	6.33
ADP K_2	4.36	3.93	3.79
AMP K_1	6.73	6.30	6.16
AMP K_2	3.99	3.77	3.71
adenosine	3.47	3.47	3.47
bisphosphoglycerate	7.96	7.10	6.83
citrate K_1	6.39	5.75	5.54
citrate K_2	4.76	4.33	4.19
isocitrate K_1	6.40	5.76	5.55
socitrate K_2	4.76	4.33	4.19
coenzyme A	8.38	8.16	8.10
HCO_3^- K_1	10.30	9.90	9.76
H_2CO_3 K_2	6.37	6.15	6.08
cysteine	8.38	8.16	8.09
dihydroxyacetone phosphate	5.70	5.27	5.13
fructose 6-phosphate K_1	6.27	5.84	5.70
fructose-1,6-biphosphate K_1	6.65	5.79	5.52
fructose-1,6-biphosphate K_2	6.05	5.41	5.20
fumarate K_1	4.60	4.17	4.03
fumarate K_2	3.09	2.88	2.81
galactose 1-phosphate K_1	6.15	5.72	5.58
glucose 6-phosphate K_1	6.42	5.99	5.85
glutathione$_{red}$	8.34	7.91	7.77
glucose 1-phosphate K_1	6.50	6.07	5.93
glyceraldehyde phosphate	5.70	5.27	5.13
glycerol 3-phosphate	6.67	6.24	6.10
malate K_1	5.26	4.83	4.69
oxalate K_1	4.28	3.85	3.71
phospho*enol*pyruvate	7.00	6.36	6.15
2-phosphoglycerate	7.64	7.00	6.79
3-phosphoglycerate	7.53	6.89	6.68
phosphate K_1	7.22	6.79	6.65
pyrophosphate K_1	9.46	8.60	8.33
pyrophosphate K_2	6.72	6.08	5.87
pyrophosphate K_3	2.26	1.83	1.69
ribose 1-phosphate K_1	6.69	6.26	6.12
ribose 5-phosphate K_1	6.69	6.26	6.12
succinate K_1	5.64	5.21	5.07
succinate K_2	4.21	3.99	3.92
succinylcoA	4.21	4.00	3.93
thioredoxin$_{red}$ K_1	8.64	8.21	8.09
thioredoxin$_{red}$ K_2	8.05	7.83	7.76

The equations and calculations described in this chapter are very useful, but so far we have not discussed thermodynamic properties other than equilibrium constants. The other properties introduced in the next three chapters provide a better understanding of the energetics and equilibria of reactions. We will consider the basic structure of thermodynamics in Chapter 2 and then to apply these ideas to chemical reactions in Chapter 3 and biochemical reactions in Chapter 4.

Structure of Thermodynamics

According to the **first law of thermodynamics**, there is a thermodynamic property U of a system, called the **internal energy**. The change in internal energy in a change in the state of a system is given by $\Delta U = q + w$, where q is the heat flow into the system and w is the work done on the system. The work can be pressure-volume work, work of transport of electric charge, chemical work (more on this later), work of stretching an elastomer, and so on.

The **second law of thermodynamics** has two parts. According to the first part there is a thermodynamic property S of a system, called the **entropy**. The change in entropy in a reversible change from one state of a system to another is given by $\Delta S = q/T$, where T is the absolute temperature. According to the second part of the second law, when a change takes place spontaneously in an isolated system, ΔS is greater than zero. This is a remarkable result because it provides a way to calculate whether a specified change in state can take place in a system on the basis of other types of measurements on the system. These conclusions apply to systems consisting of phases that are uniform in composition and do not have gradients of temperature or concentration in them.

These two laws can be combined for a system involving only pressure-volume work to obtain $dU = T\,dS - P\,dV$. This so-called **fundamental equation** shows two things: (1) thermodynamic properties of a system obey the rules of calculus and (2) the choice of independent variables (in this case S and V) plays a very important role in thermodynamics. The second law can be used to show that when S and V are held constant, the internal energy U of a system must decrease

19

in a spontaneous process. For example, the internal energy of an isolated system decreases when a spontaneous process occurs in it because S and V are constant. If thermodynamics could not provide more than this, it would be difficult to use. Fortunately mathematics provides a way to introduce intensive variables as independent variables by using **Legendre transforms**.

The concept of a Legendre transform is very important because it leads from the internal energy U to the enthalpy H and the Gibbs energy G; these thermodynamic properties are defined by $H = U + PV$ and $G = U + PV - TS = H - TS$. In this book additional Legendre transforms are used to define the transformed Gibbs energy G' and the transformed enthalpy H' at specified pH and pMg in Chapter 4 on biochemical reactions. In Chapter 6 further transformed Gibbs energies G'' and further transformed enthalpies H'' are introduced to discuss systems of biochemical reactions at specified concentrations of coenzymes. The construction of Legendre transforms is discussed in Section 2.5.

It is also important to understand that all these properties obey all the rules of calculus. As a consequence these properties are related through fundamental equations, Maxwell equations, Gibbs-Helmholtz equations, and Gibbs-Duhem equations.

The internal energy U, entropy, and the properties defined by Legendre transforms are referred to as **thermodynamic potentials** because, like the potential energy in mechanics, these extensive properties of a thermodynamic system give information about the direction of spontaneous change and the equilibrium state of the system. When work other than pressure-volume work is involved in a system, still more thermodynamic potentials can be defined by use of Legendre transforms, and these thermodynamic potentials provide criteria of equilibrium when sets of properties convenient for the experimenter are held constant. These various thermodynamic potentials are needed for the convenience of the experimenter. The thermodynamic potential used to provide a criterion of spontaneous change and equilibrium depends on the intensive variables that are held constant. Biochemical systems provide exceptional challenges to thermodynamics because the concentrations of hydrogen ions, certain metal ions, and coenzymes may be held constant, and electrical work, mechanical work, and surface work may be involved in addition to chemical work and PV work. When the equations in this chapter are applied to dilute electrolyte solutions, it is convenient to take the thermodynamic properties to be functions of the ionic strength. This is not treated in detail in this chapter, but is in the next chapter.

This chapter deals with the thermodynamics of one-phase systems, and it is understood that the phase is homogeneous and at uniform temperature. The basic structure of thermodynamics provides the tools for the treatment of more complicated systems in later chapters. This book starts with the fundamentals of thermodynamics, but the reader really needs some prior experience with thermodynamics at the level of undergraduate thermodynamics (Silbey and Alberty, 2001). Legendre transforms play an important role in this chapter, and the best single reference on Legendre transforms is Callen (1960, 1985). Other useful references for basic thermodynamics are Tisza (1966), Beattie and Oppenheim (1979), Bailyn (1994), and Greiner, Neise, and Stocker (1995).

■ 2.1 STATE OF A SYSTEM

A system that is made up of a homogeneous mass of a substance at equilibrium can be described as being in a certain **thermodynamic state** that is characterized by certain properties. If forces of various types act on the system or more of the substance is added, the system is changed to a different state. It is remarkable that only a small number of properties have to be specified to completely characterize the equilibrium state of a macroscopic system. For a system containing a single substance, three properties suffice, if they are properly chosen. For example, the state of a mass of an ideal gas is specified by its temperature T, pressure P, and amount n because the volume V is given by $V = nRT/P$, where R is the gas constant. Alternatively, V, T, P or V, P, n, or V, T, n could be specified. An ideal gas is a special case, but three properties are sufficient to define the state of a simple system (that is one substance in one phase) provided one property is extensive. If work in addition to PV work is involved, more variables have to be specified. These properties and others that characterize the state of a system are referred to as **state properties** because they determine the state of the system. The internal energy U is also a state property. It is appropriate to call them state functions because they are independent of the path of change and can be manipulated by the operations of algebra and calculus.

A thermodynamic property is said to be **extensive** if the magnitude of the property is doubled when the size of the system is doubled. Examples of extensive properties are volume V and amount of substance n. A thermodynamic property is said to be **intensive** if the magnitude of the property does not change when the size of the system is changed. Examples of intensive properties are temperature, pressure, and the mole fractions of species. The ratio of two extensive properties is an intensive property. For example, the ratio of the volume of a one-component system to its amount is the **molar volume**: $V_m = V/n$.

Experience shows that for a system that is a homogeneous mixture of N_s substances, $N_s + 2$ properties have to be specified and at least one property must be extensive. For example, we can specify T, P, and amounts of each of the N_s substances or we can specify T, P, and mole fractions x_i of all but one substance, plus the total amount in the system. Sometimes we are only interested in the intensive state of a system, and that can be described by specifying $N_s + 1$ intensive properties for a one-phase system. For example, the intensive state of a solution involving two substances can be described by specifying T, P, and the mole fraction of one of substances.

When other kinds of work are involved, it is necessary to specify more variables, but the point is that when a small number of properties are specified, all the other properties of the system are fixed. This is in contrast with the very large number of properties that have to be specified to describe the microscopic state of a macroscopic system. In classical physics the complete description of a mole of an ideal gas would require the specification of $3N_A$ components in the three directions of spatial coordinates and $3N_A$ components of velocities of molecules, where N_A is the Avogadro constant.

■ 2.2 FUNDAMENTAL EQUATION FOR THE INTERNAL ENERGY

The first and second laws of thermodynamics for a homogeneous closed system involving only PV work lead to the **fundamental equation for the internal energy** U:

$$dU = T\,dS - P\,dV \qquad (2.2\text{-}1)$$

where T is the temperature, S is the entropy of the system, P is the pressure, and V is the volume. The **first law** states that $dU = dq - P\,dV$ when only pressure-volume work is involved and the **second law** states that $dS \geqslant dq/T$, where q is heat and d indicates an inexact differential. The integral of an inexact differential depends on the path, but the integral of an exact differential does not. The test for exactness is given in Section 2.3. The greater than or equal sign indicates that the differential entropy is equal to dq/T for a reversible process and is greater than dq/T in a spontaneous process. It is important to note that all five thermodynamic properties in equation 2.2-1 have exact differentials and that the fundamental equation for U is written in the notation of calculus. This means that these properties behave like mathematical functions so that further relations can be obtained by use of the operations of calculus. The **third law** states that the entropy of each pure element or substance in a perfect crystalline form is zero at absolute zero.

The entropy provides a criterion of spontaneous change and equilibrium at constant U and V because $(dS)_{U,V} \geq 0$. Thus the entropy of an isolated system can only increase and has its maximum value at equilibrium. The internal energy also provides a criterion for spontaneous change and equilibrium. That criterion is $(dU)_{S,V} \leq 0$, which indicates that when spontaneous changes occur in a system described by equation 2.2-1 at constant S and V, U can only decrease and has its minimum value at equilibrium.

The inequalities of the previous paragraph are extremely important, but they are of little direct use to experimenters because there is no convenient way to hold U and S constant except in isolated systems and adiabatic processes. In both of these inequalities, the independent variables (the properties that are held constant) are all extensive variables. There is just one way to define thermodynamic properties that provide criteria of spontaneous change and equilibrium when intensive variables are held constant, and that is by the use of Legendre transforms. That can be illustrated here with equation 2.2-1, but a more complete discussion of Legendre transforms is given in Section 2.5. Since laboratory experiments are usually carried out at constant pressure, rather than constant volume, a new thermodynamic potential, the enthalpy H, can be defined by

$$H = U + PV \tag{2.2-2}$$

The differential of the enthalpy is given by

$$dH = dU + P\,dV + V\,dP \tag{2.2-3}$$

Substituting equation 2.2-1 yields

$$dH = T\,dS + V\,dP \tag{2.2-4}$$

The use of a Legendre transform has introduced an intensive property P as an independent variable. It can be shown that the criterion for spontaneous change and equilibrium is given by $(dH)_{S,P} \geq 0$.

The temperature can be introduced as an independent variable by defining the Gibbs energy G with the Legendre transform

$$G = H - TS \tag{2.2-5}$$

The differential of the Gibbs energy is given by

$$dG = dH - T\,dS - S\,dT \tag{2.2-6}$$

Substituting equation 2.2-4 yields

$$dG = -S\,dT + V\,dP \tag{2.2-7}$$

The use of this Legendre transform has introduced the intensive property T as an independent variable. It can be shown that the criterion for spontaneous change and equilibrium is given by $(dG)_{T,P} \geq 0$. The Gibbs energy is so useful because T and P are convenient intensive variables to hold constant and because, as we will see shortly, if G can be determined as a function of T and P, then S, V, H, and U can all be calculated.

Gibbs (1873) showed how to include the contributions of added matter to the fundamental equation by introducing the concept of the **chemical potential** μ_i of species i and writing the **fundamental equation for the internal energy of a system** involving PV work and changes in the **amounts** n_i of species as

$$dU = T\,dS - P\,dV + \sum_{i=1}^{N_s} \mu_i\,dn_i \tag{2.2-8}$$

where N_s is the number of different species. This equation really defines the
chemical potential μ_i of a species. The terms in $\mu_i dn_i$ are referred to as **chemical
work terms**. If U for a system can be determined as a function of S, V, and $\{n_i\}$,
where $\{n_i\}$ represents the set of amounts of species, then T, P, and $\{\mu_i\}$ can be
determined by taking partial derivatives of U. **Thus the intensive properties of the
system are obtained by taking derivatives of the extensive property U with respect
to extensive properties**. It may be useful to consider the internal energy to be a
function of T, P, and $\{n_i\}$ rather than S, V, and $\{n_i\}$, but when that is done, it is
not possible to calculate all the other thermodynamic properties of the system by
taking partial derivatives.

When U is expressed as a function of S, V, and $\{n_i\}$, calculus requires that the
total differential of U is given by

$$dU = \left(\frac{\partial U}{\partial S}\right)_{V,\{n_i\}} dS + \left(\frac{\partial U}{\partial V}\right)_{S,\{n_i\}} dV + \sum_{i=1}^{N_s} \left(\frac{\partial U}{\partial n_i}\right)_{S,V,n_j \neq n_i} dn_i \qquad (2.2\text{-}9)$$

Comparison of equations 2.2-8 and 2.2-9 indicates that

$$T = \left(\frac{\partial U}{\partial S}\right)_{V,\{n_i\}} = T(S, V, \{n_i\}) \qquad (2.2\text{-}10)$$

$$P = -\left(\frac{\partial U}{\partial V}\right)_{S,\{n_i\}} = P(S, V, \{n_i\}) \qquad (2.2\text{-}11)$$

$$\mu_i = \left(\frac{\partial U}{\partial n_i}\right)_{S,V,n_j \neq n_i} = \mu_i(S, V, \{n_i\}) \qquad (2.2\text{-}12)$$

The intensive variables T, P, and $\{\mu_i\}$ can be considered to be functions of S, V,
and $\{n_i\}$ because U is a function of S, V, and $\{n_i\}$. If U for a system can be
determined experimentally as a function of S, V, and $\{n_i\}$, then T, P, and $\{\mu_i\}$ can
be calculated by taking the first partial derivatives of U. Equations 2.2-10 to
2.2-12 are referred to as **equations of state** because they give relations between
state properties at equilibrium. In Section 2.4 we will see that these $N_s + 2$
equations of state are not independent of each other, but any $N_s + 1$ of them
provide a complete thermodynamic description of the system. In other words, if
$N_s + 1$ equations of state are determined for a system, the remaining equation of
state can be calculated from the $N_s + 1$ known equations of state. In the preceding
section we concluded that the intensive state of a one-phase system can be
described by specifying $N_s + 1$ intensive variables. Now we see that the determination of $N_s + 1$ equations of state can be used to calculate these $N_s + 1$ intensive
properties.

The beauty of the fundamental equation for U (equation 2.2-8) is that it
combines all of this information in one equation. Note that the $N_s + 2$ extensive
variables S, V, and $\{n_i\}$ are independent, and the $N_s + 2$ intensive variables T, P,
and $\{\mu_i\}$ obtained by taking partial derivatives of U are dependent. This is
wonderful, but equations of state 2.2-10 to 2.2-12 are not very useful because S
is not a convenient independent variable. Fortunately, more useful equations
of state will be obtained from other thermodynamic potentials introduced in
Section 2.5.

The fundamental equation for U is in agreement with the statement of the
preceding section that for a homogeneous mixture of N_s substances, the state of
the system can be specified by $N_s + 2$ properties, at least one of which is extensive.
The total number of variables involved in equation 2.2-8 is $2N_s + 5$. $N_s + 3$ of
these variables are extensive (U, S, V, and $\{n_i\}$), and $N_s + 2$ of the variables are
intensive (T, P, $\{\mu_i\}$). Note that except for the internal energy, these variables
appear in pairs, in which one property is extensive and the other is intensive; these
are referred to as **conjugate pairs**. These pairs are given later in Table 2.1 in
Section 2.7. When other kinds of work are involved, there are more than $2N_s + 5$
variables in the fundamental equation for U (see Section 2.7).

Equation 2.2-8 indicates that the internal energy U of the system can be taken to be a function of entropy S, volume V, and amounts $\{n_i\}$ because these independent properties appear as differentials in equation 2.2-8; note that these are all extensive variables. This is summarized by writing $U(S, V, \{n_i\})$. The independent variables in parentheses are called the **natural variables** of U. Natural variables are very important because when a thermodynamic potential can be determined as a function of its natural variables, all of the other thermodynamic properties of the system can be calculated by taking partial derivatives. The natural variables are also used in expressing the criteria of spontaneous change and equilibrium: For a one-phase system involving PV work, $(dU) \leqslant 0$ at constant S, V, and $\{n_i\}$.

Fundamental equation 2.2-8 has been presented as the equation resulting from the first and second laws, but thermodynamic treatments can also be based on the entropy as a thermodynamic potential. Equation 2.2-8 can alternatively be written as

$$dS = \frac{1}{T}dU + \frac{P}{T}dV - \sum_{i=1}^{N_s} \frac{\mu_i}{T}dn_i \qquad (2.2\text{-}13)$$

This **fundamental equation for the entropy** shows that S has the natural variables U, V, and $\{n_i\}$. The corresponding criterion of equilibrium is $(dS) \geqslant 0$ at constant U, V, and $\{n_i\}$. **Thus the entropy increases when a spontaneous change occurs at constant U, V, and $\{n_i\}$. At equilibrium the entropy is at a maximum.** When U, V, and $\{n_i\}$ are constant, we can refer to the system as isolated. Equation 2.2-13 shows that partial derivatives of S yield $1/T$, P/T, and μ_i/T, which is the same information that is provided by partial derivatives of U, and so nothing is gained by using equation 2.2-13 rather than 2.2-8. Since equation 2.2-13 does not provide any new information, we will not discuss it further.

Equation 2.2-8 can be integrated at constant values of the intensive properties T, P, and $\{n_i\}$ to obtain

$$U = TS - PV + \sum_{i=1}^{N_s} \mu_i n_i \qquad (2.2\text{-}14)$$

This is referred to as the **integrated form of the fundamental equation** for U.

Alternatively, equation 2.2-14 can be regarded as a result of **Euler's theorem**. A function $f(x_1, x_2, \ldots, x_N)$ is said to be homogeneous of degree n if

$$f(kx_1, kx_2, \ldots, kx_N) = k^n f(x_1, x_2, \ldots, x_N) \qquad (2.2\text{-}15)$$

For such a function, Euler's theorem states that

$$nf(x_1, x_2, \ldots, x_N) = \sum_{i=1}^{N} x_i \frac{\partial f}{\partial x_i} \qquad (2.2\text{-}16)$$

The internal energy is homogeneous of degree 1 in terms of extensive thermodynamic properties, and so equation 2.2-8 leads to equation 2.2-14. All extensive variables are homogeneous functions of the first degree of other extensive properties. All intensive properties are homogeneous functions of the zeroth degree of the extensive properties.

Since integration introduces a constant, the value of U obtained from equation 2.2-14 is uncertain by an additive constant, but that is not a problem because thermodynamic calculations always involve changes in U, that is, ΔU.

■ 2.3 MAXWELL EQUATIONS

If the differential of a function $f(x, y)$ given by

$$df = Mdx + Ndy \qquad (2.3\text{-}1)$$

and the mixed partial derivatives are equal

$$\left(\frac{\partial M}{\partial y}\right)_x = \left(\frac{\partial N}{\partial x}\right)_y \tag{2.3-2}$$

the function f is said to be **exact**, and the integral of f is independent of path. If the mixed partial derivatives are not equal, the function f is said to be **inexact**, and the integral of the function is dependent on the path. In Section 2.2 it was noted that heat q and work w are inexact differentials because they depend on the path of integration. However, all state functions are exact differentials since they are independent of path of integration. Since the internal energy has an exact differential, equation 2.3-2 applies to its fundamental equation. In thermodynamics relations like equation 2.3-2 are referred to as **Maxwell equations**.

For equation 2.2-2, the Maxwell equations are

$$\left(\frac{\partial T}{\partial V}\right)_{S,\{n_i\}} = -\left(\frac{\partial P}{\partial S}\right)_{V,\{n_i\}} \tag{2.3-3}$$

$$\left(\frac{\partial T}{\partial n_i}\right)_{S,V,n_j \neq n_i} = \left(\frac{\partial \mu_i}{\partial S}\right)_{V,\{n_i\}} \tag{2.3-4}$$

$$-\left(\frac{\partial P}{\partial n}\right)_{S,V,n_j \neq n_i} = \left(\frac{\partial \mu_i}{\partial V}\right)_{S,\{n_i\}} \tag{2.3-5}$$

$$\left(\frac{\partial \mu_i}{\partial n_j}\right)_{S,V,n_i \neq n_j} = \left(\frac{\partial \mu_i}{\partial n_i}\right)_{S,V,n_j \neq n_i} \tag{2.3-6}$$

Equations 2.2-10 to 2.2-12 and equations 2.3-3 to 2.3-6 show that the thermodynamic properties of a system are interrelated in complicated, and sometimes unexpected, ways. The next section shows that the intensive variables for a thermodynamic system are not independent of each other.

■ 2.4 GIBBS-DUHEM EQUATION AND THE PHASE RULE

The differential of the integrated form (equation 2.2-14) of the fundamental equation for the internal energy is

$$dU = T\,dS + P\,dT - P\,dV - V\,dP + \sum_{i=1}^{N_s} \mu_i\,dn_i + \sum_{i=1}^{N_s} n_i\,d\mu_i \tag{2.4-1}$$

Subtracting the fundamental equation for U (equation 2.2-8) yields

$$S\,dT - V\,dP + \sum_{i=1}^{N_s} n_i\,d\mu_i = 0 \tag{2.4-2}$$

which is referred to as the **Gibbs-Duhem equation**. This equation is important because it shows that the $N_s + 2$ intensive properties for a homogeneous system without chemical reactions are not independent. But $N_s + 1$ of them are independent. Note that this is in agreement with the experimental observation of Section 2.1 that the intensive state of a one-phase system can be specified by stating $N_s + 1$ intensive variables.

The Gibbs-Duhem equation is the basis for the phase rule of Gibbs. According to the **phase rule**, the number of **degrees of freedom** F (independent intensive variables) for a system involving only PV work, but no chemical reactions, is given by

$$F = N_s - p + 2 \tag{2.4-3}$$

where p is the number of phases. This equation is derived in a more general form later in Chapter 8 on phase equilibria. Thus for a one-phase system without chemical reaction, $F = N_s - 1 + 2 = N_s + 1$, as shown by equation 2.4-2. The independent intensive variables can be chosen to be $T, P, \mu_1, \ldots, \mu_{N_s-1}$ or $T, \{\mu_i\}$ or $P, \{\mu_i\}$.

Since F is the symbol for the number of independent intensive variables for a system, it is also useful to have a symbol for the number of natural variables for a system. To describe the extensive state of a system, we have to specify F intensive variables and in addition an extensive variable for each phase. This description of the extensive state therefore requires D variables, where $D = F + p$. Note that D is the number of natural variables in the fundamental equation for a system. For a one-phase system involving only PV work, $D = N_s + 2$, as discussed after equation 2.2-12. The number F of independent intensive variables and the number D of natural variables for a system are unique, but there are usually multiple choices of these variables. The choice of independent intensive variables F and natural variables D is arbitrary, but the natural variables must include as many extensive variables as there are phases. For example, for the one-phase system described by equation 2.2-8, the $F = N_s + 1$ intensive variables can be chosen to be $T, P, x_1, x_2, \ldots, x_{N_s-1}$ and the $D = N_s + 2$ natural variables can be chosen to be $T, P, n_1, n_2, \ldots, n_{N_s}$ or $T, P, x_1, x_2, \ldots, x_{N_s-1}$ and n (total amount in the system).

■ 2.5 LEGENDRE TRANSFORMS FOR THE DEFINITION OF ADDITIONAL THERMODYNAMIC POTENTIALS

The internal energy U has some remarkable properties and leads to many equations between the thermodynamic properties of a system, but S, V, and $\{n_i\}$ are not convenient natural variables, except for an isolated system. As shown in Section 2.2, Legendre transforms can be used to introduce other sets of $N_s + 2$ natural variables. A **Legendre transform** is a change in natural variables that is accomplished by defining a new thermodynamic potential by subtracting from the internal energy (or other thermodynamic potential) one or more products of conjugate variables. As mentioned after equation 2.2-2, examples of conjugate pairs are T and S, P and V, and μ_i and n_i. More conjugate pairs are introduced in Section 2.7. Callen (1985) emphasizes that no thermodynamic information is lost in making a Legendre transform. For reviews on Legendre transforms, see Alberty (1994d) and Alberty et al. (2001).

Legendre transforms are also used in mechanics to obtain more convenient independent variables (Goldstein, 1980). The **Lagrangian** L is a function of coordinates and velocities, but it is often more convenient to define the **Hamiltonian** H with a Legendre transform because the Hamiltonian is a function of coordinates and momenta. Quantum mechanics is based on the Hamiltonian rather than the Lagrangian.

In this section we will consider the Legendre transforms that define the **enthalpy** H, **Helmholtz energy** A, and **Gibbs energy** G.

$$H = U + PV \tag{2.5-1}$$

$$A = U - TS \tag{2.5-2}$$

$$G = U + PV - TS \tag{2.5-3}$$

If a system can be described by $dU = T\,dS - P\,dV$, there are only four thermodynamic potentials that can be defined in this way.

We will consider only the equations for the Gibbs energy and the enthalpy. According to equation 2.5-3, the differential of G is given by

$$dG = dU + P\,dV + V\,dP - T\,dS - S\,dT \tag{2.5-4}$$

Substituting equation 2.2-8 yields the **fundamental equation for the Gibbs energy for a one-phase system without chemical reactions**:

$$dG = -SdT + VdP + \sum_{i=1}^{N_s} \mu_i dn_i \qquad (2.5\text{-}5)$$

This shows that the natural variables of G for a one-phase nonreaction system are T, P, and $\{n_i\}$. The number of natural variables is not changed by a Legendre transform because conjugate variables are interchanged as natural variables. In contrast with the natural variables for U, the natural variables for G are two intensive properties and N_s extensive properties. These are generally much more convenient natural variables than S, V, and $\{n_i\}$. Thus thermodynamic potentials can be defined to have the desired set of natural variables.

It is important to understand that the change in variables provided by using a Legendre transform is quite different from the usual (much more frequent) type of change in variables. For example, in Chapter 1 functions of $[H^+]$ were converted to functions of pH by simply substituting $[H^+] = 10^{-pH}$. When a Legendre transform of a thermodynamic potential is defined, the new variable that is introduced is a partial derivative of that thermodynamic potential. For example, when U is known as a function of S, V, and $\{n_i\}$, the enthalpy is defined by use of $H = U + PV = U + V(\partial U/\partial V)_S$, and the natural variables of H are indicated by $H(S, -(\partial U/\partial V)_S)$ or $H(S, P)$. When the Gibbs energy is defined by use of $G = U + PV - TS = U + V(\partial U/\partial V)_S - S(\partial U/\partial S)_V$, and the natural variables of G are indicated by $G((\partial U/\partial S)_V, -(\partial U/\partial V)_S)$ or $G(T, P)$.

Since the natural variables of the Gibbs energy are T, P, and $\{n_i\}$, calculus yields

$$dG = \left(\frac{\partial G}{\partial T}\right)_{P,\{n_i\}} dT + \left(\frac{\partial G}{\partial P}\right)_{T,\{n_i\}} dP + \sum_{i=1}^{N_s} \left(\frac{\partial G}{\partial n_i}\right)_{T,P,n_j \neq n_i} dn_i \qquad (2.5\text{-}6)$$

Comparison with equation 2.5-5 indicates that

$$S = -\left(\frac{\partial G}{\partial T}\right)_{P,\{n_i\}} = S(T, P, \{n_i\}) \qquad (2.5\text{-}7)$$

$$V = \left(\frac{\partial G}{\partial P}\right)_{T,\{n_i\}} = V(T, P, \{n_i\}) \qquad (2.5\text{-}8)$$

$$\mu_i = \left(\frac{\partial G}{\partial n_i}\right)_{T,P,n_j \neq n_i} = \mu_i(T, P, \{n_i\}) \qquad (2.5\text{-}9)$$

Thus, if G can be determined as a function of T, P, and $\{n_i\}$, all of the thermodynamic properties of the system can be calculated. These $N_s + 2$ equations (that is equations 2.5-7 to 2.5-9) are often referred to as equations of state. Only $N_s + 1$ equations of state are independent, and so if $N_s + 1$ of them can be determined experimentally, the remaining equation of state can be calculated.

The criterion of equilibrium for this one-phase system without chemical reactions is $dG \leq 0$ at constant T, P, and $\{n_i\}$. In other words, **the Gibbs energy decreases in a spontaneous change in a system with constant T, P, and $\{n_i\}$**. For this system the $F = N_s + 1$ independent intensive variables can be chosen to be T, P, $x_1, x_2, \ldots, x_{N-1}$, and the $D = N_s + 2$ natural variables can be chosen to be T, P, $x_1, x_2, \ldots, x_{N-1}$ and n (total amount in the system), or T, P, $n_1, n_2, \ldots, n_N$.

The derivation in equations like 2.5-4 to 2.5-9 can be repeated for H and A. This shows that the natural variables for H are S, P, and $\{n_i\}$, and for A are T, V, and $\{n_i\}$. These thermodynamic potentials provide the following criteria for spontaneous change and equilibrium: $dH \leq 0$ at constant S, P, and $\{n_i\}$; $dA \leq 0$ at constant T, V, and $\{n_i\}$.

Substituting the **integrated fundamental equation** for U (equation 2.2-14) in the Legendre transforms defining H, A, and G shows that

$$H = TS + \Sigma \mu_i n_i \tag{2.5-10}$$

$$A = -PV + \Sigma \mu_i n_i \tag{2.5-11}$$

$$G = \Sigma \mu_i n_i \tag{2.5-12}$$

The interesting thing about these equations is that only the Gibbs energy G can be calculated by adding contributions from individual species. These thermodynamic potentials can be determined as functions of other variables, but **only when they are determined as functions of natural variables can all of the thermodynamic properties be obtained by taking partial derivatives.** Equations 2.5-10 to 2.5-12 can also be obtained by integrating the corresponding fundamental equations at constant values of the intensive variables.

The fundamental equation for the Gibbs energy (2.5-5) yields the following Maxwell equations:

$$-\left(\frac{\partial S}{\partial P}\right)_{T,\{n_i\}} = \left(\frac{\partial V}{\partial T}\right)_{P,\{n_i\}} \tag{2.5-13}$$

$$-\left(\frac{\partial S}{\partial n_i}\right)_{T,P,n_j \neq n_i} = \left(\frac{\partial \mu_i}{\partial T}\right)_{P,\{n_i\}} = -S_m(i) \tag{2.5-14}$$

$$\left(\frac{\partial V}{\partial n_i}\right)_{T,P,n_j \neq n_i} = \left(\frac{\partial \mu_i}{\partial P}\right)_{T,\{n_i\}} = V_m(i) \tag{2.5-15}$$

$$\left(\frac{\partial \mu_i}{\partial n_j}\right)_{T,P,n_j \neq n_i} = \left(\frac{\partial \mu_j}{\partial n_i}\right)_{T,P,n_j \neq n_i} \tag{2.5-16}$$

where $S_m(i)$ is the **molar entropy** of species i and $V_m(i)$ is its **molar volume**.

The Helmholtz energy is not very useful as a crterion for spontaneious change and equilibrium in biochemistry because experiments are not done at constant volume. However, the enthalpy is important in biochemistry because it is connected with heat evolution and the change of the equilibrium constant with temperature. The fundamental equation for the enthalpy is

$$dH = T\,dS + P\,dV + \sum_{i=1}^{N_s} \mu_i\,dn_i \tag{2.5-17}$$

Since the enthalpy is defined by $H = U + PV$, its total differential is $dH = dU + P\,dV + V\,dP$. Substituting the equation $dU = dq - P\,dV$, given earlier in Section 2.2, yields $dH = dq + V\,dP$. At constant pressure the change in enthalpy ΔH is equal to the heat q absorbed by the system in the process, which may be irreversible. Thus the change in enthalpy ΔH can be determined calorimetrically. The change in enthalpy can also be determined using the Gibbs-Helmholtz equation, which is introduced in the next paragraph, without using a calorimeter.

Equations 2.5-1 and 2.5-3 show that $G = H - TS$. Substituting the expression for S from equation 2.5-7 yields

$$H = G - T\left(\frac{\partial G}{\partial T}\right)_{P,\{n_i\}} = -T^2\left(\frac{\partial(G/T)}{\partial T}\right)_{P,\{n_i\}} \tag{2.5-18}$$

This is referred to as a **Gibbs-Helmholtz equation**, and it provides a convenient way to calculate H if G can be determined as a function of T, P, and $\{n_i\}$. There is a corresponding relation between the internal energy U and the Helmholtz energy, which is defined by equation 2.5-2:

$$U = A - T\left(\frac{\partial A}{\partial T}\right)_{P,\{n_i\}} = -T^2\left(\frac{\partial(A/T)}{\partial T}\right)_{P,\{n_i\}} \tag{2.5-19}$$

This is also referred to as a Gibbs-Helmholtz equation.

Since G is additive in terms of μ_i (equation 2.5 12), S and H are also additive in terms of partial molar entropies and partial molar enthalpies, respectively:

$$S = -\left(\frac{\partial \Sigma \mu_i n_i}{\partial T}\right)_{P,\{n_i\}} = \Sigma n_i S_m(i) \qquad (2.5\text{-}20)$$

The partial molar entropy S_{mi} was defined in equation 2.5 14. The additivity of the enthalpy can be seen by substituting equation 2.5-12 in equation 2.5-18:

$$H = -T^2\left(\frac{\partial (\Sigma \mu_i n_i/T)}{\partial T}\right)_{P,\{n_i\}} = \Sigma n_i H_{mi} \qquad (2.5\text{-}21)$$

where H_{mi} is the molar enthalpy of species i. Thus the chemical potential of a species is given by

$$\mu_i = H_{mi} - TS_{mi} \qquad (2.5\text{-}22)$$

When a system at constant pressure is heated, it absorbs heat, and the heat capacity at constant pressure C_P is defined by

$$C_P = \left(\frac{\partial H}{\partial T}\right)_P \qquad (2.5\text{-}23)$$

Substituting equation 2.5-7 in $H = G - TS$ yields

$$H = G - T\left(\frac{\partial G}{\partial T}\right)_P \qquad (2.5\text{-}24)$$

Substituting this into the definition of C_P yields

$$C_P = \left(\frac{\partial G}{\partial T}\right)_P - \left(\frac{\partial G}{\partial T}\right)_P - T\left(\frac{\partial^2 G}{\partial T^2}\right)_P = -T\left(\frac{\partial^2 G}{\partial T^2}\right)_P \qquad (2.5\text{-}25)$$

Thus we have seen that all the thermodynamic properties of a one-phase nonreaction system can be calculated from $G(T, P, \{n_i\})$.

Although Legendre transforms introducing chemical potentials of species as natural variables are not discussed until Chapter 3, there is one Legendre transform involving chemical potentials of species that needs to be given here, and that is the **complete Legendre transform** U' of the internal energy defined by

$$U' = U - TS + PV - \sum_{i=1}^{N_s} \mu_i n_i = 0 \qquad (2.5\text{-}26)$$

This transformed internal energy is equal to zero, as indicated by equation 2.2-8. The total differential of U' is

$$0 = dU - T\,dS + S\,dT + P\,dV + V\,dP - \sum_{i=1}^{N_s} \mu_i\,dn_i - \sum_{i=1}^{N_s} n_i\,d\mu_i \quad (2.5\text{-}27)$$

Substituting the fundamental equation for U yields the Gibbs-Duhem equation 2.4-2.

Now we are in a position to generalize on the number of different thermodynamic potentials there are for a system. The number of ways to subtract products of conjugate variables, zero-at-a-time, one-at-a-time, and two-at-a-time, is 2^k, where k is the number of conjugate pairs involved. In probability theory the number 2^k of ways is referred to as the number of sets of k elements. For a one-phase system involving PV work but no chemical reactions, the number of natural variables is D and the number of different thermodynamic potentials is 2^D. If $D = 2$ (as in $dU = TdS - PdV$), the number of different thermodynamic potentials is $2^2 = 4$, as we have seen with U, H, A, and G. If $D = 3$ (as in $dU = TdS - PdV + \mu\,dn$), the number of different thermodynamic potentials is $2^3 = 8$.

Thus the Gibbs-Duhem equation represents one of the 2^D thermodynamic potentials that can be defined for a system, but this thermodynamic potential is equal to zero. It should be emphasized that there is a single Gibbs-Duhem equation for a one-phase system and that it can be derived from U, H, A, or G.

When work other than PV work is involved in a system (see Section 2.7), Legendre transforms can be used to introduce intensive variables in addition to T and P as natural variables in the fundamental equation for the system. Each Legendre transform defines a new thermodynamic potential that needs a symbol and a name. Since every system has 2^D possible thermodynamic potentials, and each needs a symbol and a name, nomenclature becomes a problem. Callen (1985) showed how each possible thermodynamic potential can be given an unambiguous symbol. **Callen nomenclature** uses the symbol $U[\cdots]$, where $\cdots$ is a list of the intensive variables introduced as intensive variables in defining the particular thermodynamic potential based on the internal energy. Thus the enthalpy H can be represented by $U[P]$, the Helmholtz energy A can be represented by $U[T]$, and the Gibbs energy G can be represented by $U[T,P]$. Since all possible thermodynamic potentials can be represented in this way, this is a good method to use when there is a possibility of confusion. However, in practice, it is convenient to use symbols like U', H', A', and G' to represent transformed properties that are similar to U, H, A, and G. When this is done, it is important to specify which intensive properties have been introduced in defining these primed properties.

The number of Maxwell equations for each of the possible thermodynamic potentials is given by $D(D-1)/2$, and the number of Maxwell equations for the thermodynamic potentials for a system related by Legendre transforms is $[D(D-1)/2]2^D$. Examples are given in the following section.

■ 2.6   THERMODYNAMIC POTENTIALS FOR A SINGLE-PHASE SYSTEMS WITH ONE SPECIES

The fundamental equation for U for a single-phase system with one species is

$$dU = TdS - PdV + \mu dn \tag{2.6-1}$$

Integration of this fundamental equation at constant values of the intensive variables yields

$$U = TS + PV + \mu n \tag{2.6-2}$$

Since there are $D = 3$ natural variables, there are $2^3 - 1 = 7$ possible Legendre transforms. The Legendre transforms defining H, A, and G are given in equations 2.5-1 to 2.5-3, and the four remaining Legendre transforms are

$$U[\mu] = U - \mu n \tag{2.6-3}$$

$$U[P,\mu] = U + PV - \mu n \tag{2.6-4}$$

$$U[T,\mu] = U - TS - \mu n \tag{2.6-5}$$

$$U[T,P,\mu] = U + PV - TS - \mu n = 0 \tag{2.6-6}$$

These four Legendre transforms introduce the chemical potential as a natural variable. The last thermodynamic potential $U[T,P,\mu]$ defined in equation 2.6-6 is equal to zero because it is the complete Legendre transform for the system, and this Legendre transform leads to the Gibbs-Duhem equation for the system.

Three of the eight thermodynamic potentials for a system with one species are frequently used in statistical mechanics (McQuarrie, 2000), and there are generally accepted symbols for the corresponding partition functions: $U[T] = A = -RT \ln Q$, where Q is the **canonical ensemble partition function**;

$U[T, P] = G = -RT \ln \Delta$, where Δ is the **isothermal-isobaric partition function**; and $U[T, \mu] = -RT \ln \Xi$, where Ξ is the **grand canonical ensemble partition function**. When a system involves several species, but only one can pass through a membrane to a reservoir, $U(T, \mu_1] = -RT \ln \Psi$, where Ψ is the **semigrand ensemble partition function**. The last chapter of the book is on semigrand partition functions.

Taking the differentials of the seven thermodynamic potentials defined above and substituting equation 2.6-1 yields the fundamental equations for these seven additional thermodynamic potentials:

$$dH - TdS + VdP + \mu dn \qquad (2.6\text{-}7)$$

$$dA = -SdT - PdV + \mu dn \qquad (2.6\text{-}8)$$

$$dG = -SdT + VdP + \mu dn \qquad (2.6\text{-}9)$$

$$dU[\mu] = TdS - PdV - nd\mu \qquad (2.6\text{-}10)$$

$$dU[P, \mu] = TdS + VdP - nd\mu \qquad (2.6\text{-}11)$$

$$dU[T, \mu] = -SdT - PdV - nd\mu \qquad (2.6\text{-}12)$$

$$dU[T, P, \mu] = -SdT + VdP - nd\mu = 0 \qquad (2.6\text{-}13)$$

This last equation is the Gibbs-Duhem equation for the system, and it shows that only two of the three intensive properties (T, P, and μ) are independent for a system containing one substance. Because of the Gibbs-Duhem equation, we can say that the chemical potential of a pure substance substance is a function of temperature and pressure. The number F of independent intensive variables is $F = 1 - 1 + 2 = 2$, and so $D = F + p = 2 + 1 = 3$. Each of these fundamental equations yields $D(D - 1)/2 = 3$ Maxwell equations, and there are 24 Maxwell equations for the system.

The integrated forms of the eight fundamental equations for this system are

$$U(S, V, n) = TS - PV + \mu n \qquad (2.6\text{-}14)$$

$$H(S, P, n) = TS + \mu n \qquad (2.6\text{-}15)$$

$$A(T, V, n) = -PV + \mu n \qquad (2.6\text{-}16)$$

$$G(T, P, n) = \mu n \qquad (2.6\text{-}17)$$

$$U[\mu](S, V, \mu) = TS - PV \qquad (2.6\text{-}18)$$

$$U[P, \mu](S, P, \mu) = TS \qquad (2.6\text{-}19)$$

$$U[T, \mu](T, V, \mu) = -PV \qquad (2.6\text{-}20)$$

$$U[T, P, \mu](T, P, \mu) = 0 \qquad (2.6\text{-}21)$$

where the natural variables are shown in parentheses.

The basic question in all of thermodynamics is: A certain system is under such and such constraints, what is the equilibrium state that it can go to spontaneously? The amazing thing is that this question can be answered by making macroscopic measurements. Thermodynamics does not deal with the question as to how long it will take to reach equilibrium. We now have seven criteria for equilibrium in a one-phase system with one species and only PV work. The criteria of equilibrium provided by these thermodynamic potentials are $(dU)_{S,V,n} \leqslant 0$, $(dH)_{S,P,n} \leqslant 0$, $(dA)_{T,V,n} \leqslant 0$, $(dG)_{T,P,n} \leqslant 0$, $(dU[\mu])_{S,V,\mu} \leqslant 0$, $(dU[P, \mu])_{S,P,\mu} \leqslant 0$, and $(dU[T, \mu])_{T,V,\mu} \leqslant 0$.

The reason for going into this much detail on all of the thermodynamic potentials that can be defined for a one-phase, one-species system and the corresponding criteria for spontaneous change is to illustrate the process by which these thermodynamic potentials are defined and how they provide criteria for

Table 2.1.   Conjugate Properties Involved in Various Kinds of Work

	Extensive	Intensive	Differential Work
PV	V	$-P$	$-P\,dV$
Chemical			
non rx system	n_i	μ_i	$\mu_i\,dn_i$
rx system	n_{ci}	μ_i	$\mu_i\,dn_{ci}$
Electrical	$Q_i = Fz_in_i$	ϕ_i	$\phi_i\,dQ_i$
Mechanical	L	f	$f\,dL$
Surface	A_s	γ	$\gamma\,dA_s$
Electric polarization	$\boldsymbol{p}$	$\boldsymbol{E}$	$\boldsymbol{E}\,d\boldsymbol{p}$
Magnetic polarization	$\boldsymbol{m}$	$\boldsymbol{B}$	$\boldsymbol{B}\,d\boldsymbol{m}$

spontaneity and equilibrium under various conditions. None of these equations is immediately applicable to biochemical reactions because they are for systems containing one species. Chemical reactions are introduced in the next chapter.

■ 2.7   OTHER KINDS OF WORK

In this chapter we have discussed systems involving PV work and the transfer of species into or out of the system ($\mu_i\,dn_i$), but other kinds of work may be involved in a biochemical system. The extensive and intensive properties that are involved in various types of work are given in Table 2.1.

Table 2.1, n_{ci} is the amount of a component (see Section 3.3), ϕ_i is the electric potential of the phase containing species i, Q_i is the contribution of species i to the electric charge of a phase, z_i is the charge number, F is the Faraday constant, f is force of elongation, L is length in the direction of the force, γ is surface tension, A_s is surface area, $\boldsymbol{E}$ is electric field strength, $\boldsymbol{p}$ is the electric dipole moment of the system, $\boldsymbol{B}$ is magnetic field strength (magnetic flux density), and $\boldsymbol{m}$ is the magnetic moment of the system. Vectors are indicated by boldface type.

If a single additional work term is involved, the fundamental equation for U is

$$dU = TdS - VdP + \sum_{i=1}^{N_s} \mu_i\,dn_i + X\,dY \qquad (2.7\text{-}1)$$

where Y is an extensive variable. This shows that $D = N_s + 3$. The additional work terms should be independent of $\{n_i\}$ because natural variables must be independent. The same form of work terms appear in the fundamental equations for H, A, and G. In order to introduce the intensive properties in other kinds of work as natural variables, it is necessary to use Legendre transforms.

■ 2.8   CALCULATION OF THERMODYNAMIC PROPERTIES OF A MONATOMIC IDEAL GAS BY TAKING DERIVATIVES OF A THERMODYNAMIC POTENTIAL

The treatments in the preceding sections have been pretty abstract, and it may be hard to understand statements like: Thus, if G can be determined as a function of T, P, and $\{n_i\}$, all of the thermodynamic properties of the system can be calculated" (which appeared after equation 2.5-9). However, there is one case where this can be demonstrated in detail, and that is for a monatomic ideal gas (Greiner, Neise, and Stocker, 1995). Statistical mechanics shows that the Gibbs energy of a monatomic ideal gas without electronic excitation (Silbey and Alberty,

2001) is given as a function of T, P, and n by

$$G = -nRT \ln \left[\left(\frac{2\pi m k T}{h^2} \right)^{3/2} \frac{kT}{P} \right] \tag{2.8-1}$$

where m is the mass of the atom, k is the Boltzmann constant, and h is Planck's constant. Equations 2.5-7 to 2.5-9 show that S, V, and μ can be calculated by taking partial derivatives of G with respect to T, P, and n. Taking these partial derivatives yields

$$S = nR \left(\ln \left[\left(\frac{2\pi m k T}{h^2} \right)^{3/2} \frac{kT}{P} \right] + \frac{5}{2} \right) \tag{2.8-2}$$

$$V = \frac{nRT}{P} \tag{2.8-3}$$

$$\mu = -RT \ln \left[\left(\frac{2\pi m k T}{h^2} \right)^{3/2} \frac{kT}{P} \right] \tag{2.8-4}$$

Equation 2.8-2 is referred to as the **Sackur-Tetrode equation**. Since we have expressions for these three properties, we can calculate the properties U, H, A, and C_p:

$$H = \left(\frac{5}{2} \right) nRT \tag{2.8-5}$$

$$U = \left(\frac{3}{2} \right) nRT \tag{2.8-6}$$

$$A = -nRT \ln \left[\left(\frac{2\pi m k T}{h^2} \right)^{3/2} \frac{kT}{P} \right] - nRT \tag{2.8-7}$$

$$C_P = \left(\frac{5}{2} \right) nR \tag{2.8-8}$$

Note that U corresponds with the translational kinetic energy in three directions. Thus all the thermodynamic properties of an ideal monatomic gas can be calculated from $G(T, P, n)$.

Equations 2.8-2 and 2.8-4 can be used to derive the expressions for the standard molar entropies and standard molar Gibbs energies of a monatomic gas:

$$S_m = S_m^\circ - R \ln \left(\frac{P}{P^\circ} \right) \tag{2.8-9}$$

$$G_m = G_m^\circ + RT \ln \left(\frac{P}{P^\circ} \right) \tag{2.8-10}$$

where P° is 1 bar and

$$S_m^\circ = R \left(\ln \left[\left(\frac{2\pi m k T}{h^2} \right)^{3/2} \frac{kT}{P^\circ} \right] + \frac{5}{2} \right) \tag{2.8-11}$$

$$G_m^\circ = \mu^\circ = -RT \ln \left[\left(\frac{2\pi m k T}{h^2} \right)^{3/2} \frac{kT}{P^\circ} \right] \tag{2.8-12}$$

These are the properties of the monatomic gas at a pressure of 1 bar. It should be pointed out that this standard molar Gibbs energy is not the $\Delta_f G^\circ$ of thermodynamic tables because there the convention in thermodynamics is that the standard formation properties of elements in their reference states are set equal to zero at each temperature. However, the standard molar entropies of monatomic gases without electronic excitation calculated using equation 2.8-11 are given in thermodynamic tables.

Thus we have demonstrated the remarkable fact that equation 2.8-1 makes it possible to calculate all the thermodynamic properties for a monotomic ideal gas without electronic excitation. Here we have considered an ideal monatomic gas, but this illustrates the general conclusion that if any thermodynamic potential of a one-component system can be determined as a function of its natural variables, all of the thermodynamic properties of the system can be calculated.

Chemical Equilibrium in Aqueous Solutions

3

When a chemical reaction occurs in a closed system at constant T and P, the criterion for spontaneous change and equilibrium is no longer $dG \leqslant 0$ at constant T, P, and $\{n_i\}$ because the amounts of species change in the reaction. Therefore, the following question arises: If the amounts of species are not constant during the approach to equilibrium in a reaction system, what is? The answer is: The amounts n_{ci} of components are constant in a reaction system. When a chemical reaction occurs in a closed system, the amounts of atoms of elements and electric charge are conserved. Atoms of elements and electric charge can be taken as components, but some of these conservation equations may be redundant and are therefore not needed. Groups of atoms in molecules can also be chosen as components. This is important in biochemistry when large molecules are involved because counting atoms becomes laborious. Various choices of components can be made for a reaction system, but the number of components is independent of the set of components chosen. A particular set of components may be especially useful, depending on the objective of the calculation. In an independent set of conservation equations, no equation in the set can be obtained by adding and subtracting other equations in the set. Thus we will see in this chapter that the criterion for spontaneous change and chemical equilibrium in a closed system is

$dG \leqslant 0$ at constant T, P, and $\{n_{ci}\}$, where there are C components with amounts $\{n_{ci}\}$. Components are discussed in Section 3.3, and the various choices of components that can be used will become clearer in Chapter 5 on matrices.

In this chapter we will find that when isomers are in chemical equilibrium, it is convenient to treat isomer groups like species in order to reduce the number of terms in the fundamental equation. We will also discuss the effect of ionic strength and temperature on equilibrium constants and thermodynamic properties of species. More introductory material on the thermodynamics of chemical reactions is provided in Silbey and Alberty (2001).

■ 3.1 DERIVATION OF THE EXPRESSION FOR THE EQUILIBRIUM CONSTANT

When a chemical reaction occurs in a system, the changes in the amounts n_i of species are not independent because of the stoichiometry of the reaction that occurs. A single chemical reaction can be represented by the reaction equation

$$\sum_{i=1}^{N_s} v_i \mathbf{B}_i = 0 \tag{3.1-1}$$

where $\mathbf{B}_i$ represents species i and N_s is the number of different species. Chemical reactions balance the atoms of all elements and electric charge. The **stoichiometric numbers** v_i are positive for products and negative for reactants. The amount n_i of species i at any stage in a reaction is given by

$$n_i = n_{i0} + v_i \xi \tag{3.1-2}$$

where n_{i0} *is the initial amount of species* i and ξ is the **extent of reaction**. It is evident from this definition of ξ that it is an extensive property. Stoichiometric numbers are dimensionless, and so the extent of reaction is expressed in moles. The differential of the amount of species i is given by

$$dn_i = v_i d\xi \tag{3.1-3}$$

When a single chemical reaction occurs in a closed system, the differential of the Gibbs energy (see equation 2.5-5) is given by

$$dG = -S \, dT + V \, dP + \left(\sum_{i=1}^{N_s} \mu_i v_i \right) d\xi \tag{3.1-4}$$

This form of the fundamental equation applies at each stage of the reaction. The rate of change of G with extent of reaction for a closed system with a single reaction at constant T and P is given by

$$\left(\frac{\partial G}{\partial \xi} \right)_{T,P} = \sum_{i=1}^{N_s} \mu_i v_i = \Delta_r G \tag{3.1-5}$$

where $\Delta_r G$ is referred to as the **reaction Gibbs energy**. The Gibbs energy of the system is at a minimum at equilibrium, where $(\partial G / \partial \xi)_{T,P} = 0$. At the minimum Gibbs energy, the **equilibrium condition** is

$$\sum_{i=1}^{N_s} v_i (\mu_i)_{eq} = 0 \tag{3.1-6}$$

Notice that this relation has the same form as the chemical equation (equation 3.1-1).

To discuss equilibrium in a chemical reaction system, it is convenient to introduce the activity a_i of a species to replace the chemical potential of a species because a_i is more closely related to partial pressures and concentrations of species. The **activity** of a species is defined by

$$\mu_i = \mu_i^0 + RT \ln a_i \qquad (3.1\text{-}7)$$

where μ_i^0, which is referred to as the standard chemical potential, is the chemical potential when $a_i = 1$. A superscript zero is used to designate a standard property, and so μ_i^0 is the **standard chemical potential** of species i. For ideal mixtures of ideal gases, a_i can be replaced by P_i/P^0, where P^0 is the standard state pressure, and for ideal solutions a_i can be replaced by c_i/c^0, where c^0 is the standard concentration. Note that the activity a_i of species i is dimensionless. We will use molar concentrations, but measurements in physical chemistry are frequently based on molal concentrations (mol kg^{-1}). Molal concentrations m_i have the advantage that they do not change with temperature.

From one point of view nothing is gained by using equation 3.1-7 to define the activity a_i of a species and using it to replace the chemical potential μ_i of the species. The difference between μ_i and a_i is that μ_i of an ideal gas goes from $-\infty$ to ∞, whereas a_i goes from 0 to ∞. Thus the activity of a species in solution is more closely related to its concentration than μ_i is. However, the activity of a species in solution is directly proportional to its concentration only for ideal solutions. In general, the activity of a species in solution is given by $a_i = \gamma_i c_i$, where γ_i is the **activity coefficient** of species i. The activity coefficient of a solute is a function of the concentration, especially for ions. Strictly speaking (Mills et al., 1993), this relation should be written $a_i = \gamma_i c_i/c^0$, where c^0 is the standard concentration (1 M). However, c^0 is omitted in this book to simplify the equations. Thus equation 3.1-7 is written

$$\mu_i = \mu_i^0 + RT \ln \gamma_i c_i \qquad (3.1\text{-}8)$$

When using the molar concentration scale, the convention is that the activity coefficient of a species approaches unity as the concentration of the species approaches zero. In discussing biochemical reactions in dilute aqueous solutions, effects on activity coefficients arise primarily because of electrostatic interactions between charged species and depend on the ionic strength (see Section 1.2 and Section 3.6). Since the ionic strength is under the control of the investigator and is nearly constant during the approach to equilibrium when a biochemical reaction is carried out in dilute aqueous solution with a buffer, we can postpone discussing the effects of ionic strength to Section 3.6 by making the following observation: Equation 3.1-8 can be written

$$\mu_i = \mu_i^0 + RT \ln \gamma_i + RT \ln c_i \qquad (3.1\text{-}9)$$

In discussing biochemical thermodynamics, however, it is convenient to write this equation as

$$\mu_i = \mu_i^0 + RT \ln c_i \qquad (3.1\text{-}10)$$

where **μ_i and μ_i^0 are functions of the ionic strength**. In equations 3.1-9 and 3.1-10, μ_i and μ_i^0 have been used in two different ways, but in the rest of the book equation 3.1-10 will always be used. In other words, chemical potentials and other thermodynamic properties of species in dilute aqueous solutions will be taken to be functions of the ionic strength. This will allow us to avoid including γ_i in many places (even though the effect of ionic strength is taken into account) and to treat solutions at a specified ionic strength as "ideal solutions," that is as solutions following equation 3.1-10. We have already seen an example of this in the treatment of pH in Section 1.2.

Substituting equation 3.1-10 in 3.1-6 yields

$$\sum_{i=1}^{N_s} v_i \mu_i^0 = -RT \sum_{i=1}^{N_s} v_i \ln(c_i)_{eq} = -RT \ln \prod_{i=1}^{N_s} (c_i)_{eq}^{v_i} \tag{3.1-11}$$

Using the nomenclature of equation 3.1-5, equation 3.1-11 can be written as

$$\Delta_r G^0 = \sum_{i=0}^{N_s} v_i \Delta_f G_i^0 = -RT \ln K \tag{3.1-12}$$

where $\Delta_f G_i^0$ is the standard reaction Gibbs energy and K is the **equilibrium constant** for a chemical reaction written in terms of species:

$$K = \prod_{i=1}^{N_s} (c_i)_{eq}^{v_i} \tag{3.1-13}$$

An equilibrium constant must always be accompanied by a chemical equation. This equation is often used without the subscript "eq" that reminds us that the concentrations are equilibrium values. Strictly speaking, this equation should be written as $K = \Pi(c_i/c^0)_{eq}^{v_i}$, but the standard concentration $c^0 = 1\,M$ will be omitted, as mentioned before equation 3.1-8. Thus the equilibrium constant will be treated as a dimensionless quantity, as, of course, it must be if we are going to take its logarithm.

When H_2O is a reactant in a chemical reaction in dilute aqueous solutions, its molar concentration is not included in equation 3.1-13. The reason is that in reactions in dilute aqueous solutions the activity of water does not change significantly. The convention is that H_2O is represented in the expression for the equilibrium constant by its activity, which is essentially unity independent of the extent of reaction. However, $\Delta_f G^0(H_2O)$ is included in the calculation of $\Delta_r G^0$ using equation 3.1-12 and $\Delta_f H^0(H_2O)$ is included in the calculation of $\Delta_r H^0$ using equation 3.2-13, which is given later.

To clarify the nature of the equilibrium state of a reaction system, consider the solution reaction A = B. When one liter of ideal solution initially containing A at 1 M is considered, the Gibbs energy of the reactants at any time is given by

$$G = n_A(\mu_A^0 + RT \ln[A]) + n_B(\mu_B^0 + RT \ln[B]) \tag{3.1-14}$$

Since $n_A = 1 - \xi$ and $n_B = \xi$,

$$G = (1 - \xi)\mu_A^0 + \xi\mu_B^0 + RT[(1 - \xi)\ln(1 - (1 - \xi)) + \xi \ln \xi] \tag{3.1-15}$$

At the equilibrium state of the system, the Gibbs energy is at a minimum, and the equilibrium extent of reaction is ξ_{eq}. At $(\partial G/\partial \xi)_P = 0$,

$$\mu_B^0 - \mu_A^0 = -RT \ln(\xi_{eq}/(1 - \xi_{eq})) = -RT \ln K \tag{3.1-16}$$

Figure 3.1 shows a plot of the Gibbs energy G of a reaction system A = B as a function of the extent of reaction ξ when $\mu_A^0 = 20\,kJ\,mol^{-1}$ and $\mu_B^0 = 18\,kJ\,mol^{-1}$.

■ 3.2 CHANGES IN THERMODYNAMIC PROPERTIES IN CHEMICAL REACTIONS

In treating the fundamental equations of thermodynamics, chemical potentials of species are always used, but in making calculations when T and P are independent variables, chemical potentials are replaced by Gibbs energies of formation $\Delta_f G_i$. Therefore, we will use equation 3.1-10 in the form

$$\Delta_f G_i = \Delta_f G_i^0 + RT \ln c_i \tag{3.2-1}$$

where $\Delta_f G_i$ is the Gibbs energy of formation of species i at concentration c_i from its elements, each in its reference state. The **standard Gibbs energy of formation**

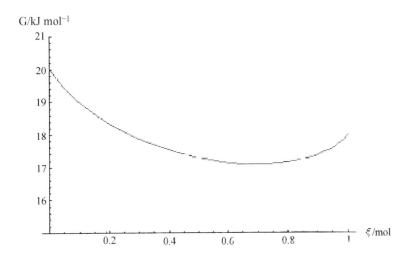

Fig. 3.1 Plot of the Gibbs energy of the reaction system A = B as a function of extent of reaction at 298.15 K (see Problem 3.2).

$\Delta_f G_i^0$ of species i is the Gibbs energy change when a mole of the species in its standard state (in the ideal gas state at 1 bar or in the ideal aqueous solution at 1 M) is formed from its elements in their reference states. The standard Gibbs energy of formation of an ion depends on the ionic strength, and the equation for $\Delta_f G_i$ has the form (equation 3.2-1) for an ideal solution at a specified ionic strength. The advantage of this procedure is that we can write equilibrium expressions in terms of concentrations and avoid the complication of dealing with activity coefficients in each calculation. The activity coefficients are taken into account in the construction of thermodynamic tables for the convenience of the user.

Substituting equation 3.2-1 in equation 3.1-5 yields

$$\Delta_r G = \sum_{i=1}^{N_s} v_i \Delta_f G_i^0 + RT \sum_{i=1}^{N_s} v_i \ln c_i = \Delta_r G^0 + RT \ln \prod_{i=1}^{N_s} c_i^{v_i} = \Delta_r G^0 + RT \ln Q$$

$$(3.2\text{-}2)$$

where Q is the **reaction quotient**:

$$Q = \prod_{i=1}^{N_s} c_i^{v_i} \qquad (3.2\text{-}3)$$

The concentrations in Q have arbitrary values. Note that the **standard reaction Gibbs energy** is given by

$$\Delta_r G^0 = \sum_{i=1}^{N_s} v_i \Delta_f G_i^0 \qquad (3.2\text{-}4)$$

where $\Delta_f G_i^0$ is the **standard Gibbs energy of formation** of species i. Thus each species in a reaction makes its own contribution to the standard Gibbs energy of reaction and to the equilibrium constant; this makes it possible to construct tables of standard thermodynamic properties of species.

The other thermodynamic properties for a reaction are related to the Gibbs energy of reaction through Maxwell equations (see Section 2.3). Because of equation 3.1-5, equation 3.1-4 can be written

$$dG = -S\,dT + V\,dP + \Delta_r G\,d\xi \qquad (3.2\text{-}5)$$

which applies at each stage of the reaction. This form of the fundamental equation for G has three Maxwell equations.

$$-\left(\frac{\partial S}{\partial \xi}\right)_{T,P} = \left(\frac{\partial \Delta_r G}{\partial T}\right)_{P,\xi} = -\Delta_r S \qquad (3.2\text{-}6)$$

$$\left(\frac{\partial V}{\partial \xi}\right)_{T,P} = \left(\frac{\partial \Delta_r G}{\partial P}\right)_{T,\xi} = \Delta_r V \qquad (3.2\text{-}7)$$

$$-\left(\frac{\partial S}{\partial P}\right)_{T,\xi} = \left(\frac{\partial V}{\partial T}\right)_{P,\xi} \qquad (3.2\text{-}8)$$

$\Delta_r S$ is the **reaction entropy** and $\Delta_r V$ is the **reaction volume**. The Legendre transforms $H = U + PV$ and $G = U + PV - TS$ lead to $G = H - TS$, and so

$$\Delta_r G = \Delta_r H - T\Delta_r S \qquad (3.2\text{-}9)$$

The relation for the entropy of reaction $\Delta_r S$ can be derived from equation 3.2-1 and equation 3.2-2. Equation 3.2-6 shows that

$$\Delta_r S = -\left(\frac{\partial \Delta_r G}{\partial T}\right)_{P,\xi} = \sum_{i=1}^{N_s} v_i \Delta_f S_i = \Delta_r S^0 - R \ln Q \qquad (3.2\text{-}10)$$

where $\Delta_f S_i$ is the **entropy of formation** of species i and $\Delta_r S^0$ is the **standard entropy of reaction** at a specified ionic strength. Thus

$$\Delta_f S_i = -\left(\frac{\partial \Delta_f G_i}{\partial T}\right)_{P,\xi} = \Delta_f S_i^0 - R \ln c_i \qquad (3.2\text{-}11)$$

where $\Delta_f S_1^0$ is the **standard entropy of formation** of species i. According to the third law of thermodynamics, absolute values of molar entropies of species can be determined, but we will be primarily concerned with the entropies of formation that can be calculated from the temperature derivative of the Gibbs energy of formation or from a combination of data on equilibrium constants and enthalpies of reaction.

The enthalpy of reaction can be calculated using the **Gibbs-Helmholtz equation** 2.5-18. Since $\Delta_r H = \Delta_r G + T\Delta_r S$ (equation 3.2-9), the enthalpy of reaction is given by

$$\Delta_r H = \Delta_r G - T\left(\frac{\partial \Delta_r G}{\partial T}\right)_P = -T^2 \left(\frac{\partial (\Delta_r G/T)}{\partial T}\right)_P \qquad (3.2\text{-}12)$$

Substituting $\Delta_r G = \Sigma v_i \Delta_f G_i$ yields

$$\Delta_r H = \sum_{i=1}^{N_s} v_i \Delta_f H_i \qquad (3.2\text{-}13)$$

where $\Delta_f H_i$ is the **enthalpy of formation** of species i. Since $H = G + TS$, it is evident that

$$\Delta_f H_i = \Delta_f G_i + T\Delta_f S_i \qquad (3.2\text{-}14)$$

and

$$\Delta_f H_i^0 = \Delta_f G_i^0 + T\Delta_f S_i^0 \qquad (3.2\text{-}15)$$

where $\Delta_f H_i^0$ is the **standard enthalpy of formation** of species i.

Taking the derivative of the enthalpy of reaction with respect to temperature yields the **heat capacity of reaction** at a constant pressure $\Delta_r C_P$:

$$\Delta_r C_P = \Delta_r C_P^0 = \sum_{i=1}^{N_s} v_i \Delta_f C_P^0(i) = \sum_{i=1}^{N_s} v_i C_{Pm}^0(i) \qquad (3.2\text{-}16)$$

$\Delta_f C_P^0(i)$ is the **standard heat capacity of formation** of species i at constant pressure and $C_{Pm}^0(i)$ is the **standard molar heat capacity** of species i at constant pressure.

Equation 3.2-12 can be written in the form

$$\left(\frac{\partial \ln K}{\partial T}\right)_P = \frac{\Delta_r H^0}{RT^2} \qquad (3.2\text{-}17)$$

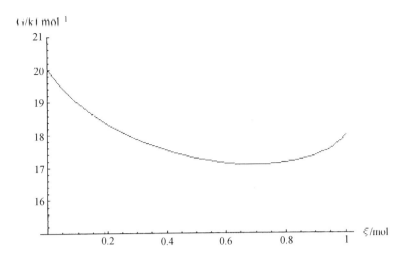

Fig. 3.2 Acid dissociation constant for acetic acid as a function of temperature (see Problem 3.4).

If $\Delta_r H^0$ is independent of temperature, integration of this equation from T_1 to T_2 yields

$$\ln \frac{K_2}{K_1} = \frac{\Delta_r H^0 (T_2 - T_1)}{R T_1 T_2} \qquad (3.2\text{-}18)$$

If $C_{Pm}^0(i)$ does not change significantly in the experimental temperature range, the enthalpy of reaction will change linearly with T and the entropy of reaction will change logarithmically:

$$\Delta_r H^0(T) = \Delta_r H^0(298.15\,\text{K}) + \Delta_r C_P^0 (T - 298.15\,\text{K}) \qquad (3.2\text{-}19)$$

$$\Delta_r S^0(T) = \Delta_r S^0(298.15\,\text{K}) + \Delta_r C_P^0 \ln \frac{T}{298.15\,\text{K}} \qquad (3.2\text{-}20)$$

Substituting these relations in $\Delta_r G^0 = -RT \ln K = \Delta_r H^0(T) - T\Delta_r S^0(T)$ yields

$$\ln K = -\frac{\Delta_r H^0(298.15)}{RT} + \frac{\Delta_r S^0(298.15)}{R} - \frac{\Delta_r C_P^0}{R} \left(1 - \frac{298.15\,\text{K}}{T} - \ln \frac{T}{298.15\,\text{K}} \right)$$

$$(3.2\text{-}21)$$

The plot in Fig. 3.2 of the acid dissociation constant for acetic acid was calculated using equation 3.2-21 and the values of standard thermodynamic properties tabulated by Edsall and Wyman (1958). When equation 3.2-21 is not satisfactory, empirical functions representing $\Delta_r C_P^0$ as a function of temperature can be used. Clark and Glew (1966) used Taylor series expansions of the enthalpy and the heat capacity to show the form that extensions of equation 3.2-21 should take up to terms in $d^3 \Delta_r C_P^0 / dT^3$.

■ 3.3 IMPORTANCE OF COMPONENTS

The role of components in reaction systems is discussed in Beattie and Oppenheim (1979) and Smith and Missen (1982). An elementary introduction to components has been provided by Alberty (1995c). In chemical reactions the atoms of each element and electric charges are conserved, but these conservation equations may not all be independent. It is only a set of independent conservation equations that provides a constraint on the equilibrium composition. The conservation equations for a chemical reaction system can also be written in terms of groups of atoms that occur in molecules. This is discussed in detail in the

Chapter 5 on matrices. In any case, the things that are conserved are referred to as **components**.

The preceding section was based on the fundamental equation for G in terms of the extent of reaction, but in order to identify the D natural variables for a one-reaction system at equilibrium, we need to apply the condition for equilibrium $\Sigma v_i \mu_i = 0$ (equation 3.1-6) that is due to the reaction. That is done by using each independent equilibrium condition to eliminate one chemical potential from equation 2.5-5. This is more easily seen for a simple reaction:

$$A + B = AB \tag{3.3-1}$$

At chemical equilibrium, equation 3.1-6 indicates that $\mu_A + \mu_B = \mu_{AB}$. Using this relation to eliminate μ_{AB} from the fundamental equation yields

$$dG = -S\,dT + V\,dP + \mu_A(dn_A + dn_{AB}) + \mu_B(dn_B + dn_{AB})$$

$$= -S\,dT + V\,dP + \mu_A\,dn_{cA} + \mu_B\,dn_{cB} \tag{3.3-2}$$

where n_{ci} is the amount of component i; $n_{cA} = n_A + n_{AB}$ and $n_{cB} = n_B + n_{AB}$. This form of the fundamental equation for G applies at chemical equilibrium. It is easy to see that $n_A + n_{AB}$ is conserved because every time a molecule of A disappears, a molecule of AB appears. These two conservation equations are constraints on the equilibrium composition. The other constraint is $K = [n_{AB}/V]/[n_A/V][n_B/V]$ where the amounts are equilibrium values; thus there are three equations and three unknowns, n_A, n_B, and n_{AB}. The natural variables for this reaction system at chemical equilibrium are T, P, n_{cA}, and n_{cB}, as shown by equation 3.3-2. Note that the number of natural variables has been decreased by one by the constraint due to reaction 3.3-1. When chemical reactions are involved in a system, μ_i and n_{ci} are conjugate variables (see Table 2.1) as indicated by equation 3.3-2.

Usually statements of problems on chemical equilibrium include the initial amounts of several species, but this doesn't really indicate the number of components. The initial amounts of all species can be used to calculate the initial amounts of components. The choice of components is arbitrary because μ_A or μ_B could have been eliminated from the fundamental equation at chemical equilibrium, rather than μ_{AB}. However, the number C of components is unique. Note that in equation 3.3-2 the components have the chemical potentials of species. This is an example of the theorems of Beattie and Oppenheim (1979) that "(1) the chemical potential of a component of a phase is independent of the choice of components, and (2) the chemical potential of a constituent of a phase when considered to be a species is equal to its chemical potential when considered to be a component." The amount of a component in a species can be negative.

The **number C of components** in a one-phase system is given by

$$C = N_s - R \tag{3.3-3}$$

where N_s is the number of different species and R is the **number of independent reactions**. The source of this equation and answers to questions about the number of components and the choice of components are clarified by the use of matrices, as described in Chapter 5. The amounts of components can be calculated from the amounts of species by use of a matrix multiplication (equation 5.1-27). When there are no reactions in a system, it is not necessary to distinguish between species and components.

Now we are in position to discuss a closed reaction system where several reactions are occuring. Equation 3.3-2 can be generalized to

$$dG = -S\,dT + V\,dP + \sum_{i=1}^{C} \mu_i\,dn_{ci} \tag{3.3-4}$$

This form of the fundamental equation for G applies to a system at chemical equilibrium. Note that the number D of natural variables of G is now $C + 2$, rather than $N_s + 2$ as it was for a nonreaction system (see Section 2.5). There are fewer independent variables because of the constraints due to the chemical reactions.

The criterion of spontaneous change and equilibrium for a nonreaction system is $dG \leqslant 0$ at constant T, P, and $\{n_i\}$, but the criterion for a system involving chemical reactions is $dG \leqslant 0$ at constant T, P, and $\{n_{vi}\}$. Therefore, to calculate the composition of a reaction system at equilibrium, it is necessary to specify the amounts of components. This can be done by specifying the initial composition because the initial reactants obviously contain all the components, but this is more information than necessary, as we will see in the chapter on matrices.

■ 3.4 GIBBS-DUHEM EQUATION AND THE PHASE RULE AT CHEMICAL EQUILIBRIUM

The Gibbs-Duhem equation and the phase rule were discussed briefly in Section 2.4, but now we want to extend those considerations to systems at chemical equilibrium. The degrees of freedom in a gaseous reaction system at chemical equilibrium was discussed by Alberty (1993b). The Gibbs-Duhem equation for a one-phase reaction system at chemical equilibrium is obtained by using the complete Legendre transform $U' = U + PV - TS - \Sigma n_{ci}\mu_i$ to interchange the roles of amounts of components and the chemical potentials of components. Thus the Gibbs-Duhem equation corresponding with equation 3.3-4 is

$$0 = -S\,dT + V\,dP - \sum_{i=1}^{c} n_{ci}\,d\mu_i \qquad (3.4\text{-}1)$$

This shows that there are $C + 2$ intensive variables for a chemical reaction system at equilibrium, but only $C + 1$ of them are independent because of this relation between them; in other words, for a one-phase system at chemical equilibrium the number F of **degrees of freedom** is given by $F = C + 1$. Since this is a one-phase system, it is evident that the **phase rule** for the reaction system is

$$F = C - p + 2 \qquad (3.4\text{-}2)$$

The number of natural variables is given by

$$D = F + p = C - p + 2 + p = C + 2 \qquad (3.4\text{-}3)$$

Since equations 3.4-2 and 3.4-3 have been introduced in a rather indirect way, more general derivations are given as follows: The composition of a phase in a system involving chemical reactions can be specified by stating $C - 1$ mole fractions, and the composition of p phases can be specified by stating $p(C - 1)$ mole fractions. If T and P are independent intensive variables, the number of independent intensive variables is equal to $p(C - 1) + 2$. The number of relationships between the chemical potentials of a single component between phases is $p - 1$. Since there are C components, there are $C(p - 1)$ equilibrium relationships. The difference F between the number of independent intensive variables and the number of relationships is given by $F = p(C - 1) + 2 - C(p - 1) = C - p + 2$. In order to describe the extensive state of the system, it is necessary to specify in addition the amounts of the p phases, and so $D = F + p = C + 2$. When special constraints are involved, the number s of these special constraints must be included in the phase rule to give $F = C - p + 2 - s$. An example of a special constraint would be taking the amounts of two reactants in the ratio of their stoichiometric numbers in a reaction. Further independent work terms in the fundamental equation increase D and F. The numbers F and D are unique, but the choices of independent intensive variables and independent extensive variables are not.

Equation 3.3-4 shows that for a chemical reaction system, the number F of intensive degrees of freedom and the number D of extensive degrees of freedom are given by

$$F = N_s - R - p + 2 \tag{3.4-4}$$

$$D = N_s - R + 2 \tag{3.4-5}$$

■ 3.5   ISOMER GROUP THERMODYNAMICS

In discussing the thermodynamics of complex reaction systems, it is helpful to have ways of reducing the complexity so that it is easier to think about the system and to make calculations. One of these ways is to aggregate isomers and make thermodynamic calculations with isomer groups, rather than species (Smith and Missen, 1982; Alberty, 1983a, 1993b). Examples of isomer groups are the butenes (3 isomers) and pentenes (5 isomers), where the numbers of isomers exclude cis-trans and stereoisomers. At higher temperatures these isomers are in equilibrium with each other, and so thermodynamic calculations can be made with butenes and pentenes. The reason this can be done is that the distribution of isomers within an isomer group is independent of the composition of a reaction system and of the other reactions that occur. The distribution of isomers within an isomer group depends only on temperature for ideal gases and ideal solutions.

The fundamental equation for G can be used to show that when isomers are in equilibrium, they have the same chemical potential. Therefore terms for isomers in the fundamental equation for G can be aggregated so that the amounts dealt with are amounts of isomer groups, rather than amounts of species. Since the number of isomers of a reactant can be significant, this can make a significant reduction in the number of chemical terms in the fundamental equation at chemical equilibrium.

There are two ways to express the Gibbs energy G_{iso} of a group of isomers at chemical equilibrium. The first method simply uses a sum of the terms for the individual isomers, and the second method utilizes the chemical potential μ_{iso} for the isomer group at equilibrium and the amount n_{iso} of the isomer group as in

$$G_{iso} = \sum_{i=1}^{N_{iso}} \mu_i n_i = \mu_{iso} n_{iso} \tag{3.5-1}$$

The number of isomers in an isomer group is represented by N_{iso}. At chemical equilibrium, all of the isomers have the same chemical potential, and this chemical potential is represented by μ_{iso}. The amount of an isomer group is represented by $n_{iso} = \Sigma n_i$. For a group of gaseous isomers at equilibrium, the **chemical potential of the isomer group** in a mixture of ideal gases is given by

$$\mu_{iso} = \mu_{iso}^\circ + RT \ln \frac{n_{iso} P}{n_t P^\circ} \tag{3.5-2}$$

where n_t is the total amount of gas in the system, P is the sum of the partial pressure of the isomers, and P^0 is the standard state pressure. At equilibrium the chemical potential of isomer i is given by

$$\mu_i = \mu_i^\circ + RT \ln \frac{n_i P}{n_t P^\circ} \tag{3.5-3}$$

These two equations can be written as

$$n_{iso} = \frac{n_t P^\circ}{P} \exp\left[\frac{\mu_{iso} - \mu_{iso}^\circ}{RT} \right] \tag{3.5-4}$$

$$n_i = \frac{n_t P^\circ}{P} \exp\left[\frac{\mu_i - \mu_i^\circ}{RT} \right] \tag{3.5-5}$$

Substituting these expressions in $n_{iso} = \Sigma n_i$ yields

$$\mu_{iso}^{\circ} = -RT \ln \sum_{i=1}^{N_{iso}} \exp\left[-\frac{\mu_i^{\circ}}{RT}\right] \qquad (3.5\text{-}6)$$

since $\mu_{iso} = \mu_i$. The corresponding expressions for the standard enthalpy, entropy, and heat capacity can be obtained by using the derivatives of equation 2.5-6 indicated by the fundamental equation for G (Alberty, 1983).

The corresponding derivation for ideal solutions is a little simpler. The chemical potential for the isomer group and for an individual isomer at chemical equilibrium are given by

$$\mu_{iso} = \mu_{iso}^{\circ} + RT \ln[B_{iso}] \qquad (3.5\text{-}7)$$

where $[B_{iso}]$ is the concentration of the isomer group. At equilibrium the chemical potential of isomer i is given by

$$\mu_i = \mu_i^{\circ} + RT \ln[B_i] \qquad (3.5\text{-}8)$$

These two equations can be written as

$$[B_{iso}] = \exp\left[\frac{\mu_{iso} - \mu_{iso}^{0}}{RT}\right] \qquad (3.5\text{-}9)$$

$$[B_i] = \exp\left[\frac{\mu_i - \mu_i^{\circ}}{RT}\right] \qquad (3.5\text{-}10)$$

Substituting these expressions in $[B_{iso}] = \Sigma[B_i]$ yields equation 3.5-6.

In making actual calculations, standard formation properties are used rather than chemical potentials, and so the **standard Gibbs energy of formation of an isomer group** is given by

$$\Delta_f G^{\circ}(\text{iso}) = -RT \ln\left\{\sum_{i=1}^{N_{iso}} \exp\left[-\frac{\Delta_f G_i^{\circ}}{RT}\right]\right\} \qquad (3.5\text{-}11)$$

Note that $\Delta_f G^{\circ}(\text{iso})$ is more negative than $\Delta_f G_i^{\circ}$ of the most stable isomer, as it must be because the isomer group has a higher mole fraction in the reaction system at equilibrium than the most stable isomer. The mole fraction r_i of the ith isomer in the isomer group at equilibrium is given by

$$r_i = \exp\left\{\frac{\Delta_f G^{\circ}(\text{iso}) - \Delta_f G_i^{\circ}}{RT}\right\} \qquad (3.5\text{-}12)$$

The summation in equation 3.5-11 has the form of a partition function, and the distribution in equation 3.5-12 has the form of a Boltzmann distribution.

The equation for the **standard enthalpy of formation of an isomer group** can be obtained by using the Gibbs-Helmholtz equation 2.5-23 in the form

$$\Delta_f H^{\circ}(\text{iso}) = -T^2 \left(\frac{\partial(\Delta_f G^{\circ}(\text{iso})/T)}{\partial T}\right)_P \qquad (3.5\text{-}13)$$

This differentiation yields

$$\Delta_f H^{0}(\text{iso}) = \sum_{i=1}^{N_{iso}} r_i \Delta_f H_i^{0} \qquad (3.5\text{-}14)$$

Thus the standard enthalpy of formation or an isomer group is the mole fraction weighted average. Equations 3.5-11 to 3.5-14 will be especially useful in the next chapter.

The equation for the **standard entropy of formation of an isomer group** can be obtained by using

$$\Delta_f S^{\circ}(\text{iso}) = -\left(\frac{\partial \Delta_f G^{\circ}(\text{iso})}{\partial T}\right)_P \qquad (3.5\text{-}15)$$

This differentiation yields

$$\Delta_f S^\circ(\text{iso}) = \sum_{i=1}^{N_{\text{iso}}} r_i \Delta_f S_i^\circ - R \sum_{i=1}^{N_{\text{iso}}} r_i \ln r_i \tag{3.5-16}$$

The same form of equation can be used to calculate the standard molar entropy $S_m^\circ(\text{iso})$ of the isomer group. The entropy of formation of the isomer group is equal to the mole-fraction-weighted entropy of formation plus the entropy of mixing the isomers.

The equation for the **standard molar heat capacity of formation of an isomer group** can be obtained by using

$$\Delta_f C_P^\circ(\text{iso}) = -\left(\frac{\partial \Delta_f H^\circ(\text{iso})}{\partial T}\right)_P \tag{3.5-17}$$

This differentiation yields

$$C_{Pm}^\circ(\text{iso}) = \sum_{i=1}^{N_{\text{iso}}} r_i C_{Pm}^\circ(i) + \frac{1}{RT^2} \left(\sum_{i=1}^{N_{\text{iso}}} r_i (\Delta_f H_i^\circ)^2 - (\Delta_f H^0(\text{iso}))^2\right) \tag{3.5-18}$$

Equation (3.5-18) has been written in terms of molar heat capacities $C_{Pm}^\circ(i)$, rather than heat capacities of formation, because the heat capacities of the elements are on both sides and cancel. The second term of this equation is always positive because the weighted average of the squares is always greater than the square of the average. Equation 3.5-18 is in accord with **LeChatelier's principle**: As the temperature is raised, the equilibrium shifts in the direction that causes the absorption of heat. Equation 3.5-18 can also be derived using $C_P = -T(\partial^2 G/\partial T^2)_P$ (equation 2.5-25).

Equations 3.5-14 and 3.5-16 can be substituted in $\Delta_f G^\circ(\text{iso}) = \Delta_f H^\circ(\text{iso}) - T\Delta_f S^\circ(\text{iso})$ to obtain another form for the standard Gibbs energy of formation of an isomer group.

$$\Delta_f G^\circ(\text{iso}) = \sum_{i=1}^{N_{\text{iso}}} r_i \Delta_f G_i^\circ + RT \sum_{i=1}^{N_{\text{iso}}} r_i \ln r_i \tag{3.5-19}$$

In other words, the standard Gibbs energy of formation of an isomer group at equilibrium is equal to the mole fraction-weighted average of standard Gibbs energies of formation of the isomers plus the Gibbs energy of mixing.

The fundamental equation for G of a system made up of isomer groups is

$$dG = -S\,dT + V\,dP + \sum_{i=1}^{N_{\text{iso}}} \mu_i(\text{iso})dn_i(\text{iso}) \tag{3.5-20}$$

where N_{iso} is the number of isomer groups. In this equation an isomer group may consist of a single species. This equation can be used to derive the equilibrium expressions for reactions written in terms of isomer groups. Since isomer groups can be treated like species in chemical thermodynamics, they can be referred to a pseudospecies. Equation 3.5-20 is based on the assumption that the species in an isomer group are in equilibrium with each other. The number of natural variables for a one-phase system consisting of N_{iso} isomer groups is $D = N_{\text{iso}} + 2$ prior to application of the constraints due to reactions between the isomer groups. If the reactions between the isomer groups are at equilibrium, the number of components replaces the number of isomer groups and $D = C + 2$.

■ **3.6   EFFECT OF IONIC STRENGTH ON EQUILIBRIA IN SOLUTION REACTIONS**

The activity coefficient γ_i of an ion depends on the ionic strength ($I = (\frac{1}{2})\Sigma z_i^2 c_i$, where z_i is the charge number) according to the Debye-Hückel theory in the limit of low ionic strengths. As discussed in Section 1.2, this equation can be extended to provide activity coefficients in the physiological range by introducing an empirical term to form the extended Debye-Hückel equation.

$$\ln \gamma_i = -\frac{\alpha z_i^2 I^{1/2}}{1 + BI^{1/2}} \tag{3.6-1}$$

where B is 1.6 $L^{1/2}$ $mol^{-1/2}$. This equation works quite well in the 0.05 to 0.25 M range of ionic strengths for a number of electrolytes for which activity coefficients have been determined accurately. It is evident from this equation that the effect of ionic strength on the thermodynamic properties of ionic species of biochemical interest are significant in the 0.05 to 0.25 M range. The effects are especially significant when ions have charges of 2, 3, or 4. The treatments of the thermodynamics of electrolyte solutions at higher concentrations require more complicated equations with more empirical parameters (Pitzer, 1991, 1995). However, there is insufficient data on the specific effects of various ions in biochemical buffers to go beyond equation 3.6-1 at present.

Thus the standard enthalpy of formation $\Delta_f H_i^\circ(I)$ and standard Gibbs energy of formation $\Delta_f G_i^\circ(I)$ of an ionic species at 298.15 K in kJ mol^{-1} can be calculated using (Clark and Glew, 1980; Goldberg and Tewari, 1991)

$$\Delta_f H_i^\circ(I) = \Delta_f H_i^\circ(I = 0) + \frac{1.4775 z_i^2 I^{1/2}}{1 + BI^{1/2}} \tag{3.6-2}$$

$$\Delta_f G_i^\circ(I) = \Delta_f G_i^\circ(I = 0) - \frac{2.91482 z_i^2 I^{1/2}}{1 + BI^{1/2}} \tag{3.6-3}$$

These equations will be very useful in the next chapter.

The standard thermodynamic properties of ions are given in tables of standard thermodynamic properties at $I = 0$. The effect of ionic strength on $\Delta_r G^\circ$ for a chemical reaction is obtained by substituting equation 3.6-3 in equation 3.1-12:

$$\Delta_r G^\circ(I) - \Delta_r G^\circ(I = 0) - \frac{2.91482 I^{1/2} \sum v_i z_i^2}{1 + BI^{1/2}} \tag{3.6-4}$$

where $\sum v_i z_i^2$ is the change in z_i^2 in the reaction. The effect of ionic strength on $\Delta_r H^\circ(I)$ for a chemical reaction is obtained by using the Gibbs-Helmholtz equation (2.5-18 and 3.2-12):

$$\Delta_r H_i^\circ(I) = \Delta_r H_i^\circ(I = 0) + \frac{1.4775 I^{1/2} \sum v_i z_i^2}{1 + BI^{1/2}} \tag{3.6-5}$$

The effect of ionic strength on the equilibrium constant for a chemical reaction at 25°C is obtained by substituting equation 3.6-4 in equation 3.1-2:

$$\ln K(I) = \ln K(I = 0) + \frac{1.17582 I^{1/2} \sum v_i z_i^2}{1 + BI^{1/2}} \tag{3.6-6}$$

■ 3.7 EFFECT OF TEMPERATURE ON THERMODYNAMIC PROPERTIES

In order to discuss thermodynamic properties in dilute aqueous solutions at temperatures other than 298.15 K, it is necessary to have the standard enthalpies of the species involved. Over narrow ranges of temperature, calculations can be based on the assumption that $\Delta_f H_i^\circ$ values are independent of temperature, but more accurate calculations can be made when $C_{Pm}^\circ(i)$ values are known. It is also necessary to take into account the temperature dependencies of the numerical coefficients in equations 3.6-4 to 3.6-6. Clarke and Glew (1980) calculated the Debye-Hückel slopes for water between 0 and 150°C. They were primarily concerned with electrostatic deviations from ideality of the solvent osmotic

Table 3.1 Debye-Hückel Constant and Limiting Slopes of $\Delta_f G_i^\circ$, $\Delta_f H_i^\circ$, and $C_{Pm}^0(i)$ as Functions of Temperature

$t/°C$	α kg$^{1/2}$ mol$^{-1/2}$	$RT\alpha$ kJ mol$^{-3/2}$ kg$^{1/2}$	$RT^2(\partial\alpha/\partial T)_P$ kJ mol$^{-3/2}$ kg$^{1/2}$	$RT[2(\partial\alpha/\partial T)_P + T(\partial^2\alpha/\partial T^2)_P]$ J mol$^{-3/2}$ kg$^{1/2}$ K^{-1}
0	1.12938	2.56494	1.075	13.255
10	1.14717	2.70073	1.213	15.41
20	1.16598	2.84196	1.3845	17.90
25	1.17582	2.91482	1.4775	19.27
30	1.18599	2.98934	1.5775	20.725
40	1.20732	3.14349	1.800	23.885

Source: With permission from R. A. Alberty, *J. Phys. Chem. B*, 105, 7865 (2001). Copyright 2001 American Chemical Society.

coefficient ϕ, and so they used the Debye-Hückel limiting law in the form $\ln\gamma = -3Am^{1/2}$, where m is the molality. The relation between these coefficients and those needed here were first discussed by Goldberg and Tewari (1991). Further discussion is to be found in Alberty (2001). The primary coefficients of interest here are those for effects of ionic strength on $\ln K$, $\Delta_f G^\circ$, $\Delta_f H^\circ$, and C_{Pm}°. These coefficients are α, $RT\alpha$, $RT^2(\partial\alpha/\partial T)_P$, and $RT^2(\partial\alpha/\partial T)_P + T(\partial^2\alpha/\partial T^2)_P]$, respectively. The third coefficient is a consequence of the Gibbs-Helmholtz equation. The fourth coefficient is a consequence of equation 2.5-25. The values of these coefficients calculated from the tables of Clark and Glew (1980) are given in Table 3.1.

In discussing the effect of temperature, it is more convenient to use the molality because molality does not change with the temperature when there are no reactions in the system. However, these values can be used in calculations based on molarities.

The calculations of standard thermodynamic properties discussed in the rest of this section are based on the assumption that the standard enthalpies of formation of species are independent of temperature: in other words, the heat capacities of species are assumed to be zero. In the future when more is known about the molar heat capacities of species, more accurate calculations can be based on the assumption that the molar heat capacities are independent of temperature. When the heat capacities of species are equal to zero, the standard entropies of formation are also independent of temperature. Under these conditions the values of $\Delta_f G_i^\circ$ at other temperatures in the neighborhood of 298.15 K can be calculated using

$$\Delta_f G_i^\circ(T) = \Delta_f H_i^\circ(298.15\ \text{K}) - T\Delta_f S_i^\circ(298.15\ \text{K}) \tag{3.7-1}$$

This equation can be written in terms of $\Delta_f G_i^\circ(298.15\ \text{K})$ and $\Delta_f H_i^\circ(298.15\ \text{K})$ by substituting the expression for the entropy of formation of the species:

$$\Delta_f G_i^\circ(T) = \left(\frac{T}{298.15}\right)\Delta_f G_i^\circ(298.15\ K) + \left(1 - \frac{T}{298.15}\right)\Delta_f H_i^\circ(298.15\ \text{K}) \tag{3.7-2}$$

In order to calculate values of $\Delta_f G_i^\circ$ at other temperatures not too far from 298.15 K, it is necessary to fit α to a power series in T. The use of Fit in *Mathematica* yields (see Problem 3.5)

$$\alpha = 1.10708 - 1.54508 \times 10^{-3}T + 5.95584 \times 10^{-6}T^2 \tag{3.7-3}$$

Clarke and Glew (1980) give an equation with more parameters to yield values of α from 0 to 150°C. When the quadratic fit is used, the coefficient $RT\alpha$ in the equation for the standard transformed Gibbs energy of formation of a species is given by

$$RT\alpha - 9.20483 \times 10^{-3}T - 1.28467 \times 10^{-5}T^2 + 4.95199 \times 10^{-8}T^3 \quad (3.7\text{-}4)$$

This equation reproduces the second column of Table 3.1 to 0.1% accuracy. The coefficient $RT^2(\partial\alpha/\partial T)_P$ in the equation for the standard transformed enthalpy of formation of a species is given by

$$RT^2\left(\frac{\partial\alpha}{\partial T}\right)_P = -1.28466 \times 10^{-5}T^2 + 9.90399 \times 10^{-8}T^3 \quad (3.7\text{-}5)$$

This equation reproduces the third column of Table 3.1 to 1% accuracy. The calculations of these three functions are shown in Problem 3.5, and they are used in the calculation of standard Gibbs energies of formation and standard enthalpies of formation of species at other temperatures in Problems 3.6 and 3.7.

Thermodynamic properties in dilute aqueous solutions are taken to be functions of ionic strength so that concentrations of reactants, rather than their activities can be used. This also means that $pH_c = -\log[H^+]$ has to be used in calculations, rather than $pH_a = -\log\{a(H^+)\}$. When the ionic strength is different from zero, this means that pH values obtained in the laboratory using a glass electrode need to be adjusted for the ionic strength and temperature to obtain the pH that is used to discuss the thermodynamics of dilute aqueous solutions. Since $pH_a = -\log\gamma(H^+) + pH_c$, the use of the extended Debye-Hückel theory yields

$$pH_a - pH_c = \frac{\alpha}{\ln(10)}\frac{I^{1/2}}{1 + 1.6I^{1/2}} \quad (3.7\text{-}6)$$

These adjustments, which are tabulated in Section 1.2, are to be subtracted from the pHa obtained with a pH meter to obtain pH_c. pH_c is lower than pH_a because the ion atmosphere of H^+ reduces its activity (see Problem 3.7). In the rest of the book, pH is taken to be pH_c.

■ 3.8 CHEMICAL THERMODYNAMIC TABLES INCLUDING BIOCHEMICAL SPECIES

A useful way to store data on equilibrium constants and enthalpies of chemical reactions is to use equations 3.2-4 and 3.2-13 to calculate standard Gibbs energies of formation and standard enthalpies of formation of species and to tabulate these values. Since there are more species than independent chemical reactions between them (remember $N_s = R + C$), this can only be done by adopting some conventions. The major convention for the construction of chemical thermodynamic tables is that $\Delta_f G_i^\circ$ and $\Delta_f H_i^\circ$ for each element in a specified reference state is taken as zero at each temperature. The reference state for the elements that are gases at room temperature is the ideal gas state at 1 bar. For each solid element, a particular state has been chosen for the reference state; this is generally the most stable state at room temperature. In order to treat the thermodynamics of electrolyte solutions, it is necessary to adopt an additional convention, and that is that $\Delta_f G_i^\circ = \Delta_f H_i^\circ = C_{Pm}^\circ(i) = 0$ at zero ionic strength for $H^+(aq)$ at each temperature. Since the thermodynamic properties of ions depend on the ionic strength, the convention of tabulating values at zero ionic strength has been adopted. These arbitrary conventions make it possible to have tables of standard Gibbs energies of formation $\Delta_f G_i^\circ$, standard enthalpies of formation $\Delta_f H_i^\circ$, and molar heat capacity $C_{Pm}^\circ(i)$ of species at 298.15 K and zero ionic strength. The NBS Tables (1982) summarize a very large body of standard thermodynamic properties of species obtained by chemical methods. However, this table does not contain very much information on biochemical metabolites because it includes only C_1 and C_2 organic molecules and ions. Cox et al. (1989) list a number of species on which there is international agreement.

As discussed in Chapter 1, Burton (1957) used equilibrium constants for enzyme-catalyzed reactions to calculate standard Gibbs energies of formation for species of biochemical interest. Wilhoit (1969) and Thauer (1977) considerably extended Burton's table. Goldberg (1984) and Goldberg and Tewari (1989) have calculated $\Delta_f G_i^\circ$ and $\Delta_f H_i^\circ$ for more species of biochemical interest from measurements of apparent equilibrium constants and enthalpies of enzyme-catalyzed reactions. When the standard Gibbs energies of formation for all the species in a system are known, the equilibrium composition can be calculated by use of a computer program that minimizes the Gibbs energy. This was illustrated for the hydrolysis of ATP, which involves 17 species (Alberty, 1991).

In making thermodynamic tables of properties of species in biochemical reactions, there are cases where it is not possible to calculate $\Delta_f G_i^\circ$ and $\Delta_f H_i^\circ$ with respect to the elements because equilibrium constants and heats of reaction have not been measured for reactions connecting the species with the elements. In these cases it is possible to assign $\Delta_f G_i^\circ = \Delta_f H_i^\circ = C_{Pm}^\circ(i) = 0$ for one species and to calculate $\Delta_f G_i^\circ$, $\Delta_f H_i^\circ$, and $C_{Pm}^\circ(i)$ for other species on the basis of this convention. Alberty and Goldberg (1992) applied this convention to adenosine because at that time it was not possible to connect ATP^{4-} with the elements in their reference states. They showed that values of $\Delta_f G_i^\circ$ and $\Delta_f H_i^\circ$ calculated in this way can be used to calculate equilibrium constants and enthalpies of reaction. But, of course, $\Delta_f G_i^\circ$ and $\Delta_f H_i^\circ$ calculated in this way cannot be used to calculate equilibrium constants and enthalpies for reactions that form adenosine from the elements. More recently Boerio-Goates et al. (2001) have determined the enthalpy of combustion of adenosine (cr), its third law entropy, and its solubility in water at 298.15 K. This makes it possible to calculate $\Delta_f H^\circ$(adenosine, aq) and $\Delta_f G^\circ$ (adenosine, aq) at 298.15 K. They have recalculated $\Delta_f H^\circ$ and $\Delta_f G^\circ$ for all of the species in the ATP series. This does not change the values of equilibrium constants and enthalpies of reaction calculated from the previous table (Alberty and Goldberg, 1992). An essential part of any table is a list of the conventions. Using values of standard formation properties from tables based on different conventions will lead to incorrect results.

The most basic data on thermodynamic properties of species at a certain temperature include $\Delta_f G_i^\circ(I = 0)$ and $\Delta_f H_i^\circ(I = 0)$, the charge number $_i$, and the number $N_H(i)$ of hydrogen atoms in the species. The number of hydrogen atoms in a species is not used in this chapter but in the next chapter. A database in computer readable form is presented in the *Mathematica* package BasicBiochemData2, which is the first item in the second part of this book. The basic data on the species of 131 reactants are given in section 2 of this package. The data for each reactant is in the form of a matrix with a row for each species of the reactant:

$$\text{namesp} = ((\{\Delta_f G_i^\circ, \Delta_f H_i^\circ, z_1, N_H(1)\}, \{\Delta_f G_2^\circ, \Delta_f H_2^\circ, z_2, N_H(2)\}, \ldots\}) \quad (3.8\text{-}1)$$

The first entry is for the species with the fewest hydrogen atoms. Only species of interest in the range pH 5 to 9 are included. No complex ions are included in this table. The advantage of this table is that, with *Mathematica* in a personal computer, it is not necessary to copy these numbers to make a calculation. Programs for making these calculations are also included in the package.

The sources of the data are described in Alberty (1998b, d, and other articles). The complications of dissolved carbon dioxide are discussed in Section 8.7. Reactions involving NAD_{ox}, NAD_{red}, $NADP_{ox}$, and $NADP_{red}$ are discussed in Chapter 9.

This data file and programs (Alberty, 2001) for using it are available on the web at *http://www.mathsource.com/cgi-bin/msitem?0211-622*. It can be read using MathReader, which is free from Wolfram Research, Inc. (100 Trade Center Drive, Champaign, IL 61820-7237): *http//www.wolfram.com*. BasicBiochemData2 is available in two forms at MathSource: a package that can be downloaded using the command ≪BasicBiochemData2* and a notebook (BasicBiochemData2.nb) that contains explanations and examples in addition to the data.

Goldberg and Akers (2001) have also published a *Mathematica* package for calculations on biochemical reactions.

BasicBiochemData2 contains a table of data on 131 reactants that is reproduced in Table 3.2 in the appendix of this chapter. It is hard to overemphasize the usefulness of this table or the importance of extending it. It can be used to calculate equilibrium constants for chemical reactions between species in the table at desired ionic strengths in the range 0 to 0.35 M. This table can also be used to calculate acid dissociation constants at desired ionic strengths, as shown in Table 1.3. When $\Delta_f H°$ is known for all of the species in a chemical reaction, Table 3.2 can be used to calculate $\Delta_r H°$. This makes it possible to calculate equilibrium constants at temperatures other than 298.15 K. Only a few $C_{p,m}°$ values are known for these species, and so equilibrium constants cannot be calculated very far from room temperature. Since $\Delta_r S° = (\Delta_r H° - \Delta_r G°)/T$, reaction entropies can also be calculated, and this is of special interest when reactant species and product species differ significantly in the disorder they introduce. For example, reactions that produce gases are generally go further to the right than reactions that do not because $\Delta_r S°$ is large and positive.

Some comments are needed about the names of species used in making calculations. Since Table 3.2 and later tables are produced using *Mathematica*, it is necessary to use short one-dimensional names that begin with lowercase letters and do not directly indicate ionic charges. The stereochemical labels are put at the end of the name, if necessary, so that they do not interfere with alphabetizing the list. Gaseous species are labeled with "g" at the end of the name, and the corresponding dissolved species are labeled "aq." The species of CO_2tot are CO_3^{2-}, HCO_3^{3-}, CO_2(aq), and H_2CO_3 (see Section 8.7). When values are given for gaseous and dissolved forms, the corresponding Henry law constants can be calculated. The distribution of CO_2 between gaseous and dissolved forms is of special interest because we will see later (Chapter 8) that the Henry law constant is also a function of pH. In Table 3.2 the chemical names of the reactants are given first, and then the name used in *Mathematica*.

Proteins that are reactants in biochemical reactions are also be included in BasicBiochemData2; examples included are cytochrome c, ferrodoxin, and thioredoxin. Later in Chapter 7 it is shown that the effect of pH on a biochemical reaction involving a protein can be calculated if the pKs of groups in the reactive site of the protein can be determined.

It is important to understand that the number of digits used in a thermodynamic table of this type does not indicate the accuracy of the measured values because the information in the table is in the differences between values. An error of 0.01 kJ mol^{-1} in the standard transformed Gibbs energy of formation of a species leads to about a 1% error in the equilibrium constant of a chemical reaction at 298.15 K. This table can be extended a good deal in the future, as indicated by the data on apparent equilibrium constants and transformed enthalpies of reaction in the critical compilations of Goldberg and Tewari (Goldberg et al., 1993; Goldberg and Tewari, 1994a, b, 1995a, b; Goldberg, 1999).

The procedure for calculating standard formation properties of species at zero ionic strength from measurements of apparent equilibrium constants is discussed in the next chapter. The future of the thermodynamics of species in aqueous solutions depends largely on the use of enzyme-catalyzed reactions. The reason that more complicated ions in aqueous solutions were not included in the NBS Tables (1992) is that it is difficult to determine equilibrium constants in systems where a number of reactions occur simultaneously. Since many enzymes catalyze clean-cut reactions, they make it possible to determine apparent equilibrium constants and heats of reaction between very complicated organic reactants that could not have been studied classically.

Table 3.2 Basic Data on Species at 298.15 K in Dilute Aqueous Solutions at Zero Ionic Strength

Reactant	Mathematica Name	$\Delta_f G°$	$\Delta_f H°$	z_i	$N_H(i)$
acetaldehyde	acetaldehyde	−139.00	−212.23	0	4
acetate	acetate	−369.31	−486.01	−1	3
		−396.45	−485.76	0	4
acetone	acetone	−159.70	−221.71	0	6
acetylcoA	acetylcoA	−188.52	—	0	3
acetylphosphate	acetylphos	−1219.39	—	−2	3
		−1268.08	—	−1	4
		−1298.26	—	0	5
cis-aconitate	aconitatecis	−917.13	—	−3	3
adenine	adenine	310.67	—	0	5
		286.70	—	1	6
adenosine	adenosine	−194.50	−621.30	0	13
		−214.28	637.70	1	14
adenosine 5′ diphosphate	adp	−1906.13	−2626.54	−3	12
		−1947.10	−2620.94	−2	13
		−1971.98	−2638.54	−1	14
alanine	alanine	−371.00	−554.80	0	7
ammonia	ammonia	−26.50	−80.29	0	3
		−79.31	−132.51	1	4
adenosine 5′-monophosphate	amp	−1040.45	−1635.37	−2	12
		−1078.86	−1629.97	−1	13
		−1101.63	−1648.07	0	14
D-arabinose	arabinose	−742.23	−1043.79	0	10
L-asparagine	asparagineL	−525.93	−766.09	0	8
L-aspartate	aspartate	−695.88	−943.41	−1	6
adenosine 5′-triphosphate	atp	−2768.10	−3619.21	−4	12
		−2811.48	−3612.91	−3	13
		−2838.18	−3627.91	−2	14
1,3-bis phosphoglycerate	bpg	−2356.14	—	−4	4
		−2401.58	—	−3	5
n-butanol	butanoln	−171.84	—	0	10
butyrate	butyrate	−352.63	—	−1	7
citrate	citrate	−1162.69	−1515.11	−3	5
		−1199.18	−1518.46	−2	6
		−1226.33	−1520.88	−1	7
isocitrate	citrateiso	−1156.04	—	−3	5
		−1192.57	—	−2	6
		−1219.47	—	−1	7
coenzyme A	coA	0	—	−1	0
		−47.83	—	0	1
glutathione-coenzyme A	coAglutathione	−35.85	—	−1	15
CO$_2$(g)	co$_2$g	−394.36	−393.50	0	0
CO$_2$tot	co$_2$tot	−527.81	−677.14	−2	0
		−586.77	−691.99	−1	1
		−623.11	−699.63	0	2
CO(aq)	coaq	−119.90	−120.96	0	0
CO(g)	cog	−137.17	−110.53	0	0
creatine	creatine	−259.20	—	0	9
creatinine	creatinine	−23.14	—	0	7
L-cysteine	cysteineL	−291.00	—	−1	6
		−338.82	—	0	7
L-cystine	cystineL	−666.51	—	0	12
cytochrome c (ox)	cytochromecox	0	—	3	0
cytochrome c (red)	cytochromecred	−24.51	—	2	0
dihydroxyacetone	dihydroxy-	−1296.26	—	−2	5

Table 3.2   *Continued*

Reactant	*Mathematica* Name	$\Delta_f G°$	$\Delta_f H°$	z_i	$N_H(i)$
phosphate	acetonephos	−1328.80	—	−1	6
ethanol	ethanol	−181.64	−288.30	0	6
ethyl acetate	ethylacetate	−337.65	−482.00	0	8
flavin-adenine dinucleotide (ox)	fadox	0	—	−2	31
flavin-adenine dinucleotide (rcd)	fadred	−38.88	—	−2	33
flavin-adenine dinucleotide-enz (ox)	fadenzox	0	—	−2	31
flavin-adenine dinucleotide-enz (red)	fadenzred	−88.60	—	−2	33
ferredoxin (ox)	ferredoxinox	0	—	1	0
ferredoxin (red)	ferredoxinred	38.07	—	0	0
flavin mononucleotide (ox)	fmnox	0	—	−2	19
flavin mononucleotide (red)	fmnred	−38.88	—	−2	21
formate	formate	−351.00	−425.55	−1	1
D-fructose	fructose	−915.51	−1259.38	0	12
D-fructose 6-phosphate	fructose6phos	−1760.80	—	−2	11
		−1796.60	—	−1	12
D-fructose 1,6-biphosphate	fructose16phos	−2601.40	—	−4	10
		−2639.36	—	−3	11
		−2673.89	—	−2	12
fumarate	fumarate	−601.87	−777.39	−2	2
		−628.14	−774.46	−1	3
		−645.80	−774.88	0	4
D-galactose	galactose	908.93	1255.20	0	12
D-galactose 1-phosphate	galactose1phos	−1756.69	—	−2	11
		−1791.77		−1	12
D-glucose	glucose	−915.90	−1262.19	0	12
D-glucose 1-phosphate	glucose1phos	−1756.87	—	−2	11
		−1793.98	—	−1	12
D-glucose 6-phosphate	glucose6phos	−1763.94	−2276.44	−2	11
		−1800.59	−2274.64	−1	12
L-glutamate	glutamate	−697.47	−979.89	−1	8
L-glutamine	glutamine	−528.02	−805.00	0	10
glutathione (ox)	glutathioneox	0	—	−2	30
glutathione (red)	glutathionered	34.17	—	−2	15
		−13.44	—	−1	16
glyceraldehyde 3-phosphate	glyceraldehyde phos	−1288.60	—	−2	5
		−1321.14	—	−1	6
glycerol	glycerol	−497.48	−676.55	0	8
sn-glycerol 3-phosphate	glycerol3phos	−1358.96	—	−2	7
		−1397.04	—	−1	8
glycine	glycine	−379.91	−523.00	0	5
glycolate	glycolate	−530.95	—	−1	3
glycylglycine	glycylglycine	−520.20	−734.25	0	8
glyoxylate	glyoxylate	−468.60	—	−1	1
H_2(aq)	h2	17.60	−4.20	0	2
H_2(g)	h2g	0	0	0	2
H_2O	h2o	−237.19	−285.83	0	2
H_2O_2(aq)	h2o2aq	−134.03	−191.17	0	2

Table 3.2   *Continued*

Reactant	*Mathematica* Name	$\Delta_f G°$	$\Delta_f H°$	z_i	$N_H(i)$
H$^+$	hydroion	0	0	1	1
hydroxypropionate	hydroxy-propionate	−518.40	—	−1	5
hypoxanthine	hypoxanthine	89.50	—	0	4
indole	indole	223.80	97.50	0	7
ketoglutarate	ketoglutarate	−793.41	—	−2	4
lactate	lactate	−516.72	−686.64	−1	5
lactose	lactose	−1567.33	−2233.08	0	22
L-isoleucine	leucineisoL	−343.90	—	0	13
L-leucine	leucineL	−352.25	−643.37	0	13
D-lyxose	lyxose	−749.14	—	0	10
L-malate	malate	−842.66	—	−2	4
		−872.68	—	−1	5
maltose	maltose	−1574.69	−2238.06	0	22
D-mannitol	mannitolD	−942.61	—	0	14
D-mannose	mannose	−910.00	−1258.66	0	12
methane(g)	methaneg	−50.72	−74.81	0	4
methane(aq)	methaneaq	−34.33	−89.04	0	4
methanol	methanol	−175.31	−245.93	0	4
L-methionine	methionineL	−502.92	—	0	11
methylamineion	methylamineion	−39.86	−124.93	1	6
N$_2$(aq)	n2aq	18.7	−10.54	0	0
N$_2$(g)	n2g	0	0	0	0
nicotinamide-adenine dinucleotide (ox)	nadox	0	0	−1	26
nicotinamide-adenine dinucleotide (red)	nadred	22.65	−31.94	−2	27
nicotinamide-adenine dinucleotide phosphate (ox)	nadpox	−835.18	0	−3	25
nicotinamide−adenine dinucleotide phosphate (red)	nadpred	−809.19	−29.18	−4	26
O$_2$(aq)	o2aq	16.40	−11.70	0	0
O$_2$(g)	o2g	0	0	0	0
oxalate	oxalate	−673.90	−825.10	−2	0
oxaloacetate	oxaloacetate	−793.29	—	−2	2
		−698.33	—	−1	1
oxalosuccinate	oxalosuccinate	−1138.88	—	−2	4
palmitate	palmitate	−259.40	—	−1	31
phosphoenolpyruvate	pep	−1263.65	—	−3	2
		−1303.61	—	−2	3
2-phospho-D-glycerate	pg2	−1496.38	—	−3	4
		−1539.99	—	−2	5
3-phospho-D-glycerate	pg3	−1502.54	—	−3	4
		−1545.52	—	−2	5
L-phenylalanine	phenylalanineL	−207.10	—	0	11
inorganic phosphate	pi	−1096.10	−1299.00	−2	1
		−1137.30	−1302.6	−1	2
2-propanol	propanol2	−185.23	−330.83	0	8
n-propanol	propanoln	−175.81	—	0	8
pyrophosphate	ppi	−1919.86	−2293.47	−4	0
		−1973.86	−2294.87	−3	1
		−2012.21	−2295.37	−2	2

Table 3.2   *Continued*

Reactant	*Mathematica* Name	$\Delta_f G°$	$\Delta_f H°$	z_i	$N_H(i)$
		−2025.11	−2290.37	−1	3
		−2029.83	−2281.17	0	4
pyruvate	pyruvate	−472.27	−596.22	−1	3
retinal	retinal	0	—	0	28
retinol	retinol	−27.91	—	0	30
D-ribose	ribose	−738.79	−1034.00	0	10
D-ribose 1-phosphate	ribose1phos	−1574.49	—	−2	9
		−1612.67	—	−1	10
D-ribose 5-phosphate	ribose5phos	−1582.57	−2041.48	−2	9
		−1620.75	−2030.18	−1	10
D-ribulose	ribulose	−735.94	−1023.02	0	10
L-serine	serineL	−510.87	—	0	7
L-sorbose	sorbose	−911.95	−1263.30	0	12
succinate	succinate	−690.44	−908.68	−2	4
		−722.62	−908.84	−1	5
		−746.64	−912.20	0	6
succinyl-coenzyme A	succinylcoA	−509.59	—	−1	4
		−533.76	—	0	5
sucrose	sucrose	−1564.70	−2199.87	0	22
thioredoxin (ox)	thioredoxinox	0	—	0	0
thioredoxin (red)	thioredoxinred	69.88	—	−2	0
		20.56	—	−1	1
		−25.37	—	0	2
L-tryptophane	tryptophaneL	−114.70	−405.20	0	12
L-tyrosine	tyrosineL	−370.70	—	0	11
ubiquinone (ox)	ubiquinoneox	0	—	0	90
ubiquinone (red)	ubiquinonered	−89.92	—	0	92
urate	urate	−325.90	—	−1	3
urea	urea	−202.80	−317.65	0	4
uric acid	uricacid	−356.90	—	0	4
L-valine	valineL	−358.65	−611.99	0	11
D-xylose	xylose	−750.49	−1045.94	0	10
D-xylulose	xylulose	−746.15	−1029.65	0	10

Note: The standard Gibbs energies of formation and standard enthalpies of formation are in kJ mol^{-1}. Conventions: $\Delta_f G° - \Delta_f H° = 0$ for elements in defined reference states, H$^+(a = 1)$, coA$^-$, FAD$_{ox}^{2-}$, FADenz$_{ox}^{2-}$, cytochromec^{3+}, ferredoxin$_{ox}^{4-}$, FMN$_{ox}^{2-}$, glutathione$_{ox}^{2-}$, NAD$_{ox}^-$, NADP$_{ox}^{3-}$, retinal^0,

Thermodynamics of Biochemical Reactions at Specified pH

As shown in Chapter 1, it is convenient to discuss the thermodynamics of biochemical reactions at specified pH in terms of reactants like ATP, which are sums of species at equilibrium at the specified pH. The apparent equilibrium constants K' for biochemical reactions in dilute aqueous solutions are functions of T, pH, and ionic strength. This chapter introduces the thermodynamics needed to discuss biochemical reactions in terms of thermodynamic properties of reactants (sums of species with different numbers of hydrogen atoms), the relations between these properties, and the relations between the properties of species and the properties of reactants. It also provides information on criteria for spontaneous change and equilibrium. An important issue in these calculations is the number of intensive degrees of freedom and the total number of degrees of freedom. The goal of this chapter is the production of functions of pH and ionic strength that make it possible to calculate the apparent equilibrium constants and transformed enthalpies of reaction of biochemical reactions at 298.15 K and desired pHs in the range 5 to 9 and ionic stregths in the range zero to 0.35 M.

When the pH is specified, we enter into a whole new world of thermodynamics because there is a complete set of new thermodynamic properties, called transformed properties, new fundamental equations, new Maxwell equations, new Gibbs-Helmholtz equations, and a new Gibbs-Duhem equation. These new equations are similar to those in chemical thermodynamics, which were discussed in the preceding chapter, but they deal with properties of reactants (sums of species) rather than species. The fundamental equations for transformed thermodynamic potentials include additional terms for hydrogen ions, and perhaps metal ions. The transformed thermodynamic properties of reactants in biochemical reactions are connected with the thermodynamic properties of species in chemical reactions by equations given here.

The relationships between the thermodynamic properties of chemical reactions and the transformed thermodynamic properties of biochemical reactions have been treated in several reviews (Alberty, 1993a, 1994c, 1997b, 2001e). *Recommendations for Nomenclature and Tables in Biochemical Thermodynamics* from an IUPAC-IUBMB Committee were published in 1994 and republished in 1996. This report is available on the Web: *http://www.chem.qmw.ac.uh/imbmb/thermod/.*

The treatment of pH as an independent variable can be extended to pMg or the free concentrations of other cations that are bound reversibly by species of a reactant.

■ 4.1 FUNDAMENTAL EQUATION FOR A BIOCHEMICAL REACTION SYSTEM AT SPECIFIED pH

In a biochemical reaction one or more reactants may be weak acids or H^+ may be produced or consumed by the reaction. Therefore the specification of the pH means that the concentration of a reacting species is held constant, and as a consequence the equilibrium composition will be different at different pHs. Actually the pH may drift during a biochemical reaction if the reaction produces or consumes H^+, but the pH is measured at equilibrium and the experimental value of K' corresponds with this pH. To find the criterion for equilibrium at specified T, P, and pH, it is necessary to use a Legendre transform (see Section 2.5) to define a **transformed Gibbs energy** G' that has the chemical potential of H^+ as a **natural variable** (see Section 2.2). This transformed Gibbs energy provides the criterion for equilibrium and spontaneous change at the specified pH. The Legendre transform of the Gibbs energy for this purpose is (Alberty, 1992a, 1992c)

$$G' = G - n_c(H)\mu(H^+) \qquad (4.1-1)$$

where $n_c(H)$ is the total amount of the hydrogen component (see Section 3.3) and

$\mu(H^+)$ is the specified chemical potential of the hydrogen ion that corresponds with the experimental pH and ionic strength. It is necessary to use the amount $n_c(H)$ of the hydrogen component in this equation because it is the conjugate variable to $\mu(H^+)$ (see Section 2.7). The transformed Gibbs energy G' plays the same role that the Gibbs energy G does when the pH is not specified. The introduction of G' leads to a transformed enthalpy H' and a transformed entropy S' for a reaction system at specified pH. Note that all of these transformed thermodynamic properties are functions of the ionic strength as well as T, P, and pH. Transformed thermodynamic properties had previously been used in connection with petroleum thermodynamics where partial pressures of molecular hydrogen, ethylene, and acetylene can be specified as independent variables (Alberty and Oppenheim, 1988, 1989, 1992, 1993a, b; Alberty, 1991c).

The amount $n_c(H)$ of the hydrogen component in a system is given by the sum of the amounts of hydrogen atoms in various species in the reaction system.

$$n_c(H) = \sum_{j=1}^{N_s} N_H(j)n_j \tag{4.1-2}$$

In this equation $N_H(j)$ is the number of hydrogen atoms in species j, and N_s is the number of different species in the system. The index number for species is represented by j so that the index number introduced later for reactants (sums of species) can be i. Substituting equation 4.1-2 and $G = \Sigma n_j \mu_j$ (equation 2.5-12) into the Legendre transform (equation 4.1-1) yields

$$G' = \sum_{j=1}^{N_s} n_j\mu_j - \sum_{j=1}^{N_s} N_H(j)\mu(H^+)n_j = \sum_{j=1}^{N_s} n_j\{\mu_j - N_H(j)\mu(H^+)\} = \sum_{j=1}^{N_s-1} n_j\mu_j' \tag{4.1-3}$$

where the **transformed chemical potential** μ_j' of species j is given by

$$\mu_j' = \mu_j - N_H(j)\mu(H^+) \tag{4.1-4}$$

Note that the transformed chemical potential of the hydrogen ion is equal to zero so that there is one less term in the last summation. Equation 4.1-3 shows that the transformed Gibbs energy G' of a system is additive in the transformed chemical potentials μ_j' of $N_s - 1$ species, just like the Gibbs energy G is additive in the chemical potentials μ_j of N_s species (see equation 2.5-12). In making the Legendre transform, the chemical potential of one species (H^+) has been changed from a **dependent variable** to an **independent variable**. The roles of $n_c(H)$ and $\mu(H^+)$ in the fundamental equation are interchanged as shown in the next paragraph.

The derivation of the fundamental equation for the transformed Gibbs energy G' starts with the fundamental equation 2.5-5 for the Gibbs energy written in terms of species:

$$dG = -S\,dT + V\,dP + \sum_{j=1}^{N_s} \mu_j\,dn_j \tag{4.1-5}$$

In order to obtain the fundamental equation for dG', it is first necessary to get the contribution for the hydrogen component into a separate term. This can be done by using equation 4.1-4 to eliminate μ_j from equation 4.1-5:

$$dG = -S\,dT + V\,dP + \sum_{j=1}^{N_s-1} \mu_j'\,dn_j + \sum_{j=1}^{N_s} N_H(j)\mu(H^+)\,dn_j \tag{4.1-6}$$

There is one less term in the first summation because $\mu'(H^+) = 0$, as is evident from equation 4.1-4. Equation 4.1-2 shows that $dn_c(H) = \Sigma N_H(j)dn_j$, and so equation 4.1-6 can be written

$$dG = -S\,dT + V\,dP + \sum_{j=1}^{N_s-1} \mu_j'\,dn_j + \mu(H^+)dn_c(H) \tag{4.1-7}$$

The number $D = N_s + 2$ of natural variables of G has not been changed by separating the term for the hydrogen component.

The differential of the transformed Gibbs energy (equation 4.1-1) is

$$dG' = dG - n_c(H)d\mu(H^+) - \mu(H^+)dn_c(H) \qquad (4.1-8)$$

and substituting equation 4.1-7 into this relation yields a form of the fundamental equation for G':

$$dG' = -S\,dT + V\,dP + \sum_{j=1}^{N_s-1} \mu'_j dn_j - n_c(H)d\mu(H^+) \qquad (4.1-9)$$

Note that the Legendre transform has interchanged the roles of the conjugate intensive $\mu(H^+)$ and extensive $n_c(H)$ variables in the last term of equation 4.1-9. The number D' of natural variables of G' is $N_s + 2$, just as it was for G, but the chemical potential of the hydrogen ion is now a natural variable instead of the amount of the hydrogen component (equation 4.1-7).

Since the chemical potential $\mu(H^+)$ depends on both the temperature and the concentration of hydrogen ions, it is not a very convenient variable when the temperature is changed. The hydrogen ion concentration can be made an independent intensive variable in the fundamental equation for G' by use of the expression for the differential of the chemical potential of H^+:

$$d\mu(H^+) = \left\{\frac{\partial\mu(H^+)}{\partial T}\right\}_{[H^+]} dT + \left\{\frac{\partial\mu(H^+)}{\partial[H^+]}\right\}_T d[H^+] \qquad (4.1-10)$$

The first partial derivative in this equation is equal to $-S_m(H^+)$, where $S_m(H^+)$ is the **molar entropy** of the hydrogen ion. To evaluate the second partial derivative in equation 4.1-10, we need to recall that the chemical potential of species B_j is given by

$$\mu_j = \mu_j^0 + RT\ln[B_j] \qquad (4.1-11)$$

where μ_j^0 is the standard chemical potential of species j. Thus the chemical potential of B_j in a 1 M solution at the specified ionic strength is given by μ_j^0. Since the thermodynamic properties are taken to be functions of the ionic strength, we do not have to deal with activity coefficients explicitly. Equation 4.1-11 indicates that $d\mu(H^+)/d[H^+] = RT/[H^+]$, and since $dpH/d[H^+] = -1/(\ln(10)[H^+])$, equation 4.1-10 can be written

$$d\mu(H^+) = -S_m(H^+)dT - RT\ln(10)dpH \qquad (4.1-12)$$

Substituting this in equation 4.1-9 yields

$$dG' = -S'dT + V\,dP + \sum_{j=1}^{N_s-1} \mu'_j dn_j + RT\ln(10)n_c(H)dpH \qquad (4.1-13)$$

where the **transformed entropy** S' of the system at a specified pH is given by

$$S' = S - n_c(H)S_m(H^+) \qquad (4.1-14)$$

Since the enthalpy H of the system is defined by $H = G + TS$, substituting equations 4.1-1 and 4.1-14 in this expression for H yields

$$H = G' + n_c(H)\mu(H^+) + T(S' + n_c(H)S_m(H^+)) = H' + n_c(H)H_m(H^+) \qquad (4.1-15)$$

where the **transformed enthalpy** H' of the system is given by

$$H' = G' + TS' \qquad (4.1-16)$$

and the **molar transformed enthalpy** of hydrogen ions is given by

$$H'_m(H^+) = \mu'(H^+) + TS'_m(H^+) \qquad (4.1-17)$$

Thus the definition of the transformed Gibbs energy for the system by $G' = G - n_c(H)\mu(H^+)$ automatically brings in the transformed enthalpy $H' = H - n_c(H)H_m(H^+)$ and the transformed entropy $S' = S - n_c(H)S_m(H^+)$ so that there is a complete set of transformed thermodynamic properties.

The summation in equation 4.1-13 can be written with fewer than $N_s - 1$ terms because when the pH is specified, groups of terms now have the same value of μ'_j. These are the terms for the different protonated species of a reactant. When a group of species differ only in the number of hydrogen atoms that they contain, these species have the same transformed chemical potential μ'_i at a specified pH, and this makes them **pseudoisomers**. Isomers have the same chemical potential at chemical equilibrium, and pseudoisomers have the same transformed chemical potential μ'_j at equilibrium at a specified pH. For example, the various protonated species (ATP^{4-}, $HATP^{3-}$, H_2ATP^{2-}) of ATP have the same transformed chemical potential μ'_j at a specified pH. This can be proved by minimizing G' at specified T, P, and pH for a system containing the three species of ATP. Since pseudoisomers have the same transformed chemical potential μ'_i, we can collect terms for pseudoisomers and use $n'_i = \Sigma n_j$ for the amount of a **pseudoisomer group**. Thus equation 4.1-13 can be rewritten as

$$dG' = -S'dT + V\,dP + \sum_{i=1}^{N'} \mu'_i\,dn'_i + RT\ln(10)n_c(H)dpH \qquad (4.1\text{-}18)$$

where N' is the number of pseudoisomer groups in the system. A pseudoisomer group may contain a single species. This is the form of the **fundamental equation for G'** that is used to treat biochemical reaction systems in a single phase. Note that this fundamental equation has a new type of term, the last one, that is proportional to dpH. The number D' of natural variables of G' is $N' + 3$, which may be considerably less than the $D = N_s + 2$ for the system described in terms of species. In writing equation 4.1-18, it is assumed that the binding of H^+ by species is at equilibrium. Acid dissociations are equilibrated much more rapidly than enzyme-catalyzed reactions.

A very important step has been taken in aggregating species in equation 4.1-18 so that the number of terms proportional to differentials in amounts is reduced from $N_s - 1$ (in equation 4.1-6) to N' (in equation 4.1-18). Aggregating groups of species makes it possible to deal with ATP as a reactant at a specified pH. This more global view makes it easier to think about systems of metabolic reactions. Within a pseudoisomer group, the transformed chemical potentials of species at equilibrium are equal, the amounts add, and the standard thermodynamic properties of the group are given by the isomer group equations discussed earlier (3.5-11 to 3.5-18). This matter will be discussed in greater detail in Section 4.3.

Equation 4.1-18 can be integrated at constant values of the intensive properties to obtain

$$G' = \sum_{i=1}^{N'} n'_i\mu'_i \qquad (4.1\text{-}19)$$

Thus the transformed Gibbs energy is additive in the transformed chemical potentials of pseudoisomer groups just like the Gibbs energy G is additive in the chemical potentials of species (equation 2.5-12).

Equation 4.1-18 shows that G' is a function of T, P, $\{n'_i\}$, and pH, and so calculus requires that

$$dG' = \left(\frac{\partial G'}{\partial T}\right)dT + \left(\frac{\partial G'}{\partial P}\right)dP + \sum\left(\frac{\partial G'}{\partial n'_i}\right)dn'_i + \left(\frac{\partial G'}{\partial pH}\right)dpH \qquad (4.1\text{-}20)$$

where the subscripts have been omitted since they are complicated. Comparison of equations 4.1-18 and 4.1-20 shows that

$$S' = -\left(\frac{\partial G'}{\partial T}\right)_{P,\mathrm{pH},\{n_i'\}} \tag{4.1-21}$$

$$V = \left(\frac{\partial G'}{\partial P}\right)_{T,\mathrm{pH},\{n_i'\}} \tag{4.1-22}$$

$$\mu_i' = \left(\frac{\partial G'}{\partial n_i'}\right)_{T,P,\mathrm{pH},\{n_j'\}} \tag{4.1-23}$$

$$RT\ln(10)n_\mathrm{c}(\mathrm{H}) = \left(\frac{\partial G'}{\partial \mathrm{pH}}\right)_{T,P,\{n_i'\}} \tag{4.1-24}$$

where $\{n_i'\}$ is the set of amounts of reactants. Equation 4.1-21 shows how the transformed entropy S' of the system can be obtained from measurements of the transformed Gibbs energy G'. Substituting equation 4.1-19 in equation 4.1-21 yields

$$S' = \sum_{i=1}^{N'} n_i' S_{\mathrm{m}i}' \tag{4.1-25}$$

where the molar transformed entropy of a reactant is given by

$$S_{\mathrm{m}i}' = -\left(\frac{\partial \mu_i'}{\partial T}\right)_{P,\mathrm{pH},\{n_j'\}} \tag{4.1-26}$$

The transformed enthalpy of the system at specified pH is given by $H' = G' + TS'$ (equation 4.1-16), and substituting equation 4.1-21 yields

$$H' = G' - T\left(\frac{\partial G'}{\partial T}\right)_{P,\mathrm{pH},\{n_i'\}} = -T^2\left(\frac{\partial(G'/T)}{\partial T}\right)_{P,\mathrm{pH},\{n_i'\}} \tag{4.1-27}$$

This is the **Gibbs-Helmholtz equation** for the system at specified pH. Substituting equation 4.1-19 in equation 4.1-27 yields

$$H' = \sum_{i=1}^{N'} n_i' H_{\mathrm{m}i}' \tag{4.1-28}$$

where the molar transformed enthalpy of a reactant is given by the Gibbs-Helmholtz equation in the form

$$H_{\mathrm{m}i}' = -T^2\left(\frac{\partial(\mu_i'/T)}{\partial T}\right)_{P,\mathrm{pH},\{n_i'\}} \tag{4.1-29}$$

If there is one reactant, equation 4.1-18 leads to $D(D-1)/2 = 4 \times 3/2 = 6$ Maxwell equations. One of these is discussed in Section 4.7 on the calculation of the average binding of hydrogen ions by a reactant.

At specified pH, equation 4.1-18 can be written

$$(\mathrm{d}G')_{\mathrm{pH}} = -S'\,\mathrm{d}T + V\,\mathrm{d}P + \sum_{i=1}^{N'} \mu_i'\,\mathrm{d}n_i' \tag{4.1-30}$$

This equation has the same form as equation 4.1-5, which applies to a chemical reaction described in terms of species. **It shows why the world of biochemical thermodynamics at specified pH looks so much like the world of chemical thermodynamics that is described by equation 4.1-5.** An important difference between these equations is that the terms in the summation on the right side of equation 4.1-30 deal with pseudoisomer groups, like ATP, rather than species.

■ 4.2 DERIVATION OF THE EXPRESSION FOR THE APPARENT EQUILIBRIUM CONSTANT

At specified pH, the biochemical reaction that corresponds with equation 4.1-18 is represented by

$$\sum_{i=1}^{N'} v_i' B_i = 0 \qquad (4.2\text{-}1)$$

where the primes on the stoichiometric numbers v_i' in a biochemical equation are used to distinguish them from the stoichiometric numbers on the underlying chemical reactions. The B_i are symbols representing pseudoisomer groups, as in $ATP + H_2O = ADP + P_i$. Biochemical reactions can produce or consume H^+, but these hydrogen ions are not shown in equation 4.2-1 because the pH is held constant. In other words, hydrogen atoms are not conserved in the reaction vessel. Figure 4.1 shows a thought experiment that corresponds with the interpretation of the determination of the apparent equilibrium constant at a specified T, P, and pH. When hydrogen ions are produced in the reaction, they diffuse into the pH reservoir through the membrane permeable to H^+ to keep the pH constant, and when hydrogen ions are consumed in the reaction, hydrogen ions diffuse into the reaction chamber to hold the pH constant. This figure shows that the pH reservoir plays the same kind of role as the heat reservoir at temperature T and the piston exerting a constant pressure P. Therefore these three independent variables have to be treated in the same way in thermodynamics; that is, they are introduced by means of Legendre transforms.

It is important to have symbols, that is, names, for reactants that are different from the symbols B_j for species, like ATP^{4-}, $HATP^{3-}$, and H_2ATP^{2-}, which are used in chemical equations. (The problems in naming are discussed later in Section 4.11.) The N' reactants in a biochemical reaction are all pseudoisomer groups; note that a pseudoisomer group is made up of one species over a wide range of pH if the reactant has no pKs in the pH range considered.

Equation 4.1-18 can be used to derive the expression for the apparent equilibrium constant K' for a biochemical reaction at a specified pH. If a single biochemical reaction is catalyzed, the amounts n_i' of the pseudoisomer groups at each stage of the reaction are given by

$$n_i' = (n_i')_0 + v_i' \xi' \qquad (4.2\text{-}2)$$

where $(n_i')_0$ is the initial amount of reactant i (pseudoisomer group i), v_i' is the stoichiometric number of reactant i in the biochemical reaction (see equation

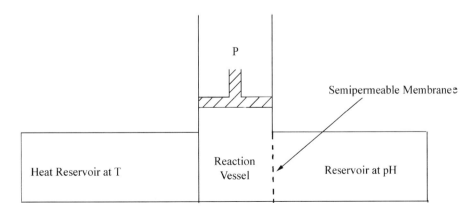

Figure 4.1 Thought experiment in which a reaction is carried out in a reaction vessel connected to a pH reservoir through a semipermeable membrane that permits H^+ to go in or out. The reaction vessel is also held at a constant temperature and pressure.

4.2-1), and ξ' is the **apparent extent of the biochemical reaction**. It is necessary to put the primes on these quantities to differentiate them from the stoichiometric numbers v_j and extents of reaction ξ of the underlying chemical reactions written in terms of species. If there is a single biochemical reaction, substituting $dn_i' = v_i' d\xi'$ in equation 4.1-18 yields

$$dG' = -S' dT + VdP + \left(\sum_{i=1}^{N'} v_i'\mu_i' \right) d\xi' + RT \ln(10) n_c(\mathrm{H}) d\mathrm{pH} \qquad (4.2\text{-}3)$$

so that

$$\Delta_r G' = \left(\frac{\partial G'}{\partial \xi'} \right)_{T,p,\mathrm{pH}} = \sum_{i=1}^{N'} v_i'\mu_i' \qquad (4.2\text{-}4)$$

where $\Delta_r G'$ is referred to as the **transformed reaction Gibbs energy**. At chemical equilibrium, $\Delta_r G'$ is equal to zero so that

$$\sum_{i=1}^{N'} v_i'(\mu_i')_{\mathrm{eq}} = 0 \qquad (4.2\text{-}5)$$

This is the **equilibrium condition**. In Chapter 3 we saw that the corresponding condition for a chemical reaction is equation 3.1-6. Note that equation 4.2-5 has the same form as biochemical equation 4.2-1.

The expression for the transformed chemical potential of a reactant is given by

$$\mu_i' = \mu_i'^0 + RT \ln[\mathrm{B}_i] \qquad (4.2\text{-}6)$$

where $\mu_i'^0$ is the standard value (that is the value for an ideal 1 M solution) and B_i represents the ith reactant (pseudoisomer group). This looks reasonable in relation to $\mu_i = \mu_j^0 + RT \ln[\mathrm{B}_j]$ for the chemical potential of species j, but we will consider equation 4.2-6 in greater detail in Section 4.4. Substituting equation 4.2-6 in equation 4.2-4 yields

$$\Delta_r G' = \sum v_i'\mu_i'^0 + RT \sum v_i' \ln[\mathrm{B}_i] \qquad (4.2\text{-}7)$$

At equilibrium, $\Delta_r G' = 0$, and so

$$\sum v_i'\mu_i'^0 = -RT \sum v_i' \ln[\mathrm{B}_i]$$

$$= -RT \sum \ln([\mathrm{B}_i]^{v_i})$$

$$= -RT \ln \Pi[\mathrm{B}_i]^{v_i} = -RT \ln K' \qquad (4.2\text{-}8)$$

where Π is the product sign. The concentrations in equation 4.2-8 are equilibrium concentrations, but it is conventional to omit the subscripts "eq" in writing expressions for equilibrium constants. The apparent equilibrium constant is given by

$$K' = \Pi[\mathrm{B}_i]^{v_i} \qquad (4.2\text{-}9)$$

This confirms that K' is written in terms of concentrations of pseudoisomer groups and that there are no terms for hydrogen ions. When dilute aqueous solutions are considered, the convention is that $[\mathrm{H_2O}]$ is omitted, but the contribution for $\mu'^0(\mathrm{H_2O})$ in equation 4.2-8 is included.

When equation 4.2-4 is substituted in equation 4.2-3, the following fundamental equation is obtained for the transformed Gibbs energy:

$$dG' = -S' dT + VdP + \Delta_r G' d\xi' + RT \ln(10) n_c(\mathrm{H}) d\mathrm{pH} \qquad (4.2\text{-}10)$$

This form of the fundamental equation has two Maxwell equations of special interest. The first Maxwell equation is

$$-\left(\frac{\partial S'}{\partial \xi'} \right)_{T,P,\mathrm{pH}} = \left(\frac{\partial \Delta_r G'}{\partial T} \right)_{P,\xi',\mathrm{pH}} = -\Delta_r S' \qquad (4.2\text{-}11)$$

where $\Delta_r S'$ is the **transformed reaction entropy**. The second Maxwell equation is

$$\left(\frac{\partial_r G'}{\partial pH}\right)_{T,P,\xi'} = RT\ln(10)\left(\frac{\partial n_c(H)}{\partial \xi'}\right)_{T,P,pH} = RT\ln(10)\Delta_r N_H \qquad (4.2\text{-}12)$$

The change in binding of hydrogen ions in the biochemical reaction is given by

$$\Delta_r N_H - \frac{1}{RT\ln(10)}\left(\frac{\partial \Delta_r G'}{\partial pH}\right)_{T,P,\xi'} \qquad (4.2\text{-}13)$$

This second Maxwell equation will be discussed in Section 4.7.

The transformed reaction entropy is of interest in its own right, but it also leads to a noncalorimetric method for determining the **transformed reaction enthalpy**.

Note that since $H' = G' + TS'$,

$$\Delta_r H' = \Delta_r G' + T\Delta_r S' \qquad (4.2\text{-}14)$$

Substituting equation 4.2-11 in this equation yields the Gibbs-Helmholtz equation at a specified pH:

$$\Delta_r H' = -T^2\left(\frac{\partial(\Delta_r G'/T)}{\partial T}\right)_{P,pH,\xi'} \qquad (4.2\text{-}15)$$

■ 4.3 TRANSFORMED THERMODYNAMIC PROPERTIES OF SPECIES AND REACTANTS

In order to learn more about the transformed chemical potential of a reactant, we consider the fundamental equation for G' (equation 4.1-18) for a system containing a single reactant

$$dG' = -S'dT + VdP + \mu'dn' + RT\ln(10)n_c(H)dpH \qquad (4.3\text{-}1)$$

This indicates that μ' is given by

$$\mu' = \left(\frac{\partial G'}{\partial n'}\right)_{T,P,pH} \qquad (4.3\text{-}2)$$

Integration of equation 4.3-1 at constant T, P, and pH yields

$$G' = \mu'n' \qquad (4.3\text{-}3)$$

so that the transformed chemical potential of a reactant is equal to its molar transformed Gibbs energy. An expression for μ' for reactant B that contains two species, which differ by one hydrogen atom, can be obtained by starting with

$$\mu' = \mu'^0 + RT\ln[B] \qquad (4.3\text{-}4)$$

The concentration of the reactant is $[B] = [B_1] + [B_2]$, where $[B_1]$ and $[B_2]$ are the concentrations of the species at the given pH. The standard transformed Gibbs energy of the reactant when the acid dissociation is at equilibrium can be calculated using

$$\mu'^0 = -RT\ln\left[\exp\left(-\frac{\mu_1'^0}{RT}\right) + \exp\left(-\frac{\mu_2'^0}{RT}\right)\right] \qquad (4.3\text{-}5)$$

or, alternatively,

$$\mu'^0 = r_1\mu_1'^0 + r_2\mu_2'^0 + RT(r_1\ln r_1 + r_2\ln r_2) \qquad (4.3\text{-}6)$$

where the equilibrium mole fractions of the species within the pseudoisomer group are given by

$$r_1 = \exp\left[\frac{\mu'^0 - \mu_1'^0}{RT}\right] \qquad (4.3\text{-}7)$$

$$r_2 = \exp\left[\frac{\mu'^0 - \mu_2'^0}{RT}\right] \qquad (4.3\text{-}8)$$

The standard transformed chemical potentials of the species at zero ionic strength are given by

$$\mu_1'^0 = \mu_1^0 - N_{H1}RT\ln(10)\text{pH} \qquad (4.3\text{-}9)$$

$$\mu_2'^0 = \mu_2^0 - N_{H2}RT\ln(10)\text{pH} \qquad (4.3\text{-}10)$$

where N_{H1} is the number of hydrogen atoms in species 1. This derivation has been made at zero ionic strength as a simplification, but the effects of ionic strength are taken into account fully in the next section. Substituting these two equations in equation 4.3-6 yields

$$\mu'^0 = r_1\mu_1^0 + r_2\mu_2^0 + RT(r_1\ln r_1 + r_2\ln r_2) - \bar{N}_H RT\ln(10)\text{pH} \qquad (4.3\text{-}11)$$

where the average number of hydrogen ions bound by the reactant is given by

$$\bar{N}_H = r_1 N_{H1} + r_2 N_{H2} \qquad (4.3\text{-}12)$$

Substituting equation 4.3-11 in equation 4.3-4 yields

$$\mu' = r_1\mu_1^0 + r_2\mu_2^0 + RT(r_1\ln r_1 + r_2\ln r_2) - \bar{N}_H RT\ln(10)\text{pH} + RT\ln[B]$$
$$(4.3\text{-}13)$$

so that the transformed chemical potential of a reactant is equal to the mole fraction average of the chemical potentials of the species, plus the Gibbs energy of mixing, minus an adjustment for the pH that is proportional to the average binding of hydrogen ions, plus $RT\ln[B]$.

■ 4.4   TRANSFORMED THERMODYNAMIC PROPERTIES OF BIOCHEMICAL REACTIONS

In the preceding three sections, μ_j has been used for a species and μ_i' has been used for a pseudoisomer group, but in treating experimental data Gibbs energies of formation $\Delta_r G_j$ and transformed Gibbs energies of formation, we use $\Delta_f G_i'$ instead because it is not possible to determine absolute values of chemical potentials. The Gibbs energies of formation of species are relative to reference states of the elements or to conventions, like $\Delta_f G^0(H^+) = 0$ at zero ionic strength at each temperature. These reference states cancel when differences are taken in discussing reactions or phase distributions.

Thus, in making calculations, we rewrite equation 4.2-4 for a biochemical reaction as

$$\Delta_r G' = \sum_{i=1}^{N'} v_i' \Delta_f G_i' \qquad (4.4\text{-}1)$$

and the **standard transformed Gibbs energy of reaction** is given by

$$\Delta_r G'^0 = \sum_{i=1}^{N'} v_i' \Delta_f G_i'^0 = -RT\ln K' \qquad (4.4\text{-}2)$$

Application of the Gibbs-Helmholtz equation to equation 4.4-1 yields the **standard transformed enthalpy of reaction**

$$\Delta_r H'^0 = \sum_{i=1}^{N'} v_i' \Delta_f H_i'^0 \qquad (4.4\text{-}3)$$

The standard transformed Gibbs energies of formation of reactants in a biochemical reaction at specified pH are very important because they can be used to calculate the value of K' of the biochemical reaction at the specified pH. This section and the next show how the standard transformed Gibbs energies of formation of biochemical reactants at specified pH and ionic strength can be calculated from the standard Gibbs energies of formation of the species they contain. This section is concerned with the standard transformed Gibbs energies of formation and standard transformed enthalpies of formation of species, and the next section is concerned with the calculation of these properties for a reactant that involves two or more species. These standard transformed properties can also be calculated from experimental data using equations 4.4-2 and 4.4-3 in the absence of information about the standard properties of the species involved.

The discussion of the standard transformed properties of a species starts with the definition of the transformed chemical potential μ_j' of the species given by $\mu_j' = \mu_j - N_H(j)\mu(H^+)$ (equation 4.1-4). This equation can be written in terms of Gibbs energies of formation and transformed Gibbs energies of formation as follows:

$$\Delta_f G_j' = \Delta_f G_j - N_H(j)\Delta_f G(H^+) \qquad (4.4\text{-}4)$$

There is a corresponding equation for the transformed enthalpy of formation of a species:

$$\Delta_f H_j' = \Delta_f H_j - N_H(j)\Delta_f H(H^+) \qquad (4.4\text{-}5)$$

Instead of $\mu_j = \mu_j^0 + RT \ln [B_j]$ (equation 4.1-11) we now use

$$\Delta_f G_j = \Delta_f G_j^0 + RT \ln[B_j] \qquad (4.4\text{-}6)$$

The corresponding relation for the enthalpy of formation of a species is

$$\Delta_f H_j = \Delta_f H_j^0 \qquad (4.4\text{-}7)$$

Using equation 4.4-6 in equation 4.4-4 yields

$$\Delta_f G_j' = \Delta_f G_j^0 + RT \ln[B_j] - N_H(j)\{\Delta_f G^0[H^+] + RT \ln(10^{-pH})\}$$
$$= \Delta_f G_j'^0 + RT \ln[B_j] \qquad (4.4\text{-}8)$$

where the standard transformed Gibbs energy of formation of species j is given by

$$\Delta_f G_j'^0 = \Delta_f G_j^0 - N_H(j)\{\Delta_f G^0[H^+] + RT \ln(10^{-pH})\} \qquad (4.4\text{-}9)$$

As we have seen in the preceding chapter, the standard thermodynamic properties of species in aqueous solutions are functions of ionic strength when they have electric charges. Substituting equation 3.6-3 for species j and for H^+ in equation 4.4-9 yields the **standard transformed Gibbs energy of formation** of species j as a function of pH and ionic strength at 298.15 K:

$$\Delta_f G_j'^0 = \Delta_f G_j^0(I=0) + N_H(j)RT \ln(10)pH - \frac{2.91482(z_j^2 - N_H(j))I^{1/2}}{1 + 1.6I^{1/2}} \qquad (4.4\text{-}10)$$

It should be noted that $\Delta_f G_j'^0$ is a function of ionic strength for uncharged species that contain hydrogen atoms, as well as charged species. There is an exception to this statement when $z_j^2 - N_H(j) = 0$ for a species. The standard transformed Gibbs energy of formation of a species is independent of ionic strength when $z_j = 0$ and $N_H(j) = 0$. Equation 4.4-10 shows how the standard transformed Gibbs energy of formation of a biochemical reactant consisting of a single species can be calculated from the standard Gibbs energy of formation of the species at zero ionic strength. Note that the coefficient of the ionic strength term is a function of temperature, as discussed in Section 3.7. The calculation of the standard transformed properties of species is discussed by Alberty (1999).

Equation 4.4-5 leads to the corresponding equation for the standard transformed enthalpy of formation of a species:

$$\Delta_f H_j'^0 = \Delta_f H_j^0 - N_H(j)\Delta_f H^0(H^+) \tag{4.4-11}$$

Substituting equation 3.6-3 yields the **standard transformed enthalpy of formation** of species j as a function of pH and ionic strength at 298.15 K:

$$\Delta_f H_j'^0 = \Delta_f H_j^0(I = 0) + \frac{1.4775(z_j^2 - N_H(j))I^{1/2}}{1 + 1.6I^{1/2}} \tag{4.4-12}$$

Thus the availability of $\Delta_f H_j^0(I = 0)$ for a species makes it possible to calculate $\Delta_f H_j'^0$, and vice versa. Note that the standard transformed enthalpy of a species is independent of pH, even when it contains hydrogen atoms. At temperatures other than 298.15 K the numerical coefficient of the ionic strength term has different values, as discussed in Section 3.7.

■ 4.5 THERMODYNAMICS OF PSEUDOISOMER GROUPS AT SPECIFIED pH

When there are two or more species in a pseudoisomer group, the standard transformed Gibbs energy of formation $\Delta_f G_i'^0$ and standard transformed enthalpy of formation $\Delta_f H_i'^0$ of the pseudoisomer group have to be calculated using isomer group thermodynamics (Section 3.5). The isomer group equations were introduced in equations 3.5-11 to 3.5-14. At a specified pH, the various forms of a reactant have the same $\Delta_f G_j'$ at chemical equilibrium, and so the **standard transformed Gibbs energy of formation of the pseuodisomer group** can be calculated using

$$\Delta_f G_i'^0 = -RT \ln \left\{ \sum_{j=1}^{N_{iso}} \exp\left[-\frac{\Delta_f G_j'^0}{RT} \right] \right\} \tag{4.5-1}$$

where N_{iso} is the number of species in the pseudoisomer group. The equilibrium mole fraction r_j of the jth pseudoisomer in the pseudoisomer group is given by

$$r_j = \exp\left\{ \frac{\Delta_f G_i'^0 - \Delta_f G_j'^0}{RT} \right\} \tag{4.5-2}$$

The **standard transformed enthalpy of formation of the pseudoisomer group** is a mole fraction weighted average and is given by

$$\Delta_f H_i'^0 = \sum_{j=1}^{N_{iso}} r_j \Delta_f H_j'^0 \tag{4.5-3}$$

Note that although $\Delta_f H_j'^0$ values for species are independent of pH, this is not true for $\Delta_f H_i'^0$ values of reactants consisting of two or more species because the r_i are functions of pH. The pseudoisomer group has a corresponding **standard transformed entropy of formation** given by

$$\Delta_f S_i'^0 = \frac{\Delta_f H_i'^0 - \Delta_f G_i'^0}{T} \tag{4.5-4}$$

The standard transformed heat capacity at constant pressure of a reactant is discussed later in Chapter 10 on calorimetry. The calculation of $\Delta_f H'^0$ using equation 4.5-3 looks simple, but note that the standard transformed Gibbs energies of formation of all of the species are involved in the calculation. These equations were applied to the ATP series by Alberty and Goldberg (1992).

Equation 4.5-1 for $\Delta_f G_i'^0$ can also be written (Alberty, 1999) in terms of the binding polynomial (partition function) P (see Section 1.3). Equation 4.5-1 can be written in the form

$$\Delta_r G_i'^0 = -RT \ln \left\{ \exp\left(\frac{\Delta_f G_1'^0}{RT} \right) \left(1 + \exp\left(\frac{\Delta_f G_1'^0 - \Delta_f G_2'^0}{RT} \right) + \cdots \right) \right\}$$

(4.5-5)

This is equivalent to

$$\Delta_f G_i'^0 = \Delta_f G_1'^0 - RT \ln P$$

(4.5-6)

where the binding polynomial is given by

$$P = 1 + \frac{[H^+]}{K_1} + \frac{[H^+]^2}{K_1 K_2} + \cdots$$

(4.5-7)

and $\Delta_f G_1'^0$ is the standard transformed Gibbs energy of formation of the species with the smallest number of dissociable hydrogen atoms. The $K_1, K_2, \ldots$ are the successive acid dissociation constants at the specified pH and ionic strength, starting with the highest pK. This equation is also useful for calculating $\Delta_f G_1'^0$ from the experimental value of $\Delta_f G_i'^0$. The value of $\Delta_f G_i'^0$ at a desired pH and ionic strength can be calculated using equation 4.4-10.

When transformed Gibbs energies of formation are used rather than chemical potentials, equation 4.3-4 can be written

$$\Delta_f G_i' = \Delta_f G_i'^0 + RT \ln[B_i]$$

(4.5-8)

From now on we will assume that $\Delta_f G_i'^0$ and $\Delta_f H_i'^0$ of biochemical reactants made up of single species have been calculated using equations 4.4-10 and 4.4-12 and that $\Delta_f G_i'^0$ and $\Delta_f H_i'^0$ of biochemical reactants with more than one species have been calculated using equations 4.5-1 and 4.5-3.

The discussion above has emphasized $\Delta_r G'^0$ and $\Delta_r H'^0$ for biochemical reactions, but it is also useful to consider $\Delta_r G'$ and $\Delta_r H'$. These quantities correspond with changes from reactants at arbitrary concentrations to products at arbitrary concentrations, rather than standard states (i.e., 1 M). Substituting equation 4.5-8 in equation 4.4-1 yields

$$\Delta_r G' = \sum_{i=1}^{N'} v_i' \Delta_f G_i'^0 + RT \ln \prod_{i=1}^{N'} [B_i]^{v_i'}$$

(4.5-9)

which can be written

$$\Delta_r G' = \Delta_r G'^0 + RT \ln Q'$$

(4.5-10)

The apparent reaction quotient Q' is given by

$$Q' = \prod_{i=1}^{N'} [B_i]^{v_i'}$$

(4.5-11)

where the concentrations of reactants can be chosen arbitrarily.

Applying equation 4.2-12 to 4.5-10 yields

$$\Delta_r S' = \Delta_r S'^0 - R \ln Q'$$

(4.5-12)

where

$$\Delta_r S'^0 = \sum_{i=1}^{N'} v_i' \Delta_f S_i'^0$$

(4.5-13)

Application of the Gibbs-Helmholtz equation derived from equation 4.2-16 to equation 4.5-10 yields

$$\Delta_r H' = \Delta_r H'^0$$

(4.5-14)

Note $\Delta_r H'^0$ does not depend on Q'.

■ 4.6   GIBBS-DUHEM EQUATION, DEGREES OF FREEDOM, AND THE CRITERION FOR EQUILIBRIUM AT SPECIFIED pH

In discussing one-phase systems in terms of species, the number D of natural variables was found to be $N_s + 2$ (where the intensive variables are T and P) and the number F of independent intensive variables was found to be $N_s + 1$ (Section 3.4). When the pH is specified and the acid dissociations are at equilibrium, a system is described in terms of N' reactants (sums of species), and the number D' of natural variables is $N' + 3$ (where the intensive variables are T, P, and pH), as indicated by equation 4.1-18. The number N' of reactants may be significantly less than the number N_s of species, so that fewer variables are required to describe the state of the system. When the pH is used as an independent variable, the Gibbs-Duhem equation for the system is

$$0 = -S' \, dT + V \, dP - \sum_{i=1}^{N'} n'_i \, d\mu'_i + \ln(10) n_c(\mathrm{H}) RT \, d\mathrm{pH} \qquad (4.6\text{-}1)$$

which indicates that there are $N' + 3$ intensive variables. However, only $N' + 2$ of them are independent for a one-phase system. This is in agreement with the phase rule in which the apparent number of independent intensive variables is given by $F' = N' - p + 3$, where the 3 refers to T, P, and pH.

The preceding paragraph applies to a system in which there are no biochemical reactions. Now we consider systems with reactions that are at equilibrium (Alberty, 1992d). For a chemical reaction system, we saw (Section 3.4) that $D = C + 2$ and $F = C + 1$ for a one-phase system. For a biochemical reaction system at equilibrium, we need the fundamental equation written in terms of apparent components to show how many natural variables there are. When the reaction conditions $\Sigma v'_i \mu'_i = 0$ for the biochemical reactions in the system are used to eliminate one μ'_i for each independent reaction from equation 4.1-18, the following fundamental equation for G' in terms of apparent components is obtained:

$$dG' = -S' \, dT + V \, dP + \sum_{i=1}^{C'} \mu'_i \, dn'_{ci} + RT \ln(10) n_c(\mathrm{H}) \, d\mathrm{pH} \qquad (4.6\text{-}2)$$

where n'_{ci} is the amount of apparent component i and C' is the number of apparent components at specified pH. This equation shows that when biochemical reactions are at equilibrium the apparent number D' of natural variables is $D' = C' + 3$. Since hydrogen is not included in the C' apparent components, $C' = C - 1$, where C is the number of components before the pH was specified. Thus the number of natural variables is the same for a system whether it is considered to be made up of species or reactants. **In other words, making the Legendre transform has not changed the number of natural variables; it has simply changed an extensive variable into an intensive variable.** The number of apparent components is given by $C' = N' - R'$, where N' is the number of pseudoisomer groups and R' is the number of independent biochemical equations. This can be compared with a chemical system that contains N_s species and whose number of components is given by $C = N_s - R$, where N_s is the number of species and R is the number of independent chemical reactions.

The Gibbs-Duhem equation that corresponds with equation 4.6-2 is

$$0 = -S' \, dT + V \, dP - \sum_{i=1}^{C'} n'_{ci} \, d\mu'_i + RT \ln(10) n_c(\mathrm{H}) \, d\mathrm{pH} \qquad (4.6\text{-}3)$$

which indicates that when enzyme-catalyzed reactions are at equilibrium, the number of apparent independent intensive variables is $F' = C' + 2$. This is in agreement with the phase rule written as $F' = C' - p + 3$. For a one-phase system, $F' = C' + 2$.

For a system of chemical reactions, the criterion for spontaneous change and equilibrium is $dG \leqslant 0$ at T, P, $\{n_{ci}\}$. When the pH is specified, the criterion for spontaneous change and equilibrium becomes $dG' \leqslant 0$ at T, P, $\{n'_{ci}\}$, pH, where $\{n'_{ci}\}$ represents the set of amounts of components other than hydrogen.

■ 4.7 CALCULATION OF THE BINDING OF HYDROGEN IONS BY REACTANTS AND THE CHANGES IN BINDING OF HYDROGEN IONS IN BIOCHEMICAL REACTIONS

When a system contains a single pseudoisomer group, equation 4.1-18 shows that

$$dG' = -S'dT + VdP + \mu'_i dn'_i + RT \ln(10) n_c(H) dpH \qquad (4.7-1)$$

The Maxwell equation involving the last two terms is

$$\left(\frac{\partial \mu'_i}{\partial pH}\right)_{T,P,n'_i} = RT \ln(10) \left(\frac{\partial n_c(H)}{\partial n'_i}\right)_{T,P,pH} \qquad (4.7-2)$$

Since the derivative on the right is the change in the amount of hydrogen atoms in the system when the amount of reactant i is changed, this equation can be written in terms of the average number $\bar{N}_H$ of hydrogen atoms bound by the reactant (Alberty, 1994c).

$$\bar{N}_H = \frac{1}{RT \ln(10)} \left(\frac{\partial \Delta_f G'^0}{\partial pH}\right)_{T,P,n_i} \qquad (4.7-3)$$

This equation is closely related to equation 4.2-13. Wyman (1964) was probably the first to use this type of equation. Note that the derivative of $\Delta_r G'$ is the same as the derivative of $\Delta_r G'^0$.

In Chapter 1 it was shown that the change in binding of hydrogen ions in a reaction can be calculated by taking the difference between the binding by products and the binding by reactants (Section 1.5). Equation 4.2-13 shows that the rate of change of $\Delta_r G'$ with pH is proportional to the change in binding of hydrogen atoms in the reaction. Since components are conserved in chemical reactions, it may be a surprise to find that the amount of the hydrogen atoms in the system is not conserved in biochemical reactions, but of course hydrogen atoms are not conserved in the system when the pH is held constant. Acid or base may have to be added to hold the pH constant. If $(\partial n_c(H)/\partial \xi')_{T,P,pH}$ is positive, it means that acid has to be added to keep the pH constant. In this case the products bind more hydrogen ions than the reactants. Since the concentration of hydrogen ions is constant, the change in the amount of the hydrogen atoms in the reaction system is equal to the change in the binding of hydrogen ions in the reaction. Since the derivative of $\Delta_r G'$ is the same as the derivative of $\Delta_r G'^0$, equation 4.2-12 can be written as

$$\Delta_r N_H = \frac{1}{RT \ln(10)} \left(\frac{\partial \Delta_r G'^0}{\partial pH}\right)_{T,P,\xi'} \qquad (4.7-4)$$

Substituting $\Delta_r G'^0 = -RT \ln K'$ yields

$$\Delta_r N_H = -\left(\frac{\partial \log K'}{\partial pH}\right)_{T,P,\xi'} \qquad (4.7-5)$$

If K' increases with pH, $\Delta_r N_H$ is negative, which indicates that less hydrogen is bound by products than reactants. In this case H^+ is produced by the reaction. The change in binding of hydrogen ions in a biochemical reaction can also be calculated from the acid dissociation constants of the reactants using

$$\Delta_r N_H = \sum_{i=1}^{N'} v_i' \bar{N}_H(i) \tag{4.7-6}$$

where $\bar{N}_H(i)$ is the average number of hydrogen atoms bound by reactant i (see equations 1.3-9 and 4.7-3).

■ 4.8 CALCULATION OF THE CHANGE IN BINDING OF MAGNESIUM IONS IN A BIOCHEMICAL REACTION

The treatment (Alberty, 1998a) of the binding of Mg^{2+}, or other metal ion that is bound reversibly by a reactant, follows the same pattern as the treatment of H^+. A term $n_c(Mg)\mu(Mg)$ can be included in the Legendre transform with the term for hydrogen as follows:

$$G' = G - n_c(H)\mu(H^+) - n_c(Mg)\mu(Mg^{2+}) \tag{4.8-1}$$

where the amount of the magnesium component is given by

$$n_c(Mg) = \sum_{i=1}^{N_s} N_{Mg}(j)n_j \tag{4.8-2}$$

$N_{Mg}(j)$ is the number of magnesium ions in species j. The inclusion of pMg as an independent variable adds a term $RT\ln(10)n_c(Mg)dpMg$ to the fundamental equation for G' (see equation 4.3-1), where $pMg = -\log[Mg^{2+}]$. For a system containing a single reactant, the fundamental equation for G' is

$$dG' = -S'dT + VdP + \mu_i'dn_i' + RT\ln(10)n_c(H)dpH + RT\ln(10)n_c(Mg)dpMg \tag{4.8-3}$$

Thus, when pMg is specified, equations 4.4-9 and 4.4-11 become

$$\Delta_f G_j'^0 = \Delta_f G_j^0 - N_H(j)\{\Delta_f G^0(H^+) + RT\ln(10^{-pH})\}$$
$$- N_{Mg}(j)\{\Delta_f G^0(Mg^{2+}) + RT\ln(10^{-pMg})\} \tag{4.8-4}$$

$$\Delta_f H_j'^0 = \Delta_f H_j^0 - N_H(j)\Delta_f H^0(H^+) - N_{Mg}(j)\Delta_f H^0(Mg^{2+}) \tag{4.8-5}$$

The equations for the average number of magnesium ions bound by a reactant $\bar{N}_{Mg}$ and the change in binding of magnesium ions in a reaction $\Delta_r N_{Mg}$ follow the development of the preceding section:

$$\bar{N}_{Mg} = \frac{1}{RT\ln(10)}\left(\frac{\partial \Delta_f G'^0}{\partial pMg}\right)_{T,P,n_i',pH} \tag{4.8-6}$$

$$\Delta_r N_{Mg} = \frac{1}{RT\ln(10)}\left(\frac{\partial \Delta_r G'^0}{\partial pMg}\right)_{T,P,\xi',pH} \tag{4.8-7}$$

The change in binding of magnesium ions in a biochemical reaction can also be calculated from the acid dissociation and magnesium complex ion dissociation constants using

$$\Delta_r N_{Mg} = \sum_{i=1}^{N'} v_i' \bar{N}_{Mg}(i) \tag{4.8-8}$$

(See equation 1.5-5.)

The derivative of $\bar{N}_{Mg}$ (equation 4.8-6) with respect to pH at T, P, and pMg, and ξ' is equal to the derivative of $\bar{N}_H$ (equation 4.7-3) with respect to pMg at T,

P, pH, and ξ', and so

$$\left(\frac{\partial \bar{N}_{Mg}}{\partial pH}\right)_{T,P,pMg} = \left(\frac{\partial \bar{N}_H}{\partial pMg}\right)_{T,P,pH} \tag{4.8-9}$$

$$\left(\frac{\partial \Delta_r N_{Mg}}{\partial pH}\right)_{T,P,pMg,\xi'} = \left(\frac{\partial \Delta_r N_H}{\partial pMg}\right)_{T,P,pH,\xi'} \tag{4.8-10}$$

Thus the effect of pH on the binding of magnesium ions by a reactant is equal to the effect of pMg on the binding of hydrogen ions. The effect of pH on the change in binding of magnesium ions in a biochemical reaction is equal to the effect of pH on the binding of magnesium ions. Thus the binding of H^+ and Mg^{2+} are linked.

■ 4.9   EFFECT OF TEMPERATURE ON TRANSFORMED THERMODYNAMIC PROPERTIES

The effect of temperature on standard transformed thermodynamic properties of species has been discussed in the preceding chapter on the assumption that $\Delta_f H^0(I = 0)$ for species are independent of temperature, or in other words, $C_{Pm}^0 = 0$. In order to make calculations at finite ionic strengths, it is necessary to adjust the Debye-Hückel coefficient α and the coefficients of the ionic strength terms in the equations for adjusting $\Delta_f G^0$ and $\Delta_f H^0$ for the effect of pH and ionic strength. As discussed in Section 3.7, Clarke and Glew (1980) gave values of the various coefficients at a series of temperatures. But in order to make calculations at arbitrary temperatures, it is necessary to fit α to an empirical equation, such as 3.7-3. The effects of temperature on $\Delta_f G'^0$ and $\Delta_f H'^0$ for biochemical reactants at specified pH and ionic strength can be calculated by calculating these effects for the species involved by use of equation 3.7-2. Alberty (2001d) calculated standard $\Delta_f G^0$ and $\Delta_f H^0$ values for 22 species of biochemical interest at 283.15 and 313.15 K and went on to calculate $\Delta_f G'^0$ and $\Delta_f H'^0$ at pH 7 and ionic strength 0.25 M for the corresponding reactants. This made it possible to calculate apparent equilibrium constants for six biochemical reactions at 283.15 and 313.15 K. *Mathematica* programs (calcdGTsp and calcdHTsp) were written to calculate $\Delta_f G'^0$ and $\Delta_f H'^0$ of species at arbitrary temperatures, pHs, and ionic strengths. In the second program, the standard transformed enthalpies of species are calculated using the Gibbs-Helmholtz equation. A biochemical reactant that consists of two or more species, $\Delta_f G'^0$ and $\Delta_f H'^0$ can be calculated for the pseudoisomer group in the usual way, but one must be careful to change the RT factor in the program for $\Delta_f G'^0(\text{iso})$. When standard enthalpies of all of the species involved in a reaction are available, K' can be calculated at desired temperatures not too far from 298.15 K. The effect of temperature on the standard transformed Gibbs energy of hydrolysis of ATP is shown in Table 4.1 (see Problem 4.6).

This discussion has not included the more accurate calculations that can be made when C_P^0 values of species are known (see equation 3.5-18). These values are not known for many species of biochemical interest. The effects of heat capacity terms are discussed in Chapter 10 because the existing information on $\Delta_r C_P'^0$ comes primarily from calorimetric data. In principle, $\Delta_r C_P'^0$ can be calculated from measurements of apparent equilibrium constants over a range of temperatures. Over short ranges of temperature, K' can be represented by

$$R \ln K' = \Delta_r S'^0 - \frac{\Delta_r H'^0}{T} \tag{4.9-1}$$

But over wider ranges of temperature, $\Delta_r S'^0$ and $\Delta_r H'^0$ are functions of temperature. Clarke and Glew (1966) have used Taylor series expansions of the enthalpy

Table 4.1 Standard Transformed Gibbs Energies in kJ mol^{-1} of Hydrolysis of ATP as a Function of Temperature, pH, and Ionic Strength

T/K	I/M	pH 5	pH 6	pH 7	pH 8	pH 9
283.15	0	-34.73	-35.38	-36.98	-41.5	-46.93
	0.10	-32.95	-33.49	-35.87	-40.50	-45.62
	0.25	-32.28	-32.87	-35.41	-40.12	-45.45
298.15	0	-35.34	-35.95	-37.64	-42.53	-48.32
	0.10	-33.33	-33.91	-36.53	-41.51	-47.13
	0.25	-32.60	-33.25	-36.07	-41.10	-46.73
313.15	0	-35.95	-36.53	-38.31	-43.56	-49.71
	0.10	-33.71	-34.32	-37.16	-42.51	-48.43
	0.25	-32.91	-33.63	-36.72	-42.08	-48.01

Source: With permission from R. A. Alberty, *J. Phys. Chem. B* 105, 7865–7870 (2001). Copyright 2001 American Chemical Society.

and entropy to show the form that extensions of equation 4.9-1 should take up to $d^3\Delta_r C_P^0/dT^3$. However, it takes very accurate measurements to determine the curvature.

■ 4.10 CALCULATION OF STANDARD TRANSFORMED GIBBS ENERGIES OF SPECIES FROM EXPERIMENTAL MEASUREMENTS OF APPARENT EQUILIBRIUM CONSTANTS

Apparent equilibrium constants have been measured for about 500 biochemical reactions involving about 1000 reactants. In principle, this makes it possible to put the species of all these reactants in the table BasicBiochemData2 described in Section 3.8. As indicated in Sections 4.4 and 4.5, the calculation of thermodynamic properties of species from experimental measurements of K' is rather complicated, and so it is important to look at this process from a broader viewpoint. The preceding discussions were based on the assumption that species properties are known and properties of reactants are to be calculated. But, in calculating species data from K' and $\Delta_r H'$, we are interested in the inverse process (Alberty, 2002c). Callen (1985) discussed the Legendre transform to go from a function of (X, Y) to a function of (P, ϕ) and pointed out that "the relationship between (X, Y) and (P, ϕ) is symmetrical with its inverse except for a change in sign in the equation for the Legendre transform." The **inverse Legendre transform** used here is the definition of the Gibbs energy G in terms of the transformed Gibbs energy G':

$$G = G' + n_c(\mathrm{H})\mu(\mathrm{H}^+) \qquad (4.10\text{-}1)$$

The following derivation provides guidance in writing computer programs to calculate standard Gibbs energies of formation and standard enthalpies of formation of organic species in dilute aqueous solutions from K' and $\Delta_r H'$ values for enzyme-catalyzed reactions. The first step is to see how S' and n_i' in the fundamental equation for G' (equation 4.1-18) can be divided up into contributions of species. The partial derivative of the transformed Gibbs energy with respect to temperature is equal to $-S'$, and so equation 4.1-14 shows that

$$\left(\frac{\partial G'}{\partial T}\right)_{P,\{n_i'\},\mathrm{pH}} = -S' = -S - n_c(\mathrm{H})\left(\frac{\partial \mu(\mathrm{H}^+)}{\partial T}\right)_{P,\{n_i'\},\mathrm{pH}} \qquad (4.10\text{-}2)$$

Thus the fundamental equation for the transformed Gibbs energy can be written as

$$dG' = -S\,dT + V\,dP + \sum_{i=1}^{N'} \mu_i'\,dn_i'$$

$$- n_c(H)\left(\left(\frac{\partial \mu(H^+)}{\partial T}\right)_{P,\{n_i'\},pH}\,dT - RT\ln(10)\,dpH\right) \quad (4.10\text{-}3)$$

$$= -S\,dT + V\,dP + \sum_{i=1}^{N'} \mu_i'\,dn_i' - n_c(H)\,d\mu(H^+)$$

where the term in parentheses is the total differential of the chemical potential of hydrogen ions (see equation 4.1-10).

The summation in equation 4.10-3 can be written in terms of species exclusive of the hydrogen ion because when species in a pseudoisomer group are in equilibrium at a specified pH, these species have the same transformed chemical potential.

$$dG' = -S\,dT + V\,dP + \sum_{j=1}^{N_s-1} \mu_i'\,dn_j - n_c(H)\,d\mu(H^+) \quad (4.10\text{-}4)$$

where N_s is the number of different species.

Now the inverse Legendre transform given in equation 4.10-1 is needed. The differential of the Gibbs energy is given by

$$dG = dG' + n_c(H)\,d\mu(H^+) + \mu(H^+)\,dn_c(H) \quad (4.10\text{-}5)$$

Substituting equation 4.10-4 into this equation yields

$$dG = -S\,dT + V\,dP + \sum_{j=1}^{N_s-1} \mu_j'\,dn_j + \mu(H^+)\,dn_c(H) \quad (4.10\text{-}6)$$

The amount of the hydrogen component $n_c(H)$ in the system is given by equation 4.1-2, and so equation 4.10-6 can be written as

$$dG = -S\,dT + V\,dP + \sum_{j=1}^{N_s} \{\mu_j' + N_H(j)\mu(H^+)\}\,dn_j \quad (4.10\text{-}7)$$

The term in braces is the chemical potential of ion j:

$$\mu_j = \mu_j' + N_H(j)\mu(H^+) \quad (4.10\text{-}8)$$

and so equation 4.1-5 is obtained as expected.

If the apparent equilibrium constant K' for an enzyme-catalyzed reaction has been determined at 298.15K and $\Delta_r G'^0$ values can be calculated at the experimental pH and ionic strength using known functions of pH and ionic strength for all the reactants but one, the $\Delta_f G'^0$ of that reactant under the experimental conditions can be calculated using equation 4.4-2. So far functions of pH and ionic strength that yield $\Delta_f G'^0$ are have been published for 131 reactants at 298.15 K (Alberty, 2001f).

When the reactant of interest consists of a single species, $\Delta_f G^0(I = 0)$ for this species at 298.15 K can be calculated using equation 4.10-8 in the following form (see equation 4.4-10):

$$\Delta_f G_j^0(I = 0) = \Delta_f G'^0(pH, I) - N_H(j)RT\ln(10)pH$$

$$+ \frac{2.91482(z_j^2 - N_H(j))I^{1/2}}{1 + 1.6I^{1/2}} \quad (4.10\text{-}9)$$

A program calcGef1sp has been written to produce output in the form of equation 3.8-1 for a reactant made up of one species. It is given in the package Basic-BiochemData2.

When the reactant of interest consists of two species with different numbers of hydrogen atoms, the pK of the weak acid is needed to calculate $\Delta_f G'^0(I = 0)$ of the two species, and the calculation is more complicated. The standard transformed Gibbs energy of formation of a pseudoisomer group containing two species is given by

$$\Delta_f G'^0 = \Delta_f G_1'^0 - RT \ln(1 + 10^{pK_1 - pH}) \tag{4.10-10}$$

where pK_1 is the value at the experimental ionic strength at 298.15 K calculated using

$$pK_1(I) = pK_1(I = 0) + \frac{0.510651(\Sigma v_i z_i^2)I^{1/2}}{1 + 1.6I^{1/2}} \tag{4.10-11}$$

Now equation 4.10-9 is used to adjust $\Delta_f G_1'^0(I)$ to $\Delta_f G_1^0(I = 0)$. After $\Delta_f G_1^0(I = 0)$ has been calculated, $\Delta_f G_2^0(I = 0)$ can be calculated using

$$\Delta_f G_2^0(I = 0) = \Delta_f G_1^0(I = 0) - RT \ln(10)pK_1(I = 0) \tag{4.10-12}$$

A *Mathematica* program calcGef2sp has been written to produce output in the form of equation 3.8-1 for a reactant made up of two species. It is given in the package BasicBiochemData2. This output can be added to the database in BasicBiochemData2 and can be used to calculate $\Delta_f G'^0$ of the reactant at 298.15 K, pH 5 to 9, and ionic strengths 0 to 0.35 M.

When the reactant consists of three species with different numbers of hydrogen atoms, equation 4.10-10 becomes

$$\Delta_f G'^0 = \Delta_f G_1'^0 - RT \ln(1 + 10^{pK_1 pH} + 10^{pK_1 + pK_2 - 2pH}) \tag{4.10-13}$$

$\Delta_f G_1'^0(I = 0)$ can be calculated by using equation 4.10-9, and equation 4.10-10 can be used to calculate $\Delta_f G_2'^0(I = 0)$. Then $\Delta_f G_3'^0(I = 0)$ can be calculated using

$$\Delta_f G_3^0(I = 0) = \Delta_f G_2^0(I = 0) - RT \ln(10)pK_2(I = 0) \tag{4.10-14}$$

A *Mathematica* program calcGef3sp has been written to produce output in the form of equation 3.8-1, and it is given in BasicBiochemData2.

The species matrix for a reactant can be verified by use of the programs calcdGmat and calckprime, which are also given in BasicBiochemData2. The program calcdGmat yields the function of pH and ionic strength for $\Delta_r G'^0$ of the reactant. The program calckprime can then be used to calculate K' for the reaction used at the experimental pH and ionic strength.

A good deal of work will have to be done to extract species information from the apparent equilibrium constants that have been reported for about 500 reactions. Beyond that, use can be made of analogies with known reactions; for example, the various ribonucleotide phosphates (AMP, GMP, CMP, UMP, and dTMP) are believed to have the same hydrolysis constants and pKs. Beyond that, the group additivity method (Alberty, 1998c) can be used to estimate thermodynamic properties.

■ 4.11 TABLES OF STANDARD TRANSFORMED THERMODYNAMIC PROPERTIES AT 298.15 K FOR BIOCHEMICAL REACTANTS AT SPECIFIED pH AND IONIC STRENGTH

Table 4.2 provides $\Delta_f G^0$ and $\Delta_f H^0$ for species of 131 biochemical reactants at 298.15 K in dilute aqueous solutions at zero ionic strength. These values are available in the package BasicBiochemData2 (Alberty, 2002d), which is the first item in the second part of this book. These values can be used to calculate $\Delta_f G_i'^0$ and $\Delta_f H_i'^0$ for biochemical reactants at desired pHs in the range 5 to 9 and desired ionic strengths in the range 0 to about 0.35 M, as described in this chapter.

Table 4.2   Values of $\Delta_f G_i'^0$ in kJ mol^{-1} at 298.15 K and pHs 5, 6, 7, 8, and 9 at Ionic Strength 0.25 M

Reactant	pH 5	pH 6	pH 7	pH 8	pH 9
acetaldehyde	−21.60	1.23	24.06	46.90	69.73
acetate	282.71	265.02	247.83	230.70	−213.57
acetone	16.40	50.65	84.90	119.14	153.39
acetylcoA	−100.47	−83.35	−66.22	−49.10	−31.97
acetylphos	−1153.77	−1129.84	−1107.02	−1085.39	−1066.49
aconitatecis	−836.37	−819.24	−802.12	−785.00	−767.87
adenine	457.06	485.92	514.50	543.04	571.58
adenosine	187.48	261.25	335.46	409.66	483.87
adp	−1569.05	−1495.55	−1424.70	−1355.78	−1287.24
alanine	−165.55	−125.60	−85.64	−45.68	5.73
ammonia	37.28	60.11	82.93	105.64	127.51
amp	−698.40	−625.22	−554.83	−486.04	−417.51
arabinose	−448.73	−391.65	−334.57	−277.49	−220.41
asparagineL	−291.13	−245.47	−199.80	−154.14	−108.47
aspartate	−520.59	−486.34	−452.09	−417.85	−383.60
atp	−2437.46	−2363.76	−2292.50	−2223.44	−2154.88
bpg	−2262.15	−2233.92	−2207.30	−2183.36	−2160.38
butanoln	121.66	178.74	235.82	292.90	349.98
butyrate	−147.99	−108.03	−68.08	−28.12	11.83
citrate	−1027.23	−995.44	−966.23	−937.62	−909.07
citrateiso	−1020.58	−988.80	−959.58	−930.97	−902.42
coA	−18.48	−12.79	−7.26	−2.82	−1.10
coAglutathione	403.59	489.21	574.83	660.45	746.07
co2g	−394.36	−394.36	−394.36	−394.36	−394.36
co2tot	−564.61	−554.49	−547.10	−541.18	−535.80
coaq	−119.90	−119.90	−119.90	−119.90	−119.90
cog	−137.17	−137.17	−137.17	−137.17	−137.17
creatine	4.95	56.32	107.69	159.07	210.44
creatinine	182.31	222.27	262.22	302.18	342.13
cysteineL	−133.37	−93.43	−53.65	−14.97	20.99
cystineL	−314.31	−245.82	−177.32	−108.82	−40.33
cytochromecox	−7.29	−7.29	−7.29	−7.29	−7.29
cytochromecred	−27.75	−27.75	−27.75	−27.75	−27.75
dihydroxy-acetonephos	−1154.88	−1124.53	−1095.70	−1067.13	−1038.59
ethanol	−5.54	28.71	62.96	97.20	131.45
ethylacetate	−102.85	−57.19	−11.52	34.14	79.81
fadox	906.61	1083.56	1260.51	1437.46	1614.40
fadred	926.43	1114.79	1303.16	1491.53	1679.89
fadenzox	906.61	1083.56	1260.51	1437.46	1614.40
fadenzred	876.71	1065.07	1253.44	1441.80	1630.17
ferredoxinox	−0.81	−0.81	−0.81	−0.81	−0.81
ferredoxinred	38.07	38.07	38.07	38.07	38.07
fmnox	554.41	662.86	771.32	879.77	988.22
fmnred	574.23	694.10	813.97	933.84	1053.70
formate	−322.46	−316.75	−311.04	−305.34	−299.63
fructose	−563.31	−494.82	−426.32	−357.82	−289.33
fructose6phos	−1445.66	−1379.42	−1315.74	−1252.84	−1190.04
fructose16phos	−2326.42	−2264.57	−2206.78	−2149.62	−2092.54
fumarate	−546.67	−535.02	−523.58	−512.16	−500.75
galactose	−556.73	−488.24	−419.74	−351.24	−282.75
galactosephos	−1440.96	−1375.09	−1311.60	−1248.72	−1185.93
glucose	−563.70	−495.21	−426.71	−358.21	−289.72
glucose1phos	−1442.86	−1376.01	−1311.89	−1248.92	−1186.11
glucose6phos	−1449.53	−1382.88	−1318.92	−1255.98	−1193.18
glutamate	−463.48	−417.82	−372.15	−326.49	−280.82
glutamine	−234.52	−177.44	−120.36	−63.28	−6.20

Table 4.2 *Continued*

Reactant	pH 5	pH 6	pH 7	pH 8	pH 9
glutathioneox	877.26	1048.50	1219.74	1390.98	1562.22
glutathionered	455.34	546.64	637.62	726.89	813.52
glyceraldehyde phos	−1147.22	−1116.87	−1088.04	−1059.47	−1030.93
glycerol	−262.68	−217.02	−171.35	−125.68	−80.02
glycerol3phos	−1163.24	−1118.83	−1077.13	−1036.91	−996.93
glycine	−233.16	−204.62	−176.08	−147.54	−119.00
glycolate	−443.71	−426.59	−409.46	−392.34	−375.21
glycylglycine	−285.40	−239.74	−194.07	−148.41	−102.74
glyoxylate	−440.06	−434.35	−428.64	−422.94	−417.23
h2aq	76.30	87.72	99.13	110.55	121.96
h2g	58.70	70.12	81.53	92.95	104.36
h2o	−178.49	−167.07	−155.66	−144.24	−132.83
h2o2aq	−75.33	−63.91	−52.50	−41.08	−29.67
hydroxy propionateb	−372.46	−343.92	−315.38	−286.84	−258.30
hypoxanthine	206.90	229.73	252.56	275.40	298.23
indole	429.25	469.21	509.16	549.12	589.07
ketoglutarate	−679.25	−656.42	−633.59	−610.75	−587.92
lactate	−370.78	−342.24	−313.70	−285.16	−256.62
lactose	−921.63	−796.06	−670.48	−544.90	−419.33
leucineisoL	37.65	111.85	186.06	260.26	334.47
leucineL	29.30	103.50	177.71	251.91	326.12
lyxose	−455.64	−398.56	−341.48	−284.40	−227.32
malate	−729.49	−705.79	−682.85	−660.00	−637.17
maltose	−928.99	−803.42	−677.84	−552.26	−426.69
mannitolD	−531.71	−451.80	−371.89	−291.97	−212.06
mannose	−557.80	−489.31	−420.81	−352.31	−283.82
methaneaq	83.07	105.90	128.73	151.57	174.40
methaneg	66.68	89.51	112.34	135.18	158.01
methanol	−57.91	−35.08	−12.25	10.59	33.42
methionineL	−180.07	−117.28	−54.49	8.29	71.08
methylamineion	135.43	169.68	203.93	238.17	272.42
n2aq	18.70	18.70	18.70	18.70	18.70
n2g	0	0	0	0	0
nadox	762.29	910.696	1059.11	1207.51	1355.92
nadred	811.86	965.98	1120.09	1274.21	1428.33
nadpox	−108.72	33.98	176.68	319.38	462.08
nadpred	−59.05	89.36	237.77	386.18	534.59
o2aq	16.40	16.40	16.40	16.40	16.40
o2g	0	0	0	0	0
oxalate	−677.26	−677.15	−677.14	−677.14	−677.14
oxaloacetate	−737.83	−726.41	−715.00	−703.58	−692.17
oxalosuccinate	−1024.72	−1001.89	−979.06	−956.22	−933.39
palmitate	649.64	826.59	1003.54	1180.48	1357.43
pep	−1218.97	−1203.00	−1189.73	−1178.02	−1166.58
pg2	−1396.52	−1368.31	−1341.79	−1317.92	−1294.95
pg3	−1402.06	−1373.94	−1347.73	−1324.05	−1301.11
phenylalanineL	115.75	178.54	241.33	304.11	366.90
pi	−1079.46	−1068.49	−1059.49	−1052.97	−1047.17
ppi	−1957.07	−1947.46	−1940.66	−1935.64	−1933.29
propanol2	49.57	95.23	140.90	186.56	232.23
propanoln	58.99	104.65	150.32	195.98	241.65
pyruvate	−385.03	−367.91	−350.78	−333.66	−316.53
retinal	821.80	981.62	1141.45	1301.27	1461.10
retinol	852.59	1023.83	1195.07	1366.31	1537.55
ribose	−445.29	−388.21	−331.13	−274.05	−216.97

Table 4.2   *Continued*

Reactant	pH 5	pH 6	pH 7	pH 8	pH 9
ribose1phos	−1320.16	−1264.30	−1211.14	−1159.50	−1108.09
ribose5phos	−1328.24	−1272.38	−1219.22	−1167.58	−1116.17
ribulose	−442.44	−385.36	−328.28	−271.20	−214.12
serineL	−305.42	−265.47	−225.51	−185.55	−145.60
sorbose	−559.75	−491.26	−422.76	−354.26	−285.77
succinate	−578.32	−553.73	−530.64	−507.79	−484.95
succinylcoA	−393.33	−370.32	−347.47	−324.63	−301.80
sucrose	−919.00	−793.43	−667.85	−542.27	−416.70
thioredoxinox	0	0	0	0	0
thioredoxinred	33.33	44.70	55.74	64.03	66.35
tryptophaneL	237.50	306.00	374.49	442.99	511.49
tyrosineL	−47.85	14.94	77.73	140.51	203.30
ubiquinoneos	2641.49	3155.21	3668.94	4182.66	4696.38
ubiquinonered	2610.27	3135.41	3660.55	4185.69	4710.83
urate	−238.66	−221.54	−204.41	−187.29	−170.16
urea	−85.40	−62.57	−39.74	−16.90	5.93
uricacid	−239.50	−216.67	−193.84	−171.00	−148.17
valineL	−35.80	26.99	89.78	152.56	215.35
xylose	−456.99	−399.91	−342.83	−285.75	−228.67
xylulose	−452.65	−395.57	−338.49	−281.41	−224.33

Various types of tables can be constructed, and the package produces table 1 ($\Delta_f G'^0$ of reactants at pH 7 and ionic strengths of 0, 0.10, and 0.25 M), table 2 ($\Delta_f G'^0$ of reactants at pHs 5, 6, 7, 8, and 9 and ionic strengths 0.25 M), table 3 ($\Delta_f H'^0$ of reactants at pH 7 and ionic strengths of 0, 0.10, and 0.25 M), and table 4 ($\Delta_f H'^0$ of reactants at pHs 5, 6, 7, 8, and 9 and ionic strengths 0.25 M). A table of standard transformed Gibbs energies of formation at 298.15 K, pH 7, and ionic strengths of 0, 0.10, and 0.25 M for about 100 reactants was provided earlier (Alberty, 2000a). Table 4.2 gives the standard transformed Gibbs energies of formation of all of the reactants in BasicBiochemData2 at 298.15 K , pHs 5, 6, 7, 8, and 9, and ionic strength 0.25 M. The values in Table 4.2 can be used to calculate the apparent equilibrium constant for any reaction between reactants in this table that balances atoms of elements other than hydrogen. Part of the dependence of $\Delta_f G_i'^0$ on pH may cancel between reactants and products in a biochemical reaction. Note that the functions used to calculate Table 4.2 can also be used to calculate $\bar{N}_H$ for a reactant as a function of pH and ionic strength. As we will see later in Chapter 9, Table 4.2 can be used to calculate standard transformed reduction potentials for redox half-reactions and electromotive forces for galvanic cells at specified pH and ionic strength.

In discussing chemical reactions, note that the names of species show ionic charges so that charge balance can be checked. Names of reactants in biochemical reactions should not show electric charges because, in general, they are not integers, and biochemical reactions do not balance electric charges. The names of biochemical reactants used in writing reactions should, to the extent possible, indicate that they represent sums of species. Names of biochemical reactants should not contain H because that suggests that hydrogen atoms should balance. These naming problems are readily solved with ATP but are more difficult with NAD^+ and NADH, as they are generally represented. Therefore NAD_{ox} and NAD_{red} are used in writing reactions, and nadox and nadred are used in *Mathematica* because names of functions should start with lowercase letters.

Table 4.3 gives $\Delta_f H_i'^0$ for the biochemical reactants at 298.15 K and pHs 5, 6, 7, 8, 9 at ionic strength 0.25 M.

Table 4.3   Values of $\Delta_f H_i'^0$ in kJ mol^{-1} at 298.15 K, pH 5, 6, 7, 8, and 9 at ionic strength 0.25 M

	pH 5	pH 6	pH 7	pH 8	pH 9
acetaldehyde	-213.87	-213.87	-213.87	-213.87	-213.87
acetate	-486.96	-486.85	-486.83	-486.83	-486.83
acetone	-224.17	-224.17	-224.17	-224.17	-224.17
adenosine	-590.92	-622.97	-626.27	-626.60	-626.63
adp	-2625.82	-2625.74	-2627.24	-2627.71	-2627.76
alanine	-557.67	-557.67	-557.67	-557.67	-557.67
ammonia	-133.74	-133.71	-133.45	-130.97	-115.00
amp	-1635.98	-1636.50	-1638.19	-1638.60	-1638.65
arabinose	-1047.89	-1047.89	-1047.89	-1047.89	-1047.89
asparagineL	-769.37	-769.37	-769.37	-769.37	-769.37
aspartate	-945.46	-945.46	-945.46	-945.46	-945.46
atp	-3615.67	-3615.43	-3616.92	-3617.49	-3617.56
citrate	-1518.50	-1514.96	-1513.66	-1513.49	-1513.47
co2g	-393.50	-393.50	-393.50	-393.50	-393.50
co2tot	-699.80	-696.59	-692.86	-691.80	-689.51
coaq	-120.96	-120.96	-120.96	-120.96	-120.96
cog	-110.53	-110.53	-110.53	-110.53	-110.53
ethanol	-290.76	-290.76	-290.76	-290.76	-290.76
ethylacetate	-485.28	-485.28	-485.28	-485.28	-485.28
formate	-425.55	-425.55	-425.55	-425.55	-425.55
fructose	-1264.31	-1264.31	-1264.31	-1264.31	-1264.31
fumarate	-776.45	-776.56	-776.57	-776.57	-776.57
galactose	-1260.13	-1260.13	-1260.13	-1260.13	-1260.13
glucose	-1267.12	-1267.12	-1267.12	-1267.12	-1267.12
glucose6phos	-2279.17	-2279.25	-2279.30	-2279.31	-2279.31
glutamate	-982.76	-982.76	-982.76	-982.76	-982.76
glutamine	-809.10	-809.10	-809.10	-809.10	-809.10
glycerol	-679.83	-679.83	-679.83	-679.83	-679.83
glycine	-525.05	-525.05	-525.05	-525.05	-525.05
glycylglycine	-737.53	-737.53	-737.53	-737.53	-737.53
h2aq	-5.02	-5.02	-5.02	-5.02	-5.02
h2g	-0.82	-0.82	-0.82	-0.82	-0.82
h2o	-286.65	-286.65	-286.65	-286.65	-286.65
h2o2aq	-191.99	-191.99	-191.99	-191.99	-191.99
indole	94.63	94.63	94.63	94.63	94.63
lactate	-688.28	-688.28	-688.28	-688.28	-688.28
lactose	-2242.11	-2242.11	-2242.11	-2242.11	-2242.11
leucineL	-648.71	-648.71	-648.71	-648.71	-648.71
maltose	-2247.09	-2247.09	-2247.09	-2247.09	-2247.09
mannose	-1263.59	-1263.59	-1263.59	-1263.59	-1263.59
methaneg	-76.45	-76.45	-76.45	-76.45	-76.45
methaneaq	-90.68	-90.68	-90.68	-90.68	-90.68
methanol	-247.57	-247.57	-247.57	-247.57	-247.57
methylamineion	-126.98	-126.98	-126.98	-126.98	-126.98
n2aq	-10.54	-10.54	-10.54	-10.54	-10.54
n2g	0.00	0.00	0.00	0.00	0.00
nadox	-10.26	-10.26	-10.26	-10.26	-10.26
nadpox	-6.57	-6.57	-6.57	-6.57	-6.57
nadpred	-33.28	-33.28	-33.28	-33.28	-33.28
nadred	-41.38	-41.38	-41.38	-41.38	-41.38
o2aq	-11.70	-11.70	-11.70	-11.70	-11.70
o2g	0.00	0.00	0.00	0.00	0.00
pi	-1302.89	-1302.03	-1299.36	-1297.99	-1297.79
ppi	-2294.18	-2292.80	-2291.57	-2290.05	-2287.69
propanol2	-334.11	-334.11	-334.11	-334.11	-334.11
pyruvate	-597.04	-597.04	-597.04	-597.04	-597.04
ribose	-1038.10	-1038.10	-1038.10	-1038.10	-1038.10

Table 4.3   *Continued*

	pH 5	pH 6	pH 7	pH 8	pH 9
ribose5phos	−2034.57	−2038.10	−2042.43	−2043.41	−2043.52
ribulose	−1027.12	−1027.12	−1027.12	−1027.12	−1027.12
sorbose	1268.23	1268.23	−1268.23	−1268.23	−1268.23
succinate	−909.85	−908.87	−908.70	908.68	908.68
sucrose	−2208.90	−2208.90	−2208.90	−2208.90	−2208.90
tryptophaneL	−410.13	−410.13	−410.13	−410.13	−410.13
urea	−319.29	−319.29	−319.29	−319.29	−319.29
valineL	−616.50	−616.50	−616.50	−616.50	−616.50
xylose	−1050.04	−1050.04	−1050.04	−1050.04	−1050.04
xylulose	−1033.75	−1033.75	−1033.75	−1033.75	−1033.75

The values of $\Delta_r H'^0$ calculated using this table can be used to calculate apparent equilibrium constants at other temperatures not too far from 298.15 K. Note that standard transformed enthalpies of reactants that consist of a single species are not a function of pH (see equation 4.4-12). The standard transformed enthalpies of reactants are functions of pH when there are more than two species because r_i depends on pH. As indicated by the pH dependencies in Table 4.3, these differences are often small.

These tables apply to single sets of values of pH and ionic strength. A more general approach is to use the functions of ionic strength and pH for each reactant that give the values of standard transformed thermodynamic properties at 298.15 K. For reactants for which $\Delta_f H^0$ is known for all species, functions of temperature, pH, and ionic strength can be used to calculate standard transformed thermodynamic properties at temperatures in the approximate range 273.15 to 313.15 K, as discussed in Section 4.9.

The database BasicBiochemData2 (Alberty, 2002d) contains functions of pH and ionic strength that give $\Delta_f G'^0$ and $\Delta_f H'^0$ at 298.15 K for reactants for which species information has been tabulated; for example, the function for $\Delta_f G'^0(\text{ATP})$ is obtained by typing "atp" and the function for $\Delta_f H'^0(\text{ATP})$ is obtained by typing "atph." This makes it very convenient to calculate $\Delta_r G'^0$ and $\Delta_r H'^0$ for any reaction between reactants in the table at specified pH and ionic strength. The program calctrGerx[eq_,pHlist_,islist_] can be used to calculate these properties for the reaction typed in as an argument, as illustrated by calctrGerx[atp+H$_2$O+de == adp+pi,{5,6,7,8,9},.25] (see Problem 4.6). The corresponding values of $\Delta_f H'^0$ are obtained by appending "h" to each of the reactant names, but the number of reactions for which this can be done is significantly less than for $\Delta_r G'^0$ and K'. The values of K' at 298.15 K can be calculated using the program calckprime. Values of $\Delta_r N_H$ can be calculated by using calcNHrx. This is illustrated by the following tables for the reactions of glycolysis, gluconeogenesis, and the citric acid cycle. Table 4.4 gives the standard transformed reaction Gibbs energies of the 10 reactions of glycolysis and the net reaction for glycolysis. Added information on the effects of ionic strength is given in Problem 4.8.

There is a difference between the way these biochemical reactions for glycolysis are written here and in most biochemistry textbooks, which include H$^+$ in reactions 1, 3, 6, and 10 and 2H$^+$ in the net reaction. These H$^+$ are wrong, in principle, because at constant pH, hydrogen atoms in a reaction system are not conserved, and they are stoichiometrically incorrect because integer amounts of hydrogen ions are not consumed or produced, except under special conditions (see Table 4.6).

Table 4.5 gives the corresponding apparent equilibrium constants for the reactions in glycolysis.

Table 4.4 Standard Transformed Reaction Gibbs Energies (in kJ mol^{-1}) for the Reactions of Glycolysis at 298.15 K and 0.25 M Ionic Strength

	pH 5	pH 6	pH 7	pH 8	pH 9
1. Glucose + ATP = glucose 6-phosphate + ADP	−17.41	−19.47	−24.42	−30.11	−35.82
2. Glucose 6-phosphate = fructose 6-phosphate	3.87	3.36	3.19	3.15	3.14
3. Fructose 6-phosphate + ATP = fructose 1,6-biphosphate + ADP	−12.46	−16.95	−23.25	−29.12	−34.86
4. Fructose 1,6-biphosphate = dihydroxyacetone phosphate + glyceraldehyde phosphate	24.33	23.18	23.03	23.02	23.02
5. Dihydroxyacetone phosphate = glyceraldehyde phosphate	7.66	7.66	7.66	7.66	7.66
6. Glyceraldehyde phosphate + P_i + NAD_{ox} = 1,3-bisphosphoglycerate + NAD_{red}	14.10	6.71	1.12	−4.22	−9.88
7. 1,3-Bisphosphoglycerate + ADP = 3-phosphoglycerate + ATP	−8.31	−8.22	−8.22	−8.34	−8.37
8. 3-phosphoglycerate = 2-phosphoglycerate	5.54	5.63	5.94	6.13	6.16
9. 2-phosphoglycerate = phosphoenolpyruvate + H_2O	−0.94	−1.76	−3.60	−4.35	−4.45
10. Phosphoenolpyruvate + ADP = pyruvate + ATP	−34.47	−33.01	−28.85	−23.29	−17.60
Net reaction: Glucose + $2P_i$ + 2ADP + $2NAD_{ox}$ = 2pyruvate + 2ATP + $2NAD_{red}$ + $2H_2O$	−41.90	−63.4	−80.6	−93.4	−104.9

Note: See Problem 4.8.

Table 4.6 gives the changes in binding of hydrogen ions in these reactions that are calculated using equation 4.7-4. These changes in the binding of hydrogen ions can be viewed as the causes of the pH dependencies. If $\Delta_r N_H$ is positive, products bind more hydrogen ions than reactants; therefore, raising the pH reduces the apparent equilibrium constant. If $\Delta_r N_H$ is negative, reactants bind more hydrogen ions than products; therefore, raising the pH increases the apparent equilibrium constant.

Gluconeogenesis uses seven of the reactions in glycolysis, but three are replaced by the sum of the pyruvate carboxylase and phospho*enol*pyruvate carboxykinase reactions, the fructose 1,6-biphosphatase reaction, and the glucose 6-phosphatase reaction. Tables 4.7, 4.8, and 4.9 give the thermodynamic properties of these reactions and the net reaction for gluconeogenesis.

There is a difference between the ways the first two reactions are written in Tables 4.7, 4.8, and 4.9 and in biochemistry textbooks. Textbooks give the reactions in terms of gaseous carbon dioxide, but here CO_2tot is used because in thinking about living cells it is of more interest to know the equilibrium concentration of total carbon dioxide in the solution (see Section 8.7). When $CO_2(g)$ is replaced by CO_2tot in a biochemical reaction, it is necessary to insert a H_2O on the other side of the equation to balance oxygen atoms. At pH 7 and ionic strength 0.25 M, $\Delta_f G_i'^0(CO_2tot) - \Delta_f G_i'^0(H_2O) - \Delta_f G_i'^0(CO_2(g)) = 2.93$ kJ mol^{-1}, and so the standard transformed reaction Gibbs energies for reactions 1 and 2 at pH 7 and 0.25 M ionic strength would be −5.99 kJ mol^{-1} and −1.40 kJ mol^{-1} if they were balanced with $CO_2(g)$. Thus, if $CO_2(g)$ is on the left side of the reaction, replacing it with CO_2tot (and adding H_2O to the other side) makes $\Delta_r G_i'^0$ more negative by 2.92 kJ mol^{-1}. If $CO_2(g)$ is on the right side, the change makes $\Delta_r G_i'^0$ more positive by 2.92 kJ mol^{-1}. The effect of changing from

Table 4.5   Apparent equilibrium constants for glycolysis at 298.15 K and 0.25 M ionic strength (see Problem 4.8)

	pH 5	pH 6	pH 7	pH 8	pH 9
1. Glucose + ATP − glucose 6-phosphate + ADP	1.12×10^3	2.57×103	1.900×10^4	1.89×10^5	1.89×10^6
2. Glucose 6-phosphate = fructose 6-phosphate	0.210	0.248	0.276	0.281	0.282
3. Fructose 6-phosphate + ATP = fructose 1,6-biphosphate + ADP	146	933	1.18×10^4	1.27×10^4	1.28×10^6
4. Fructose 1,6-biphosphate = dihydroxyacetone phosphate + glyceraldehyde phosphate	5.46×10^{-5}	8.68×10^{-5}	9.23×10^{-5}	9.28×10^{-5}	9.28×10^{-5}
5. Dihydroxyacetone phosphate = glyceraldehyde phosphate	0.0455	0.0455	0.0455	0.0455	0.0455
6. Glyceraldehyde phosphate + P_i + NAD_{ox} − 1,3-bisphosphoglycerate + NAD_{red}	0.034	0.067	0.610	5.49	53.9
7. 1,3-Bisphosphoglycerate + ADP = 3-phosphoglycerate + ATP	28.65	27.57	27.59	29.0	29.2
8. 3-phosphoglycerate = 2-phosphoglycerate	0.107	0.103	0.091	0.084	0.083
9. 2-phosphoglycerate = phosphoenolpyruvate + H_2O	1.46	2.04	4.27	5.77	6.02
10. Phosphoenolpyruvate + ADP = pyruvate + ATP	1.05×10^6	6.06×10^5	1.09×10^5	1.16×10^4	1.16×10^3
Net reaction: Glucose + $2P_i$ + 2ADP + $2NAD_{ox}$ = 2pyruvate + 2ATP + $2NAD_{red}$ + $2H_2O$	2.4×10^7	1.4×10^{11}	1.4×10^{14}	2.5×10^{16}	2.6×10^{18}

Note: See Problem 4.8.

Table 4.6   Change in the Binding of Hydrogen Ions for Glycolysis at 298.15 K and 0.25 M Ionic Strength

	pH 5	pH 6	pH 7	pH 8	pH 9
1. Glucose + ATP = glucose 6-phosphate + ADP	− 0.14	− 0.65	− 0.98	− 1.00	− 1.00
2. Glucose 6-phosphate = fructose 6-phosphate	− 0.04	− 0.08	− 0.02	0.00	0.00
3. Fructose 6-phosphate + ATP = fructose 1,6-biphosphate + ADP	− 0.41	− 1.09	− 1.07	− 1.01	− 1.00
4. Fructose 1,6-biphosphate = dihydroxyacetonephosphate + glyceraldehyde phosphate	− 0.29	− 0.07	− 0.01	0.00	0.00
5. Dihydroxyacetone phosphate = glyceraldehyde phosphate	0.00	0.00	0.00	0.00	0.00
6. Glyceraldehyde phosphate + P_i + NAD_{ox} = 1,3-bisphosphoglycerate + NAD_{red}	− 1.57	− 1.07	− 0.92	− 0.98	− 1.00
7. 1,3-Bisphosphoglycerate + ADP = 3-phosphoglycerate + ATP	− 0.01	− 0.02	− 0.27	− 0.01	0.00
8. 3-phosphoglycerate = 2-phosphoglycerate	0.00	0.03	0.06	0.01	0.00
9. 2-phosphoglycerate = phosphoenolpyruvate + H_2O	− 0.05	− 0.27	− 0.26	− 0.04	0.00
10. Phosphoenolpyruvate + ADP = pyruvate + ATP	0.08	0.48	0.93	0.99	1.00
Net reaction: Glucose + $2P_i$ + 2ADP + $2NAD_{ox}$ = 2pyruvate + 2ATP + $2NAD_{red}$ + $2H_2O$	− 3.92	− 3.30	2.32	2.07	2.00

Note: See Problem 4.8.

Table 4.7   Standard Transformed Reaction Gibbs Energies for the New Reactions in Gluconeogenesis at 298.15 K and 0.25 M Ionic Strength

	pH 5	pH 6	pH 7	pH 8	pH 9
Pyruvate + CO2tot + ATP = oxaloacetate + ADP + P_i	0.75	− 4.31	− 8.81	− 14.06	− 19.36
Oxaloacetate + ATP + H_2O = phospho*enol*phosphate + ADP + CO_2tot	1.15	4.20	1.62	− 3.72	− 9.74
Pyruvate + 2ATP + H_2O = phosphoenolphosphate + 2ADP + P_i	1.91	− 0.11	− 7.19	− 17.78	− 29.10
Fructose 1,6-biphosphate + H_2O = fructose 6-phosphate + P_i	− 20.21	− 16.26	− 12.79	− 11.95	− 11.84
Glucose 6-phosphate + H_2O = glucose + P_i	− 15.15	− 13.75	− 11.62	− 10.96	− 10.88
Net reaction: 2Pyruvate + 6ATP + $2NAD_{red}$ + $6H_2O$ = glucose + 6ADP + $6P_i$ + $2NAD_{ox}$	− 88.8	− 69.3	− 63.3	− 70.7	− 81.7

See Problem 4.9

Table 4.8   Apparent Equilibrium Constants for the New Reactions in Gluconeogenesis at 298.15 K and 0.25 M Ionic Strength

	pH 5	pH 6	pH 7	pH 8	pH 9
Pyruvate + CO_2tot + ATP = oxaloacetate + ADP + P_i	0.74	5.69	34.9	291	2.47×10^3
Oxaloacetate + ATP + H_2O = phospho*enol*phosphate + ADP + CO_2tot	0.63	0.18	0.52	4.49	50.9
Pyruvate + 2ATP + H_2O = phospho*enol*phosphate + 2ADP + P_i	0.46	1.05	18.2	1.30×10^3	1.25×10^5
Fructose 1,6-biphosphate + H_2O = fructose 6-phosphate + P_i	3.47×10^3	7.07×10^2	1.74×10^2	1.24×10^2	1.19×10^2
Glucose 6-phosphate + H_2O = glucose + P_i	4.51×10^2	2.56×10^2	1.08×10^2	0.83×10^2	0.81×10^2
Net reaction: 2Pyruvate + 6ATP + 2NAD$_{red}$ + 6H_2O = glucose + 6ADP + 6P_i + 2NAD$_{ox}$	2.8×10^{15}	1.4×10^{12}	1.2×10^{11}	2.5×10^{12}	2.0×10^{14}

Note: See Problem 4.9

Table 4.9   Changes in the Binding of Hydrogen Ions for the New Reactions in Gluconeogenesis at 298.15 K and 0.25 M Ionic Strength

	pH 5	pH 6	pH 7	pH 8	pH 9
Pyruvate + CO_2tot + ATP = oxaloacetate + ADP + P_i	−0.96	−0.80	−0.85	−0.96	−0.85
Oxaloacetate + ATP + H_2O = phosphoenolphosphate + ADP + CO_2tot	0.84	.07	−0.82	−1.00	−1.15
Pyruvate + 2ATP + H_2O = phosphoenolphosphate + 2ADP + P_i	−0.12	−0.73	−1.67	−1.96	−2.00
Fructose 1,6-biphosphate + H_2O = fructose 6-phosphate + P_i	0.37	0.84	0.32	0.05	0.01
Glucose 6-phosphate + H_2O = glucose + P_i	0.10	0.40	0.24	0.04	0.00
Net reaction: 2Pyruvate + 6ATP + 2NAD$_{red}$ + 6H_2O = glucose + 6ADP + 6P_i + 2NAD$_{ox}$	3.77	2.51	-0.45	-1.79	−1.98

See Problem 4.9

CO_2(g) to CO_2tot is smaller at higher pH and larger at lower pH. Also note that GTP and GDP have been replaced with ATP and ADP in the phospho*enol*-pyruvate carboxykinase reaction because because the correct result can be obtained in this way.

Tables 4.10, 4.11, and 4.12 are the corresponding tables for pyruvate dehydrogenase, the citric acid cycle, the net reaction for the citric acid cycle, the net

Table 4.10   Standard Transformed Reaction Gibbs Energies for Pyruvate Dehydrogenase, the Citric Acid Cycle, and Net Reactions at 298.15 K and 0.25 M Ionic Strength

	pH 5	pH 6	pH 7	pH 8	pH 9
PDH: Pyruvate + CoA + NAD$_{ox}$ + H_2O = totCO$_2$ + NAD$_{red}$ + acetylCoA	−25.34	−26.62	−30.48	−44.70	−36.75
1. AcetylCoA + oxaloacetate + H_2O = CoA + citrate	−37.08	−39.56	−44.77	−51.67	−61.36
2. Citrate = cis-aconitate + H_2O	12.36	9.12	8.46	8.38	8.37
3. cis-Aconitate + H_2O = iso-citrate	−5.72	−2.48	−1.81	−1.73	−1.72
4. iso-Citrate + NADox + H_2O = ketoglutarate + CO_2tot + NADred	4.79	0.25	−4.46	−10.03	−16.07
5. Ketoglutarate + NAD$_{ox}$ + CoA + H_2O = succinylCoA + CO_2tot + NAD$_{red}$	−2.15	−33.24	−37.08	−41.31	−43.35
6. SuccinylCoA + P_i + ADP = succinate + ATP + CoA	7.58	4.09	1.26	−0.65	−4.72
7. Succinate + FADenz$_{ox}$ = fumarate + FADenz$_{red}$	1.75	0.22	0.00	−0.03	−0.03
8. Fumarate + H_2O = L-malate	−4.34	−3.69	−3.61	−3.60	−3.60
9. L-Malate + NAD$_{ox}$ = oxaloacetate + NAD$_{red}$	41.23	34.65	28.84	23.12	17.41
CAC: AcetylCoA + 3NAD$_{ox}$ + FADenz$_{ox}$ + ADP + P_i + 4H_2O = 2CO_2tot + 3NAD$_{red}$ + FADez$_{red}$ + ATP + CoA	−11.56	−30.64	−53.18	−77.52	−105.07
PDH⁺CAC: Pyruvate + 4NAD$_{ox}$ + FADenz$_{ox}$ + ADP + P_i + 5H_2O = 3CO_2tot + 4NAD$_{red}$ + FADenz$_{red}$ + ATP	−36.90	−57.26	−83.65	−112.22	−141.82
GLY + PDH⁺CAC: Glucose + 10NAD$_{ox}$ + 2FADenz$_{ox}$ + 4ADP + 4P_i + 8H_2O = 6CO_2tot + 10NAD$_{red}$ + 2FADenz$_{red}$ + 4ATP	−115.90	−178.14	−248.12	−318.01	−388.78

Note: See Problem 4.10

Table 4.11 Apparent Equilibrium Constants for Pyruvate Dehydrogenase, the Citric Acid Cycle, and Net Reactions at 298.15 K and 0.25 M Ionic Strength

	pH 5	pH 6	pH 7	pH 8	pH 9
PDH: Pyruvate + CoA + NAD_{ox} + H_2O − totCO_2 + NAD_{red} + acetylCoA	2.75×10^4	4.62×10^4	2.19×10^5	1.20×10^6	2.74×10^6
1. AcetylCoA + oxaloacetate + H_2O = CoA + citrate	3.13×10^6	8.51×10^6	6.97×10^7	1.13×10^9	5.63×10^{10}
2. Citrate = cis-aconitate + H_2O	0.68×10^{-2}	2.52×10^{-2}	3.30×10^{-2}	3.41×10^{-2}	3.42×10^{-2}
3. cis-Aconitate + H_2O = iso-citrate	10.1	2.72	2.07	2.01	2.00
4. iso-Citrate + NAD_{ox} + H_2O = ketoglutarate + CO_2tot + NAD_{red}	0.144	0.91	6.04	57.1	653
5. Ketoglutarate + NAD_{ox} + CoA + H_2O − succinylCoA + CO_2tot + NAD_{red}	4.28×10^5	6.67×10^5	3.14×10^6	1.72×10^7	3.92×10^7
6. SuccinylCoA + Pi + ADP = succinate + ATP + CoA	0.046	0.192	0.601	1.30	6.72
7. Succinate + FADenzox = fumarate + $FADenz_{red}$	0.493	0.915	1.00	1.01	1.01
8. Fumarate + H_2O = L-malate	5.74	4.43	4.29	4.27	4.27
9. L-Malate + NAD_{ox} = oxaloacetate + NAD_{red}	5.98×10^{-8}	8.50×10^{-7}	8.87×10^{-6}	8.91×10^{-5}	8.91×10^{-4}
CAC: AcetylCoA + $3NAD_{ox}$ + FADenz$_{ox}$ + ADP + P_i + $4H_2O$ = $2CO_2$tot + $3NAD_{red}$ + FADez$_{red}$ + ATP + CoA	1.06	2.33×10^5	2.07×10^9	3.8×10^{13}	2.6×10^{18}
PDH^+CAC: Pyruvate + $4NAD_{ox}$ + FADcnz$_{ox}$ + ADP + Pi + $5H_2O$ = $3CO_2$tot + $4NAD_{red}$d + FADenz$_{red}$ + ATP	2.92×10^6	1.07×10^{10}	4.53×10^{14}	4.57×10^{19}	7.01×10^{24}
GLY + PDH^+CAC: Glucose + $10NAD_{ox}$ + 2FADenz$_{ox}$ + 4ADP + $4P_i$ + $8H_2O$ = $6CO_2$tot + $10NAD_{red}$ + 2FADenz$_{red}$ + 4ATP	2.01×10^{20}	1.62×10^{31}	2.94×10^{43}	5.15×10^{55}	1.29×10^{68}

Note: See Problem 4.10

reaction for pyruvate dehydrogenase and the citric acid cycle, and the net reaction from glucose to CO_2tot. The sixth reaction in the citric acid cycle involves GTP and GDP, but ATP and ADP have been substituted for these calculations because the apparent equilibrium constant fot the reaction ATP + GTP = ADP + GTP is expected to be very close to unity. There is a problem with FAD_{ox} and FAD_{red} in the seventh reaction. The values for fadoxsp

Table 4.12 Changes in the Binding of Hydrogen Ions for Pyruvate Dehydrogenase, the Citric Acid Cycle, and Net Reactions at 298.15 K and 0.25 M Ionic Strength

	pH 5	pH 6	pH 7	pH 8	pH 9
PDH: Pyruvate + CoA + NAD_{ox} + H_2O = totCO_2 + NAD_{red} + acetylCoA	−0.08	−0.44	−0.82	0.56	−0.26
1. AcetylCoA + oxaloacetate + H_2O = CoA + citrate	0.09	−0.74	−1.04	−1.44	−1.89
2. Citrate = cis-aconitate + H_2O	−0.91	−0.27	−0.03	0.00	0.00
3. cis-Aconitate + H_2O = iso-citrate	0.90	0.27	0.03	0.00	0.00
4. iso-Citrate + NAD_{ox} + H_2O = ketoglutarate + CO_2tot + NAD_{red}	−0.98	−0.72	−0.93	−1.01	−1.15
5. Ketoglutarate + NAD_{ox} + CoA + H_2O = succinylCoA + CO_2tot + NAD_{red}	−0.08	−0.44	−0.82	−0.56	−0.26
6. SuccinylCoA + Pi + ADP = succinate + ATP + CoA	−0.44	−0.66	−0.32	−0.48	−0.89
7. Succinate + FADenz$_{ox}$ = fumarate + FADenz$_{red}$	−0.50	−0.10	−0.01	0.00	0.00
8. Fumarate + H_2O = L-malate	0.23	0.04	0.00	0.00	0.00
9. L-Malate + NAD_{ox} = oxaloacetate + NAD_{red}	−1.33	−1.05	−1.01	−1.00	−1.00
CAC: AcetylCoA + $3NAD_{ox}$ + FADenz$_{ox}$ + ADP + P_i + $4H_2O$ = $2CO_2$tot + $3NAD_{red}$ + FADez$_{red}$ + ATP + CoA	−3.11	−3.66	−4.11	−4.49	−5.19
PDH^+CAC: Pyruvate + $4NAD_{ox}$ + FADenz$_{ox}$ + ADP + Pi + $5H_2O$ = $3CO_2$tot + $4NAD_{red}$ + FADenz$_{red}$ + ATP	−3.19	−4.11	−4.93	−5.05	−5.44
GLY + PDH^+CAC: Glucose + $10NAD_{ox}$ + 2FADenz$_{ox}$ + 4ADP + 4Pi + $8H_2O$ = $6CO_2$tot + $10NAD_{red}$ + 2FADenz$_{red}$ + 4ATP	−10.30	−11.72	−12.40	−12.17	−12.89

Note: See Problem 4.10

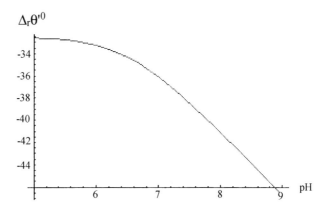

Figure 4.2 Standard transformed Gibbs energy of reaction in kJ mol^{-1} for ATP + H$_2$O = ADP + P$_i$ at 298.15 K and $I = 0.25$ M (see Problem 4.3).

and fadredsp in BasicBiochemData2 are based on the standard apparent reduction potential at pH 7. However, the seventh reaction involves FAD$_{ox}$ and FAD$_{red}$ bound to the enzyme. Since K' for the seventh reaction is believed to be essentially zero, $\Delta_f G^0$ of the species of FAD bound by the enzyme have been adjusted to give this value. The value of $\Delta_f G^0$(FADenz$_{ox}$) has been taken to be the same as for FAD$_{ox}$, but $\Delta_f G^0$(FADenz$_{red}$) has been taken to be different from $\Delta_f G^0$(FAD$_{red}$). Replacing CO$_2$tot with CO$_2$(g) would increase $\Delta_r G'^0$ for the pyruvate dehydrogenase reaction, reaction 4, and reaction 5 by 2.92 kJ mol^{-1} and for the last three net reactions by 2, 3, and 6 times this much.

■ 4.12   PLOTS OF THERMODYNAMIC PROPERTIES OF BIOCHEMICAL REACTIONS VERSUS pH

The functions of pH and ionic strength that yield $\Delta_r G_i'^0$, $\Delta_r H_i'^0$, and $\Delta_r N_H$ can also be used to plot these properties in terms of pH at a chosen ionic strength and in terms of ionic strength at a chosen pH. Figure 4.2 shows the dependence of the standard transformed Gibbs energy of the hydrolysis of ATP to ADP on pH.

Figure 4.3 shows the dependence of the standard transformed enthalpy of this reaction on pH. Figure 4.4 shows the dependence of $\Delta_r S'^0$ on pH. The enthalpy does not vary much with pH, but the entropy increases significantly above pH 6. This causes the equilibrium to shift further in the direction of hydrolysis at higher pHs.

Figure 4.5 shows the dependence of $\log K'$ on pH. The change in binding of hydrogen ions in the hydrolysis of ATP to ADP and P$_i$ is shown in Fig. 4.6 at

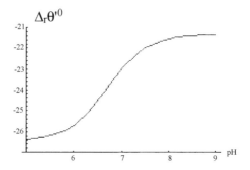

Figure 4.3 Standard transformed enthalpy of reaction in kJ mol^{-1} for ATP + H$_2$O = ADP + P$_i$ at 298.15 K and $I = 0.25$ M (see Problem 4.3).

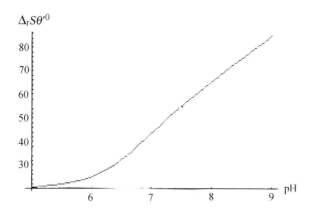

Figure 4.4 Standard transformed entropy of reaction in J K^{-1} mol^{-1} for ATP + H$_2$O = ADP + P$_i$ at 298.15 K and $I = 0.25$ M (see Problem 4.3).

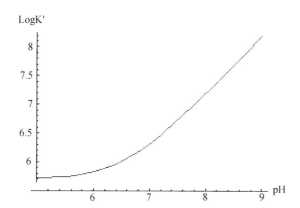

Figure 4.5 Logarithm base 10 of the apparent equilibrium constant K' for ATP + H$_2$O = ADP + P$_i$ at 298.15 K and $I = 0.25$ M.

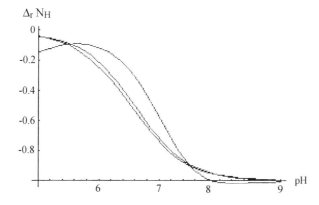

Figure 4.6 Plot of the change in binding of hydrogen ions in the reaction ATP + H$_2$O = ADP + P$_i$ at $I = 0, 0.10$ and 0.25 M at 298.15 K. The more curved plot is at zero ionic strength. (See Problem 4.5.)

298.15 K as a function of pH at ionic strengths of 0, 0.10, and 0.25 M (see Problem 4.5).

As a test of the values of the thermodynamic properties that can be calculated using BasicBiochemData2, it has been used to calculate the apparent equilibrium constants at 298.15 K at the experimental pH and ionic strength of a number of reactions in the critical compilations of Goldberg and Tewari (Goldberg et al., 1992; Goldberg and Tewari, 1994a, b, 1995a, b; Goldberg, 1999). The table in Problem 4.7 is a first step toward identifying errors in BasicBiochemData2. Since the values of $\Delta_f G^0$ for species can be obtained from the existing data in different ways, this redundancy can be used to find errors in experimental measurements of K' and in the calculations to produce a thermodynamic database.

Matrices in Chemical and Biochemical Thermodynamics

- **5.1** Chemical Equations as Matrix Equations
- **5.2** Biochemical Equations as Matrix Equations
- **5.3** Coupling of Biochemical Reactions
- **5.4** Matrix Forms of the Fundamental Equations for Chemical Reaction Systems
- **5.5** Matrix Forms of the Fundamental Equations for Biochemical Reaction Systems
- **5.6** Operations of Linear Algebra

When a system involves more than one chemical reaction, it is convenient to represent the conservation equations and reaction equations by matrices. Chemical reactions balance atoms of elements and electric charges, and that means that there is a set of conservation equations. The coefficients in these conservation equations are related to the stoichiometric numbers in the reaction equations (Alberty, 1991b). For larger systems of reactions, it is very convenient to use conservation matrices and stoichiometric number matrices because linear algebra provides mathematical operations for changing and interconverting matrices. Conservation matrices and stoichiometric number matrices are related mathematically and actually contain the same information. But for some purposes it is better to use conservation matrices, and for other purposes it is better to use stoichiometric number matrices. Conservation matrices are very helpful in identifying components and showing how noncomponents are made up of components.

Biochemical reactions balance the atoms of all elements except for hydrogen, or of metals when they are bound reversibly and their ionic concentrations are held constant. Thus a system of biochemical reactions can be represented by an apparent conservation matrix or an apparent stoichiometric number matrix. The adjective "apparent" is used because hydrogen ions are omitted in the apparent conservation matrix since they are not conserved. Hydrogen ions are also omitted in the apparent stoichiometric number matrix since they do not appear in biochemical reactions. The conservation and stoichiometric number matrices for a system of biochemical reactions can be derived from the conservation matrix for the underlying chemical reactions. As in the case of chemical reactions, the apparent conservation matrix is related mathematically to the apparent stoichiometric number matrix. Matrix notation is also useful in writing fundamental equations and Gibbs-Duhem equations and in calculating equilibrium compositions. There will be more applications of matrix operations in subsequent chapters. More information on matrices is to be found in Smith and Missen (1982) and in a textbook on linear algebra, such as Strang (1988).

■ 5.1 CHEMICAL EQUATIONS AS MATRIX EQUATIONS

The conservation relationships in chemical reactions can be represented by reaction equations or by conservation equations. When using reaction equations in thermodynamics, it is important to remember that a reaction equation can be multiplied by any positive or negative integer without changing the equilibrium composition that will be calculated. Of course, the expression for the equilibrium constant K of the reaction must be changed appropriately. When an equilibrium calculation is made on a multireaction system, only an independent set of reactions is used. An independent set of reaction equations is one in which no equation in the set can be obtained by adding or subtracting other reactions in the set. We will find that linear algebra provides a much more practical test of independence. The number R of independent reactions in a set is unique, but the particular reactions in the set are not. Any two reactions in a set of independent reactions can be added, and this reaction can be used to replace one of the two reactions without changing the equilibrium concentration that will be calculated. These remarks apply in thermodynamics, but not in discussing rates of reactions.

The corresponding conservation equations are less familiar, but they contain the same information as a set of independent chemical reactions. The conservation equations for a system containing N_s species are given by

$$\sum_{j=1}^{N_s} N_{ij}n_j = n_{ci} \tag{5.1-1}$$

where n_{ci} is the amount of **component** i, N_{ij} is the number of units of component i in species j, and n_j is the amount of species j. For chemical reactions the conservation equations are usually written in terms of amounts of elements and electric charge, but they can also be written in terms of specified groups of atoms. The things that are conserved are referred to as components. **The amounts of components in a closed system are not changed by chemical reactions.** The conservation equations for the components in a reaction system must be independent; that is, no conservation equation in the set can be calculated by adding and subtracting the other equations in the set. The number C of components for a chemical reaction system is unique, but the components that are chosen are not.

Equation 5.1-1 for a reaction system can be written in matrix form as

$$An = n_c \tag{5.1-2}$$

where A is the **conservation matrix** made up of the N_{ij} values, with a row for each component and a column for each species. In equation 5.1-2, n is the column matrix of amounts of species and n_c is the column matrix of amounts of components. The matrix product of the $C \times N_s$ conservation matrix A and the $N_s \times 1$ amount of species matrix n is equal to the $C \times 1$ matrix n_c of amounts of components. Equation 5.1-2 can be used to calculate amounts of components in more complicated systems (see equation 5.1-27). The number N_s of different species in a system of chemical reactions is given by

$$N_s = C + R \tag{5.1-3}$$

where R is the number of independent reactions. This equation can be interpreted by pointing out that the number N_s of unknown concentrations of species in an equilibrium calculation is equal to the number of C components plus the number R of independent reactions. Note that there is a conservation equation for each component and an equilibrium constant expression for each reaction. The species in a chemical reaction system can be divided into C components and R noncomponents. Various choices of components and non-components can be made, but the numbers C and R are unique for a system of chemical reactions.

Consider a gaseous reaction system in which the only reaction is

$$CO + 3H_2 = CH_4 + H_2O \tag{5.1-4}$$

When stoichiometric numbers are taken to be signed quantities, this chemical equation can be written as

$$-CO - 3H_2 + CH_4 + H_2O = 0 \tag{5.1-5}$$

This may not look like a matrix equation, but it actually is. When we replace the chemical formulas with column vectors that give the numbers of C, H, and O atoms in each species, equation 5.1-5 can be written as

$$-\begin{pmatrix} 1 \\ 0 \\ 1 \end{pmatrix} - 3\begin{pmatrix} 0 \\ 2 \\ 0 \end{pmatrix} + \begin{pmatrix} 1 \\ 4 \\ 0 \end{pmatrix} + \begin{pmatrix} 0 \\ 2 \\ 1 \end{pmatrix} = \begin{pmatrix} 0 \\ 0 \\ 0 \end{pmatrix} \tag{5.1-6}$$

This equation can be written as a matrix multiplication:

$$\begin{pmatrix} 1 & 0 & 1 & 0 \\ 0 & 2 & 4 & 2 \\ 1 & 0 & 0 & 1 \end{pmatrix} \begin{pmatrix} -1 \\ -3 \\ 1 \\ 1 \end{pmatrix} = \begin{pmatrix} 0 \\ 0 \\ 0 \end{pmatrix} \tag{5.1-7}$$

where the conservation matrix A is given by

$$
A = \begin{array}{c} \\ C \\ H \\ \\ \end{array}
\begin{array}{cccc}
CO & H_2 & CH_4 & H_2O \\
1 & 0 & 1 & 0 \\
0 & 2 & 4 & 2 \\
1 & 0 & 0 & 1
\end{array}
\tag{5.1-8}
$$

In the A matrix there is a column for each species and a row for each component. Note that the components are taken to be atoms of C, H, and O. (In equation 5.1-15 we will see that other choices of components can be made.) The stoichiometric number matrix corresponding with equation 5.1-5 is

$$
v = \begin{array}{c} \\ CO \\ H_2 \\ CH_4 \\ H_2O \end{array}
\begin{array}{c}
rx5.1\text{-}4 \\
-1 \\
-3 \\
1 \\
1
\end{array}
\tag{5.1-9}
$$

In the **v** matrix there is a column for each reaction and a row for each species, with the species in the same order as in the columns of the A matrix. The matrix multiplication in equation 5.1-7 is represented in general by

$$A\mathbf{v} = \mathbf{0} \tag{5.1-10}$$

Note that the matrix product of the $C \times N_s$ conservation matrix and the $N \times R$ stoichiometric number matrix is a $C \times R$ zero matrix.

Equation 5.1-10 applies to a multireaction system. For example, if in addition to reaction 5.1-4, the following reaction occurs:

$$CO + H_2O = CO_2 + H_2 \qquad (5.1-11)$$

We can add a column to the conservation matrix for CO_2 and a column to the stoichiometric number matrix for this reaction to obtain

$$\begin{pmatrix} 1 & 0 & 1 & 0 & 1 \\ 0 & 2 & 4 & 2 & 0 \\ 1 & 0 & 0 & 1 & 2 \end{pmatrix} \begin{pmatrix} -1 & -1 \\ -3 & 1 \\ 1 & 0 \\ 1 & -1 \\ 0 & 1 \end{pmatrix} = \begin{pmatrix} 0 & 0 \\ 0 & 0 \\ 0 & 0 \end{pmatrix} \qquad (5.1-2)$$

The new conservation matrix A is given by

$$A = \begin{matrix} & CO & H_2 & CH_4 & H_2O & CO_2 \\ C & 1 & 0 & 1 & 0 & 1 \\ H & 0 & 2 & 4 & 2 & 0 \\ O & 1 & 0 & 0 & 1 & 2 \end{matrix} \qquad (5.1-13)$$

The new stoichiometric number matrix v is given by

$$v = \begin{matrix} & rx5.1\text{-}4 & rx5.1\text{-}11 \\ CO & -1 & -1 \\ H_2 & -3 & 1 \\ CH_4 & 1 & 0 \\ H_2O & 1 & -1 \\ CO_2 & 0 & 1 \end{matrix} \qquad (5.1-14)$$

Now we want to show that the conservation matrix can be written in terms of other components. The easiest way to obtain an equivalent A matrix is to make a row reduction (Gaussian elimination) to obtain the **canonical form** of the matrix with an **identity matrix** in the left side. An identity matrix is a square matrix that has ones on the diagonal with the other positions occupied by zeros. Subtracting the first row of matrix 5.1-13 from the last row and using the difference with a change in sign to replace the last row, subtracting the last row from the first, and subtracting two times the third row from the second yields the canonical form of the conservation matrix:

$$A = \begin{matrix} & CO & H_2 & CH_4 & H_2O & CO_2 \\ CO & 1 & 0 & 0 & 1 & 2 \\ H_2 & 0 & 1 & 0 & 3 & 2 \\ CH_4 & 0 & 0 & 1 & -1 & -1 \end{matrix} \qquad (5.1-15)$$

The canonical form of a matrix is readily obtained using RowReduce in *Mathematica*. In equation 5.1-15 the conservation equations are for the conservation of CO, H_2, and CH_4 rather than for the atoms of C, H, and O; in other words, the components have been chosen to be CO, H_2, and CH_4. The last two columns show how the noncomponents H_2O and CO_2 are made up of the components. They show that H_2O is made up of $CO + 3H_2 - CH_4$, and CO_2 is made up of $2CO + 2H_2 - CH_4$. If one of the conservation equations were redundant, it would yield a row of zeros that would be dropped. Since there are three rows in this A matrix that are not all zeros after row reduction, the A matrix has a rank of 3, and so the number of components is given by

$$C = \text{rank } A \qquad (5.1-16)$$

In matrix 5.1-15 the three components are CO, H_2, and CH_4. However, if the order of the columns were changed, other components would be chosen. Thus the conservation matrix is not unique. A set of components must contain all the elements that are not redundant. The rank of the stoichiometric number matrix is equal to the number of independent reactions.

$$R - \text{rank } A \qquad (5.1\text{-}17)$$

Thus equation 5.1-3 ($N_s = C + R$) can be written

$$N_s = \text{rank } A + \text{rank } \mathbf{v} \qquad (5.1\text{-}18)$$

Next we want to consider the fact that a stoichiometric number matrix can be calculated from the conservation matrix, and vice versa. Since $A\mathbf{v} = \mathbf{0}$, the A matrix can be used to calculate a basis for the stoichiometric number matrix $\mathbf{v}$. The stoichiometric number matrix $\mathbf{v}$ is referred to as the **null space** of the A matrix. When the conservation matrix has been row reduced it is in the form $A = [I_C, Z]$, where I_C is an identity matrix with rank C. A basis for the null space is given by

$$v = \begin{pmatrix} -Z \\ I_R \end{pmatrix} \qquad (5.1\text{-}19)$$

where Z is $C \times R$ and I_R is an identity matrix with rank R. It is necessary to say that the null space calculated using equation 5.1-19, or using a computer, is a **basis** for the null space because the stoichiometric number matrix for a system of reactions can be written in many different ways, as mentioned before (equation 5.1-1). All of these forms of the $\mathbf{v}$ matrix satisfy equation 5.1-10.

Equation 5.1-19 shows that the stoichiometric number matrix corresponding with conservation matrix 5.1-15 is

$$v = \begin{array}{c} \\ CO \\ H_2 \\ CH_4 \\ H_2O \\ CO_2 \end{array} \begin{array}{cc} \text{rx } 5.1\text{-}14 & \text{rx } 5.1\text{-}21 \\ -1 & -2 \\ -3 & -2 \\ 1 & 1 \\ 1 & 0 \\ 0 & 1 \end{array} \qquad (5.1\text{-}20)$$

Note that reaction 5.1-11 has been replaced by

$$2CO + 2H_2 = CH_4 + CO_2 \qquad (5.1\text{-}21)$$

which is stoichiometrically correct and independent. The correct equilibrium composition can be calculated with either set of reactions. Stoichiometric number matrices 5.1-14 and 5.2-20 have the same row-reduced form, and so they are equivalent. This is an example of the fact that the stoichiometric number matrix for a system is not unique. It is important to realize that the equilibrium composition that is calculated for a system of reactions is valid for all possible reactions that can be obtained by adding and subtracting the reactions used in the calculation of the equilibrium composition.

Equation 5.1-10 provides the means for calculating a basis for the stoichiometric number matrix that corresponds with the conservation matrix. Similarly the transposed stoichiometric number matrix provides the means for calculating a basis for the transposed conservation matrix. This is done by using the following equation, which is equivalent to equation 5.1-10:

$$\mathbf{v}^T A^T = \mathbf{0} \qquad (5.1\text{-}22)$$

where "T" indicates the transpose. The transpose of a matrix is obtained by exchanging rows and columns. Thus a set of independent reactions for a reaction system can be used to calculate a set of conservation equations for the system.

This can be illustrated by starting with the transposed stoichiometric number matrix 5.1-14, which is

$$
\mathbf{v}^\mathrm{T} = \begin{array}{cc}
 & \begin{array}{ccccc} \text{CO} & \text{H}_2 & \text{CH}_4 & \text{H}_2\text{O} & \text{CO}_2 \end{array} \\
\begin{array}{c} \text{rx } 5.1\text{-}4 \\ \text{rx } 5.1\text{-}11 \end{array} & \begin{array}{ccccc} -1 & -3 & 1 & 1 & 0 \\ -1 & 1 & 0 & -1 & 1 \end{array}
\end{array}
\qquad (5.1\text{-}23)
$$

Row reduction of matrix 5.1-23 yields

$$
\mathbf{v}^\mathrm{T} = \begin{array}{ccccc}
\text{CO} & \text{H}_2 & \text{CH}_4 & \text{H}_2\text{O} & \text{CO}_2 \\
1 & 0 & -\frac{1}{4} & \frac{1}{2} & -\frac{3}{4} \\
0 & 1 & -\frac{1}{4} & -\frac{1}{2} & \frac{1}{4}
\end{array}
\qquad (5.1\text{-}24)
$$

The transpose of the conservation matrix can be obtained by using

$$
A^\mathrm{T} = \begin{pmatrix} -Z \\ I_\text{c} \end{pmatrix}
\qquad (5.1\text{-}25)
$$

This yields

$$
A = \begin{array}{c}
 \\ \text{CH}_4 \\ \text{H}_2\text{O} \\ \text{CO}_2
\end{array}
\begin{array}{ccccc}
\text{CO} & \text{H}_2 & \text{CH}_4 & \text{H}_2\text{O} & \text{CO}_2 \\
\frac{1}{4} & \frac{1}{4} & 1 & 0 & 0 \\
-\frac{1}{2} & \frac{1}{2} & 0 & 1 & 0 \\
\frac{3}{4} & -\frac{1}{4} & 0 & 0 & 1
\end{array}
\qquad (5.1\text{-}26)
$$

This looks different from equation 5.1-8, but it yields 5.1-15 on row reduction, which shows that the two matrices are equivalent.

Mathematica is very useful for carrying out these matrix operations. The operation for row reduction is RowReduce, and the operation for calculating a basis for the null space is NullSpace. Row reduction is also used to determine whether the equations in a set of conservation equations or reaction equations are independent. Rows that are dependent come out as all zeros when this is done, and they must be deleted because they do not provide any useful information.

We return to equation 5.1-1 for the system we have been discussing:

$$
A\boldsymbol{n} = \begin{array}{c}
 \\ \text{C} \\ \text{H} \\ \text{O}
\end{array}
\begin{array}{ccccc}
\text{CO} & \text{H}_2 & \text{CH}_4 & \text{H}_2\text{O} & \text{CO}_2 \\
1 & 0 & 1 & 0 & 1 \\
0 & 2 & 4 & 2 & 0 \\
1 & 0 & 0 & 1 & 2
\end{array}
\begin{pmatrix} n(\text{CO}) \\ n(\text{H}_2) \\ n(\text{CH}_4) \\ n(\text{H}_2\text{O}) \\ n(\text{CO}_2) \end{pmatrix}
= \begin{pmatrix} n_\text{c}(\text{C}) \\ n_\text{c}(\text{H}) \\ n_\text{c}(\text{O}) \end{pmatrix}
\qquad (5.1\text{-}27)
$$

The product of a $C \times N_\text{s}$ matrix and a $N_\text{s} \times 1$ matrix is a $C \times 1$ matrix; note that N_s disappears as one of the dimensions of the resultant matrix. The amounts of components in a reaction system are independent variables and consequently do not change during a chemical reaction. The amounts of species are dependent variables because their amounts do change during chemical reactions. Equation 5.1-27 shows that A is the transformation matrix that transforms amounts of species to amounts of components. The order of the columns in the A matrix is arbitrary, except that it is convenient to include all of the elements in the species on the left so that the canonical form can be obtained by row reduction. When the row-reduced form of A is used, the amounts of the components CO, H_2, and CH_4 can be calculated (see Problem 5.1).

As an example of a set of chemical reactions in aqueous solution that are of biochemical interest, consider the hydrolysis of adenosine triphosphate to adenosine diphosphate and inorganic phosphate in the neighborhood of pH 7. The chemical reactions involved are

$$ATP^{4-} + H_2O = ADP^{3-} + HPO_4^{2-} + H^+ \qquad (5.1\text{-}28)$$

$$HATP^{3-} = H^+ + ATP^{4-} \qquad (5.1\text{-}29)$$

$$HADP^{2-} = H^+ + ADP^{3-} \qquad (5.1\text{-}30)$$

$$H_2PO_4^- = H^+ + HPO_4^{2-} \qquad (5.1\text{-}31)$$

This set of chemical reactions is not unique; for example, the reference reaction can be written with $H_2PO_4^-$. Additional reactions are involved if Mg^{2+} or other cations are bound reversibly by these species. The conservation matrix for this system is

		ATP^{4-}	H^+	H_2O	HPO_4^{2-}	ADP^3	$HATP^3$	$HADP^{2-}$	$H_2PO_4^-$
	C	10	0	0	0	10	10	10	0
$A =$	H	12	1	2	1	12	13	13	2
	O	13	0	1	4	10	13	10	4
	P	3	0	0	1	2	3	2	1

$$(5.1\text{-}32)$$

The rows for nitrogen and electric charges are redundant, and therefore are omitted. The row-reduced conservation matrix is

		ATP^{4-}	H^+	H_2O	HPO_4^{2-}	ADP^{3-}	$HATP^{3-}$	$HADP^{2-}$	$H_2PO_4^-$
	ATP^{4-}	1	0	0	0	1	1	1	0
$A =$	H^+	0	1	0	0	-1	1	0	1
	H_2O	0	0	1	0	1	0	1	0
	HPO_4^{2-}	0	0	0	1	-1	0	-1	1

$$(5.1\text{-}33)$$

The last four species can be considered to be made up of the four components labeling the rows.

A basis for the stoichiometric number matrix can be calculated using *Mathematica*.

	ATP^{4-}	H^+	H_2O	HPO_4^{2-}	ADP^{3-}	$HATP^{3-}$	$HADP^{2-}$	$H_2PO_4^-$
	0	-1	0	-1	0	0	0	1
$\mathbf{v}^T =$	-1	0	-1	1	0	0	1	0
	-1	-1	0	0	0	1	0	0
	-1	1	-1	1	1	0	0	0

$$(5.1\text{-}34)$$

This does not correspond with reactions 5.1-28 to 5.1-32, but it is equivalent because the row-reduced form of equation 5.1-34 is identical with the row-reduced form of the stoichiometric number matrix for reactions 5.1-28 to 5.1-32 (see Problem 5.2). The application of matrix algebra to electrochemical reactions is described by Alberty (1993d).

■ 5.2 BIOCHEMICAL EQUATIONS AS MATRIX EQUATIONS

As discussed in Chapter 4, biochemists are generally more interested in reactions at specified pH. At specified pH, hydrogen is not conserved, and so this row and column of matrix 5.1-32 are omitted to obtain the following conservation matrix

(Alberty, 1992b):

$$
A' = \begin{array}{c|ccccccc}
 & \text{ATP}^{4-} & \text{ADP}^{3-} & \text{HPO}_4^{2-} & \text{H}_2\text{O} & \text{HATP}^{3-} & \text{HADP}^{2-} & \text{H}_2\text{PO}_4^{-} \\
\hline
\text{C} & 10 & 10 & 0 & 0 & 10 & 10 & 0 \\
\text{O} & 13 & 10 & 4 & 1 & 13 & 10 & 4 \\
\text{P} & 3 & 2 & 1 & 0 & 3 & 2 & 1
\end{array}
\tag{5.2-1}
$$

where the prime on A' indicates that the pH has been specified so that hydrogen atoms are not conserved. This matrix has three pairs of redundant columns. Since the columns for ATP^{4-} and HATP^{3-} are the same, we can delete one and label the remaining column as ATP, where this abbreviation refers to the sum of ATP^{4-} and HATP^{3-}. The abbreviations ADP and P_i are introduced in the same way to obtain

$$
A' = \begin{array}{c|cccc}
 & \text{ATP} & \text{H}_2\text{O} & \text{ADP} & \text{P}_i \\
\hline
\text{C} & 10 & 0 & 10 & 0 \\
\text{O} & 13 & 1 & 10 & 4 \\
\text{P} & 3 & 0 & 2 & 1
\end{array}
\tag{5.2-2}
$$

This is referred to as an **apparent conservation matrix** to distinguish it from the conservation matrix in equation 5.1-32. Thus specifying the pH has the effect of simplifying the conservation matrix of the system by reducing the number of rows by one and the number of columns by four. The matrix in equation 5.2-2 is not unique. An equivalent apparent conservation matrix can be obtained more simply by conserving adenosine groups; this leads to

$$
A' = \begin{array}{c|cccc}
 & \text{ATP} & \text{H}_2\text{O} & \text{ADP} & \text{P}_i \\
\hline
\text{aden} & 1 & 0 & 0 & 1 \\
\text{O} & 9 & 1 & 6 & 4 \\
\text{P} & 3 & 0 & 2 & 1
\end{array}
\tag{5.2-3}
$$

The row-reduced forms of matrices 5.2-2 and 5.2-3 are the same, and so they are equivalent.

$$
A' = \begin{array}{c|cccc}
 & \text{ATP} & \text{H}_2\text{O} & \text{ADP} & \text{P}_i \\
\hline
\text{ATP} & 1 & 0 & 0 & 1 \\
\text{H}_2\text{O} & 0 & 1 & 0 & 1 \\
\text{ADP} & 0 & 0 & 1 & -1
\end{array}
\tag{5.2-4}
$$

Apparent conservation matrices A' and apparent stoichiometric number matrices $\mathbf{v}'$ at specified pH have the properties indicated by equations 5.1-10 and 5.1-22 so that

$$
A'\mathbf{v}' = \mathbf{0} \tag{5.2-5}
$$

$$
(\mathbf{v}')^{\text{T}}(A')^{\text{T}} = \mathbf{0} \tag{5.2-6}
$$

The rank of the A' matrix is the number C' of apparent components, and the rank of the apparent stoichiometric number matrix is the number R' of independent biochemical reactions.

$$
C' = \text{rank } A' \tag{5.2-7}
$$

$$
R' = \text{rank } \mathbf{v}' \tag{5.2-8}
$$

$$
N' = C' + R' \tag{5.2-9}
$$

A basis for the null space $\mathbf{v}'$ of conservation matrix 5.2-5 at specified pH obtained with equation 5.1-19 or with a computer is

$$\mathbf{v}' = \begin{array}{cc} \text{ATP} & -1 \\ \text{H}_2\text{O} & -1 \\ \text{ADP} & 1 \\ \text{P}_i & 1 \end{array} \qquad (5.2\text{-}10)$$

which is referred to as an **apparent stoichiometric matrix** because it is made up of the stoichiometric numbers for a reactants at specified pH, rather than species. Matrix 5.2-10 corresponds with

$$\text{ATP} + \text{H}_2\text{O} = \text{ADP} + \text{P}_i \qquad (5.2\text{-}11)$$

which is referred to as a **biochemical reaction** to distinguish it from the underlying chemical reactions. This equation does not balance hydrogen atoms because the chemical potential of hydrogen ion is specified (see Problem 5.3). This conversion of a set of chemical equations to a single biochemical equation is discussed by Alberty (1992b).

The product of the apparent conservation matrix $\mathbf{A}'$ and the column vector $\mathbf{n}'$ of **amounts of reactants** (pseudoisomer groups) gives the column vector $\mathbf{n}'_c$ of the **amounts of the apparent components**:

$$\mathbf{A}'\mathbf{n}' = \mathbf{n}'_c \qquad (5.2\text{-}12)$$

This is like the product of the conservation matrix $\mathbf{A}$ and the amounts $\mathbf{n}$ of species, which gives the amounts of components $\mathbf{n}_c$ (equation 5.1-12). The apparent components in equation 5.2-4 are ATP, H_2O, and ADP.

In summary, the linear algebra of the hydrolysis of ATP at specified pH is very much like the linear algebra of chemical reactions, even though hydrogen atoms are not conserved in the biochemical reaction and the reactants are sums of species.

■ 5.3   COUPLING OF BIOCHEMICAL REACTIONS

Some enzyme-catalyzed reactions are sums of biochemical reactions that could in principle occur separately. This is important in considering conservation equations because the mechanisms of such reactions may introduce additional conservation equations, in other words, additional components. When two biochemical reactions without a common reactant are coupled together by an enzymatic mechanism, the number of biochemical reactions in the system is decreased by one, but the number of reactants is unchanged: $\Delta C' = \Delta N' - \Delta R' = 0 - (-1) = 1$. There is then one more apparent component. In discussing enzyme-catalyzed reactions, it is convenient to use EC numbers (Webb, 1992).

Glutamate-ammonia ligase (EC 6.3.1.2) couples the following two reactions:

$$\text{Glutamate} + \text{ammonia} = \text{glutamine} + \text{H}_2\text{O} \qquad (5.3\text{-}1)$$

$$\text{ATP} + \text{H}_2\text{O} = \text{ADP} + \text{P}_i \qquad (5.3\text{-}2)$$

so that the reaction catalyzed is

$$\text{Glutamate} + \text{ATP} + \text{ammonia} = \text{glutamine} + \text{ADP} + \text{P}_i \qquad (5.3\text{-}3)$$

The transposed stoichiometric number matrix for this reaction is

$$(\mathbf{v}')^{\text{T}} = \begin{array}{cccccc} \text{Glutamate} & \text{ATP} & \text{Amm} & \text{ADP} & \text{P}_i & \text{Glutamine} \\ -1 & -1 & -1 & 1 & 1 & 1 \end{array} \qquad (5.3\text{-}4)$$

The use of NullSpace yields the following row reduced conservation matrix:

$$A' = \begin{array}{c} \\ \end{array} \begin{array}{cccccc} \text{Glutamate} & \text{ATP} & \text{Amm} & \text{ADP} & \text{P}_i & \text{Glutamine} \\ 1 & 0 & 0 & 0 & 0 & 1 \\ 0 & 1 & 0 & 0 & 0 & 1 \\ 0 & 0 & 1 & 0 & 0 & 1 \\ 0 & 0 & 0 & 1 & 0 & -1 \\ 0 & 0 & 0 & 0 & 1 & -1 \end{array} \qquad (5.3\text{-}5)$$

This shows that five apparent components are in agreement with $C' = N' - R' = 6 - 1 = 5$, but only four elements (C, O, N, and P) are to be conserved. The fifth conservation equation is needed to tie reactions 5.3-1 and 5.3-2 one to one. This conservation equation can be written in a number of ways, but one way is

$$n(\text{ATP}) + n(\text{glutamine}) = \text{const.} \qquad (5.3\text{-}6)$$

Thus the apparent conservation matrix is given by

$$A' = \begin{array}{c} \\ \text{C} \\ \text{O} \\ \text{N} \\ \text{P} \\ \text{con1} \end{array} \begin{array}{cccccc} \text{Glutamate} & \text{ATP} & \text{Amm} & \text{ADP} & \text{P}_i & \text{Glutamine} \\ 5 & 10 & 0 & 10 & 0 & 5 \\ 4 & 13 & 0 & 10 & 4 & 3 \\ 1 & 5 & 1 & 5 & 0 & 2 \\ 0 & 3 & 0 & 2 & 1 & 0 \\ 0 & 1 & 0 & 0 & 0 & 1 \end{array} \qquad (5.3\text{-}7)$$

Where con1 is the component represented by equation 5.3-6. Row reduction yields equation 5.3-5, which shows that the stoichiometric number matrix and conservation matrix are equivalent. The last column of equation 5.3-5 shows that there is a single reaction and that it agrees with equation 5.3-3. When coupling introduces additional conservation equations, components can be chosen in such a way that the conservation relations are all expressed in terms of conservations of reactants that are chosen as components. Thus equation 5.3-5 utilizes the five components glutamate, ATP, ammonia, ADP, and P_i.

Coupling does not necessarily involve constraints in addition to element balances. For example, glucokinase (EC 2.7.1.2) couples the hydrolysis of ATP to ADP with the phosphorylation of glucose to G6P. The reaction catalyzed is

$$\text{ATP} + \text{Glc} = \text{ADP} + \text{G6P} \qquad (5.3\text{-}8)$$

The apparent conservation matrix is

$$A' = \begin{array}{c} \\ \text{aden} \\ \text{glc} \\ \text{P} \end{array} \begin{array}{cccc} \text{ATP} & \text{Glc} & \text{ADP} & \text{G6P} \\ 1 & 0 & 1 & 0 \\ 0 & 1 & 0 & 1 \\ 3 & 0 & 2 & 2 \end{array} \qquad (5.3\text{-}9)$$

Note that here the reactions that are coupled share a reactant that is not a reactant in reaction 5.3-8. It is usually more convenient to count groups rather than atoms. Row reduction and use of the analogue of equation 5.1-19 show that reaction 5.3-8 is obtained.

An extreme example of additional constraints introduced by the enzymatic mechanism of a biochemical reaction is the NAD synthase (glutamine-hydrolyzing) reaction (EC 6.3.5.1) (Alberty, 1994b):

$$\text{ATP} + \text{deamido-NAD}_{\text{ox}} + \text{L-glutamine} + \text{H}_2\text{O}$$

$$= \text{AMP} + \text{PP}_i + \text{NAD}_{\text{ox}} + \text{L-glutamate} \qquad (5.3\text{-}10)$$

This reaction involves eight reactants, and so $C' = N' - R' = 8 - 1 = 7$, but only four elements are involved. So there are three additional constraints. Reaction 5.3-10 can be considered to be made up of the following two reactions:

$$ATP + H_2O = AMP + PP_i \qquad (5.3-11)$$

$$\text{deamido-NAD}_{ox} + \text{L-glutamine} = NAD_{ox} + \text{L-glutamate} \qquad (5.3-12)$$

The constraints in the urea cycle are discussed in Alberty (1997c).

When water is a reactant in a system of reactions, there is a sense in which oxygen atoms are not conserved because they can be brought into reactants or expelled from reactants without altering the activity of water in the solution. When water is not involved as a reactant in a system of reactions, this problem does not arise, and oxygen atoms are conserved in the reactants. When water is a reactant, there are problems with equilibrium calculations, as are discussed later in connection with the calculation of the equilibrium composition in the next chapter in Section 6.3. The problem encountered in using A' and v' matrices can be avoided by using a Legendre transform to define a further transformed Gibbs energy G'' that takes advantage of the fact that oxygen atoms are available to a reaction system from an essentially infinite reservoir when dilute aqueous solutions are considered at specified pH (Alberty, 2001a, 2002b).

■ 5.4 MATRIX FORMS OF THE FUNDAMENTAL EQUATIONS FOR CHEMICAL REACTION SYSTEMS

In treating systems of biochemical reactions it is convenient to use the fundamental equation for G' in matrix form (Alberty, 2000b). The extent of reaction ξ for a chemical reaction was discussed earlier in Section 2.1. For a system of chemical reactions, the extent of reaction vector ξ is defined by

$$n = n_0 + v\xi \qquad (5.4-1)$$

where n is the $N_s \times 1$ column matrix of amounts of species, n_0 is the $N_s \times 1$ column matrix of initial amounts of species, v is the $N_s \times R$ matrix of stoichiometric numbers, and ξ is the $R \times 1$ column vector of extents of the R independent reactions. The differential of the matrix for amounts of species is

$$dn = vd\xi \qquad (5.4-2)$$

Thus the fundamental equation for the Gibbs energy of a chemical reaction system can be written as

$$dG = -SdT + VdP + \mu dn \qquad (5.4-3)$$

where μ is the $1 \times N_s$ chemical potential matrix. We will see later that this equation can also be applied to phase equilibria (Chapter 8). Substituting equation 5.4-2 yields

$$dG = -SdT + VdP + \mu v d\xi \qquad (5.4-4)$$

This equation is useful for setting up the fundamental equation for consideration of a chemical reaction system described by a particular stoichiometric number matrix.

As an example of the usefulness of equation 5.4-4, consider the fumarase reaction (fumarate + H_2O = L-malate) in the range pH 5 to 9 where the chemical reactions are

$$\text{fum}^{2-} + H_2O = \text{mal}^{2-} \qquad (5.4-5)$$

$$\text{Hfum}^- = H^+ + \text{fum}^{2-} \qquad (5.4-6)$$

$$\text{Hmal}^- = H^+ + \text{mal}^{2-} \qquad (5.4-7)$$

The fundamental equation is

$$dG = -SdT + VdP + \mu(\text{fum}^{2-})dn(\text{fum}^{2-}) + \mu(\text{mal}^{2-})dn(\text{mal}^{2-})$$
$$+ \mu(\text{Hfum}^-)dn(\text{Hfum}^-) + \mu(\text{Hmal}^-)dn(\text{Hmal}^-) + \mu(\text{H}^+)dn(\text{H}^+)$$
$$+ \mu(\text{H}_2\text{O})dn(\text{H}_2\text{O}) \tag{5.4-8}$$

It is hard to divide the six chemical work terms into three terms for three chemical reactions, but this can be done using equation 5.4-4 with the following stoichiometric number matrix:

$$
\mathbf{v} =
\begin{array}{c}
 \\
\text{fum}^{2-} \\
\text{mal}^{2-} \\
\text{Hfum}^- \\
\text{Hmal}^- \\
\text{H}^+ \\
\text{H}_2\text{O}
\end{array}
\begin{array}{ccc}
\text{rx } 5.4\text{-}5 & \text{rx } 5.4\text{-}6 & \text{rx } 5.4\text{-}7 \\
-1 & 1 & 0 \\
1 & 0 & 1 \\
0 & -1 & 0 \\
0 & 0 & -1 \\
0 & 1 & 1 \\
1 & 0 & 0
\end{array}
\tag{5.4-9}
$$

The row matrix of chemical potentials of species is $\mathbf{\mu} = \{\{\mu(\text{fum}^{2-}), \mu(\text{mal}^{2-}), \mu(\text{Hfum}^-), \mu(\text{Hmal}^-), \mu(\text{H}^+), \mu(\text{H}_2\text{O})\}\}$, and the column matrix of extents of reaction is $\mathbf{\xi} = \{\{\xi_1\}, \{\xi_2\}, (\xi_3)\}$. Therefore the last term in equation 5.4-4 is given by

$$\mathbf{\mu v}d\mathbf{\xi} = (-\mu(\text{fum}^{2-}) - \mu(\text{H}_2\text{O}) + \mu(\text{mal}^{2-}))d\xi_1 + (\mu(\text{fum}^{2-}) + \mu(\text{H}^+)$$
$$- \mu(\text{Hfum}^-))d\xi\xi_2 + (\mu(\text{mal}^{2-}) + \mu(\text{H}^+) - \mu(\text{Hmal}^-))d\xi_3 \tag{5.4-10}$$

This shows how the fundamental equation for a system of chemical reactions can be written in terms of the three extents of reaction (see Problem 5.5.)

It is important to be able to write the fundamental equation for a system of chemical reactions in terms of components because components are involved in the criterion for spontaneous change and equilibrium. We have seen earlier (Section 2.3) that this is done by eliminating one chemical potential from the fundamental equation with each independent equilibrium condition of the form $\Sigma v_i \mu_i = 0$ to obtain

$$dG = -SdT + VdP + \sum_{i=1}^{c} \mu_{ci} dn_{ci} \tag{5.4-11}$$

where μ_{ci} is the chemical potential of the species that corresponds with component i. This equation can be written in terms of $\mathbf{\mu}_c$, which is the $1 \times C$ matrix of chemical potentials of components, and $\mathbf{n}_c$, which is the $C \times 1$ column matrix of amounts of components.

$$dG = -SdT + VdP + \mathbf{\mu}_c d\mathbf{n}_c \tag{5.4-12}$$

This one-phase system has $C + 2$ natural variables in agreement with $D = F + 1$, where $F = C - p + 2$ is the number of intensive degrees of freedom. Thus $D = C - 1 + 2 + 1 = C + 2$. Since the amounts of components are given by $\mathbf{n}_c = A\mathbf{n}$ (equation 5.1-2), the fundamental equation can also be written in the form

$$dG = -SdT + VdP + \mathbf{\mu}_c A d\mathbf{n} \tag{5.4-13}$$

where A is the transformation matrix that converts the matrix $\mathbf{n}$ into the matrix $\mathbf{n}_c$. This form of the fundamental equation is useful for setting up the fundamental equation for consideration of a reaction system described by a particular set of components.

Equation 5.4-12 indicates that the corresponding **Gibbs-Duhem equation** for a system of chemical reactions is

$$-SdT + VdP - (d\boldsymbol{\mu}_c)\boldsymbol{n}_c = 0 \qquad (5.4\text{-}14)$$

Because of this relation between the $C + 2$ intensive variables, the number of intensive degrees of freedom is $F - C + 1$.

■ 5.5 MATRIX FORMS OF THE FUNDAMENTAL EQUATIONS FOR BIOCHEMICAL REACTION SYSTEMS

For a biochemical reaction system at specified pH, equations 5.4-1 and 5.4-2 become

$$\boldsymbol{n'} = \boldsymbol{n'_0} + \boldsymbol{v'}\boldsymbol{\xi'} \qquad (5.5\text{-}1)$$

$$d\boldsymbol{n'} = \boldsymbol{v'}\,d\boldsymbol{\xi'} \qquad (5.5\text{-}2)$$

Therefore equation 4.2-3 can be written in matrix form:

$$dG' = -S'dT + VdP + \sum_{i=1}^{N'} \mu_i'\,dn_i' + RT\ln(10)\,n_c(\mathrm{H})d\mathrm{pH} \qquad (5.5\text{-}3)$$

$$= -S'\,dT + V\,dP + \boldsymbol{\mu'}\,d\boldsymbol{n'} + \boldsymbol{RT}\ln(10)\,n_c(\mathrm{H})d\mathrm{pH}$$

$$= -S'\,dT + V\,dP + \boldsymbol{\mu'}\boldsymbol{v'}\,d\boldsymbol{\xi'} + RT\ln(10)\,n_c(\mathrm{H})d\mathrm{pH}$$

The primes on the amounts are needed to indicate that they are amounts of reactants, which are sums of species that are pseudoisomers at specified pH. The primes on the stoichiometric number matrices and extents of reaction column matrices are needed to indicate that these matrices are for biochemical reactions written in terms reactants (sums of species). The primes are needed on the transformed chemical potentials to distinguish them from chemical potentials of species.

The biochemical analogues of equations 5.4-11, 5.4-12, and 5.4-13 are

$$dG' = -S'\,dT + V\,dP + \sum_{i=1}^{C'} \mu_{ci}'\,dn_{ci}' + RT\ln(10)\,n_c(\mathrm{H})d\mathrm{pH} \qquad (5.5\text{-}4)$$

$$= -S'\,dT + V\,dP + \boldsymbol{\mu'_c}\,d\boldsymbol{n'_c} + RT\ln(10)\,n_c(\mathrm{H})d\mathrm{pH}$$

$$= -S'dT + V\,dP + \boldsymbol{\mu'_c}\boldsymbol{A'}d\boldsymbol{n'} + RT\ln(10)\,n_c(\mathrm{H})d\mathrm{pH}$$

The prime on the amount of a component indicates that these are the components other than the hydrogen component. The corresponding Gibbs-Duhem equation is

$$-S'\,dT + V\,dP - (d\boldsymbol{\mu'_c})\boldsymbol{n'_c} + RT\ln(10)\,n_c(\mathrm{H})d\mathrm{pH} = 0 \qquad (5.5\text{-}5)$$

Since the thermodynamics of a biochemical reaction system is considered at specific pH, we need to consider equation 5.5-4 in the form

$$(dG')_{\mathrm{pH}} = -S'\,dT + V\,dP + \boldsymbol{\mu'_c}\,d\boldsymbol{n'_c} \qquad (5.5\text{-}6)$$

and equation 5.5-5 in the form

$$-S'\,dT + V\,dP - (d\,\boldsymbol{\mu'_c})\boldsymbol{n'_c} = 0 \qquad (5.5\text{-}7)$$

These equations look like equations 5.4-13 and 5.4-14, where C' components play the role of C components in equations 5.4-13 and 5.4-14.

The number D' of natural variables for a system and the number F' of intensive degrees of freedom for a one-phase system at equilibrium were discussed in Section 4.6, but now we can discuss these numbers in a more general way. Table 5.1 gives these numbers for three descriptions of a one-phase reaction

Table 5.1 Numbers of Natural Variables and Numbers of Intensive Degrees of Freedom for One-Phase Reaction Systems at Equilibrium

	Numbers of Natural Variables	Numbers of Intensive Degrees of Freedom
Chemical reaction system	$D = C + 2$	$F = C + 1$
Biochemical reaction system	$D' = C' + 3$	$F' = C' + 2$
	$= C + 2$	$= C + 1$
Biochemical reaction system after specification of pH	$D' = C' + 2$	$F' = C' + 1$
	$= C + 1$	$= C$

system at equilibrium. The first line of the table describes a chemical reaction system at equilibrium (see equations 5.4-13 and 5.4-14). The second line of the table describes a biochemical reactin system at equilibrium before the pH is held constant (see equations 5.5-4 and 5.4-5). Since $C' = C - 1$, the number of natural variables and intensive degrees of freedom are not changed in making the Legendre transform and separating out a term in dpH. However, after the specification that the pH is constant, the third line of the table shows that the number D' of natural variables and the number F' of intensive degrees of freedom have each been reduced by one. Since $C' = C - 1$, the number of extensive degrees of freedom is reduced to C by holding the pH constant. In the next chapter we will see that it may be useful to make further Legendre transforms, but the maximum number of these further Legendre transforms is $C - 1$, because at least one component must remain.

When the chemical potentials of several species are held constant, it may be useful to write the Legendre transform in terms of matrices. For example, when the chemical potentials of several species are specified, the Legendre transform can be written as

$$G' = G - \mu_c N_c n \qquad (5.5\text{-}8)$$

where μ_c is a $1 \times C$ row matrix of the components for which the chemical potentials are specified. The number of components for which chemical potentials are specified has to be at least one less than the number of components in the reaction system. N_c is a $C \times N_s$ matrix of the numbers of specified components in the N_s species. This N_c matrix is like the conservation matrix A, except that the rows correspond to the specified components, and not all of the components. The $N_s \times 1$ column matrix n gives the amounts of all species in the system. The differential of G' is given by

$$dG' = dG - \mu_c N_c dn - (d\mu_c)N_c n \qquad (5.5\text{-}9)$$

Equation 4.1-6 can be written

$$dG = -S\,dT + V\,dP + \mu'_{nc}\,dn_{nc} + \mu_c dn_c \qquad (5.5\text{-}10)$$

where $\mu'_{nc}dn_{nc}$ is the term for noncomponents after the chemical potentials μ_c have been specified for certain components. Substituting equation 5.5-11 in 5.5-10 yields

$$dG' = -SdT + V\,dP + \mu'_{nc}\,dn_{nc} - (d\mu_c)N_c n \qquad (5.5\text{-}11)$$

since $dn_c = N_c dn$ (see equation 5.1-2). The transformed chemical potentials μ'_{nc} of noncomponents are given by

$$\mu'_{nc} = \mu'_{nc} - \mu_c N_c \qquad (5.5\text{-}12)$$

Note that μ_c is $1 \times C$ and N_c is $C \times N_{nc}$.

■ 5.6 OPERATIONS OF LINEAR ALGEBRA

Although matrix multiplications, row reductions, and calculation of null spaces can be done by hand for small matrices, a computer with programs for linear algebra are needed for large matrices. *Mathematica* is very convenient for this purpose. More information about the operations of linear algebra can be obtained from textbooks (Strang, 1988), but this section provides a brief introduction to making calculations with *Mathematica* (Wolfram, 1999).

Matrix multiplication. The first dimension of a matrix is the number of rows, and the second dimension of a matrix is the number of columns. In order to multiply matrix *b* by matrix *a*, it is necessary that the second dimension of matrix *a* be the same as the first dimension of matrix *b*. The product *ab* has the first dimension of *a* and the second dimension of *b*. In *Mathematica* this product is calculated by putting a period between *a* and *b* or by using Dot[*a*, *b*].

Gaussian reduction. The rows of a matrix can be multiplied by integers and be added and subtracted to produce zero elements. This can be done to obtain the matrix in row-reduced canonical form in which there is a identity matrix on the left. An identity matrix is a square matrix of zeros with ones along the diagonal. In *Mathematica* the row-reduced canonical form of *a* is obtained by using RowReduce[*a*]. If, after row reduction, one of the rows is made up of zeros, one of the rows is not independent, and should be deleted. If two matrices have the same row-reduced form, they are equivalent. We say that a matrix is not unique because it can be written in different forms that are equivalent.

Null space. If the product of two matrices is a zero matrix (all zeros), *ax* = **0** is said to be a homogeneous equation. The matrix *x* is said to be the null space of *a*. In *Mathematica* a basis for the null space of *a* can be calculated by use of NullSpace[*a*]. There is a degree of arbitrariness in the null space in that it provides a basis, and alternative forms can be calculated from it, that are equivalent. See Equation 5.1-19 for a method to calculate a basis for the null space by hand. When a basis for the null space of a matrix needs to be compared with another matrix of the same dimensions, they are both row reduced. If the two matrices have the same row-reduced form, they are equivalent.

Solution of linear equations. A set of linear equations is represented by *ax* = *b*. The solution *x* can be obtained in *Mathematica* by use of LinearSolve[*a*, *b*]. Matrix *a* can be square or rectangular.

Transposition. In *Mathematica* the Transpose[*a*] transposes the first two levels of *a*. Equations 5.1-14 and 5.1-23 give a matrix and its transpose.

Transformation matrix. When the conservation matrix *a* for a system is written in terms of elemental compositions, the elements are used as components. But we can change the choice of components (change the basis) by making a matrix multiplication that does not change the row-reduced form of the *a* matrix or its null space. Since components are really coordinates, we can shift to a new coordinate system by multiplying by the inverse of the transformation matrix between the two coordinate systems. A new choice of components can be made by use of a component transformation matrix *m*, which gives the composition of the new components (columns) in terms of the old components (rows). The following matrix multiplication yields a new a matrix in terms of the new components.

$$a(\text{new}) = m^{-1}a(\text{old}) \tag{5.5-1}$$

The inverse of the transformation matrix *m* is represented by m^{-1}. In *Mathematica*, the inverse of *m* is calculated with Inverse[*m*].

Systems of Biochemical Reactions

Systems of biochemical reactions like glycolysis, the citric acid cycle, and larger and smaller sequential and cyclic sets of enzyme-catalyzed reactions present challenges to make calculations and to obtain an overview. The calculations of equilibrium compositions for these systems of reactions are different from equilibrium calculations on chemical reactions because additional constraints, which arise from the enzyme mechanisms, must be taken into account. These additional constraints are taken into account when the stoichiometric number matrix is used in the equilibrium calculation via the program equcalcrx, but they must be explicitly written out when the conservation matrix is used with the program equcalcc. The stoichiometric number matrix for a system of reactions can also be used to calculate net reactions and pathways.

Since concentrations of ATP, ADP, NAD$_{ox}$, and NAD$_{red}$ may be in steady states, it is of interest to calculate equilibrium compositions that correspond with these steady state concentrations. These calculations are referred to as level 3 equilibrium calculations because they are based on the introduction of [ATP], [ADP], and the like, as natural variables by use of a Legendre transform.

■ 6.1 CALCULATION OF NET REACTIONS USING MATRIX MULTIPLICATION

In dealing with large systems of biochemical reactions it is important to find ways to obtain a more global view, and one way is to use net reactions because they show what is accomplished by a set of reactions. In calculating a net reaction, the intermediates are eliminated because they are produced and consumed in equal amounts. In order to prevent accumulation of intermediates, some of the reactions in the set have to run at 2, 3,... times the rates of other reactions. These integers are referred to as the **stoichiometric numbers of steps**, which are represented by s_i' for step i. The **pathway matrix** (column vector) for a set of biochemical reactions is represented by s'. The pathway matrix is $R' \times 1$. The advantage of using matrices is that linear algebra and computers can be used.

The relation between a stoichiometric number matrix v' for a set of R' reactions involving N' reactants and the **stoichiometric number matrix v_{net}' for a net reaction** is a system of linear equations that is represented by the following matrix multiplication (Alberty, 1996):

$$
\begin{array}{ccccc}
v_{11}' & v_{12}' & \cdots & v_{1R}' & s_1' \\
v_{21}' & v_{22}' & \cdots & v_{2R}' & s_2' \\
\cdots & \cdots & \cdots & \cdots & \cdots \\
\cdots & \cdots & \cdots & \cdots & s_R' \\
v_{N1}' & v_{N2}' & \cdots\cdots & v_{NR}' &
\end{array}
=
\begin{array}{c}
v_{net1}' \\
v_{net2}' \\
\cdots \\
\\
v_{netN}'
\end{array}
\qquad (6.1\text{-}1)
$$

Equation 6.1-1 can be written in the form

$$
s_1'
\begin{array}{c}
v_{11}' \\ v_{21}' \\ \cdots \\ v_{N1}'
\end{array}
+ s_2'
\begin{array}{c}
v_{12}' \\ v_{22}' \\ \cdots \\ v_{N2}'
\end{array}
+ \cdots + s_R'
\begin{array}{c}
v_{1R}' \\ v_{2R}' \\ \cdots \\ v_{NR}'
\end{array}
=
\begin{array}{c}
v_{net1}' \\ v_{net2}' \\ \cdots \\ v_{netN}'
\end{array}
\qquad (6.1\text{-}2)
$$

This shows that the solution s' to the system of linear equations represented by equation 6.1-1 is made up of the stoichiometric numbers s_i' that give the number of times the various biochemical reactions have to occur to accomplish the net reaction. Equation 6.1-1 is conveniently written in matrix notation as

$$ v' \cdot s' = v_{net}' \qquad (6.1\text{-}3) $$

The stoichiometric number matrix v' for the system is $N' \times R'$, the pathway matrix s' is $R' \times 1$, and the stoichiometric number matrix v_{net}' for the net reaction is $N' \times 1$. When the pH (and perhaps the free concentrations of cations that are bound reversibly) is specified, a prime is used on the symbols in equation 6.1-3 to distinguish the stoichiometric numbers of the biochemical reactions from those of the underlying chemical reactions. Since it is easy to make errors in typing a stoichiometric number matrix into a computer, it is useful to check the matrix by using it to print out the reactions. This can be done using the programs mkeqn (Alberty, 1996a) and nameMatrix (Alberty, 2000c), which are given in Problem 6.1. The use of these programs is illustrated in Problem 6.1.

The net reaction for the 10 steps of **glycolysis** is

$$ \text{glucose} + 2P_i + 2ADP + 2NAD_{ox} = 2\text{pyruvate} + 2ATP + 2NAD_{red} + 2H_2O \qquad (6.1\text{-}4) $$

This net reaction is obtained by multiplying the first five reactions of glycolysis by 1, the second five reactions by 2, and adding. This causes the intermediates to cancel. Alternatively, this net reaction can be calculated by multiplying the stoichiometric number matrix v' for the 10 reactions of glycolysis by the pathway matrix s', where $(s')^T = \{1, 1, 1, 1, 1, 2, 2, 2, 2, 2\}$, according to equation 6.1-3.

A matrix multiplication can be used to calculate the $\Delta_r G'^0$ values for a series of reactions from a vector of $\Delta_f G'^0$ values for the reactants involved.

$$[\Delta_f G_1'^0, \Delta_f G_2'^0, \ldots, \Delta_f G_N'^0] \cdot \mathbf{v}' = [\Delta_r G_1'^0, \Delta_r G_2'^0, \ldots, \Delta_r G_R'^0] \qquad (6.1\text{-}5)$$

(Note that $(1 \times N')(N' \times R') = 1 \times R'$ and see Problem 6.2.) The $\Delta_r G_{net}'^0$ for a particular path s' is obtained by multiplying both sides of this equation by s':

$$[\Delta_f G_1'^0, \Delta_f G_2'^0, \ldots, \Delta_f G_N'^0] \cdot \mathbf{v}' \cdot s' = \Delta_r G_{net}'^0 \qquad (6.1\text{-}6)$$

Note the first matrix is $1 \times N'$, the second is $N' \times R'$, and the third is $R' \times 1$, and so the result is 1×1.

■ 6.2 CALCULATION OF PATHWAYS BY SOLVING LINEAR EQUATIONS

Since equation 6.1-2 represents a set of linear equations, the path can be calculated from the stoichiometric number matrix and a particular net reaction by solving the set of linear equations (Alberty,1996a). In *Mathematica* this can be done with LinearSolve:

$$s' = \text{LinearSolve}[\mathbf{v}', \mathbf{v}_{net}'] \qquad (6.2\text{-}1)$$

This calculation can be made for chemical reactions, biochemical reactions at specified pH, or at steady state concentrations of reactants like ATP and ADP, as is discussed in Section 6.6. The advantage of the matrix formulation of this calculation is that very large matrices can be handled.

Problem 6.2 illustrates the use of equation 6.2-1 by applying it to four net reactions that represent the oxidation of glucose to carbon dioxide and water: (1) the net reaction for glycolysis, (2) the net reaction catalyzed by the pyruvate dehydrogenase complex, (3) the net reaction for the citric acid cycle, and (4) the net reaction for oxidative phosphorylation. The $\mathbf{v}'$ in equation 6.2-1 is the apparent stoichiometric number matrix for these four reactions. The net reaction is

$$\text{glucose} + 6O_2 + 40ATP + 40P_i = 46H_2O + 6CO_2 + 40ATP \qquad (6.2\text{-}2)$$

The nunet in equation 6.1-1 is the list of stoichiometric coefficients for this reaction for the order of the reactants in $\mathbf{v}'$. The use of equation 6.2-1 yields the following path: $\{1, 2, 2, 12\}$. This means that reaction 1 has to occur once, reaction 2 has to occur twice, reaction 3 has to occur twice, and reaction 4 has to occur 12 times in order to oxidize a mole of glucose to carbon dioxide and water.

■ 6.3 USE OF A LEGENDRE TRANSFORM FOR REACTIONS INVOLVING WATER AS A REACTANT

When water is a reactant, the calculation of K' using $\Delta_r G'^0 = -RT \ln K'$ is based on the convention that $\Delta_f G'^0(H_2O)$ is involved in calculating $\Delta_r G'^0$, but that the activity of H_2O in the expression for K' is taken to be unity. This is a practical convention in treating a single reaction, but in treating systems of reactions, it is almost a necessity to use matrices, linear algebra, and a computer. Linear algebra can be used to convert sets of stoichiometric equations to sets of conservation equations, and vice versa, but these operations are incompatible when H_2O is a reactant and the convention that $a(H_2O)$ is equal to unity is used (Alberty, 2001a, 2002b). In considering systems of reactions, it is advantageous to use apparent conservation matrices and apparent stoichiometric matrices that are interconvertible using the operations of linear algebra. In dilute aqueous solutions, the solvent provides an essentially infinite reservoir for H_2O, and so a Legendre transform can be used to define a **further transformed Gibbs energy** G'' that provides the criterion for spontaneous change and equilibrium at specified pH and $a(H_2O) = 1$. When this is done, the apparent conservation matrix A'' that does not include the conservation of H_2O becomes consistent with the apparent stoichiometric number matrix v'' that does not include the stoichiometric number for H_2O.

The Legendre transform that defines the further transformed Gibbs energy G'', which provides the criterion for spontaneous change and equilibrium in dilute aqueous solutions, is

$$G'' = G' - n_c(O)\mu'^0(H_2O) \tag{6.3-1}$$

The amount of the oxygen component in the system is given by $n_c(O) = \Sigma N_O(i)n_i$, where $N_O(i)$ is the number of oxygen atoms in reactant i. $\mu'^0(H_2O)$ is the standard transformed chemical potential for H_2O at the specified pH and ionic strength. The standard further transformed Gibbs energy of formation of reactant i is given by

$$\Delta_f G''^0_i = \Delta_f G'^0_i - N_O(i)\Delta_f G'^0(H_2O) \tag{6.3-2}$$

where $\Delta_f G'^0(H_2O)$ is given by equation 4.4-10. Note that $\Delta_f G''^0(H_2O) = 0$. When this adjustment of the standard transformed Gibbs energy of formation of reactant i is made, this reactant becomes a pseudoisomer of other reactants that differ from it only with respect to the number of oxygen atoms they contain, and so the standard further transformed Gibbs energy of formation of the pseudoisomer group has to be calculated using the analogue of equation 4.5-1. The apparent equilibrium constant K'' for a biochemical reaction at specified pH and $a(H_2O) = 1$ is given by

$$\Delta_r G''^0 = -RT \ln K'' = \sum v''_i \Delta_f G''^0_i \tag{6.3-3}$$

There is no term for H_2O in the summation. When the pH is specified and $a(H_2O) = 1$, the criterion for spontaneous change and equilibrium is $dG'' \leqslant 0$ at specified T, P, pH, $a(H_2O) = 1$, and amounts of apparent components. Note that oxygen is no longer a component.

Thus the inconsistency between A' and v'' is eliminated by using A'' and v''. The number C'' of apparent components can be determined by row reduction of A'' since $C'' = \text{rank } A''$. The number R'' of independent reactions can be determined by row reduction of v'' because $R'' = \text{rank } v''$. Note that $N'' = C'' + R''$. These two types of matrices can be interconverted by use of

$$A''v'' = 0 \quad \text{and} \quad (v'')^T(A'')^T = 0 \tag{6.3-4}$$

The apparent stoichiometric number matrix v'' can be obtained from the row-reduced form of A'' by use of the analogue of equation 5.1-19 or by calculating a basis for the null space using a computer program.

Further transformed Gibbs energies of formation are especially useful in calculating equilibrium compositions by computer programs that accept conservation matrices and vectors of initial amounts, as discussed in the next section.

■ 6.4 CALCULATIONS OF EQUILIBRIUM COMPOSITIONS FOR SYSTEMS OF BIOCHEMICAL REACTIONS

One of the important things that thermodynamics can tell us about a system of reactions is the composition at equilibrium for given initial amounts of reactants. For a single reaction there is an analytic solution for this problem, but for a system consisting of two or more reactions, an iteration using the Newton-Raphson method is required to find the composition of the system that yields the lowest possible transformed Gibbs energy, given the conservation equations and equilibrium expressions. Computer programs for doing that were written by

Krambeck (1978, 1991) in APL and in *Mathematica* and by Smith and Missen (1982) in *Fortran*. Krambeck wrote the program equcalc for use on gaseous mixtures involved in petroleum processing and also adapted it to solution reactions as equcalcc and equcalcrx. These latter programs, which are given in BasicBiochemData2, have the advantage that they operate with conservation matrices and stoichiometric number matrices, respectively, so that they can be used with systems of any size.

Equcalcc was written to calculate compositions of reaction systems in terms of species, given the conservation matrix A, the vector of standard Gibbs energies of formation, and initial amounts of species, but since the fundamental equation for G' at specified pH has the same form as the fundamental equation for G, equcalcc can be used with the apparent conservation matrix A', the vector of standard transformed Gibbs energies of formation of pseudoisomer groups, and initial amounts of pseudoisomer groups. However, there is a problem when H_2O is a reactant, as was discussed in the preceding section. When equcalcc is used for reactions involving water as a reactant, further transformed Gibbs energies of formation (based on equation 6.3-2) and conservation matrices omitting oxygen have to be used.

Equcalcrx was written to calculate the equilibrium composition in terms of species, given the stoichiometric number matrix, but it can be used with the apparent stoichiometric number matrix at a specified pH. Apparent equilibrium constants have to be known for a set of independent biochemical reactions for the system. This program has the advantage over equcalcc that further transformed Gibbs energies of formation (based on equation 6.3-2) do not have to be calculated when water is involved as a reactant. Actually equcalcrx obtains A' by calculating the null space of $(v')^T$ and then using equcalcc. Although equcalcc and equcalcrx appear to require the vector of initial amounts, it is really only the vector of initial amounts of components that is used.

The inputs for these programs are designated by the following terminology:

1. When equcalcc[as,lnk,no] is applied to a system of R independent chemical reactions, it requires a $C \times N$ conservation matrix as, a list lnk of standard Gibbs energies of formation of species multiplied by $(-1/RT)$, and a list no of the initial concentrations of species. It can be used at specified pH by using a $C' \times N'$ conservation matrix as, a list of standard transformed Gibbs energies of pseudoisomer groups multiplied by $(-1/RT)$, and a list no of initial concentrations of pseudoisomer groups.

2. When equcalcrx[nt,lnkr,no] is applied to a system of R independent chemical reactions, it requires a $R \times N$ transposed stoichiometric number matrix nt, a vector of natural logarithms of the equilibrium constants of independent reactions, and a vector no of the initial concentrations. It can be used at a specified pH by using a $R' \times N'$ transposed stoichiometric number matrix nt, a vector lnkr of natural logarithms of the apparent equilibrium constants of independent biochemical reactions, and a vector no of the initial concentrations.

The use of these programs is illustrated in Problems 6.4 to 6.8.

The calculation of the equilibrium composition of a system of chemical reactions with equcalcc is based on minimizing the Gibbs energy subject to the conservation condition $An = n_c$. This is accomplished by using a **Lagrangian L** defined by

$$L = G - \lambda(An - n_c) = (\mu - \lambda A)n + \lambda n_c \qquad (6.4\text{-}1)$$

where λ is the $1 \times C$ vector of **Lagrange multipliers**. At equilibrium the rates of change of L with respect to the amounts of each of the species must be equal to zero. Thus at equilibrium,

$$\mu = \lambda A \qquad (6.4\text{-}2)$$

This is equivalent to the equilibrium condition $\boldsymbol{\mu v} = 0$, and so the objective is to calculate the Lagrange multipliers. The number of Lagrange multipliers is equal to the number of components. Once the Lagrange multipliers have been obtained, the chemical potentials can be calculated using equation 5.3-2 and the equilibrium mole fractions can be calculated using

$$x = \exp\left[\frac{\lambda A - \boldsymbol{\mu}^*}{RT}\right] \tag{6.4-3}$$

where the elements of $\boldsymbol{\mu}^*$

$$\mu_j^* = \mu_j^0 + RT \ln[B_j] \tag{6.4-4}$$

are the chemical potentials of the species.

■ 6.5 THREE LEVELS OF CALCULATIONS OF COMPOSITIONS FOR SYSTEMS OF BIOCHEMICAL REACTIONS

In Chapter 4 we saw how specifying the pH and using the transformed Gibbs energy G' provides a more global view of a biochemical reaction. This process of making Legendre transforms can be continued by specifying the concentrations of coenzymes like ATP, ADP, NAD_{ox}, and NAD_{red} (Alberty,1993c, 2000b, c, 2002a). These reactants are produced and consumed by many reactions, and so, in a living cell, their concentrations are in steady states. Thus the thermodynamics of a system of enzyme-catalyzed reactions can be discussed at three levels. Description of a reaction system in terms of species is referred to as level 1, discussion in terms of reactants (sums of species) at specified pH (e.g., or specified pH and pMg) is referred to as level 2, and discussion in terms of reactants at specified steady state concentrations of coenzymes is referred to as level 3. For the purpose of derivations we can imagine that the system at level 3 is connected with reservoirs of coenzymes at their steady state concentrations by means of semiper-meable membranes. The maximum number of components that can be specified in this way is one less than the number of components in the system. Table 6.1 shows the criteria for spontaneous change and equilibrium for various specifica-tion of independent variables.

When the concentrations of ATP and ADP are in a steady state, these concentrations can be made natural variables by use of a Legendre transform that defines a **further transformed Gibbs energy** G'' as follows.

$$G'' = G' - n'_c(\text{ATP})\mu'(\text{ATP}) - n'_c(\text{ADP})\mu'(\text{ADP}) \tag{6.5-1}$$

In this equation $n'_c\text{ATP})$ is the amount of the ATP component in the system, that is, the total amount of ATP free and bound. Thus $n'_c(\text{ATP})$ and $\mu'(\text{ATP})$ are conjugate variables, and $n'_c(\text{ADP})$ and $\mu'(\text{ADP})$ are conjugate variables. It may seem remarkable that ATP and ADP at a specified pH can be taken as components, but any group or combination of atoms can be taken as a

Table 6.1   Levels of Thermodynamic Treatment

Level	Independent Variables	Criterion for spontaneous change and equilibrium at constant independent variables
1	$T, P, \{n_c\}$	$dG \leqslant 0$
2	$T, P, \text{pH}, \{n'_c\}$	$dG' \leqslant 0$
3	$T, P, \text{pH}, [\text{ATP}], [\text{ADP}], \ldots, \{n''_c\}$	$dG'' \leqslant 0$

component. As mentioned in the discussion of matrices in Section 5.1, the reactants listed first in the conservation matrix will be taken as components if they are sufficiently different. Since $G' = \Sigma \mu_i' n_i'$, $G'' = \Sigma \mu_i'' n_i''$, $n_c'(ATP) = \Sigma N_{ATP}(i) n_i'$, and $n_c'(ADP) = \Sigma N_{ADP}(i) n_i'$, equation 6.5-1 shows that the further transformed chemical potential of reactant i (sum of species) is given by

$$\mu_i'' - \mu_i' - N_{ATP}(i)\mu'(ATP) \quad N_{ADP}(i)\mu'(ADP) \tag{6.5-2}$$

where $N_{ATP}(i)$ and $N_{ADP}(i)$ are the numbers of ATP and ADP molecules required to make up the ith reactant; note that these numbers may be positive or negative. The values of $N_{ATP}(i)$ and $N_{ADP}(i)$ can be obtained from the apparent conservation matrix (see equation 6.5-21).

The differential of the Legendre transform in equation 6.5-1 is

$$dG'' = dG' - n_c'(ATP)d\mu'(ATP)) - \mu'(ATP)dn_c'(ATP) - n_c'(ADP)d\mu'(ADP)$$
$$- \mu'(ADP)dn_c'(ADP) \tag{6.5-3}$$

The general equation for dG' at a given pH is equation 4.1-18. This equation can be written in terms of the amounts of components $n_c'(ATP)$ and $n_c'(ADP)$ by use of equation 6.5-2. This yields

$$dG' = -S'dT + VdP + \sum_{i=1}^{N'-2} \mu_i'' dn_i' + \mu'(ATP)dn_c'(ATP) + \mu'(ADP)dn_c'(ADP)$$
$$+ n_c(H)RT\ln(10)dpH \tag{6.5-4}$$

where $dn_c'(ATP) = \Sigma N_{ATP}(i)dn_i'$ and $dn_c'(ADP) = \Sigma N_{ADP}(i)dn_i'$. Substituting equation 6.5-4 into equation 6.5-3 yields the following fundamental equation for G'':

$$dG' = -S'dT + VdP + \sum_{i=1}^{N'-2} \mu_i'' dn_i' - n_c'(ATP)d\mu'(ATP) + n_c'(ADP)d\mu'(ADP)$$
$$+ n_c(H)RT\ln(10)dpH \tag{6.5-5}$$

However, $\mu'(ATP)$ and $\mu'(ADP)$ are not convenient independent variables because they depend on temperature as well as concentration. To eliminate $d\mu'(ATP)$ and $d\mu'(ADP)$ from equation 6.5-5, the following equations are used:

$$d\mu'(ATP) = \left[\frac{\partial\mu'(ATP)}{\partial T}\right]_{P,pH,[ATP]} dT + \left[\frac{\partial\mu'(ATP)}{\partial[ATP]}\right]_{T,P,pH} d[ATP] \tag{6.5-6}$$

$$d\mu'(ADP) = \left[\frac{\partial\mu'(ADP)}{\partial T}\right]_{P,pH,[ADP]} dT + \left[\frac{\partial\mu'(ADP)}{\partial[ADP]}\right]_{T,P,pH} d[ADP] \tag{6.5-7}$$

The derivatives in the first terms of these equations are $-S_m'(ATP)$ and $-S_m'(ADP)$, and the derivatives in the second terms are calculated using $\mu'(ATP) = \mu'^0(ATP) + RT \ln[ATP]$. Thus

$$d\mu'(ATP) = -S_m'(ATP)dT + RT \, d\ln[ATP] \tag{6.5-8}$$

$$d\mu'(ADP) = -S_m'(ADP)dT + RT \, d\ln[ADP] \tag{6.5-9}$$

Substituting these equations in equation 6.5-5 yields

$$dG'' = -S''dT + VdP + \sum_{i=1}^{N''} \mu_i'' dn_i'' + RT\ln(10)n_c(H)dpH$$
$$- n_c'(ATP)RT \, d\ln[ATP] - n_c'(ADP)RT \, d\ln[ADP] \tag{6.5-10}$$

It can be shown that when the reactions of ATP and ADP with reactants in the system are at equilibrium, the further transformed chemical potentials of some of the reactants are equal; these reactants form a pseudoisomer group with amount n_i''. Thus holding [ATP] and [ADP] constant makes it possible to reconceptualize the system into a smaller set of pseudoisomer groups; specifically, the number of pseudo isomer groups is reduced from $N' + 2$ to N''. A double prime is used on the amounts in the summation to indicate that these are amounts of pseudoisomer groups at specified [ATP] and [ADP] as well as pH. The further transformed entropy of the system is given by

$$S'' = S' - n'_c(\text{ATP})S'_m(\text{ATP}) - n'_c(\text{ADP})S'_m(\text{ADP}) \qquad (6.5\text{-}11)$$

where $S'_m(\text{ATP})$ is the molar transformed entropy of ATP at the specified pH and ionic strength. There is also a further transformed enthalpy given by $H'' = G'' + TS''$.

The fundamental equation for G'' given in equation 6.5-10 leads to several new types of relations between properties. First consider the equation for dG'' for a system containing a single pseudoisomer group; that is, the summation is replaced with $\Delta_f G''_{iso} dn''_{iso}$. The Maxwell equation between this term and the term in dpH is

$$\left(\frac{\partial \Delta_f G''_{iso}}{\partial \text{pH}}\right)_{T,P,n''_{iso},[\text{ATP}],[\text{ADP}]} = RT\ln(10)\bar{N}_H \qquad (6.5\text{-}12)$$

where $n_c(\text{H})$ is replaced by $\bar{N}_H$, which is a more useful symbol for the average binding of hydrogen atoms by the pseudoisomer group containing different reactants. This equation is like equation 4.7-3. It gives the average binding of hydrogen atoms by the pseudoisomer group as a function of the pH at specified concentrations of ATP and ADP.

The Maxwell equations between $\Delta_f G''_{iso} dn''_{iso}$ and the terms in dln[ATP] and dln[ADP] are

$$\left(\frac{\partial \Delta_f G''_{iso}}{\partial \ln[\text{ATP}]}\right)_{T,P,n''_{iso},\text{pH},[\text{ADP}]} = -RT\,\bar{N}_{\text{ATP}} \qquad (6.5\text{-}13)$$

$$\left(\frac{\partial \Delta_f G''_{iso}}{\partial \ln[\text{ADP}]}\right)_{T,P,n''_{iso},\text{pH},[\text{ATP}]} = -RT\,\bar{N}_{\text{ADP}} \qquad (6.5\text{-}14)$$

In these equations $\bar{N}_{\text{ATP}}$ is equal to the rate of change of $n_c(\text{ATP})$ with respect to the amount of the pseudoisomer group, and $\bar{N}_{\text{ADP}}$ is equal to the rate of change of $n_c(\text{ADP})$ with respect to the amount of the pseudoisomer group. Thus these equations give $\bar{N}_{\text{ATP}}$ and $\bar{N}_{\text{ADP}}$ as functions of the independent variables. The binding of a component can be negative, as pointed out in Section 3.3.

Taking the derivative of equation 6.5-13 with respect to ln[ADP] yields the same result as taking the derivative of equation 6.5-14 with respect to ln[ATP]; therefore,

$$\left(\frac{\partial \bar{N}_{\text{ATP}}}{\partial \ln[\text{ADP}]}\right)_{T,P,n_{iso},\text{pH},[\text{ATP}]} = \left(\frac{\partial \bar{N}_{\text{ADP}}}{\partial \ln[\text{ATP}]}\right)_{T,P,n_{iso},\text{pH},[\text{ADP}]} \qquad (6.5\text{-}15)$$

We have seen this type of reciprocal relation twice earlier: see equation 1.3-17 and equation 4.8-9. There are also reciprocal relations between the binding of ATP and hydrogen ions and between ADP and hydrogen ions.

The complete Legendre transform for the system we are discussing yields the Gibbs-Duhem equation for the system:

$$0 = -S''\,dT + V\,dP + \sum_{i=1}^{C''} n''_{ci}\,d\mu''_i + RT\ln(10)n_c(\text{H})\text{dpH} - n'_c(\text{ATP})RT\,d\ln[\text{ATP}]$$

$$- n'_c(\text{ADP})RT\,d\ln[\text{ADP}] \qquad (6.5\text{-}16)$$

This relation between the $C'' + 5$ intensive properties of the system shows that the number of independent intensive degrees of freedom is $F'' = C'' + 4$. Since this is a one-phase system, the total number of degrees of freedom is $D'' = C'' + 5$.

The expressions for apparent equilibrium constants K'' are written in terms of concentrations of the N'' pseudoisomer groups; thus [ATP] and [ADP] do not appear explicitly in equilibrium constant expressions for the system. The criterion for spontaneous change and equilibrium is $dG'' \leqslant 0$ at specified T, P, pH, [ATP], [ADP], and amounts of remaining components.

To see the effects of specifying [ATP] and [ADP], consider the first three reactions of glycolysis:

$$\text{Glucose} + \text{ATP} = \text{glucose 6-phosphate} + \text{ADP} \qquad (6.5\text{-}17)$$

$$\text{Glucose 6-phosphate} = \text{fructose 6-phosphate} \qquad (6.5\text{-}18)$$

$$\text{Fructose 6-phosphate} + \text{ATP} = \text{fructose 1,6-biphosphate} + \text{ADP} \qquad (6.5\text{-}19)$$

These three biochemical reactions are catalyzed by hexokinase (EC 2.7.1.1), glucose-6 phosphate isomerase (EC 5.3.1.9), and 6-phosphofructokinase (EC 2.7.1.11), respectively. The EC numbers are from *Enzyme Nomenclature* (Webb, 1992). The first step is to write the conservation matrix for this reaction system at specified pH because that will show how to calculate the further transformed Gibbs energies of formation at specified [ATP] and [ADP].

At specified pH the apparent conservation matrix for this system is

$$
A' = \begin{array}{c} \\ \text{Aden} \\ \text{P} \\ \text{Glu} \end{array}
\begin{array}{cccccc}
\text{ATP} & \text{ADP} & \text{Glu} & \text{G6P} & \text{F6P} & \text{F16BP} \\
1 & 1 & 0 & 0 & 0 & 0 \\
3 & 2 & 0 & 1 & 1 & 2 \\
0 & 0 & 1 & 1 & 1 & 1
\end{array}
\qquad (6.5\text{-}20)
$$

where the C' components are the adenine group, phosphorus atoms, and the glucose framework. Other components can be chosen, but the number of components is 3 because $C' = N' - R' = 6 - 3 = 3$. Row reduction yields

$$
A' = \begin{array}{c} \\ \text{ATP} \\ \text{ADP} \\ \text{Glu} \end{array}
\begin{array}{cccccc}
\text{ATP} & \text{ADP} & \text{Glu} & \text{G6P} & \text{F6P} & \text{F16BP} \\
1 & 0 & 0 & 1 & 1 & 2 \\
0 & 1 & 0 & -1 & -1 & -2 \\
0 & 0 & 1 & 1 & 1 & 1
\end{array}
\qquad (6.5\text{-}21)
$$

Note that the C' components are now ATP, ADP, and glucose. Matrix 6.5-21 shows the amounts of ATP and ADP in the four pseudoisomers Glu, G6P, F6P, and F16BP (see last row). G6P and F6P can each be considered to contain 1ATP and -1ADP. F16BP can be considered to contain 2ATP and -2ADP. When the rows and columns for ATP and ADP are deleted, this conservation matrix shows that the remaining four reactants are pseudoisomers. The reactions between these four pseudoisomer groups can be represented by

$$\text{Glucose} = \text{glucose 6-phosphate} \qquad (6.5\text{-}22)$$

$$\text{Glucose 6-phosphate} = \text{fructose 6-phosphate} \qquad (6.5\text{-}23)$$

$$\text{Fructose 6-phosphate} = \text{fructose 1,6-biphosphate} \qquad (6.5\text{-}24)$$

The standard further transformed Gibbs energies of formation of the pseudoisomers can be calculated using equation 6.5-2 and can be written in the form

$$\Delta_f G''^0(i) = \Delta_f G'^0(i) - N_{\text{ATP}}(i)\Delta_f G'(\text{ATP}) - N_{\text{ADP}}(i)\Delta_f G'(\text{ADP}) \quad (6.5\text{-}25)$$

where $\Delta_f G'(i) = \Delta_f G'^0(i) + RT\ln[i]$. The numbers of ATP molecules and ADP molecules involved in these four reactions are shown in the row-reduced conservation matrix (equation 6.5-21). When the equilibrium concentration of ATP is 0.0001 M and the equilibrium concentration of ADP is 0.01 M, the standard further transformed Gibbs energies of formation in kJ mol^{-1} of the remaining four reactants at 298.15 K and $I = 0$ are as follows: glucose, -426.71 (note that this value is not changed); glucose 6-phosphate, -439.73; fructose 6-phosphate, -436.55; fructose 1,6-biphosphate, -449.98. Since these reactants are pseudo-isomers, the **standard further transformed Gibbs energy of formation of the**

pseudoisomer group can be calculated by using

$$\Delta_f G''^0(\text{iso}) = -RT \ln \sum_{i=1}^{N_{\text{iso}}} \exp\left(-\frac{\Delta_f G''^0(i)}{RT}\right) = -450.03 \text{ kJ mol}^{-1} \quad (6.5\text{-}26)$$

where N_{iso} is the number of pseudoisomers in the group, which is 4. The equilibrium mole fractions of glucose, glucose 6-phosphate, fructose 6-phosphate, and fructose 1,6-biphosphate can be calculated by using

$$r_i = \exp\left[\frac{\Delta_f G''^0(\text{iso}) - \Delta_f G''^0(i)}{RT}\right] \quad (6.5\text{-}27)$$

Thus the equilibrium concentrations of glucose, glucose 6-phosphate, fructose 6-phosphate, and fructose 1,6-biphosphate are 8.21×10^{-7}, 1.56×10^{-4}, 4.34×10^{-5}, and 9.8×10^{-2} M. Note how much the specification of the equilibrium concentrations of ATP and ADP has simplified this equilibrium calculation. The level 2 discussion of reactions 6.5-17 to 6.5-19 at specified pH involves 6 reactants and 3 apparent equilibrium constants. The level 3 discussion at specified pH and specified concentrations of ATP and ADP involves 4 reactants, but since these reactants are pseudoisomers under these conditions, the system consists of a single pseudoisomer group, and so equations 6.5-26 and 6.5-27 can be used.

Apparent equilibrium constants K'' at specified concentrations of coenzymes for a system larger than 6.5-17 to 6.5-19 can be calculated using

$$\Delta_r G''^0 = -RT \ln K'' \quad (6.5\text{-}28)$$

■ 6.6 CONSIDERATION OF GLYCOLYSIS AT SPECIFIED [ATP],[ADP],[NAD$_{\text{ox}}$], [NAD$_{\text{red}}$], AND [P$_i$]

Glycolysis involves 10 biochemical reactions and 16 reactants. Water is not counted as a reactant in writing the stoichiometric number matrix or the conservation matrix for reasons described in Section 6.3. Thus there are six components because $C' = N' - R' = 16 - 10 = 6$. From a chemical standpoint this is a surprise because the reactants involve only C, H, O, N, and P. Since H and O are not conserved at specified pH in dilute aqueous solution, there are only three conservation equations based on elements. Thus three additional conservation relations arise from the mechanisms of the enzyme-catalyzed reactions in glycolysis. Some of these conservation relations are discussed in Alberty (1992a). At specified pH in dilute aqueous solutions the reactions in glycolysis are represented by

$$\text{Glc} + \text{ATP} = \text{G6P} + \text{ADP} \quad (6.6\text{-}1)$$

$$\text{G6P} = \text{F6P} \quad (6.6\text{-}2)$$

$$\text{F6P} + \text{ATP} = \text{FBP} + \text{ADP} \quad (6.6\text{-}3)$$

$$\text{FBP} = \text{DHAP} + \text{GAP} \quad (6.6\text{-}4)$$

$$\text{DHAP} = \text{GAP} \quad (6.6\text{-}5)$$

$$\text{GAP} + \text{P}_i + \text{NAD}_{\text{ox}} = \text{BPG} + \text{NAD}_{\text{red}} \quad (6.6\text{-}6)$$

$$\text{BPG} + \text{ADP} = \text{PG3} + \text{ATP} \quad (6.6\text{-}7)$$

$$\text{PG3} = \text{PG2} \quad (6.6\text{-}8)$$

$$\text{PG2} = \text{PEP} (+\text{H}_2\text{O}) \quad (6.6\text{-}9)$$

$$\text{PEP} + \text{ADP} = \text{Pyr} + \text{ATP} \quad (6.6\text{-}10)$$

where the following abbreviations are used: glucose (Glc), adenosine triphosphate (ATP), adenosine diphosphate (ADP), glucose 6-phosphate (G6P), fructose 6-

	1	2	3	4	5	6	7	8	9	10
Glc	-1	0	0	0	0	0	0	0	0	0
ATP	-1	0	-1	0	0	0	1	0	0	1
ADP	1	0	1	0	0	0	-1	0	0	-1
NAD_{ox}	0	0	0	0	0	-1	0	0	0	0
NAD_{red}	0	0	0	0	0	1	0	0	0	0
P_i	0	0	0	0	0	-1	0	0	0	0
G6P	1	-1	0	0	0	0	0	0	0	0
F6P	0	1	-1	0	0	0	0	0	0	0
FBP	0	0	1	-1	0	0	0	0	0	0
DHAP	0	0	0	1	-1	0	0	0	0	0
13BPG	0	0	0	0	0	1	-1	0	0	0
3PG	0	0	0	0	0	0	1	-1	0	0
2PG	0	0	0	0	0	0	0	1	-1	0
PEP	0	0	0	0	0	0	0	0	1	-1
GAP	0	0	0	1	1	-1	0	0	0	0
Pyr	0	0	0	0	0	0	0	0	0	1

Figure 6.1 Apparent stoichiometric number matrix v' for the 10 reactions of glycolysis at specified pH in dilute aqueous solutions. (see Problem 6.3) [With permission from R. A. Alberty, *J. Phys. Chem. B* 104, 4807–4814 (2000). Copyright 2000 American Chemical Society.]

phosphate (F6P), fructose 1,6-biphosphate (FBP), D-glyceraldehyde 3-phosphate (GAP), dihydroxyacetone phosphate (DHAP), 1,3-bisphosphoglycerate (3-phospho-D-glycerol phosphate)(BPG), nicotinamide adenine dinucleotide-oxidized (NAD_{ox}), nicotinamide adenine dinucleotide-reduced (NAD_{red}), 3-phospho-D-glycerate (PG3), 2-phospho-D-glycerate (PG2), phospho*enol*pyruvate (PEP), and pyruvate (Pyr). If reactions 6.6-6 to 6.6-10 are each multiplied by 2 and the reactions are added, the net reaction is

$$Glc + 2P_i + 2ADP + 2NAD_{ox} = 2Pyr + 2ATP + 2NAD_{red}(+2H_2O) \qquad (6.6\text{-}11)$$

When using a computer, a net reaction is obtained more conveniently by use of a matrix multiplication (see Section 6.1). H_2O is put in parentheses because its stoichiometric number is not used in the stoichiometric number matrix, but it is involved in the calculation of K' for this net reaction using $\Delta_r G'^0 = -RT \ln K'$.

In writing the stoichiometric number matrix for glycolysis, there is a choice as to the order of the reactants. To make Glc, ATP, ADP, NAD_{ox}, NAD_{red}, and P_i components, they are put first in the rows for reactants in the apparent stoichiometric number matrix, followed by the rest of the reactants ending with Pyr. The stoichiometric number matrix for glycolysis is shown in Fig. 6.1. To check that these 10 reactions are indeed independent, a row reduction of the transposed stoichiometric number matrix can be used. Another way to test the correctness of this matrix is to calculate the net reaction using equation 6.1-3.

Conservation matrix A' that corresponds to this stoichiometric matrix is obtained by calculating the null space of $(v')^T$, as indicated by equation 6.3-4. In order to obtain a conservation matrix with identifiable rows, RowReduce is used again and the result is shown in Fig. 6.2. The figure shows that Glu, ATP, ADP, NAD_{ox}, NAD_{red}, and P_i can be taken as the six components for glycolysis. This

	Glc	ATP	ADP	NAD_{ox}	NAD_{red}	P_i
Glc	1	0	0	0	0	0
ATP	0	1	0	0	0	0
ADP	0	0	1	0	0	0
NAD_{ox}	0	0	0	1	0	0
NAD_{red}	0	0	0	0	1	0
P_i	0	0	0	0	0	1
G6P	1	1	-1	0	0	0
F6P	1	1	-1	0	0	0
F6BP	1	2	-2	0	0	0
DHAP	1/2	1	-1	0	0	0
13BPG	1/2	1	-1	1	-1	1
3PG	1/2	0	0	1	-1	1
2PG	1/2	0	0	1	-1	1
PEP	1/2	0	0	1	-1	1
GAP	1/2	1	-1	0	0	0
Pyr	1/2	-1	1	1	-1	1

Figure 6.2 Transposed apparent conservation matrix $(A')^T$ for glycolysis at specified pH in dilute aqueous solution, calculated from the apparent stoichiometric number matrix in the previous figure. This conservation matrix shows the composition of the noncomponents (the last 10 rows) in terms of components (see Problem 6.3). [With permission from R. A. Alberty, *J. Phys. Chem. B* 104, 4807–4814 (2000). Copyright 2000 American Chemical Society.]

is not the only possible choice for components, but it is the one to use to obtain a global view of the thermodynamics of the reaction system at specified concentrations of coenzymes. The transposed conservation matrix in Fig. 6.2 is another illustration of the statement that conservation equations can be expressed in terms of reactants, rather than elements and constraints arising from enzyme mechanisms.

When the concentration of a component is held constant in an equilibrium calculation, its row and column in the conservation matrix A' are deleted. When the rows and columns for ATP, ADP, NAD_{ox}, NAD_{red}, and P_i are deleted, the remaining apparent conservation matrix is dramatically reduced, in fact it is reduced to a vector, namely $\{\{1,1,1,1,\frac{1}{2},\frac{1}{2},\frac{1}{2},\frac{1}{2},\frac{1}{2},\frac{1}{2},\frac{1}{2}\}\}$, which applies to the conservation of the glucose component (Alberty, 2000c). Under these conditions the reactants in glycolysis consist of two pseudoisomer groups. The first contains Glc, G6P, F6P, and FBP, and this group will be referred to as C_6, where the C refers to the element carbon. The second contains DHAP, 13BPG, 3PG, 2PG, PEP, GAP, and Pyr and will be referred to as C_3. Deleting the redundant columns in the conservation matrix for glycolysis yields the apparent conservation matrix $A'' = \{\{1,\frac{1}{2}\}\}$, where the 1 is for C_6 and the $\frac{1}{2}$ is for C_3. This is the conservation matrix that applies when [ATP], [ADP], $[NAD_{ox}]$, $[NAD_{red}]$, and $[P_i]$ are held constant. Calculating the null space for this conservation matrix yields $(v'')^T = \{\{-\frac{1}{2},1\}\}$, which indicates that there is a single reaction $(\frac{1}{2}) C_6 = C_3$ or

$$C_6 = 2C_3 \qquad (6.6\text{-}12)$$

at level 3. Thus specifying the concentrations of 5 coenzymes has reduced the system of 10 reactions and 16 reactants to 1 reaction and 2 reactants. The expression for the apparent equilibrium constant K''_{GLY} under these conditions can be written

$$K''_{GLY} = [C_3]^2/[C_6] \qquad (6.6\text{-}13)$$

where K''_{GLY} is a function of T, P, pH, [ATP], [ADP], [NAD$_{ox}$], [NAD$_{red}$], [P$_i$], and ionic strength. Equilibrium constants are dimensionless, but the reference concentration $c^0 = 1$ M is omitted in the denominator of equation 6.6-13 as a simplification. This equilibrium expression provides the most global view of the thermodynamics of glycolysis. The value of the apparent equilibrium constant K''_{GLY} can be calculated when $\Delta_f G_i'^0$ values are known for the 16 reactants at the desired T, P, pH, and ionic strength. These values can be used in equations like 6.5-25 to calculate the $\Delta_f G_i''^0$ values of the reactants at the desired concentrations of the coenzymes.

Figure 6.2 shows the content of each of these reactants in terms of components; in other words, the rows give the values of $N_c(i)$ for each of the components held constant. Equation 6.5-26 can then be used to calculate the $\Delta_f G''^0(\text{iso})$ values of the C_6 and C_3 pseudoisomer groups. In order to calculate the numerical value for K''_{GLY} using equation 6.5-28, the reaction between pseudoisomer groups should be written $C_6 = 2C_3(+2H_2O)$ because $\Delta_f G''^0(H_2O)$ is involved in calculating K''_{GLY}. Note that $\Delta_f G''^0(H_2O) = \Delta_f G'^0(H_2O)$ because H_2O does not contain coenzymes. The equilibrium concentrations of C_6 and C_3 can be calculated using equation 6.6-13, and then the concentrations of C_6 and C_3 can be divided into the equilibrium concentrations of each of the reactants (noncomponents) by use of equation 6.5-27. Thus K''_{GLC} provides the means to calculate the equilibrium concentrations of the 11 reactants for which concentrations have not been specified.

This more global view of the thermodynamics of a system of biochemical reactions provides different information than the net reaction (equation 6.6-11) for the system because it deals with the pseudoisomer groups C_6 and C_3. The expression for the apparent equilibrium constant for the net reaction for glycolysis can be used to calculate the equilibrium value of $[\text{Pyr}]^2/[\text{Glu}]$ by setting the concentrations of the coenzymes equal to their steady state values. The apparent equilibrium constant for the net reaction at 298.15 K, pH 7, and 0.25 M ionic strength calculated from $\Delta_f G'^0$ values is 1.41×10^{14}. When the steady state concentrations of ATP and NAD$_{red}$ are 0.01 M and of ADP and NAD$_{ox}$ are 10^{-5} M, and that of P$_i$ is 0.001 M, the equilibrium concentration of glucose will exceed that of pyruvate. The advantage of the equilibrium expression in equation 6.5-13 is that it yields the equilibrium concentrations of pseudoisomer groups C_6 and C_3; that is, it accounts for all of the reactants in glycolysis."The equilibrium calculations given here show how calculations become simpler as the values of more intensive variables are held constant.

■ 6.7 CALCULATION OF THE EQUILIBRIUM COMPOSITION FOR GLYCOLYSIS

The standard transformed Gibbs energies of formation of the reactants in the first five reactions of glycolysis are known at pH 7 and 0.25 M ionic strength (see the first column of Table 6.2), and so the equilibrium composition can be calculated for specified steady state concentrations of ATP and ADP (Alberty, 2001g). The standard further transformed Gibbs energies of formation at two different sets of concentrations of ATP and ADP are given in the last two columns.

There is no adjustment for glucose, and the adjustment for G6P is given by

$$\Delta_f G''^0(\text{G6P}) = \Delta_f G'^0(\text{G6P}) - \{\Delta_f G'^0(\text{ATP}) + RT\ln[\text{ATP}]\}$$

$$+ \{\Delta_f G'^0(\text{ADP}) + RT\ln[\text{ADP}]\} \qquad (6.7\text{-}1)$$

Table 6.2   Standard Transformed Gibbs Energies of Formation at 298.15 K, pH 7, and 0.25 M Ionic Strength, Standard Further Transformed Gibbs Energies of Formation at $[ATP] = 10^{-4}$ M and $[ADP] = 10^{-2}$ M, and Standard Further Transformed Gibbs Energies of Formation at $[ATP] = 10^{-2}$ M and $[ADP] = 10^{-2}$ M

	$\Delta_f G'^0$/kJ mol^{-1}	$\Delta_f G''^0$/kJ mol^{-1}	$\Delta_f G''^0$/kJ mol^{-1}
Glc	-426.71	-426.71	-416.71
G6P	-1318.92	-439.74	-451.16
F6P	-1315.74	-436.56	-447.97
FBP	-2206.78	-448.42	-471.25
GAP	-1088.04	-208.86	-220.28
GlycP	-1095.70	-216.52	-227.94
ATP	-2097.89		
ADP	-1230.12		

Source: Reprinted from R. A. Alberty, *Biophys. Chem.* 93, 1–10 (2001), with permission from Elsevier Science.

The same adjustment is applied to the other reactants, except for FBP where the adjustment terms are both multiplied by 2 because it contains two phosphate groups.

When [ATP] and [ADP] are specified, the four reactants with six carbon atoms become pseudoisomers and the two reactants with three carbon atoms become pseudoisomers. The standard transformed Gibbs energies of formation of these two pseudoisomer groups at the two sets of concentrations are given in Table 6.3 of the article "Systems of biochemical reactions from the point of view of a semigrand partition function" (Alberty, 2001g) along with the apparent equilibrium constants $K'' = 0.00213$ and $K'' = 0.0021$ for the two sets of conditions. The equilibrium extent of reaction ξ'' can be calculated for any desired initial concentration of glucose by use of the quadratic formula. The equilibrium values of ξ'' are given for an initial concentration of glucose of 0.01 M. This makes it possible to calculate $[C_6]_{eq}$ and $[C_3]_{eq}$; then the equilibrium concentrations of the various reactants can be calculated using equation 5.6-27.

It is perhaps surprising that raising the concentration of ATP by a factor of 100 makes so little difference, but of course it does make a big difference for the first three reactants. The concentration of fructose 1,6-biphosphate cannot increase very much because it already dominates, and that limits the effects on GAP and GlycP. These calculations can be applied to larger systems and can include the specification of the concentrations of other coenzymes like NAD_{ox} and NAD_{red}.

As mentioned in the Preface, Callen (1985) pointed out that "The choice of variables in terms of which a given problem is formulated, while a seemingly innocuous step is often the most critical step in the solution." This calculation is

Table 6.3   Standard Further Transformed Gibbs Energies of Formation of C_6 and C_3 at pH 7 Ionic Strength 0.25 M for Different Specified Concentrations of ATP

	$[ATP] = 10^{-4}$ M $[ATP] = 10^{-2}$ M	$[ADP] = 10^{-2}$ M $[ADP] = 10^{-2}$ M
$\Delta_f G''^0(C_6)$/kJ mol^{-1}	-448.51	-471.25
$\Delta_f G''^0(C_3)$/kJ mol^{-1}	-216.63	-228.05
$\Delta_r G''^0$(rx 6.6-12)/kJ mol^{-1}	15.25	15.15
K''	0.00213	0.00221
ξ''	0.00205	0.00209

Source: With permission from R. A. Alberty, *Biophys. Chem.* 93, 1–10 (2001).

Table 6.4 Equilibrium Compositions (M) for the First
Five Reactions of Glycolysis at 298.15 K, pH 7, Ionic Strength
0.25 M, and Specified Concentrations of ATP and ADP

	$[ATP] = 10^{-4}$ M $[ATP] = 10^{-2}$ M	$[ADP] = 10^{-2}$ M $[ADP] = 10^{-2}$ M
[Glc]/M	1.20×10^{-9}	1.24×10^{-10}
[G6P]/M	2.31×10^{-1}	2.39×10^{-6}
[F6P]/M	6.39×10^{-5}	6.61×10^{-7}
[FBP]/M	7.64×10^{-3}	7.90×10^{-3}
[GAP]M	1.79×10^{-4}	1.82×10^{-4}
[GlycP]M	3.94×10^{-3}	4.00×10^{-3}

Source: With permission from R. A. Alberty, *Biophys. Chem.* 93, 1–10
(2001).

an example of that and emphasizes the fact that there is no loss of information in
making a Legendre transform to introduce new intensive variables. Note that
there are further transformed enthalpies H'' and entropies S'' that are not
discussed here. For larger reaction systems, further coenzyle concentrations can
be specified, but note that there is a maximum number of intensive variables that
can be specified for a given reaction system because one component must remain
unspecified.

Thermodynamics of the Binding of Ligands by Proteins

The binding of oxygen by hemoglobin is an important example of the binding of a ligand by a protein, and so it is of interest to consider this series of reactions from the point of view of the transformed Gibbs energy at a specified pH. The experimental determination of the oxygen binding by the tetramer is complicated by the partial dissociation of the tetramer into dimers. In view of the fact that it is not possible to connect either the tetramer or dimer to its elements in the standard state, the standard transformed Gibbs energy of the tetramer can be set equal to zero. This convention has already been used for other reactants that cannot be connected to the elements by reactions with known equilibrium constants. This chapter shows how all seven apparent equilibrium constants for the binding of oxygen by hemoglobin at specified pH can be determined by measuring the fractional saturation of heme as a function of the concentration of molecular oxygen and the concentration of heme at a specified pH.

A number of biochemical reactions involve proteins as reactants, and so it is important to be able to determine the standard transformed Gibbs energies of formation of their reactive sites at specified pH. The standard transformed Gibbs energies of formation of the active sites of ferredoxin, cytochrome c, and thioredoxin are given in tables discussed earlier in Chapter 4.

The effect of pH on protein-ligand equilibria is discussed and the equations are applied to the binding of succinate, D-tartrate, L-tartrate, and meso-tartrate by the catalytic site of fumarase.

The binding of oxygen by hemoglobin is discussed by Wyman and Gill (1990) and ligand-receptor energetics are discussed by Klotz (1997).

■ 7.1 THE BINDING OF OXYGEN BY HEMOGLOBIN TETRAMERS

Since a protein is a weak acid, its transformed thermodynamic properties are functions of pH, and that is discussed specifically in Section 7.6. However, it is not necessary to examine the pH dependence of the binding of a ligand first. This is illustrated by the consideration of the binding of oxygen by hemoglobin at specified pH. Since the pH is an independent variable, the criterion for equilibrium is provided by the transformed Gibbs energy G'. Hemoglobin is more complicated than a weak acid in that its binding properties are affected by chloride ion and perhaps other ligands in the buffer used. If necessary, the Legendre transform to define G' at a specified pH can include specification of the concentrations of chloride ions and other ligands that affect the binding of molecular oxygen. Since the **tetramers** of hemoglobin ($\alpha_2\beta_2$, represented here as T) can combine with 1 to 4 molecules of oxygen, the fundamental equation for the transformed Gibbs energy G' for the binding by tetramer at a specified pH is

$$dG' = -S'\,dT + V\,dP + \sum_{i=0}^{4} \mu'(T(O_2)_i)dn'(T(O_2)_i)$$

$$+ \mu'(O_2)dn'(O_2) + RT\ln(10)n_c(H)d\text{pH} \qquad (7.1\text{-}1)$$

where $\mu'(T(O_2)_i)$ is the transformed chemical potential of the sum of various protonated species binding i molecules of molecular oxygen and $n'(T(O_2)_i)$ is the amount of species binding i molecules of molecular oxygen. When equation 7.1-1 is integrated at constant values of the intensive variables,

$$G' = \sum \mu_i' n_i' \qquad (7.1\text{-}2)$$

is obtained.

The biochemical equations for the binding reactions can be written in different ways, but the usual way is

$$T + O_2 = T(O_2) \qquad K_{41}' = \frac{[T(O_2)]}{[T][O_2]} = 4.397 \times 10^4 \qquad (7.1\text{-}3)$$

$$T(O_2) + O_2 = T(O_2)_2 \qquad K_{42}' = \frac{[T(O_2)_2]}{[T(O_2)][O_2]} = 1.221 \times 10^4 \qquad (7.1\text{-}4)$$

$$T(O_2)_2 + O_2 = T(O_2)_3 \qquad K_{43}' = \frac{[T(O_2)_3]}{[T(O_2)_2][O_2]} = 4.049 \times 10^5 \qquad (7.1\text{-}5)$$

$$T(O_2)_3 + O_2 = T(O_2)^4 \qquad K_{44}' = \frac{[T(O_2)_4]}{[T(O_2)_3][O_2]} = 6.644 \times 10^5 \qquad (7.1\text{-}6)$$

where the values for the apparent equilibrium constants are those determined by Mills, Johnson, and Ackers (1976) for human hemoglobin at 21.5°C, 1 bar, pH 7.4, $[Cl^-] = 0.2$ M, and 0.2 M ionic strength. Molar concentrations are used, but the apparent equilibrium constants are considered to be dimensionless.

The following equilibrium conditions for the four reactions (see 7.1-3 to 7.1-6) can be derived using equation 7.1-1:

$$\mu'(T) + \mu'(O_2) = \mu'(TO_2) \qquad (7.1\text{-}7)$$

$$\mu'(T(O_2)) + \mu'(O_2) = \mu'(T(O_2)_2) \qquad (7.1\text{-}8)$$

$$\mu'(T(O_2)_2) + \mu'(O_2) = \mu'(T(O_2)_3) \qquad (7.1\text{-}9)$$

$$\mu'(T(O_2)_3) + \mu'(O_2) = \mu'(T(O_2)_4) \qquad (7.1\text{-}10)$$

Substituting these equilibrium conditions in equation 7.1-1 yields

$$dG' = -S'dT + VdP + \mu'(\text{TotT})dn'_c(\text{TotT}) + \mu'(O_2)dn'_c(O_2)$$
$$+ RT\ln(10)n_c(\text{H})d\text{pH} \tag{7.1-11}$$

where $\mu'(\text{TotT})$ is the transformed chemical potential of the T component and $n'_c(\text{TotT})$ is the amount of the T component, namely

$$n'_c(\text{TotT}) = n'(T) + n'(T(O_2)) + n'(T(O_2)_2) + n'(T(O_2)_3) + n'(T(O_2)_4) \tag{7.1-12}$$

The amount $n'_c(O_2)$ of the molecular oxygen component is given by

$$n'_c(O_2) = n'(O_2) + n'(T(O_2)) + 2n'(T(O_2)_2)$$
$$+ 3n'(T(O_2)_3) + 4n'(T(O_2)_4) \tag{7.1-13}$$

Equation 7.1-12 shows that the natural variables for G' are T, P, $n'_c(T)$, $n'_c(O_2)$, and pH, and so the criterion for spontaneous change and equilibrium is $dG' \leqslant 0$ at constant T, P, $n'_c(\text{TotT})$, $n'_c(O_2)$, and pH. The number of natural variables is five, $D' = 5$. The number of independent intensive properties is $F' = 4$, and they can be taken to be T, P, pH, and $[O_2]$.

It is convenient to use the fundamental equation in matrix form (see Chapter 5). The stoichiometric number matrix $\mathbf{v'}$ for reactions 7.1-3 to 7.1-6 is

$$\mathbf{v'} = \begin{array}{c} \\ T \\ T(O_2) \\ T(O_2)_2 \\ T(O_2)_3 \\ T(O_2)_4 \\ O_2 \end{array} \begin{array}{cccc} \text{rx 7.1-3} & \text{rx 7.1-4} & \text{rx 7.1-5} & \text{rx 7.1-6} \\ -1 & 0 & 0 & 0 \\ 1 & -1 & 0 & 0 \\ 0 & 1 & -1 & 0 \\ 0 & 0 & 1 & -1 \\ 0 & 0 & 0 & 1 \\ -1 & -1 & -1 & -1 \end{array} \tag{7.1-14}$$

Substituting this matrix in equation 5.5-3 yields

$$dG' = -S'dT + VdP + \sum_{i=1}^{4} \Delta_r G'_i d\xi'_i + RT\ln(10)n_c(\text{H})d\text{pH} \tag{7.1-15}$$

where

$$\Delta_r G'_i = \mu'(T(O_2)_i) - \mu'(T(O_2)_{i-1}) - \mu'(O_2), \quad i = 1, 2, 3, 4 \tag{7.1-16}$$

Since at equilibrium $\Delta_r G'_i = 0$, these four equations can be used to derive the expressions for the apparent equilibrium constants K' for the four reactions that are given in equations 7.1-3 to 7.1-6.

In the absence of experimental methods for distinguishing experimentally between the five forms of the tetramer, the fractional saturation of hemoglobin is measured. The **fractional saturation of tetramer** Y_T is defined by

$$Y_T = \frac{[T(O_2)] + 2[T(O_2)_2] + 3[T(O_2)_3] + 4[T(O_2)_4]}{4\{[T] + [T(O_2)] + [T(P_2)_2] + [T(O_2)_3] + [T(O_2)_4]\}} \tag{7.1-17}$$

Substituting the equilibrium expressions defined in equations 7.1-13 to 7.1-16 yields

$$Y_T =$$

$$\frac{K'_{41}[O_2] + 2K'_{41}K'_{42}[O_2]^2 + 3K'_{41}K'_{42}K'_{43}[O_2]^3 + 4K'_{41}K'_{42}K'_{43}K'_{44}[O_2]^4}{4(1 + K'_{41}[O_2] + K'_{41}K'_{42}[O_2]^2 + K'_{41}K'_{42}K'_{43}[O_2]^3 + K'_{41}K'_{42}K'_{43}K'_{44}[O_2]^4)}$$
$$\tag{7.1-18}$$

This is often referred to as the **Adair equation**. A plot of the fractional saturation for tetramer, which shows the cooperative effect, is given in Fig. 7.1.

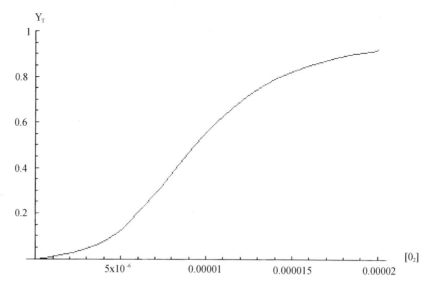

Figure 7.1   Fractional saturation Y_T of tetramer with molecular oxygen calculated with the binding constants in equations 7.1-3 to 7.1-6 (see Problem 7.3).

The four apparent equilibrium constants can be calculated from measurements of Y_T as as function of $[O_2]$ in a particular buffer by use of the method of least squares. The determination of accurate values for the four equilibrium constants is difficult because of the cooperative effect that causes molecular oxygen to be bound more strongly after some is bound. A further complication is that the tetramer is partially dissociated into dimers (see Section 7.3). Values of these four apparent equilibrium constants for reactions 7.1-3 to 7.1-6 are available in the literature for different hemoglobins and various buffers. By assigning $\Delta_f G'^0 = 0$ to the tetramer T without bound oxygen molecules, we can calculate standard transformed Gibbs energies of formation for the other four forms using

$$\Delta_r G'^0 = \sum v'_i \Delta_f G'^0_i = -RT \ln K' \qquad (7.1\text{-}19)$$

In order to calculate the standard transformed Gibbs energies of formation of the four oxygenated forms of hemoglobin, we need the value of $\Delta_f G^0$ for molecular oxygen in aqueous solution at 21.5°C. The NBS Table (1992) indicates that $\Delta_f G^0(O_2(ao)) = 16.1$ kJ mol^{-1} at 21.5°C. The value of $\Delta_f G'^0(T(O_2))$ is calculated using

$$\Delta_r G'^0 = -(8.31451 \times 10^{-3}\text{ kJ K}^{-1}\text{ mol}^{-1})(294.65\text{ K}) \ln 4.397 \times 10^4$$

$$= \Delta_f G'^0(T(O_2)) - \Delta_f G'^0(T) - \Delta_f G'^0(O_2)$$

$$= \Delta_f G'^0(T(O_2)) - 0 - 16.1 \qquad (7.1\text{-}20)$$

which shows that $\Delta_f G'^0(T(O_2)) = -10.0922$ kJ mol^{-1}. Experimental errors are usually large enough that values like this can be rounded to 0.01 kJ mol^{-1}, but sometimes in making calculations, it is a good idea to keep more digits. The standard transformed Gibbs energies of formation of $T(O_2)_2$, $T(O_2)_3$, and $T(O_2)_4$ are -17.045, -32.577, and -49.321 kJ mol^{-1}, respectively.

The values of the standard transformed Gibbs energies of formation of the five forms of hemoglobin at specified T, P, and buffer can be used to calculate the equilibrium constants for other reactions that can be written between these forms, such as $T + 4O_2 = T(O_2)_4$. But it is also of interest to consider the tetramer of hemoglobin as an entity at a specified pressure of molecular oxygen, just as ATP is considered as an entity at a specified pH. This is discussed in the next section.

■ 7.2 FURTHER TRANSFORMED GIBBS ENERGY AT SPECIFIED OXYGEN CONCENTRATION

In order to introduce the chemical potential of molecular oxygen as a natural variable, the following **Legendre transform** is used to define a further transformed Gibbs energy G'' (Alberty, 1996b):

$$G'' = G' - n'_c(O_2)\mu'(O_2) \tag{7.2-1}$$

where $n'_c(O_2)$ is the amount of molecular oxygen in the system, free and bound. Substituting $G'' = \Sigma \mu''_i n''_i$ and $G' - \Sigma \mu'_i dn'_i$ (equation 7.1-2) yields the following expression for the **further transformed chemical potential** μ''_i of reactant i ($i = 0 - 4$):

$$\mu''_i = \mu'_i - N_i(O_2)\mu'(O_2) \tag{7.2-2}$$

where $N_i(O_2)$ is the number of O_2 molecules bound by i. Note that $\mu''(O_2) = 0$. Using equation 7.2-2 to eliminate $\mu'(T)$, $\mu'(T(O_2))$, $\mu'(T(O_2)_2)$, $\mu'(T(O_2)_3)$, and $\mu'(T(O_2)_4)$ from equation 7.1-1 yields

$$\begin{aligned} dG' = &-S'\,dT + V\,dP + \mu''(T)dn'(T) + \mu''(T(O_2))dn'(T(O_2)) \\ &+ \mu''(T(O_2)_2)dn'(T(O_2)_2) + \mu''(T(O_2)_3)dn'((T(O_2)_3) \\ &+ \mu''(T(O_2)_4)dn'(T(O_2)_4) + \mu'(O_2)dn'_c(O_2) + RT\ln(10)n_c(H)dpH \end{aligned} \tag{7.2-3}$$

Taking the differential of G'' in equation 7.2-1 yields

$$dG'' = dG' - n'_c(O_2)d\mu'(O_2) - \mu'(O_2)dn'_c(O_2) \tag{7.2-4}$$

Substituting equation 7.2-3 yields

$$\begin{aligned} dG'' = &-S'dT + V\,dP + \mu''(T)dn'(T) + \mu''(T(O_2))dn'(T(O_2)) \\ &+ \mu''(T(O_2)_2)dn'(T(O_2)_2) + \mu''(T(O_2)_3)dn'((T(O_2)_3) \\ &+ \mu''(T(O_2)_4)dn'(T(O_2)_4) - n'_c(O_2)d\mu'(O_2) \\ &+ RT\ln(10)n_c(H)dpH \end{aligned} \tag{7.2-5}$$

At specified T, P, $\mu'(O_2)$, and pH,

$$\begin{aligned} (dG'')_{T,P,\mu'(O_2),pH} = &\mu''(T)dn'(T) + \mu''(T(O_2))dn'(T(O_2)) + \mu''(T(O_2)_2)dn'(T(O_2)_2) \\ &+ \mu''(T(O_2)_3)dn'((T(O_2)_3) + \mu''(T(O_2)_4)dn'(T(O_2)_4) \end{aligned} \tag{7.2-6}$$

The four reactions at specified $\mu'(O_2)$ can be written as

$$T = T(O_2) \qquad K''_{41} = \frac{[T(O_2)]}{[T]} \tag{7.2-7}$$

$$T(O_2) = T(O_2)_2 \qquad K''_{42} = \frac{[T(O_2)_2]}{[T(O_2)]} \tag{7.2-8}$$

$$T(O_2)_2 = T(O_2)_3 \qquad K''_{43} = \frac{[T(O_2)_3]}{[T(O_2)_2]} \tag{7.2-9}$$

$$T(O_2)_3 = T(O_2)_4 \qquad K''_{44} = \frac{[T(O_2)_4]}{[T(O_2)_3]} \tag{7.2-10}$$

These reactions do not balance oxygen because its chemical potential is specified. At specified $[O_2]$, these five forms of hemoglobin are pseudoisomers, and they have the same further transformed chemical potential: $\mu''(T) = \mu''(T(O_2)) = $

Table 7.1   Standard Transformed Gibbs Energy of Formation $\Delta_f G'^0$ and Standard Further Transformed Gibbs Energies of Formation $\Delta_f G''^0$ of Hemoglobin Tetramer at 21.5°C, 1 bar, pH 7.4, $[Cl^-] = 0.2$ M, and 0.2 M Ionic Strength

		$\Delta_f G''^0$/kJ mol^{-1}		
	$\Delta_f G'^0$/kJ mol^{-1}	$[O_2] = 5 \times 10^{-6}$ M	$[O_2] = 10^{-5}$ M	$[O_2] = 2 \times 10^{-5}$ M
T	0	0	0	0
T(O$_2$)	$-10.092\ 2$	3.711 09	2.012 97	0.314 85
T(O$_2$)2	$-17.045\ 5$	10.561 10	7.164 84	3.768 59
T(O$_2$)$_3$	-32.5768	8.833 13	3.738 77	$-1.355\ 60$
T(O$_2$)4	-49.3213	5.891 89	$-0.900\ 59$	$-7.693\ 07$
$\Delta_f G''^0$(TotT)/kJ mol^{-1}	$-0.736\ 51$	-0.73651	$-2.814\ 90$	$-8.069\ 06$

Source: Reprinted from R. A. Alberty, *Biophys. Chem.* 62, 141–159 (1996), with permission from Elsevier Science.
Note: See Problem 7.1.

$\mu''(T(O_2)_2) = \mu''(T(O_2)_3) = \mu''(T(O_2)_4)$. Therefore equation 7.2-5 can be written

$$dG'' = -S'dT + V\,dP + \mu''(T)dn'_c(TotT) - n'_c(O_2)d\mu'(O_2) + RT\ln(10)n_c(H)dpH$$

(7.2-11)

This shows that the natural variables for the further transformed Gibbs energy G'' are T, P, $n'_c(T)$, $\mu'(O_2)$, and pH, and so the criterion for spontaneous change and equilibrium is $dG'' \leqslant 0$ at constant T, P, $n'_c(T)$, $\mu''(O_2)$, pH. There are $D'' = 5$ natural variables and $F'' = 4$ independent intensive variables, the same as for G'.

The integrated form of equation 7.2-11 at constant values of the intensive variables is

$$G'' = \mu''(T)n'_c(TotT)$$

(7.2-12)

so this system behaves like a one-component system. Under these conditions the entity TotT, which is made up of the various forms of the tetramer, has a set of further transformed thermodynamic properties. As we have seen before, the standard further transformed Gibbs energy of formation $\Delta_f G''^0$(TotT) of the tetramer pseudoisomer group at a specified concentration of molecular oxygen can be calculated by using equation 4.5-1 for an isomer group.

Derivations are carried out with chemical potentials, but calculations are carried out with Gibbs energies of formation, and so equation 7.2-2 is used in the form

$$\Delta_f G''^0_i = \Delta_f G'^0_i + N_i(O_2)(\Delta_f G^0(O_2) + RT\ln[O_2])$$

(7.2-13)

where $\Delta_f G'^0$ values are given after equation 7.1-20. The values of $\Delta_f G'^0_i$ calculated using equation 7.2-13 for the five forms at three $[O_2]$ are given in Table 7.1. The $\Delta_f G''^0_i$ values for T are independent of $[O_2]$ because T does not contain O_2. The other $\Delta_f G''^0_i$ values decrease as $[O_2]$ is raised because the oxygenated forms become more stable relative to T. The value of $\Delta_f G''^0$(TotT) is calculated using the partition function in equation 4.5-1. $\Delta_f G''^0$(TotT) is more negative than $\Delta_f G''^0_i$ for any of the pseudoisomers and becomes more negative as the concentration of oxygen is increased. The entries in Table 7.1 were calculated with the *Mathematica* program calctgfT (Alberty, 1996b).

The equilibrium mole fractions of the various forms of the tetramer can be calculated using the following analog of equation 4.5-2:

$$r_i = \exp\left\{\frac{\Delta_f G''^0(TotT) - \Delta_f G''^0_i}{RT}\right\}$$

(7.2-14)

These equilibrium mole fractions are given in Table 7.2.

Table 7.2   Equilibrium Mole Fractions of Forms of the Tetramer at 21.4°C, 1 bar, pH 7.4, $[Cl^-] = 0.2$ M, and 0.2 M Ionic Strength at $[heme] = 10^{-5}$ M

	$[O_2] = \times 10^{-6}$ M	$[O_2] = 10^{-5}$ M	$[O_2] = 2 \times 10^{-5}$ M
T	0.763	0.317	0.037
$T(O_2)$	0.163	0.139	0.033
$T(O_2)_2$	0.010	0.017	0.008
$T(O_2)_3$	0.020	0.069	0.065
$T(O_2)_4$	0.067	0.458	0.858

Note. See Problem 7.1.

For these five forms of the tetramer, $\Delta_f G''^0(\text{TotT})$ is the same as the **binding potential** Π defined by Wyman (1948,1964) as $\Pi = RT \ln P$, except for the difference in sign. The **binding polynomial** P is defined as (see Section 1.3)

$$P = \frac{[M] + [ML] + [ML_2] + \cdots}{[M]} \tag{7.2-15}$$

The binding polynomial for the binding of oxygen by the tetramer is given by

$$
\begin{aligned}
P_T &= 1 + K'_{41}[O_2] + K'_{41}K'_{42}[O_2]^2 + K'_{41}K'_{42}K'_{43}[O_2]^3 \\
&\quad + K'_{41}K'_{42}K'_{43}K'_{44}[O_2]^4 \\
&= 1 + K'_{41}[O_2](1 + K'_{42}[O_2](1 + K'_{43}[O_2](1 + K'_{44}[O_2]))) \tag{7.2-16}
\end{aligned}
$$

Binding potentials Π become more positive with increasing stability, in contrast to Gibbs energies of formation which become more negative. The values of $\Delta_f G''^0$ and Π agree for the tetramer because of the convention that $\Delta_f G'^0(T) = 0$. But for the dimer D, these two physical quantities are not equal as shown in the next section.

In this section we have seen that in addition to the $\Delta_f G'^0$ for the various forms of the tetramer at a specified pH, the sum of various forms of the tetramer have $\Delta_f G''^0$ values that are function of $[O_2]$.

■ 7.3 PARTIAL DISSOCIATION OF TETRAMERS INTO DIMERS

The tetramer ($\alpha_4\beta_4$) of hemoglobin is partially dissociated into dimers ($\alpha\beta$). Mills, Johnson, and Ackers (1976) give the following value for the apparent equilibrium constant $^\circ K'_2$ (their symbol) for the association reaction in the absence of oxygen:

$$2D = T \quad ^\circ K'_2 = \frac{[T]}{[D]^2} = 4.633 \times 10^{10} \tag{7.3.1}$$

Therefore, since the standard transformed Gibbs energy of formation of T is taken as zero, the standard transformed Gibbs energy of formation of D is 30.083 kJ mol^{-1}. The equilibrium constants for the dimer are given by

$$D + O_2 = D(O_2) \qquad K'_{21} = \frac{[D(O_2)]}{[D][O_2]} = 3.253 \times 10^6 \tag{7.3-2}$$

$$D(O_2) + O_2 = D(O_2)^2 \qquad K'_{22} = \frac{[D(O_2)2]}{[D(O_2)][O_2]} = 8.155 \times 10^5 \tag{7.3-3}$$

Since $\Delta_f G'^0(D) = 30.08$ kJ mol^{-1}, $\Delta_f G''^0(D)$ is given by (Alberty, 1996b)

$$\Delta_f G''^0(D) = 30.08 - RT \ln P_D \tag{7.3-4}$$

Table 7.3 Standard Transformed Gibbs Energy of Formation $\Delta_f G'^0$ and Standard Further Transformed Gibbs Energies of Formation $\Delta_s G''^0$ of Hemoglobin Dimer at 21.5°C, 1 bar, pH 7.4, $[Cl^-] = 0.2$ M, and 0.2 M Ionic Strength

	$\Delta_f G'^0$(TotD)/kJ mol^{-1}	$\Delta_f G''^0$(TotD)/kJ mol^{-1}		
		$[O_2] = 5 \times 10^{-6}$ M	$[O_2] = 10^{-5}$ M	$[O_2] = 2 \times 10^{-5}$ M
D	30.083 25	30.083 25	30.083 25	30.083 25
D(O$_2$)	9.447 23	23.250 50	21.552 40	19.854 30
D(O$_2$)$_2$	−7.799 32	19.807 30	16.411 00	13.014 80
$\Delta_f G''^0$(TotD)/kJ mol^{-1}		19.240 4	16.119 4	12.866 8
$\Delta_f G''^0$(eq 7.3-4)/kJ mol^{-1}	−33.772 6*	−39.217 3	−35.053 7	−33.802 7
K''(eq. 7.3-7)	9.508×10^5*	8.956×10^6	1.637×10^6	0.982×10^6

Source: Reprinted from R. A. Alberty, *Biophys. Chem.* 62, 141–159 (1996), with permission from Elsevier Science.
Note: See Problem 7.2.

The corresponding binding polynomial for dimer is given by

$$P_D = 1 + K'_{21}[O_2] + K'_{22}[O_2]^2 \qquad (7.3\text{-}5)$$

Thermodynamic properties of dimmers are summarized in Table 7.3.

The **fractional saturation of dimer** Y_D is given by the Adair equation (see equation 7.1-18)

$$Y_D = \frac{K'_{21}[O_2] + 2K_{22}[O_2]^2}{2(1 + K'_{21}[O_2] + K_{22}[O_2]^2)} \qquad (7.3\text{-}6)$$

The binding curve for dimer, which does not have a cooperative effect, is shown in Fig. 7.2.

The fundamental equations for the dimer are similar to those for the tetramer. Table 7.3 gives $\Delta_f G''^0$ and $\Delta_f G''^0$(TotD) at the same three oxygen concentrations as Table 7.1. The standard transformed Gibbs energies of formation of the three forms of the dimer are based on the convention that $\Delta_f G'^0(T) = 0$.

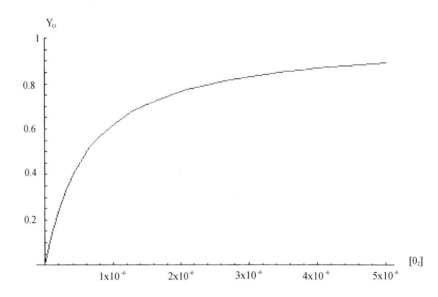

Figure 7.2 Fractional saturation Y_D of dimer with molecular oxygen calculated with the binding constants in equations 7.3-2 and 7.3-3 (see Problem 7.4).

The values of $\Delta_f G''^0(\text{TotT})$ in Table 7.1 and $\Delta_f G''^0(\text{TotD})$ in Table 7.2 make it possible to calculate the apparent equilibrium constant for the reaction

$$2\text{TotD} = \text{TotT} \tag{7.3-7}$$

at three concentrations of molecular oxygen. The apparent equilibrium constant K'', which is a function of $[\text{O}_2]$, is defined by

$$K'' = \frac{[\text{TotT}]}{[\text{TotD}]^2} \tag{7.3-8}$$

This apparent equilibrium constant can be written in terms of the binding polynomials of tetramer and dimer and $^\circ K_2''$:

$$K'' = \frac{^\circ K_2'' P_\text{T}}{(P_\text{D})^2} \tag{7.3-9}$$

As $[\text{O}_2]$ approaches infinity, this apparent equilibrium constant approaches a limiting value because the reaction becomes

$$2\text{D}(\text{O}_2)_2 = \text{T}(\text{O}_2)_4 \tag{7.3-10}$$

in the limit of infinite $[\text{O}_2]$. The value of this apparent equilibrium constant at very high oxygen concentrations is given by

$$K'' = \exp\left\{-\frac{[\Delta_f G'^0(\text{TotT}(\text{O}_2)_4) - 2\Delta_f G'^0(\text{D}(\text{O}_2)_2)]}{RT}\right\} = 9.50814 \times 10^5 \tag{7.3-11}$$

Note that this equilibrium constant is 5×10^4 times smaller than for $2\text{D} = \text{T}$ (see Problem 7.3). This is not unexpected because of the cooperative effect in the tetramer.

■ 7.4 EXPERIMENTAL DETERMINATION OF SEVEN APPARENT EQUILIBRIUM CONSTANTS

The fractional saturation of tetramer Y_T and the fractional saturation of dimer Y_D are functions only of $[\text{O}_2]$ at specified T, P, pH, etc., as shown by equations 7.1-18 and 7.3-6. However, since the tetramer form is partially dissociated into dimers, the **fractional saturation of heme** Y is a function of both $[\text{O}_2]$ and [heme]. Ackers and Halvorson (1974) derived an expression for the function $Y([\text{O}_2], [\text{heme}])$. When Legendre transforms are used, a simpler form of this function is obtained, and it can be used to derive limiting forms at high and low [heme]. These limiting forms are of interest because they show that if data can be obtained in regions where Y is linear in some function of [heme], extrapolations can be made to obtain Y_T and Y_D. These fractional saturations can be analyzed separately to obtain the Adair constants for the tetramer and the dimmer (Alberty, 1997a).

When the tetramer and dimer are in equilibrium, the fractional saturation of heme is given by

$$Y = f_\text{D} Y_\text{D} + f_\text{T} Y_\text{T} \tag{7.4-1}$$

where $f_\text{D} = 2[\text{TotD}]/[\text{heme}]$ is the fraction of the heme in the dimer and $f_\text{T} = 4[\text{TotT}]/[\text{heme}]$ is the fraction of the heme in the tetramer. Since $f_\text{T} = 1 - f_\text{D}$, equation 7.4-1 can be written

$$Y = Y_\text{T} + f_\text{D}(Y_\text{D} - Y_\text{T}) \tag{7.4-2}$$

The concentration of heme in the solution is given by

$$[\text{heme}] = 2[\text{TotD}] + 4[\text{TotT}] = 2[\text{TotD}] + 4K''[\text{TotD}]^2 \tag{7.4-3}$$

where equation 7.3-8 has been used in writing the last form. Applying the quadratic formula shows that the equilibrium concentration of TotD is given by

$$[\text{TotD}] = \frac{-2 + (4 + 16K''[\text{heme}])^{1/2}}{8K''} \tag{7.4-4}$$

The fraction of heme in the dimer is given by

$$f_{\text{D}} = \frac{2[\text{TotD}]}{2[\text{TotD}] + 4[\text{TotT}]} = \frac{1}{1 + 2K''[\text{TotD}]} \tag{7.4-5}$$

Substituting equation 7.4-4 into equation 7.4-5 yields

$$f_{\text{D}} = \frac{2}{1 + (1 + 4K''[\text{heme}])^{1/2}} \tag{7.4-6}$$

Substituting this equation into equation 7.4-2 yields

$$Y = Y_{\text{T}} + \frac{2(Y_{\text{D}} - Y_{\text{T}})}{1 + (1 + 4K''[\text{heme}])^{1/2}} \tag{7.4-7}$$

At specified $[O_2]$, Y is a function only of [heme] in a way that is described by three parameters, Y_{T}, Y_{D}, and K''. This equation is the same as that derived by Ackers and Halvorson (1974), although it has a rather different form.

If it were possible to titrate hemoglobin with oxygen at sufficiently high [heme], Y_{T} could be obtained directly. However, for the values of the seven equilibrium constants obtained by Mills, Johnson, and Ackers (1976), the tetramer is partially dissociated at the highest practical heme concentrations of about 5 mM. Equation 7.4-7 indicates that if Y can be determined at several high [heme], a linear extrapolation is possible (see equation 7.4-12). As is evident from equation 7.4-7, the question as to whether [heme] is high or low depends on whether $[\text{heme}] > \frac{1}{4}K''$ or $[\text{heme}] < \frac{1}{4}K''$. Of course, this criterion depends on $[O_2]$. In considering plots of Y versus some function of [heme], $[\text{heme}] = \frac{1}{4}K''$ can be used to divide the dependence of Y on [heme] into high-heme and low-heme regions. If $4K''[\text{heme}] \gg 1$, equation 7.4-7 reduces to

$$Y = Y_{\text{T}} + \frac{(Y_{\text{D}} - Y_{\text{T}})}{(K'')^{1/2}[\text{heme}]^{1/2}} \tag{7.4-8}$$

Thus a plot of Y versus $[\text{heme}]^{-1/2}$ at a specified $[O_2]$ must approach linearity as [heme] is increased. The intercept of the limiting slope of a plot of Y versus $[\text{heme}]^{-1/2}$ at $[\text{heme}]^{-1/2} = 0$ is Y_{T}, and the limiting slope is $(Y_{\text{D}} - Y_{\text{T}})/(K'')^{1/2}$. This slope is determined by two factors, $(Y_{\text{D}} - Y_{\text{T}})$ and K''. The slope will be low at high $[O_2]$ because $Y_{\text{D}} - Y_{\text{T}}$ is small. The slope will be low at low $[O_2]$ because K'' is large. Once Y_{T} has been determined at a series of $[O_2]$ by use of extrapolations of this type, K_{41}', K_{42}', K_{43}', and K_{44}' can be calculated by the method of nonlinear least squares.

The values of Y_{D} at various $[O_2]$ can be determined by extrapolations at low [heme]. If $4K''[\text{heme}] \ll 1$, the square root term in equation 7.4-7 can be rewritten using

$$(1 + x)^{1/2} \approx 1 + \frac{x}{2} \qquad (x \ll 1) \tag{7.4-9}$$

to obtain

$$Y = Y_{\text{T}} + \frac{(Y_{\text{D}} - Y_{\text{T}})}{1 + K''[\text{heme}]} \tag{7.4-10}$$

Since $4K''[\text{heme}] \ll 1$ is satisfied, we can use

$$\frac{1}{1 + x} \approx 1 - x \qquad (x \ll 1) \tag{7.4-11}$$

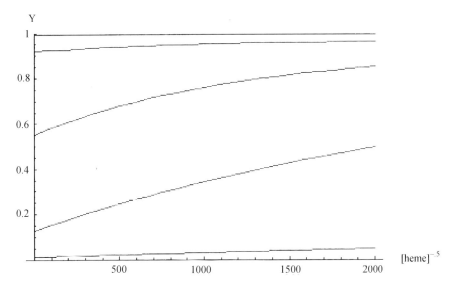

Figure 7.3 Calculated plots of Y versus $[\text{heme}]^{-1/2}$ at five $[O_2]$, expressed as molar concentrations. Starting at the top the oxygen conentrations are 10^{-4}, 2×10^{-5}, 10^{-5}, 5×10^{-6}, and 10^{-6} M. The intercepts give values of Y_T. [Reprinted from R. A. Alberty, *Biophys. Chem.* 63, 119–132 (1997), with permission from Elsevier Science.]

to obtain

$$Y = Y_D - (Y_D - Y_T)K''[\text{heme}] \qquad (7.4\text{-}12)$$

Thus as $[\text{heme}] \to 0$, Y becomes a linear function of $[\text{heme}]$, and Y approaches Y_D in the limit of $[\text{heme}] = 0$. The condition that $4K''[\text{heme}] \ll 1$ is hard to satisfy experimentally because low concentrations of heme have to be used. There is a steep slope at low $[O_2]$ because $(Y_D - Y_T)$ and K'' are both large. Note that when the slope of a plot is large, the determination of the intercept is more uncertain. At high $[O_2]$ the slope will be smaller because $(Y_D - Y_T)$ and K'' are smaller. The determination of Y_D at a series of $[O_2]$ yields K'_{21} and K'_{22}.

Once Y_T and Y_D have been determined by extrapolation, the slope of each plot at specified $[O_2]$ yields K''. It is also possible to calculate K'' from any measured Y value by use of equation 7.4-7 written in the form

$$\frac{[2(Y_D - Y_T)/Y - Y_T]^2 - 1}{4[\text{heme}]} = K'' \qquad (7.4\text{-}13)$$

The determination of K'' at a series of $[O_2]$ and knowledge of Y_T and Y_D as a function of $[O_2]$ makes it possible to calculate $^{\circ}K'_2$ (see equation 7.3-9), the apparent association constant for the reaction $2D = T$ at the specified T, P, pH, $[Cl^-]$, ionic strength, etc.

To show how the limiting forms of Y as a function of $[\text{heme}]$ can be used to determine all the equilibrium constants for the binding of oxygen by hemoglobin that is partially dissociated into dimers, values of Y were calculated with the parameters at $21.4°C$, pH 7.4, $[Cl^-] = 0.2$ M, and 0.2 M ionic strength. These calculated values of Y were then plotted versus $[\text{heme}]^{-1/2}$. In Fig. 7.3, the intercepts at the Y axis corrrespond to bindings at very high heme concentrations where the dissociation into dimers is negligible. Thus the intercepts can be used to calculate the four equilibrium constants for the tetramer. These plots show that the extrapolation becomes linear as $[\text{heme}]^{-1/2}$ is reduced. Since the limiting slope is $(Y_D - Y_T)/(K'')^{1/2}$, the value of K'' at a particular $[O_2]$ can be calculated from the limiting slope after Y_T has been determined.

To determine the binding constants for the dimer, Y is plotted versus $[\text{heme}]$ at the lowest possible heme concentrations, as shown in Fig. 7.4. The intercepts

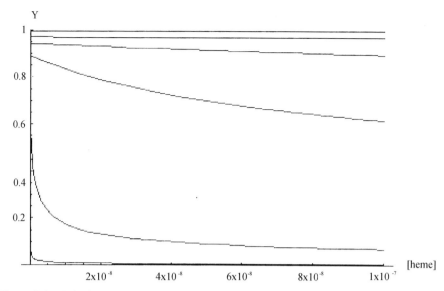

Figure 7.4   Calculated values of Y versus [heme] at heme concentrations starting at the top of 10^{-4}, 2×10^{-5}, 10^{-5}, 5×10^{-6}, 10^{-6}, and $10^{-7}\,M$. The intercepts give the values of Y_D. (see Problem 7.5). [Reprinted from R. A. Alberty, *Biophys. Chem.* **63**, 119–132 (1997), with permission from Elsevier Science.]

at the Y-axis correspond to the bindings by the dimer. Thus the intercepts can be used to calculate the two equilibrium constants for the binding of oxygen by the dimer. The plots show that the extrapolation becomes linear as [heme] is reduced to low values, but these have to be very low values, especially at $[O_2]$ that half saturate the dimer. The direct determination of K'_{21} and K'_{22} from oxygen binding experiments will require very low [heme], which has not yet been achieved in oxygen binding experiments. This may be achievable using long absorption cells, multipath cells, or a Fourier transform spectrometer. Since the limiting slope is $(Y_D - Y_T)/8.2426(K'')^{1/2}$, the value of K'' at a particular $[O_2]$ can be determined in this way. A check on the values of the seven apparent equilibrium constants is that they can be used to calculate the shapes of both of these plots, including the nonlinear regions.

■ 7.5   DISSOCIATION OF A DIPROTIC ACID

Before discussing the effect of pH on protein-ligand equilibria, it is necessary to discuss an aspect of acid dissociations that was too advanced for Chapter 1. Consider a protein A that has two acid groups. The acid dissociation constants are defined by

$$HA^- = H^+ + A^{2-} \qquad K_1 \tag{7.5-1}$$

$$H_2A = H^+ + HA^- \qquad K_2 \tag{7.5-2}$$

The binding polynomial for this system is

$$P = 1 + \frac{[H^+]}{K_1} + \frac{[H^+]^2}{K_1 K_2} \tag{7.5-3}$$

The average binding of hydrogen ions is given by

$$\bar{N}_H = \frac{[H^+]dP}{Pd[H^+]} = \frac{[H^+]d\ln P}{d[H^+]} \tag{7.5-4}$$

The plot of $\bar{N}_H$ versus pH is the titration curve for H_2A. Note that $\ln P$ can be calculated by integrating $(\bar{N}_H/[H^+])d[H^+]$. Applying equation 7.5-4 to equation 7.5-3 yields

$$\bar{N}_H = \frac{([H^+]/K_1) + (2[H^+]^2/K_1K_2)}{1 + ([H^+]/K_1) + ([H^+]^2/K_1K_2)} \tag{7.5-5}$$

At very high pH, the binding of H^+ approaches zero, and at very low pH it approaches 2.

There is another way to look at this binding, and that is to assume that the two groups are independent. In this case the dissociation reactions are written

$$HA^- = H^+ + A^{2-} \qquad \kappa_1 \tag{7.5-6}$$

$$HAH = H^+ + HA^- \qquad \kappa_2 \tag{7.5-7}$$

In other words, κ_1 is the dissociation constant for the left hydrogen atom and κ_2 is the dissociation constant for the right hydrogen atom. In this case the binding is simply the sum of the bindings at the two sites:

$$\bar{N}_H = \frac{[H^+]/\kappa_1}{1 + ([H^+]/\kappa_1)} + \frac{[H^+]/\kappa_2}{1 + ([H^+]/\kappa_2)} \tag{7.5-8}$$

This equation can be rearranged to

$$\bar{N}_H = \frac{[H^+]((1/\kappa_1) + (1/\kappa_2)) + 2[H^+]^2/\kappa_1\kappa_2}{1 + [H^+]((1/\kappa_1) + (1/\kappa_2)) + [H^+]^2/\kappa_1\kappa_2} \tag{7.5-9}$$

Thus

$$K_1 = \frac{\kappa_1\kappa_2}{\kappa_1\kappa_2} \tag{7.5-10}$$

$$K_2 = \kappa_1 + \kappa_2 \tag{7.5-11}$$

If $\kappa_1 \ll \kappa_2$, K_1 is essentially equal to κ_1 and K_2 is essentially equal to κ_2. When the groups are independent, the binding polynomial is given by

$$P = 1 + [H^+]\left(\frac{1}{\kappa_1} + \frac{1}{\kappa_2}\right) + \frac{[H^+]^2}{\kappa_1\kappa_2} = \left(1 + \frac{[H^+]}{\kappa_1}\right)\left(1 + \frac{[H^+]}{\kappa_2}\right) \tag{7.5-12}$$

K_1 and K_2 can be evaluated by curve fitting the plot of P (equation 7.5-3) versus $[H^+]$, or better $\log P$ versus $\log[H^+]$. If these two dissociation constants have nearly the same magnitude, a quadratic has to be solved to evaluate κ_1 and κ_2:

$$\kappa_2 = \frac{K_2 \pm \sqrt{K_2^2 - 4K_1K_2}}{2} \tag{7.5-13}$$

Note that K_1 cannot be equal to K_2. If the two groups are identical with dissociation constants κ (in equations 7.5-6 and 7.5-7), $K_1 = \kappa/2$ and $K_2 = 2\kappa$.

If $\kappa_1 = \kappa_2$,

$$P = \left(1 + \frac{[H^+]}{\kappa}\right)^2 \tag{7.5-14}$$

$$\ln P = 2\ln\left(1 + \frac{[H^+]}{\kappa}\right) \tag{7.5-15}$$

Equation 7.5-4 yields

$$\bar{N}_H = 2\frac{[H^+]/\kappa}{1 + ([H^+]/\kappa)} \tag{7.5-16}$$

■ 7.6 EFFECT OF pH ON PROTEIN-LIGAND EQUILIBRIA

When the binding of a ligand by a protein is accompanied by the production or consumption of hydrogen ions, the apparent dissociation constant K' for the protein-ligand complex will be a function of the pH. The apparent dissociation constant is defined by

$$PL_{tot} = P_{tot} + L_{tot} \qquad K' = \frac{[P_{tot}][L_{tot}]}{[PL_{tot}]} \qquad (7.6\text{-}1)$$

The abbreviations for reactants represent sums of species at a specified pH, and the expression for the equilibrium constant is written in terms of concentrations because K' is taken to be a function of ionic strength as well as pH. The pH dependence of K' can be written in terms of the binding polynomials of the three reactants (see equation 1.4-8):

$$K' = \frac{K_{ref} P(P_{tot}) P(L_{tot})}{P(PL_{tot})} \qquad (7.6\text{-}2)$$

where $K_{ref} = [P][L]/[PL]$ is written in terms of species for a reaction that is independent of pH. The binding polynomials in equation 7.6-2 include all weak acid groups in the three reactants.

If the ligand does not have pKs in the pH range studied, $P(L_{tot}) = 1$. In this case hydrogen ions are produced or consumed when there are acid groups in the binding site that have pKs in the pH range studied and the pKs of these groups are changed by the binding of the ligand. If the various acid dissociations of the protein are independent, the binding polynomials are written as products of terms of the form $(1 + 10^{-pH+pK_1})^{n_1}(1 + 10^{-pH+pK_2})^{n_2}$, when there are n_1 groups with pK_1 and n_2 groups with pK_2. However, it is not necessary to go this far in making assumptions. If the acid groups in the binding site are independent of the acid groups in the rest of the protein molecule, the binding polynomial for the protein is given by (Alberty, 2000d)

$$P(P_{tot}) = P(P_{nonsite}) P(P_{site}) \qquad (7.6\text{-}3)$$

where $P(P_{nonsite})$ is the binding polynomial for the acid groups outside of the binding site and $P(P_{site})$ is the binding polynomial for the acid groups in the unoccupied binding site. In this equation the acid groups in the binding site are defined as the acid groups in the protein that undergo a shift in pK when the ligand is bound. This major step in the treatment of the effect of the binding of a ligand by a protein is possible if the binding of the ligand changes the pKs of only some of the acid dissociations of the protein.

The corresponding binding polynomial for the protein with the binding site occupied (PL$_{tot}$) is given by

$$P(PL_{tot}) = P(P_{nonsite}) P(PL_{site}) \qquad (7.6\text{-}4)$$

where $P(P_{nonsite})$ is the same polynomial that is in equation 7.6-3 and $P(PL_{site})$ has the same number of terms as $P(P_{site})$. The nonsite groups can interact with each other, the site groups in the protein can interact with each other, and the site groups in the protein-ligand complex can interact with each other.

Substituting equations 7.6-3 and 7.6-4 in 7.6-2 when the ligand does not have pKs in the pH range considered yields a simpler equation for the dependence of K' on pH:

$$K' = \frac{K_{ref} P(P_{site})}{P(PL_{site})} \qquad (7.6\text{-}5)$$

This equation is important because it shows that the pH dependence of K' is determined entirely by the pKs of the acid groups in the binding site when it is unoccupied and in the binding site when it is occupied by ligand. If the number of acid groups in the site is small, their pKs in the unoccupied and occupied site can be calculated from the dependence of K' on pH. Thus the determination of the pH dependence of K' can yield important information about the acid groups in the binding site. If the effect of temperature on K' is studied, the standard transformed enthalpies and entropies of the site can also be determined.

As an application of equation 7.6-5, consider the effect of pH on the inhibition constants of fumarase, which have been determined for succinate, D-tartrate, L-tartrate, and meso-tartrate inhibitors (Wigler and Alberty, 1960). The kinetics of the conversion of fumarate to L-malate and the inhibition by these competitive inhibitors indicate that there are two acid groups in the catalytic site that affect the binding:

$$
\begin{array}{ccc}
PL_{site} = & P_{site} + L \\
\parallel & \parallel \\
HL_{site} & HP_{site} + L & \qquad (7.6\text{-}6) \\
\parallel & \parallel \\
H_2PL_{site} & H_2P_{site} + L
\end{array}
$$

The binding by a competitive inhibitor can be represented by equation 7.6-5.

The binding polynomial for the unoccupied site in the protein is given by

$$ P(P_{site}) = 1 + 10^{-pH + pK_{PS1}} + 10^{-2pH + pK_{PS1} + pK_{PS2}} \qquad (7.6\text{-}7) $$

The binding polynomial for the site when it is occupied by a competitive inhibitor is given by

$$ P(PL_{site}) = 1 + 10^{-pH + pK_{PLS1}} + 10^{-2pH + pK_{PLS1} + pK_{PLS2}} \qquad (7.6\text{-}8) $$

Substituting equations 7.5-7 and 7.5-8 in equation 7.5-5 yields

$$ K' = K_{ref} \frac{1 + 10^{-pH + pK_{PS1}} + 10^{-2pH + pK_{PS1} + pK_{PS2}}}{1 + 10^{-pH + pK_{PLS1}} + 10^{-2pH + pK_{PLS1} + pK_{PLS2}}} \qquad (7.6\text{-}9) $$

The pKs of the acid groups in the catalytic site of fumarase at 298 K and an ionic strength of 0.01 M are given in Table 7.4 along with the equilibrium constants for the reference reactions. More information on the experimental determination of these parameters is available in Wigler and Alberty (1960). The pKs for the two acid groups in the unoccupied catalytic site of fumarase are $pK_{PS1} = 6.9$ and $pK_{PS2} = 6.3$. These two acid groups can be considered to be identical and independent because their difference is $0.6 = \log 4$. The pKs for site-L-tartrate are $pK_{PLS1} = 7.5$ and $pK_{PLS2} = 7.4$, indicating there is a **cooperative effect** because they are closer than 0.6 pH units (see Section 1.2). The acid dissociations of the ligands are ignored in these calculations because we are primarily concerned with what happens in the neighborhood of pH 7. With these values of the pKs, the pH dependence of the apparent equilibrium constant for

Table 7.4 pKs of Acid Groups in the Catalytic Site in Fumarase at 298.15 K and an Ionic Strength of 0.01 M

Ligand	pK$_1$	pK$_2$	K$_{ref}$
Unoccupied	6.9	6.3	
succinate	7.5	6.5	1.2×10^{-3}
D-tartrate	7.8	6.9	2.5×10^{-3}
L-tartrate	7.5	7.4	4.1×10^{-3}
meso-tartrate	7.1	5.7	4.6

Source: With permission from P. W. Wigler and R. A. Alberty, *J. Am. Chem. Soc.* 82, 5482–5488(1960). Copyright 1960 American Chemical Society.

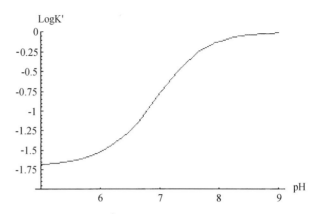

Figure 7.5 Plot of log K' for the reaction site-L-tartrate = site + L-tartrate over a range of pH at 25°C and an ionic strength of 0.01 M. [With permission from R. A. Alberty, *J. Phys. Chem. B* 104, 9929–9934 (2000). Copyright 2000 American Chemical Society.]

the dissociation of L-tartrate from the complex is given by

$$K' = K_{\text{ref}} \frac{1 + 10^{-\text{pH}+6.9} + 10^{-2\text{pH}+13.2}}{1 + 10^{-\text{pH}+7.5} + 10^{-2\text{pH}+14.9}} \tag{7.6-10}$$

The base 10 logarithm of K' calculated using equation 7.6-10 is given as a function of pH in Fig. 7.5. The constant $K_{\text{ref}} = 4.1 \times 10^{-3}$ has been omitted in making Fig. 7.5 because it is not involved in the pH dependence.

The change in binding of hydrogen ions in the dissociation of the site-L-tartrate complex can be calculated by taking the derivative of log K' with respect to pH (equation 4.7-5). The pH dependence of $\Delta_r N_{\text{H}}$ is shown in Fig. 7.6.

Since the products (unoccupied site plus L-tartrate) bind hydrogen ions less strongly than the complex, $\Delta_r N_{\text{H}}$ is negative at all pH values. Another way to express this is that hydrogen ions are produced in the dissociation, except in the limit of very high and very low pH values where $\Delta_r N_{\text{H}} = 0$.

The preceding discussion has been concerned with the apparent dissociation constant of a protein-ligand complex and the change in binding of H^+ in the dissociation of the ligand. Now we return to the discussion of the hydrogen ion binding (Chapter 1) of the unoccupied site of fumarase and especially the site occupied by L-tartrate. The average number of hydrogen ions bound by the

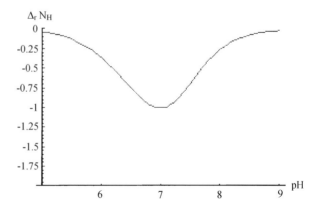

Figure 7.6 Plot of $\Delta_r N_{\text{H}}$ for the reaction site-L-tartrate = site + L-tartrate over a range of pH at 25°C and an ionic strength of 0.01 M. [With permission from R. A. Alberty, *J. Phys. Chem. B* 104, 9929–9934 (2000). Copyright 2000 American Chemical Society.]

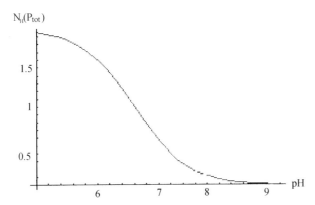

Figure 7.7 Plot of the average number of hydrogen ions bound by the unoccupied catatytic site of fumarase at 25°C and an ionic strength of 0.10 M. [With permission from R. A. Alberty, *J. Phys. Chem. B* 104, 9929–9934 (2000). Copyright 2000 American Chemical Society.]

catalytic site is given by $\bar{N}_H = - (1/\ln 10) d \ln P(P_{site})/pH$. This quantity for the unoccupied site is plotted in Fig. 7.7.

This plot has the same shape as the titration curve of a single site because $K_2 = 4K_1$, but with the ordinate multiplied by 2.

The **binding capacity** for hydrogen ions is defined by

$$\frac{d\bar{N}_H}{dpH} = -\frac{1}{\ln(10)} \frac{d^2 \ln P}{dpH^2} \qquad (7.6\text{-}11)$$

Di Cera, Gill, and Wyman (1988) adopted this name because this quantity is analogous to the heat capacity, which is given by the second derivative of the Gibbs energy G with respect to temperature (equation 2.5-25). They pointed out that the binding capacity is a measure of cooperativity.

The binding capacity for the unoccupied site, which is calculated using equation 7.6-11, is plotted versus pH in Fig. 7.8. The number of hydrogen ions bound by the catalytic site in the fumarase-L -tartrate complex is plotted in Fig. 7.9. This is steeper than the titration curve of a diprotic acid with identical and independent groups. The binding capacity for the site occupied by meso-tartrate is shown in Fig. 7.10. The slope of the binding curve is steeper than for the unoccupied site shown in Fig. 7.6, as expected since the binding is cooperative.

The preceding example of the determination of the pKs of acid groups in the binding site of a protein and the binding capacity is based on the study of the

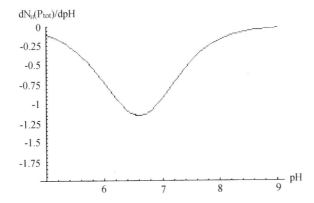

Figure 7.8 Plot of the binding capacity (see equation 7.6-11) for an unoccupied catalytic site of fumarase at 25°C and an ionic strength of 0.01 M. [With permission from R. A. Alberty, *J. Phys. Chem. B* 104, 9929–9934 (2000). Copyright 2000 American Chemical Society.]

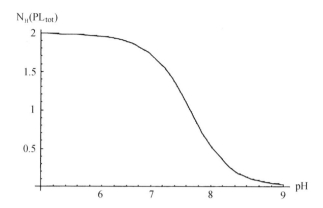

Figure 7.9   Plot of the average number of hydrogen ions bound $\bar{N}_H$ by the catalytic site of fumarase occupied by L-tartrate at 25°C and an ionic strength of 0.01 M. [With permission from R. A. Alberty, *J. Phys. Chem. B* 104, 9929–9934 (2000). Copyright 2000 American Chemical Society.]

competitive inhibition of an enzyme, but the method of curve fitting the pH dependence of K' (see equation 7.6-9) can be used when the apparent equilibrium constant can be measured spectrophotometrically or by equilibrium dialysis (Klotz, 1997).

■ 7.7   CALCULATION OF STANDARD TRANSFORMED GIBBS ENERGIES OF FORMATION OF THE CATALYTIC SITE OF FUMARASE

The apparent equilibrium constant for a biochemical reaction at a specified pH can be calculated from the standard transformed Gibbs energies of formation of the reactants, and the standard transformed Gibbs energy of formation of the reactants are calculated using isomer group thermodynamics (see Section 4.5-1). Alberty (1999a) has shown that $\Delta_f G_i'^0$ for a biochemical reactant is given by

$$\Delta_f G_1'^0 = \Delta_f G_1'^0 - RT \ln P \tag{7.7-1}$$

where $\Delta_f G_1'^0$ is the standard transformed Gibbs energy of formation for the species with the fewest hydrogen atoms and P is the binding polynomial for the reactant. This equation can be applied to the enzymatic site for fumarase and to the complexes formed with competitive inhibitors (Alberty, 2000d).

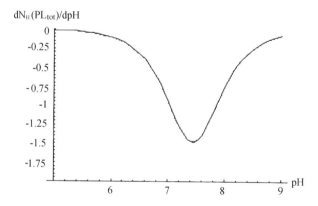

Figure 7.10   Plot of the binding capacity (see equation 7.5-11) for the catalytic site of fumarase occupied by L-tartrate at 25°C and an ionic strength of 0.01 M. [With permission from R. A. Alberty, *J. Phys. Chem. B* 104, 9929–9934 (2000). Copyright 2000 American Chemical Society.]

Table 7.5 Standard Transformed Gibbs Energies of Formation for the Catalytic Site of Fumarase in kJ mol^{-1} at 25°C and Ionic Strength 0.01 M

	pH 5	pH 6	pH 7	pH 8	pH 9
site	−18.39	−7.96	−1.66	−0.19	−0.02
succinate	−576.28	−553.45	−530.62	−507.78	−484.95
site-succinate	−615.86	−582.23	−551.35	−525.15	−501.70
D-tartrate	114.16	136.99	159.83	182.66	205.49
site-D−tartrate	72.45	106.43	138.75	166.52	190.48
L-tartrate	114.16	136.99	159.83	205.49	
site-L-tartrate	72.56	106.71	140.02	168.21	191.78
meso-tartrate	114.16	136.99	159.83	182.66	205.49
site meso-tartrate	101.51	133.36	161.52	186.15	209.24

Source: [With permission from R. A. Alberty, *J. Phys. Chem. B* 104, 9929–9934 (2000). Copyright 2000 American Chemical Society.]
*This table is based on the convention that $\Delta_f G^0 = 0$ at 25°C and zero ionic strength for the doubly charged ions of D-tartrate, L-tartarate, and meso-tartrate. In addition, the convention is that $\Delta_f G'^0 = 0$ for the binding site at high pH.

The apparent dissociation constant of the fumarase site-succinate complex to yield unoccupied site and succinate is represented by the following function of pH:

$$K' = 1.2 \times 10^{-3} \frac{1 + 10^{6.9 - \mathrm{pH}} + 10^{13.2 - 2\mathrm{pH}}}{1 + 10^{7.5 - \mathrm{pH}} + 10^{14.0 - 2\mathrm{pH}}} \qquad (7.7\text{-}2)$$

According to equation 7.1-19, this K' is given by

$$-RT \ln K' = \Delta_f G'^0(\text{site}) + \Delta_f G'^0(\text{succ}) - \Delta_f G'^0(\text{site-succ}) \qquad (7.7\text{-}3)$$

The value of $\Delta_f G^0(\text{succ}^{2-})$ at 25°C and zero ionic strength is -690.44 kJ mol^{-1}, and the pH dependence of $\Delta_f G'^0(\text{succ})$ is given by $-690.44 - 4RT \ln 10^{-\mathrm{pH}}$ (see equation 4.4-10), neglecting the effect of the binding hydrogen ions at lower pH values. This value is independent of the ionic strength because $z_i^2 = N_\mathrm{H}(i)$. $\Delta_f G'^0(\text{site})$ is taken as zero in the limit of high pH by convention so that

$$\Delta_f G'^0(\text{site}) = -RT \ln(1 + 10^{6.9 - \mathrm{pH}} + 10^{13.3 - 2\mathrm{pH}}) \qquad (7.7\text{-}4)$$

Equation 7.7-3 can be written as

$$-RT \ln(1 + 10^{6.9 - \mathrm{pH}} + 10^{13.3 - 2\mathrm{pH}}) K'$$
$$= -RT \ln - 960.44 - 4RT \ln(10^{-\mathrm{pH}}) - \Delta_f G'^0(\text{site-succ}) \qquad (7.7\text{-}5)$$

Substituting equation 7.6-2 yields

$$\Delta_f G'^0(\text{site-succ}) = -707.11 - 4RT \ln(10^{-\mathrm{pH}})$$
$$- RT \ln(1 + 10^{7.5 - \mathrm{pH}} + 10^{14.0 - 2\mathrm{pH}}) \qquad (7.7\text{-}6)$$

The values of $\Delta_f G'^0$ for the catalytic site, succinate, and site-succinate calculated in this way are shown in Table 7.5. Similar calculations have been made for D-tartrate, L-tartrate, and meso-tartrate using data from Table 7.4. Since the $\Delta_f G^0$ values for these three reactants are not known, the convention has been adopted that they are equal to zero. Table 7.5 shows that standard transformed Gibbs energies of formation at specified pH values can be calculated for an unoccupied binding site and the binding site occupied by a ligand.

Phase Equilibrium in Aqueous Systems

When a system involves two or more phases, there is a single fundamental equation for the Gibbs energy that is the sum of the fundamental equations for the separate phases: $dG = dG_\alpha + dG_\beta + \cdots$. The fundamental equation for the system provides the criterion for spontaneous change and equilibrium. However, there is a separate Gibbs-Duhem equation for each phase because any intensive property of a phase is related to the other intensive properties of that phase. In the treatments here the amount of material in the interface is ignored on the assumption that the amounts there are negligible compared with the amounts in the bulk phases. The effects of small pressure differences between the phases are also ignored. New phases may form spontaneously under certain circumstances, but phases can also be separated by membranes with specified permeabilities.

Phase equilibrium across semipermeable membranes is of special interest in biological applications. First, we will consider two-phase aqueous systems without chemical reactions, then introduce reactions, and finally electric potential differences between phases. The numbers of intensive degrees of freedom F and extensive degrees of freedom D have been discussed in Chapter 3, and F' and D' at specified pH have been discussed in Chapter 4. That discussion is continued here. The distribution of carbon dioxide between the gas phase and aqueous solution is discussed as a function of pH and ionic strength.

■ 8.1 TWO-PHASE SYSTEMS WITHOUT CHEMICAL REACTIONS

One Species, Two Phases

This system is not useful for representing a biochemical system but is needed as a foundation. The fundamental equation for G for a system containing alpha and beta phases

$$dG = -SdT + VdP + \mu_{A\alpha}dn_{A\alpha} + \mu_{A\beta}dn_{A\beta} \qquad (8.1\text{-}1)$$

This shows that the natural variables for G for this system before phase equilibrium is established are T, P, $n_{A\alpha}$, and $n_{A\beta}$. When A is transferred from one phase to the other, $dn_{A\alpha} = -dn_{A\beta}$. Substituting this conservation relation into equation 8.1-1 yields

$$dG = -SdT + VdP + (\mu_{A\alpha} - \mu_{A\beta})dn_{A\alpha} \qquad (8.1\text{-}2)$$

which shows that $\mu_{A\alpha} = \mu_{A\beta} = \mu_A$ at phase equilibrium. Substituting this equilibrium condition in equation 8.1-1 yields

$$dG = -SdT + VdP + \mu_A(dn_{A\alpha} + dn_{A\beta}) = -SdT + VdP + \mu_A dn_{cA} \qquad (8.1\text{-}3)$$

where n_{cA} is the amount of component A in the two-phase system. This fundamental equation shows that the system at phase equilibrium has $D = 3$ natural variables, which seems to suggest T, P, and n_{cA}. However, we will see in the next paragraph that this is not a suitable choice of natural variables. The Gibbs-Duhem equations for the two phases at phase equilibrium are

$$0 = -S_\alpha dT + V_\alpha dP - n_{A\alpha}d\mu_A \qquad (8.1\text{-}4)$$

$$0 = -S_\beta dT + V_\beta dP - n_{A\beta}d\mu_A \qquad (8.1\text{-}5)$$

Eliminating $d\mu_A$ between these two equations yields the **Clapeyron equation**

$$\frac{dP}{dT} = \frac{\Delta H_{mA}}{T\Delta V_{mA}} \qquad (8.1\text{-}6)$$

where $\Delta H = H_{mA\alpha} - H_{mA\beta} = T(S_{mA\alpha} - S_{mA\beta})$ is the change in molar enthalpy and ΔV_m is the change in molar volume in the phase change. Thus the pressure can be taken to be a function of the temperature, or the temperature can be taken to a function of the pressure. This indicates that a two-phase system with a single species has a single independent intensive variable, in agreement with the phase rule $F = C - p + 2 = 1 - 2 + 2 = 1$.

Using the Clapeyron equation to eliminate dP from equation 8.1-1 and substituting $\mu_{A\alpha} = \mu_{A\beta}$ yields

$$dG = \left(-S + V\frac{\Delta H_{mA}}{T\Delta V_{mA}}\right)dT + \mu_{A\alpha}dn_{A\alpha} + \mu_{A\alpha}dn_{A\beta} \qquad (8.1\text{-}7)$$

This form of the fundamental equation, which applies at equilibrium, indicates that the natural variables for this system are T, $n_{A\alpha}$, and $n_{A\beta}$. Alternatively, P, $n_{A\alpha}$, and $n_{A\beta}$ could be chosen. Specification of the natural variables gives a complete description of the extensive state of the system at equilibrium, and so the criterion of spontaneous change and equilibrium is $dG \leq 0$ at constant T, $n_{A\alpha}$, and $n_{A\beta}$ or

$\mathrm{d}G \leqslant 0$ at constant P, $n_{A\alpha}$, and $n_{A\beta}$. The natural variables for a multiphase system must include extensive variables that are related to the sizes of all the phases, since the amounts of the various phases are independent variables. The number D of natural variables is in agreement with $D = F + p$ because this yields $D = 1 + 2 = 3$. In summary, a suitable choice of natural variables includes F intensive variables and p extensive variables, which may be taken as the amounts in the p phases.

Two Species, Two Phases

The fundamental equation for G is

$$\mathrm{d}G = -S\mathrm{d}T + V\mathrm{d}P + \mu_{A\alpha}\mathrm{d}n_{A\alpha} + \mu_{B\alpha}\mathrm{d}n_{B\alpha} + \mu_{A\beta}\mathrm{d}n_{A\beta} + \mu_{B\beta}\mathrm{d}n_{B\beta} \quad (8.1\text{-}8)$$

This can be used to derive two equilibrium expressions that convert this fundamental equation to its form at phase equilibrium, which is

$$\mathrm{d}G = -S\mathrm{d}T + V\mathrm{d}P + \mu_A\mathrm{d}n_{cA} + \mu_B\mathrm{d}n_{cB} \quad (8.1\text{-}9)$$

where n_{cA} is the amount of the A in the system and n_{cB} is the amount of B in the system. This shows that the system at phase equilibrium has $D = 4$ natural variables, which can be taken to be T, P, n_{cA} and n_{cB}. The Gibbs-Duhem equations for the two phases at equilibrium are

$$0 = -S_\alpha\mathrm{d}T + V_\alpha\mathrm{d}P - n_{A\alpha}\mathrm{d}\mu_A - n_{B\alpha}\mathrm{d}\mu_B \quad (8.1\text{-}10)$$

$$0 = -S_\beta\mathrm{d}T + V_\beta\mathrm{d}P - n_{A\beta}\mathrm{d}\mu_A - n_{B\beta}\mathrm{d}\mu_B \quad (8.1\text{-}11)$$

$\mathrm{d}\mu_B$ can be eliminated between these two equations to obtain μ_A as a function of T and P. Thus $F = 2$. The relation $D = F + p$ is satisfied, and the natural variables can be taken to be T, P, n_{cA}, and n_{cB}, although it might be more convenient to use T, P, n_α, and n_β, where $n_\alpha = n_{A\alpha} + n_{B\alpha}$.

N_s Species, Two Phases

When there are N_s species the fundamental equation for G can be written

$$\mathrm{d}G = -S\mathrm{d}T + V\mathrm{d}P + \sum_{i=1}^{N_s} \mu_{i\alpha}\mathrm{d}n_{i\alpha} + \sum_{i=1}^{N_s} \mu_{i\beta}\mathrm{d}n_{i\beta} \quad (8.1\text{-}12)$$

where there is a term for each species in each phase. This equation can be used to derive the N_s equilibrium conditions $\mu_{i\alpha} = \mu_{i\beta}$. Substituting these equilibrium conditions in the fundamental equation yields

$$\mathrm{d}G = -S\mathrm{d}T + V\mathrm{d}P + \sum_{i=1}^{C} \mu_i\mathrm{d}n_{ci} \quad (8.1\text{-}13)$$

The number C of components is equal to the number of terms in the summations in equation 8.1-12 minus the number N_s of independent equilibria between phases, that is, $C = 2N_s - N_s = N_s$. Equation 8.1-13 shows that there are $D = C + 2 = N_s + 2$ natural variables. The Gibbs-Duhem equations for the two phases are

$$0 = -S_\alpha\mathrm{d}T + V_\alpha\mathrm{d}P - \sum_{i=1}^{N_s} n_{i\alpha}\mathrm{d}\mu_i \quad (8.1\text{-}14)$$

$$0 = -S_\beta\mathrm{d}T + V_\beta\mathrm{d}P - \sum_{i=1}^{N_s} n_{i\beta}\mathrm{d}\mu_i \quad (8.1\text{-}15)$$

Since there are $N_s + 2$ intensive variables and two relations between them, $F = N_s$. This is in agreement with $D = F + p$. The criterion for spontaneous change and equilibrium for this system is $\mathrm{d}G \leqslant 0$ at constant T, P, and $\{n_{ci}\}$.

■ 8.2 TWO-PHASE SYSTEM WITH A CHEMICAL REACTION AND A SEMIPERMEABLE MEMBRANE

Consider an aqueous two-phase system containing A, B, C, and solvent H_2O in which the reaction $A + B = C$ occurs. The two phases are separated by a membrane, and the membrane is permeable to all four species. The fundamental equation for the Gibbs energy of the α phase is

$$dG_\alpha = -S_\alpha dT + V_\alpha dP + \mu_{A\alpha}dn_{A\alpha} + \mu_{B\alpha}sn_{B\alpha} + \mu_{C\alpha}dn_{C\alpha} + \mu_{H_2O\alpha}dn_{H_2O\alpha} \quad (8.2\text{-}1)$$

The corresponding equation for the β phase is

$$dG_\beta = -S_\beta dT + V_\beta dP + \mu_{A\beta}dn_{A\beta} + \mu_{B\beta}dn_{B\beta} + \mu_{C\beta}dn_{C\beta} + \mu_{H_2O\beta}dn_{H_2O\beta} \quad (8.2\text{-}2)$$

These equations can be used to show that at chemical equilibrium $\mu_{A\alpha} + \mu_{B\alpha} = \mu_{C\alpha}$ and $\mu_{A\beta} + \mu_{B\beta} = \mu_{C\beta}$. Substituting these equilibrium conditions into equation 8.2-1 yields

$$dG_\alpha = -S_\alpha dT + V_\alpha dP + \mu_{A\alpha}dn_{cA\alpha} + \mu_{B\alpha}dn_{cB\alpha} + \mu_{H_2O\alpha}dn_{H_2O\alpha} \quad (8.2\text{-}3)$$

where the amount of the A component is $n_{cA\alpha} = n_{A\alpha} + n_{C\alpha}$ and the amount of the B component is $n_{cB\alpha} = n_{B\alpha} + n_{C\alpha}$. The corresponding fundamental equation for the β phase is

$$dG_\beta = -S_\beta dT + V_\beta dP + \mu_{A\beta}dn_{cA\beta} + \mu_{B\beta}dn_{cB\beta} + \mu_{H_2O\beta}dn_{H_2O\beta} \quad (8.2\text{-}4)$$

The fundamental equation for the Gibbs energy of the system at chemical equilibrium in each phase is the sum of equations 8.2-3 and 8.2-4:

$$dG = -SdT + VdP + \mu_{A\alpha}dn_{cA\alpha} + \mu_{B\alpha}dn_{cB\alpha} + \mu_{H_2O\alpha}dn_{H_2O\alpha}$$
$$+ \mu_{A\beta}dn_{cA\beta} + \mu_{B\beta}dn_{cB\beta} + \mu_{H_2O\beta}dn_{H_2O\beta} \quad (8.2\text{-}5)$$

This equation can be used to show that $\mu_{A\alpha} = \mu_{A\beta} = \mu_A$, $\mu_{B\alpha} = \mu_{B\beta} = \mu_B$, and $\mu_{H_2O\alpha} = \mu_{H_2O\beta} = \mu_{H_2O}$. When these phase equilibrium conditions are inserted in equation 8.2-5, it becomes

$$dG = -SdT + VdP + \mu_A dn_{cA} + \mu_B dn_{cB} + \mu_{H_2O}dn_{cH_2O} \quad (8.2\text{-}6)$$

where n_{ci} represents the amount of component i in the system. The amount of component A in the system is represented by $n_{cA} = n_{cA\alpha} + n_{cA\beta}$, and the amount of the solvent is represented by $n_{cH_2O} = n_{H_2O\beta} + n_{H_2O\beta}$. Equation 8.2-6 indicates that there are $D = 5$ natural variables, and that they might be taken to be T, P, n_{cA}, n_{cB}, and n_{cH_2O} in the criterion of spontaneous change and equilibrium: $dG \leqslant 0$ at constant T, P, n_{cA}, n_{cB}, and n_{cH_2O}.

The Gibbs-Duhem equations for the two phases at equilibrium can be derived from equations 8.2-3 and 8.2-4:

$$0 = -S_\alpha dT + V_\alpha dP - n_{cA\alpha}d\mu_A - n_{cB\alpha}d\mu_B - n_{H_2O\alpha}d\mu_{H_2O} \quad (8.2\text{-}7)$$

$$0 = -S_\beta dT + V_\beta dP - n_{cA\beta}d\mu_A - n_{cB\beta}d\mu_B - n_{H_2O\beta}d\mu_{H_2O} \quad (8.2\text{-}8)$$

where the phase subscripts on the chemical potentials have been dropped because of phase equilibrium. $d\mu_{H_2O}$ can be eliminated between these two equations, and the resulting equation can be solved for $d\mu_B$, which is a function of T, P, and μ_A. Thus there are three independent intensive properties, in agreement with $F = C - p + 2 = 3 - 2 + 2 = 3$. This is in agreement with $D = F + p = 3 + 2 = 5$. The natural variables for the expression of the criterion for spontaneous change and equilibrium based on G might be taken to be T, P, n_{cA}/n_{cB}, n_α, and n_β, where $n_\alpha = n_{A\alpha} + n_{B\alpha} + n_{C\alpha} + n_{S\alpha}$. Then the equilibrium condition would be $dG \leqslant 0$ at constant T, P, n_{cA}/n_{cB}, n_α, and n_β.

If the phases are both dilute aqueous solutions and the membrane separating the phases is permeable to all species, the species will have the same concentrations in the two phases, and an equilibrium relation of the form

$$K_\alpha = \frac{[C]_\alpha}{[A]_\alpha[B]_\alpha} = K_\beta = \frac{[C]_\beta}{[A]_\beta[B]_\beta} \qquad (8.2\text{-}9)$$

will be satisfied in each phase and in the system as a whole.

■ 8.3 TWO-PHASE SYSTEM WITH A MEMBRANE PERMEABLE BY A SINGLE ION

When two different phases are separated by a membrane permeable by a single ion and that ion has different activities on the two sides of the membrane, an electric potential difference will be set up at equilibrium (Alberty, 1995a, d, 1997). We first consider a two-phase system with an aqueous solution of a single salt on both sides of a membrane that is permeable only to cation C. When electrolytes are involved, it is necessary that **counterions** be present because bulk phases are electrically neutral. When cation C diffuses through the membrane into the β phase, the β phase becomes positively charged with respect to the α phase. Diffusion stops when a sufficient difference in electric potential has been established. When a conductor is charged, the charge migrates to the surface, and for an aqueous solution of a salt this occurs in the **charge relaxation time** of about one nanosecond. Thus a positively charged layer is formed at the surface of the membrane toward the β phase and a negatively charged layer is formed at the surface of the membrane toward the α phase. The thickness of the layer in each aqueous phase is the **Debye length** of about 1 nm at an ionic strength of 0.1 M. The amount of charge required to set up a significant potential difference between the phases is very small. Many biological membranes have **capacitances** of about one microfarad per square centimeter (Weiss, 1996). The charge transfer per square centimeter required to set up a potential difference of 0.1 V is therefore 10^{-12} mol of singly charged ions. As the electric potential of the β phase increases due to the diffusion of cation C, the process of diffusion slows and an equilibrium potential difference is reached.

The fundamental equations for G for the phases on either side of the center of the membrane are

$$dG_\alpha = -S_\alpha dT + V_\alpha dP + \mu_{C\alpha} dn_{C\alpha} + \phi_\alpha dQ_\alpha \qquad (8.3\text{-}1)$$

$$dG_\beta = -S_\beta dT + V_\beta dP + \mu_{C\beta} dn_{C\beta} + \phi_\beta dQ_\beta \qquad (8.3\text{-}2)$$

where $n_{C\alpha}$ is the amount of cation C in the α phase and Q is the amount of charge transferred across the center of the membrane. No term is included for the monovalent anion A because its concentration in the bulk phase is equal to that of the monovalent cation C. Since $dn_{A\alpha} = dn_{C\alpha}$, the inclusion of a term for A in equation 8.3-1 would yield $(\mu_{C\alpha} + \mu_{A\alpha})dn_{C\alpha}$. However, since the chemical potential can be defined for an arbitrary reference potential (cf, $\Delta_f G_i^0$), $\mu_{A\alpha}$ can be set equal to zero.

Since $-dQ_\alpha = dQ_\beta = dQ$, the fundamental equation for G for this two-phase system prior to the establishment of phase equilibrium is the sum of equation 8.3-1 and 8.3-2:

$$dG = -SdT + VdP + \mu_{C\alpha} dn_{C\alpha} + \mu_{C\beta} dn_{C\alpha} + (\phi_\beta - \phi_\alpha)dQ \qquad (8.3\text{-}3)$$

There is difference in electric potential across the membrane. So there is an electric field in the membrane, but there are no electric fields in the two bulk phases. The electrical work required to move charge dQ across the center of the membrane is $(\phi_B - \phi_\alpha)dQ$. The polarization of the membrane does not change in this process of charge transport because the potential difference is constant. Since $dn_{C\alpha} = -dn_{C\beta}$, equation 8.3-3 can be used to show that at phase equilibrium,

$\mu_{C\alpha} = \mu_{C\beta} = \mu_C$. Substitution of this relation in equation 8.3-3 yields

$$dG = -SdT + VdP + \mu_C dn_{cC} + (\phi_\beta - \phi_\alpha)dQ \qquad (8.3\text{-}4)$$

where n_{cC} is the amount of the component C. This indicates that there are four natural variables for this system, $D = 4$. Integration of equation 8.3-4 at constant values of the intensive variables yields

$$G = \mu_C n_{cC} + (\phi_\beta - \phi_\alpha)Q \qquad (8.3\text{-}5)$$

The Gibbs-Duhem equations for the two phases at phase equilibrium are

$$0 = -S_\alpha dT + V_\alpha dP - n_{C\alpha}d\mu_C - Q_\alpha d\phi_\alpha \qquad (8.3\text{-}6)$$

$$0 = -S_\beta dT + V_\beta dP - n_{C\beta}d\mu_C C - Q_\beta d\phi_\beta \qquad (8.3\text{-}7)$$

This looks like there are five intensive variables, but there are not because only the difference in electric potentials between the phases is important. We can take $\phi_\alpha = 0$ and delete the electric work term in equation 8.3-6. Since there are four intensive variables and two equations, $F = 2$, in agreement with $F = C - p + 2 = 2 - 2 + 2 = 2$. Note that Q_β is taken as a component. This leads to $D = F + p = 2 + 2 = 4$ in agreement with equation 8.3-4.

In considering the thermodynamics of systems in which there are electric potential differences, the **activity** a_i of an ion can be defined in terms of its chemical potential μ_i and the electric potential ϕ_i of the phase it is in by

$$\mu_i = \mu_i^0 + RT\ln a_i + Fz_i\phi_i \qquad (8.3\text{-}8)$$

where μ_i^0 is the **standard chemical potential** in a phase where the electric potential is zero, F is the Faraday constant, and z_i is the charge number. This is the arbitrary introduction of a property of a species, the activity, that is more convenient in making calculations than the chemical potential of the species. According to equation 8.3-8 the chemical potential of an ion is a function of ϕ_i as well as a_i. The activity here has the same functional dependence on intensive variables in the presence of electric potential differences as in the absence of electric potential differences. When equation 8.3-8 is substituted in $\mu_{C\alpha} = \mu_{C\beta}$, we obtain

$$\frac{a_{C\beta}}{a_{C\alpha}} = \exp\left(-\frac{Fz_C\phi_\beta}{RT}\right) \qquad (8.3\text{-}9)$$

or

$$-\frac{RT}{Fz_C}\ln\frac{a_{C\beta}}{a_{C\alpha}} = \phi_\beta \qquad (8.3\text{-}10)$$

based on the convention that $\phi_\alpha = 0$. This is referred to as the **membrane equation**, and it has been very useful in research on ion transport and nerve conduction. It is really a form of the Nernst equation (equation 9.1-4).

■ 8.4 TWO-PHASE SYSTEM WITH A CHEMICAL REACTION AND A MEMBRANE PERMEABLE BY A SINGLE ION

In this system the reaction A + B = C occurs in both phases, but only C can diffuse through the membrane (Alberty, 1997d). The fundamental equation for G is

$$dG = -SdT + VdP + \mu_{A\alpha}dn_{A\alpha} + \mu_{B\alpha}dn_{B\alpha} + \mu_{C\alpha}dn_{C\alpha} + \mu_{A\beta}dn_{A\beta} + \mu_{B\beta}dn_{B\beta}$$
$$+ \mu_{C\beta}dn_{C\beta} + (\phi_\beta - \phi_\alpha)dQ \qquad (8.4\text{-}1)$$

The reactions in the system are represented as

$$A_\alpha + B_\alpha = C_\alpha \tag{8.4-2}$$

$$A_\beta + B_\beta = C_\beta \tag{8.4-3}$$

$$C_\alpha = C_\beta \tag{8.4-4}$$

Note that the phase transfer is treated like a chemical reaction. When the equilibrium conditions for these three reactions are introduced into equation 8.4-1, we obtain

$$dG = -SdT + VdP + \mu_{A\alpha}dn_{cA\alpha} + \mu_{A\beta}dn_{cA\beta} + \mu_C dn_{cC} + (\phi_\beta - \phi_\alpha)dQ \tag{8.4-5}$$

The partial derivative of the Gibbs energy with respect to the amount of a component yields the chemical potential of a species (Beattie and Oppenheim, 1979).

$$\left(\frac{\partial G}{\partial n_{cA\alpha}}\right)_{T,P,n_{cA\beta},n_{cC},Q} = \mu_{A\alpha} \tag{8.4-6}$$

The criterion for equilibrium based on G is $dG \leqslant 0$ at constant T, P, $n_{cA\alpha}$, $n_{cA\beta}$, n_{cC}, and Q. Integration of equation 8.4-1 at constant values of the intensive variables yields

$$G = \mu_{A\alpha}n_{cA\alpha} + \mu_{A\beta}n_{cA\beta} + \mu_C n_{cC} + (\phi_B - \phi_\alpha)Q \tag{8.4-7}$$

The Gibbs-Duhem equations for the two phases are

$$0 = -S_\alpha dT + V_\alpha dP - n_{cA\alpha}d\mu_{A\alpha} - n_{cC\alpha}d\mu_C \tag{8.4-8}$$

$$0 = -S_\beta dT + V_\beta dP - n_{cA\beta}d\mu_{A\beta} - n_{cC\beta}d\mu_C - Qd\phi_\beta \tag{8.4-9}$$

where ϕ_α has been taken equal as zero. Since there are six variables and two equations, $F = 4$, which can be taken to be T, P, $\mu_{A\alpha}$, and ϕ_B. The number of independent intensive variables can also be calculated using the phase rule: $F = C - p + 3 = 3 - 2 + 3 = 4$, where the 3 is for T, P, and ϕ_B. The number D of natural variables is given by $D = F + p = 4 + 2 = 6$.

The equilibrium expressions for reactions 8.4-2 to 8.4-4, which are derived from the equilibrium conditions using equation 8.3-8, are

$$K_\alpha = \frac{a_{c\alpha}}{a_{A\alpha}a_{B\alpha}} = \exp[-\mu_C^0 - \mu_A^0 - \mu_B^0)/RT] \tag{8.4-10}$$

$$K_\beta = \frac{a_{C\beta}}{a_{A\beta}a_{B\beta}} = \exp[-(\mu_C^0 - \mu_A^0 - \mu_B^0)/RT] \tag{8.4-11}$$

$$K_C = \frac{a_{C\beta}}{a_{C\alpha}} = \exp\left(-\frac{Fz_C\phi_\beta}{RT}\right) \tag{8.4-12}$$

The effect of the electric potential cancels in a chemical reaction in a phase. Note that $a_{C\beta}$ and $a_{C\alpha}$ are not independent variables in the chemical reaction system. Substituting the equilibrium concentrations of C from equations 8.4-10 and 8.4-11 in equation 8.4-12 yields

$$\frac{a_{A\beta}a_{B\beta}}{a_{A\alpha}a_{B\beta}} = \exp\left(-\frac{Fz_C\phi_\beta}{RT}\right) \tag{8.4-13}$$

or

$$-\frac{RT}{Fz_C}\ln\frac{a_{A\beta}a_{B\beta}}{a_{A\alpha}a_{B\alpha}} = \phi_\beta \tag{8.4-14}$$

This shows how a chemical reaction can establish an electric potential difference between phases. This potential difference can then be used to transport other ions between the phases against their concentration gradients.

■ 8.5 TRANSFORMED GIBBS ENERGY OF A TWO-PHASE SYSTEM WITH A CHEMICAL REACTION AND A MEMBRANE PERMEABLE BY A SINGLE ION

The equilibrium relations of the preceding section were derived on the assumption that the charge transferred Q can be held constant, but that is not really practical from an experimental point of view. It is better to consider the potential difference between the phases to be a natural variable. That is accomplished by use of the Legendre transform (Alberty, 1995c; Alberty, Barthel, Cohen, Ewing, Goldberg, and Wilhelm, 2001)

$$G' = G - \phi_\beta Q \tag{8.5-1}$$

that defines the transformed Gibbs energy G'. Since

$$dG' = dG - \phi_\beta dQ - Q d\phi_\beta \tag{8.5-2}$$

substituting equation 8.4-5 with $\phi_\alpha = 0$ yields

$$dG' = -SdT + VdP + \mu_{A\alpha}dn_{cA\alpha} + \mu_{A\beta}dn_{cA\beta} + \mu_C dn_{cC} - Q d\phi_\beta \tag{8.5-3}$$

To learn more about the derivatives of the transformed Gibbs energy, the chemical potentials of species are replaced by use of equation 8.3-8 to obtain

$$dG' = -SdT + VdP + (\mu_A^0 + RT \ln a_{A\alpha})dn_{cA\alpha} + (\mu_A^0 + RT \ln a_{A\beta} + Fz_A\phi_\beta)dn_{cA\beta}$$
$$+ (\mu_C^0 RT \ln a_{C\alpha})dn_{cC} - Q d\phi_\beta \tag{8.5-4}$$

Thus

$$\left(\frac{\partial G'}{\partial n_{cA\alpha}}\right)_{T,P,n_{cA\beta},n_{cC},\phi_\beta} = \mu_A^0 + RT \ln a_{A\alpha} = \mu_{A\alpha}' \tag{8.5-5}$$

where $\mu_{A\alpha}'$ is the **transformed chemical potential** of A in the α phase. This corresponds with writing equation 8.3-8 as

$$\mu_i = \mu_i' + Fz_i\phi_i \tag{8.5-6}$$

Thus μ_i' is equal to the chemical potential of i in a phase where $\phi_i = 0$.

Equation 8.5-3 indicates that the number of natural variables for the system is 6, $D = 6$. Thus the number D of natural variables is the same for G and G', as expected, since the Legendre transform interchanges conjugate variables. The criterion for equilibrium is $dG' \leqslant 0$ at constant $T, P, n_{cA\alpha}$, $n_{cA\beta}$, n_{cC}, and ϕ_β. The Gibbs-Duhem equations are the same as equations 8.4-8 and 8.4-9, and so the number of independent intensive variables is not changed. Equation 8.5-3 yields the same membrane equations (8.4-13 and 8.4-14) derived in the preceding section.

The integration of equation 8.5-3 at constant values of the intensive variables yields

$$G' = \mu_{A\alpha}n_{cA\alpha} + \mu_{A\beta}n_{cA\beta} + \mu_C n_{cC} \tag{8.5-7}$$

which agrees with equations 8.4-7 and 8.5-1.

■ 8.6 EFFECTS OF ELECTRIC POTENTIALS ON MOLAR PROPERTIES OF IONS

The fundamental equation for G for a two-phase system with a potential difference can be written

$$dG = -SdT + VdP + \sum \mu_{i\alpha}dn_{i\alpha} + \sum \mu_{i\beta}dn_{i\beta} + \phi_\beta dQ \tag{8.6-1}$$

where $\phi_\alpha = 0$. Integration at constant values of intensive variables yields

$$G = \sum \mu_{i\alpha} n_{i\alpha} + \sum \mu_{i\beta} n_{i\beta} + \phi_\beta Q \qquad (8.6\text{-}2)$$

The entropy of the system can be obtained by use of the following derivative:

$$S = -\left(\frac{\partial G}{\partial T}\right)_{P,\{n_{i\alpha}\},\{n_{i\beta}\},Q} \qquad (8.6\text{-}3)$$

Taking this derivative of G yields

$$S = \sum n_{i\alpha} S_{mi\alpha} + \sum n_{i\beta} S_{mi\beta} \qquad (8.6\text{-}4)$$

where S_{mi} is the **molar entropy** of i. Substituting $\mu_i = \mu_i' + Fz_i\phi_i$ in equation 8.6-2 yields

$$G = \sum \mu_{i\alpha}' n_{i\alpha} + \sum \mu_{i\beta}' n_{i\beta} + F\phi_\beta \sum z_i n_{i\beta} + \phi_\beta Q \qquad (8.6\text{-}5)$$

Taking the derivative in equation 8.6-3 yields

$$S = \sum n_{i\alpha} S_{mi\alpha}' + \sum n_{i\beta} S_{mi\beta}' \qquad (8.6\text{-}6)$$

where S_{mi}' is the **transformed molar entropy** of i. Comparing this equation with equation 8.6-4 shows that the molar entropy if a species is not affected by the electric potential of a phase, thus $S_{mi} = S_{mi}'$ and $S = S'$.

The corresponding **molar enthalpy** is obtained by use of the Gibbs-Helmholtz equation: $H = -T^2[(\partial(G/T)/\partial T]_P$. Applying the Gibbs-Helmholtz equation to equations 8.6-2 and 8.6-5 yields

$$H = \sum n_{i\alpha} H_{mi\alpha} + \sum n_{i\beta} H_{mi\beta} + \phi_\beta Q \qquad (8.6\text{-}7)$$

where H_{mi} is the molar enthalpy of i, and

$$H = \sum n_{i\alpha} H_{mi\alpha}' + \sum n_{i\beta} H_{mi\beta}' + F\phi_\beta \sum z_i n_i + \phi_\beta Q \qquad (8.6\text{-}8)$$

where H_{mi}' is the **transformed molar enthalpy**. Comparing equation 8.6-7 and 8.6-8 shows that

$$H_{mi} = H_{mi}' + Fz_i\phi_\beta \qquad (8.6\text{-}9)$$

Thus the molar enthalpy of an ion is affected by the electric potential of the phase in the same way as the chemical potential.

In 1974 IUPAC (Parsons, 1974) recommended that the electrochemical potential $\tilde{\mu}_i$ be defined by

$$\tilde{\mu}_i = \mu_i + Fz_i\phi \qquad (8.6\text{-}10)$$

where μ_i was referred to as the chemical potential. There are several problems with this recommendation. The electrochemical potential is actually the chemical potential defined by equation 8.4-6. The quantity represented by μ_i on the right-hand side of equation 8.6-10 is the transformed chemical potential μ_i' defined by equation 8.5-5. According to the equations presented here, the chemical potential for an ion should be defined by equation 8.3-8, rather than by equation 8.6-10. Guggenheim (1967) wrote

$$\mu_i = RT \ln l_i^0 + RT \ln a_i + Fz_i\phi_i \qquad (8.6\text{-}11)$$

and pointed out that l_i^0 is independent of the electric potential of a phase. Thus his equation is the same as equation 8.3-8 with a different symbol for the constant term. Physicists consider that the chemical potential μ_i of an electron in a metal includes a contribution due to the electrostatic energy of the electron and is constant throughout the system, so their use of the chemical potential μ_i is the same as that recommended here.

■ 8.7 EQUILIBRIUM DISTRIBUTION OF CARBON DIOXIDE BETWEEN THE GAS PHASE AND AQUEOUS SOLUTION

The distribution of carbon dioxide between the gas phase and aqueous solution is much more complex than the distributions of H_2, O_2, and N_2 because in the aqueous phase, carbon dioxide is distributed between $CO_2(aq)$, H_2CO_3, HCO_3^-, and CO_3^{2-}. This equilibrium can be treated with data in the NBS Tables (1982), even though it only provides data on $CO_2(g)$, $H_2CO_3(ao)$, $HCO_2^-(ao)$, and $CO_3^{2-}(ao)$. The "ao" designates undissociated molecules in water. The standard formation properties from the NBS Tables are given in Table 8.1. The entry for $H_2CO_3(ao)$ is simply the sum of the entries for $CO_2(ao)$ and $H_2O(ao)$. The introductory material of the NBS Tables explains that some species in aqueous solution are listed with two or more formulas that differ only in the number of molecules of water contained in them. These forms are referred to as being equivalent in the sense that the thermodynamic properties of each pair are connected by the formal chemical equation

$$A(aq) + nH_2O(l) = A \cdot nH_2O(aq) \tag{8.7-1}$$

for which $\Delta_r H^0 = \Delta_r G^0 = \Delta_r S^0 = 0$ by convention. Thus the arbitrary convention is that $K = 1$ at each temperature for this reaction. This convention is necessary when there is no way to distinguish between $A(aq)$ and $A \cdot nH_2O(aq)$ in dilute aqueous solutions. The number n of water molecules bound cannot be determined by equilibrium measurements because the concentration of H_2O cannot be changed (Alberty, 2002b). This convention is also used in the NBS Tables (1982) for SO_2 and H_2SO_3, NH_3 and NH_4OH, FeO_2^{2-} and $Fe(OH)_4^{2-}$, and so on. Thus the entry for $H_2CO_3(ao)$ in the NBS Table is simply the sum of the entries for $CO_2(ao)$ and $H_2O(ao)$.

Dissolved carbon dioxide is different from species like SO_3 and NH_3 in aqueous solutions in that the hydration reaction is slow enough ($t_{1/2} = 15$ seconds at pH 7 and 298 K) so that the rate constants involved can be determined and can be used to calculate the hydrolysis equilibrium constant (Edsall, 1969) at 298.15 K in terms of species for

$$CO_2(sp) + H_2O(l) = H_2CO_3(sp) \qquad K_h = 2.584 \times 10^{-3} \tag{8.7-2}$$

The slowness of this reaction leads to a fading end point when a solution containing bicarbonate is titrated with sodium hydroxide using methyl orange as an indicator. The rate of liberation of CO_2 from carbonate buffers in the neutral range is so slow that this reaction has to be catalyzed in our lungs by carbonic anhydrase.

Table 8.1   Standard Formation Properties at 298.15 K from the NBS Tables (1982)

	$\Delta_f H^0/kJ\ mol^{-1}$	$\Delta_f G^0/kJ\ mol^{-1}$
$CO_2(g)$	-373.51	-394.36
$CO_2(ao)$	-413.80	-385.98
$CO_3^{2-}(ao)$	-677.14	-527.81
$HCO_3^-(ao)$	-691.99	-586.77
$H_2CO_3(ao)$	-699.63	-623.08
$H_2O(ao)$	-285.83	-237.13

Table 8.2   Standard Formation Properties of Species at 298.15 K

	$\Delta_f H^0$/kJ mol^{-1}	$\Delta_f G^0$/kJ mol^{-1}
$CO_2(sp)$	413.81	-385.97
$CO_3^{2-}(sp)$	-677.14	-527.81
$HCO_3^-(sp)$	-691.99	-586.77
$H_2CO_3(sp)$	-694.91	-606.33

Reprinted with permission from R. A. Alberty, *J. Phys. Chem.*, 99, 11028 (1995). Copyright 1995 American Chemical Society.

The acid dissociation constant for the species H_2CO_3 is given by

$$H_2CO_3(sp) = H^+(ao) + HCO_3^-(sp) \quad K(H_2CO_3) = K_1 \left(\frac{1+1}{K_h} \right) = 1.668 \times 10^{-4}$$

$$(8.7\text{-}3)$$

where $K_1 = 4.300 \times 10^{-7}$. Knowing this equilibrium constant makes it possible to make a table of thermodynamic properties of species (Alberty, 1995b), as shown in Table 8.2.

Thus there are four terms in the calculation of $\Delta_f G'^0$(iso) and $\Delta_f H'^0$(iso) for TotCO$_2$, which is the sum of the four species. Table 8.2 make it possible to calculate the standard transformed Gibbs energies of formation and standard transformed enthalpies of formation of the equilibrium mixture of species of carbon dioxide in dilute aqueous solution as a function of pH and ionic strength by the methods discussed earlier in Section 3.4. The standard transformed Gibbs energies of formation are given as a function of pH and ionic strength in Table 8.3.

Later this table was also calculated using equilibrium constants (Alberty, 1997). A third way to calculate this table is to use the properties of $H_2CO_3(ao)$, $HCO_3^-(ao)$, and $CO_3^{2-}(ao)$ directly from the NBS Tables (Alberty, 1998b). The reason is that when only dilute aqueous solutions are considered, the thermodynamic properties of TotCO$_2$ are independent of the value of K_h.

The values of $\Delta_f G'^0$(TotCO$_2$) make it possible to calculate the apparent Henry's law constant for carbon dioxide as a function of pH and ionic strength. This constant is the equilibrium constant for the reaction

$$TotCO_2(aq) = CO_2(g) + H_2O \quad K'_H = \frac{P(CO_2, g)}{[TotCO_2]} \quad (8.7\text{-}4)$$

where $P(CO_2, g)$ is in bars. The H_2O on the right-hand side is required to balance oxygen atoms. As the pH is decreased, the apparent Henry's law constant approaches the equilibrium constant for the reaction

$$CO_2(aq) = CO_2(g) \quad K = \frac{P(CO_2)}{[CO_2(aq)]} \quad (8.8.5)$$

Table 8.3 $\Delta_f G'^0$(TotCO$_2$) in kJ mol^{-1} in Dilute Aqueous Solution at 298.15 K

pH	$I = 0$ M	$I = 0.10$ M	$I = 0.25$ M
5	-566.19	-565.28	-564.67
6	-555.56	-554.93	-554.55
7	-547.39	-547.24	-547.16
8	-541.23	-541.23	-541.24
9	-535.58	-535.70	-535.85

Table 8.4   Apparent Henry's Law Constants for Carbon Dioxide at 298.15 K

pH	$I = 0$ M	$I = 0.10$ M	$I = 0.25$ M
5	28.18	27.64	27.15
6	20.55	17.99	16.09
7	5.544	4.005	3.169
8	0.6650	0.4524	0.3456
9	0.0652	0.0421	0.0303

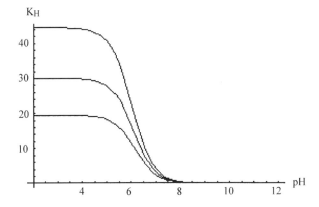

Fig. 8.1 The apparent Henry's law constant for carbon dioxide in water at 283.15 K, 298.15 K, and 313.15 K and ionic strength 0.25 M as a function of pH. The value of the Henry's law constant increases with the temperature (see Problem 8.1).

This reaction balances oxygen atoms, so it is not necessary to write a H_2O on the right-hand side of the equation. Apparent Henry's law constants for carbon dioxide are given as a function of pH and ionic strength in Table 8.4.

The effect of temperature on the apparent Henry's law constant at ionic strength 0.25 M is shown by Fig. 8.1. The effect of ionic strength on the Henry's law constant at 298.15 K and 0.25 M ionic strength is shown by Fig 8.2. The equilibrium constant expression for a biochemical reaction involving carbon dioxide at a specified pH can be written as

$$A = B + CO_2(g) \qquad K' = \frac{[B]P(CO_2)}{[A]} \qquad (8.7\text{-}6)$$

or

$$A + H_2O = B + TotCO_2 \quad K' = \frac{[B][TotCO_2]}{[A]} \qquad (8.7\text{-}7)$$

where A and B represent sums of species. The H_2O in reaction 8.7-7 is required to balance oxygen atoms. These two equilibrium constants have different values. The second has the advantage it yields the concentrations of species of carbon dioxide in the cell.

■ 8.8   PHASE SEPARATION IN AQUEOUS SYSTEMS CONTAINING HIGH POLYMERS

Aqueous systems containing high polymers such as polyethylene glycols and dextrans may separate spontaneously into two phases (Guan et al., 1993). These

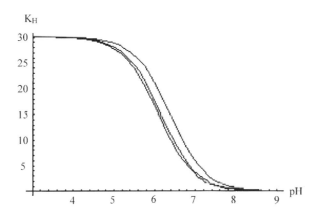

Fig. 8.2 The apparent Henry's law constant for carbon dioxide in water at 298.15 K and ionic strengths 0, 0.10, and 0.25 M as a function of pH. The value of the Henry's law constant decreases with increasing ionic strength (see Problem 8.2).

systems may separate at as low as 5% by weight of the water-soluble polymers. This suggests that phase separation may occur spontaneously in a living cell where water-soluble polymers are formed. If the contents of a living cell separate into two phases, the interface will adsorb molecules and tend to orient them with respect to the two phases. Is it possible that such an interface might become a cell wall?

Oxidation-Reduction Reactions

9

When electrons are transferred in a chemical reaction or a biochemical reaction, the reactions can, in principle, be carried out separately in two half-cells of a galvanic cell. Each half-cell reaction makes its independent contribution to the equilibrium electromotive force of the galvanic cell. When chemical half-reactions in terms of species are considered, their reduction potentials and standard reduction potentials are represented by E and E^0, respectively. When biochemical half-reactions in terms of reactants (sums of species) are considered at a specified pH, their apparent reduction potentials and standard apparent reduction potentials are represented by E' and E'^0, respectively. Reduction potentials for chemical half-reactions are measured in relation to the standard hydrogen electrode, which by convention has a standard reduction potential of zero volts at each temperature at zero ionic strength. The standard hydrogen electrode consists of molecular hydrogen at 1 bar bubbling over a platinum electrode immersed in a solution with hydrogen ions at unit activity. For a chemical half-reaction the standard reduction potential E^0 is obtained when the concentrations of the species are 1 M; this potential and the corresponding thermodynamic properties are taken to be functions of the ionic strength. Remember that in equation 3.1-10 the convention was adopted that the Gibbs energy of formation of a species is expressed by $\Delta_f G_j = \Delta_f G_i^0 + RT \ln[B_j]$, where $\Delta_f G_j$ and $\Delta_f G_j^0$ are taken to be functions of ionic strength. Chemical half-reactions balance all atoms and charges. At specified pH, biochemical half reactions do not balance hydrogen atoms or charges. Standard apparent reduction potentials are obtained when the reactants (sums of species) are all at 1 M. The measurement of electromotive force at specified pH yields the same type of information as the measurement of apparent equilibrium constants, namely $\Delta_f G_i'^0$ values for reactants. Conversely, $\Delta_f G_i'^0$ values calculated from

155

measurements of K' can be used to calculate E'^0 values. The relation between E and $\Delta_r G$ for a chemical half reaction is $E = -\Delta_r G/|v_e|F$, where $|v_e|$ is the number of electrons involved and F is the Faraday constant (the product of the Avogadro constant and the proton charge, 96 485.309 C mol^{-1}). The relation between E' and $\Delta_r G'$ for a biochemical half-reaction at specified pH is $E' = -\Delta_r G'/|v_e|F$. The classical book in this field is W. M. Clarke, *Oxidation-Reduction Potentials of Organic Systems* (1961). The use of transformed thermodynamic properties in this field was introduced by Alberty (1993c, 1998d, 2001b).

When a biochemical half-reaction involves the production or consumption of hydrogen ions, the electrode potential depends on the pH. When reactants are weak acids or bases, the pH dependence may be complicated, but this dependence can be calculated if the pKs of both the oxidized and reduced reactants are known. Standard apparent reduction potentials E'^0 have been determined for a number of oxidation-reduction reactions of biochemical interest at various pH values, but the E'^0 values for many more biochemical reactions can be calculated from $\Delta_r G'^0$ values of reactants from the measured apparent equilibrium constants K'. Some biochemical redox reactions can be studied potentiometrically, but often reversibility cannot be obtained. Therefore a great deal of the information on reduction potentials in this chapter has come from measurements of apparent equilibrium constants.

Since tables of standard apparent reduction potentials and standard transformed Gibbs energies of formation contain the same basic information, there is a question as to whether this chapter is really needed. However, the consideration of standard apparent reduction potentials provides a more global view of the driving forces in redox reactions. There are two contributions to the apparent equilibrium constant for a biochemical redox reaction, namely the standard apparent reduction potentials of the two half-reactions. Therefore it is of interest to compare the standard apparent reduction potentials of various half reactions.

■ 9.1   BASIC EQUATIONS

An enzyme-catalyzed redox reaction can be divided into two half-reactions, one producing electrons and the other consuming electrons. The standard apparent reduction potentials $E_R'^0$ and $E_L'^0$ for the two half-reactions in an enzyme-catalyzed redox reaction at a specified pH and ionic strength determine E'^0 for the overall reaction, which is positive for a reaction that can occur spontaneously. A biochemical redox reaction at a specified pH can be represented schematically by

$$\text{Ox} + \text{Red}' = \text{Red} + \text{Ox}' \quad E'^0 = E_R'^0 - E_L'^0 \tag{9.1-1}$$

Where Ox, Ox', Red, and Red' are reactants (sums of species). The subscripts are abbreviations for right and left, but the two half-reactions could be distinguished in other ways. The half-reactions and their standard apparent reduction potentials at a specified pH are represented by

$$\text{Ox} + |v_e|e^- = \text{Red} \quad E_R'^0 \tag{9.1-2}$$

$$\text{Ox}' + |v_e|e^- = \text{Red}' \quad E_L'^0 \tag{9.1-3}$$

Of course, these reactions may be very much more complicated. Since the pH is specified, H$^+$ is not included as a reactant, and a reactant may be a sum of species if the reactant has pKs in the pH region of interest. These biochemical reactions do balance atoms of elements other than hydrogen, but they do not balance electric charges. When the half-reactions occur in half-cells connected by a KCl salt bridge, the difference E' in electric potential between the metallic electrodes at equilibrium at a specified pH is given by (Silbey and Alberty, 2001)

$$E' = E'^0 - \frac{RT}{|v_e|F} \ln \frac{[\text{Red}][\text{Ox}']}{[\text{Red}'][\text{Ox}]} \tag{9.1-4}$$

This is usually referred to as the Nernst equation (see equations 8.3-10 and 8.4-14). It is assumed that the salt bridge contributes a negligible junction potential. This equation can be written as

$$E' = \left(E'^0_R - \frac{RT}{|v_e|F} \ln \frac{[\text{Red}]}{[\text{Ox}]} \right) - \left(E'^0_L - \frac{RT}{|v_e|F} \ln \frac{[\text{Red}']}{[\text{Ox}']} \right) \tag{9.1-5}$$

Where the two terms correspond with half-reactions 9.1-2 and 9.1-3.

When the pH is specified, the change in the transformed Gibbs energy G' in a biochemical redox reaction like equation 9.1-1 is given by

$$\Delta_r G' = \Delta_r G'_R - \Delta_r G'_L \tag{9.1-6}$$

where $\Delta_r G'_R$ and $\Delta_r G'_L$ are the transformed Gibbs energies for half-reactions 9.1-2 and 9.1-3. Equation 9.1-6 can be written as

$$\Delta_r G' = \left(\Delta_r G'^0_R + RT \ln \frac{[\text{Red}]}{[\text{Ox}]} \right) - \left(\Delta_r G'^0_L + RT \ln \frac{[\text{Red}']}{[\text{Ox}']} \right) \tag{9.1-7}$$

When the reactants are at their standard concentrations (1 M) or standard pressures (1 bar), the logarithmic terms disappear and this equation becomes

$$\Delta_r G'^0 = \Delta_r G'^0_R - \Delta_r G'^0_L \tag{9.1-8}$$

Comparison of equations 9.1-5 and 9.1-7 shows that the standard apparent reduction potentials for the half-reactions at specified pH are given by

$$E'^0_R = -\frac{\Delta_r G'^0_R}{|v_e|F} = -\frac{\Sigma (v'_i \Delta_f G'^0_i)_R}{|v_e|F} \tag{9.1-9}$$

$$E'^0_L = -\frac{\Delta_r G'^0_L}{|v_e|F} = -\frac{\Sigma (v'_i \Delta_f G'^0_i)}{|v_e|F} \tag{9.1-10}$$

where v'_i is the stoichiometric number of reactant i. The prime is needed to distinguish these stoichiometric numbers from the stoichiometric numbers of the underlying chemical reactions. The $\Delta_f G'^0_i$ are the standard transformed Gibbs energies of formation of reactants (sums of species). In calculating the standard transformed Gibbs energies for half-reactions, we take the standard transformed Gibbs energies of the formal electrons in equation 9.1-2 and 9.1-3 to be zero. Substituting equation 9.1-9 and 9.1-10 in equation 9.1-8 yields

$$\Delta_r G'^0 = -|v_e|FE'^0_R + |v_e|FE'^0_L = |v_e|FE'^0 \tag{9.1-11}$$

Note that when two half-reactions are added, their $\Delta_r G'^0$ values add but their E'^0 values do not.

The apparent equilibrium constant for a biochemical reaction (like equation 9.1-1) at specified pH can be calculated using

$$K' = \exp \left[\frac{-\Delta_r G'^0}{RT} \right] = \exp \left[-\Sigma \frac{v'_i \Delta_f G'^0_i}{RT} \right] \tag{9.1-12}$$

Equations 9.1-8 to 9.1-10 show that the apparent equilibrium constant can also be calculated using

$$K' = \exp \left[\frac{|v_e|F(E'^0_R - E'^0_L)}{RT} \right] = \exp \left[\frac{|v_e|FE'^0}{RT} \right] \tag{9.1-13}$$

Table 9.1   Standard Reduction Potentials E^0 at 298.15 K, 1 bar, and Zero Ionic Strength

Half-reaction	E^0/V
$F_2(g) + 2e^- = 2F^-$	2.87
$O_2(g) + 4H^+ + 4e^- = 2H_2O$	1.2288
$Fe^{3+} + e^- = Fe^{2+}$	0.771
$O_2(g) + 2H^+ + 2e^- = H_2O_2$	0.70
$N_2(g) + 8H^+ + 6e^- = 2NH_4^+$	0.274
$C_2H_4O(acetaldehyde) + 2H^+ + 2e^- = C_2H_6O(ethanol)$	0.221
$Cu_2^+ + e^- = Cu^+$	0.153
$2H^+ + 2e^- = H_2(g)$	0
$Fe^{2+} + 2e^- = Fe(s)$	-0.440
$Li^+ + e^- = Li(s)$	-3.045

This equation is the basis for the statement that knowledge of $E_R^{\prime 0}$ and $E_L^{\prime 0}$ at a specified pH for the two half-reactions determines the direction of spontaneous reaction for the overall redox reaction.

I. H. Segel (1976) discussed oxidation-reduction reactions very clearly and has written: "When any two of the half-reactions are coupled, the one with the greater tendency to gain electrons (the one with the more positive reduction potential) goes as written (as a reduction). Consequently, the other half-reaction (the one with the lesser tendency to gain electrons as shown by the less positive reduction potential) is driven backwards (as an oxidation). The reduced forms of those substances with highly negative reduction potentials are good reducing agents (and are easily oxidized). The oxidized forms of those substances with highly positive reduction potentials are good oxidizing agents (and are easily reduced)."

A few standard reduction potentials of chemical half reactions at 298.15 K, 1 bar, and zero ionic strength are given in Table 9.1. Note that the half-reactions balance atoms and charges and that the half-reactions are arbitrarily written in such a way that there are no fractional stoichiometric numbers. It is interesting to consider the extremes in this table. The fluorine molecule-fluoride ion electrode has a very high affinity for electrons, and this is indicated by its very positive standard reduction potential. Thus gaseous fluorine is a powerful oxidizing agent. At the other extreme, the lithium ion-metallic lithium electrode has a very low affinity for electrons, and this is indicated by its very negative standard reduction potential. Thus lithium metal is a powerful reducing agent. Oxidizing agents with reduction potentials above 1.2 V tend to oxidize H_2O to $O_2(g)$, and metals with negative reduction potentials tend to reduce H_2O to $H_2(g)$. Note that half-reactions with more positive reduction potentials than 1.229 V tend to produce oxygen gas in aqueous solutions and half-reactions with negative reduction potentials tend to produce molecular hydrogen.

■ 9.2   OXIDATION-REDUCTION REACTIONS INVOLVING SINGLE SPECIES AT SPECIFIED pH

When reactants in half-reactions involve only single species in a range of pH, the dependence of $E^{\prime 0}$ on pH is linear in this range. The standard apparent reduction potentials of a number of half-reactions have been calculated at 298.15 K at a series of pHs and ionic strengths and are given in Table 9.2. This table gives both the chemical form of the half reaction and the biochemical form of the half-reaction. The chemical form is useful for understanding the pH and ionic strength dependencies of the standard apparent reduction potentials, but the standard apparent reduction potentials apply at a specified pH and ionic strength and

Table 9.2 Standard Apparent Reduction Potentials E'^0 in Volts at 298.15 K and 1 bar as a Function of pH and Ionic Strength

	pH 5	pH 6	pH 7	pH 8	pH 9
	1. $O_2(aq) + 4H^+ + 4e^- = 2H_2O$				
	$O_2(aq) + 4\ e^- = 2H_2O$				
I = 0 M	0.9759	0.9167	0.8575	0.7984	0.7392
I = 0.10 M	0.9695	0.9103	0.8512	0.7920	0.7329
I = 0.25 M	0.9675	0.9083	0.8491	0.7900	0.7308
	2. $O_2(g) + 4H^+ + 4e^- = 2H_2O$				
	$O_2(g) + 4\ e^- = 2H_2O$				
I = 0 M	0.9334	0.8742	0.8150	0.7559	0.6967
I = 0.10 M	0.9270	0.8679	0.8087	0.7495	0.6904
I = 0.25 M	0.9250	0.8658	0.8066	0.7475	0.6883
	3. $O_2(aq) + 2H^+ + 2e^- = H_2O_2$				
	$O_2(aq) + 2\ e^- = H_2O_2$				
I = 0 M	0.4838	0.4246	0.3654	0.3063	0.2471
I = 0.10 M	0.4774	0.4183	0.3591	0.2999	0.2408
I = 0.25 M	0.4754	0.4162	0.3570	0.2979	0.2387
	4. $O_2(g) + 2H^+ + 2e^- = H_2O_2$				
	$O_2(g) + 2\ e^- = H_2O_2$				
I = 0 M	0.3988	0.3396	0.2805	0.2213	0.1621
I = 0.10 M	0.3924	0.3333	0.2741	0.2149	0.1558
I = 0.25 M	0.3904	0.3312	0.2721	0.2129	0.1537
	5. cytochrome $c(Fe^{3+}) + e^- =$ cytochrome $c(Fe^{2+})$				
	cytochrome $c_{ox} + e^- =$ cytochrome c_{red}				
I = 0 M	0.2540	0.2540	0.2540	0.2540	0.2540
I = 0.10 M	0.2223	0.2223	0.2223	0.2223	0.2223
I = 0.25 M	0.2121	0.2121	0.2121	0.2121	0.2121
	6. pyruvate$^-$ + 2H$^+$ + 2e$^-$ = lactate				
	pyruvate + 2 e$^-$ = lactate				
I = 0 M	−0.0655	−0.1246	−0.1838	−0.2429	−0.3021
I = 0.10 M	−0.0718	−0.1310	−0.1901	−0.2493	−0.3084
I = 0.25 M	−0.0739	−0.1330	−0.1922	−0.2513	−0.3105
	7. acetaldehyde + 2H$^+$ + 2e$^-$ = ethanol				
	acetaldehyde + 2e$^-$ = ethanol				
I = 0 M	−0.0748	−0.1340	−0.1932	−0.2523	−0.3115
I = 0.10 M	−0.0812	−0.1403	−0.1995	−0.2587	−0.3179
I = 0.25 M	−0.0832	−0.1424	−0.2015	−0.2607	−0.3199
	8. $FMN^{2-} + 2H^+ + 2e^- = FMNH_2^{2-}$				
	$FMN_{ox} + 2e^- = FMN_{red}$				
I = 0 M	−0.0943	−0.1535	−0.2126	−0.2718	−0.3310
I = 0.10 M	−0.1007	−0.1598	−0.2190	−0.2781	−0.3373
I = 0.25 M	−0.1027	−0.1619	−0.2210	−0.2801	−0.3394
	9. retinal + 2H$^+$ + 2e$^-$ = retinol				
	retinal + 2 e$^-$ = retinol				
I = 0 M	−0.1512	−0.2103	−0.2695	−0.3287	−0.3878
I = 0.10 M	−0.1575	−0.2167	−0.2758	−0.3350	−0.3942
I = 0.25 M	−0.1596	−0.2187	−0.2779	−0.3370	−0.3962

Table 9.2   *Continued*

	pH 5	pH 6	pH 7	pH 8	pH 9
			10. acetone $+ 2H^+ + 2e^- = $ 2-propanol		
			acetone $+ 2\,e^- = $ 2-propanol		
$I = 0$ M	-0.1635	-0.2227	-0.2818	-0.3410	-0.4001
$I = 0.10$ M	-0.1698	-0.2290	-0.2882	-0.3473	-0.4065
$I = 0.25$ M	-0.1719	-0.2311	-0.2902	-0.3493	-0.4085
			11. $NAD^- + H^+ + 2e^- = NADH^{2-}$-		
			$NAD_{ox} + 2\,e^- = NAD_{red}$		
$I = 0$ M	-0.2653	-0.2949	-0.3244	-0.3540	-0.3836
$I = 0.10$ M	-0.2589	-0.2885	-0.3181	-0.3477	-0.3773
$I = 0.25$ M	-0.2569	-0.2865	-0.3160	-0.3456	-0.3752
			12. $NADP^{3-} + H^+ + 2e^- = NADPH^{4-}$		
			$NADP_{ox} + 2e^- = NADP_{red}$		
$I = 0$ M	-0.2826	-0.3122	-0.3417	-0.3713	-0.4009
$I = 0.10$ M	-0.2636	-0.2931	-0.3227	-0.3523	-0.3819
$I = 0.25$ M	-0.2574	-0.2870	-0.3166	-0.3462	-0.3757
			13. $Fd^+ + e^- = Fd^0$		
			ferredoxin$_{ox} + e^- = $ ferredoxin$_{red}$		
$I = 0$ M	-0.3946	-0.3946	-0.3946	-0.3946	-0.3946
$I = 0.10$ M	-0.4009	-0.4009	-0.4009	-0.4009	-0.4009
$I = 0.25$ M	-0.4030	-0.4030	-0.4030	-0.4030	-0.4030
			14. Acetyl-CoA$^{4-} + 2H^+ + 2e^- = $ CoA$^{4-} + $ acetaldehyde		
			Acetyl-CoA $+ 2e^- = $ CoA $+ $ acetaldehyde		
$I = 0$ M	-0.2950	-0.3541	-0.4133	-0.4725	-0.5316
$I = 0.10$ M	-0.313	-0.3605	-0.4196	-0.4788	-0.5380
$I = 0.25$ M	-0.3034	-0.3625	-0.4217	-0.4808	-0.5400
			15. $2H^+ + 2e^- = H_2(g)$		
			$2e^- = H_2(g)$		
$I = 0$ M	-0.2958	-0.3550	-0.4141	-0.4733	-0.5324
$I = 0.10$ M	-0.3021	-0.3613	-0.4205	-0.4796	-0.5388
$I = 0.25$ M	-0.3042	-0.3634	-0.4225	-0.4817	-0.5408
			16. $2H^+ + 2e^- = H_2(aq)$		
			$2e^- = H_2(aq)$		
$I = 0$ M	-0.3870	-0.4462	-0.5053	-0.5645	-0.6237
$I = 0.10$ M	-0.3934	-0.4525	-0.5117	-0.5708	-0.6399
$I = 0.25$ M	-0.3954	-0.4546	-0.5137	-0.5729	-0.6320

Source: With permission from R. A. Alberty, *Arch. Biochem. Biophys.* 389, 94–109 (2001). Copyright Academic Press.

correspond with the biochemical form of the half-reaction. Note that standard apparent reduction potentials always decrease with increasing pH or are independent of pH. The numerical calculations in this chapter have been made using a file of *Mathematica* (BasicBiochemData2) functions that give the dependencies of the standard transformed Gibbs energies of formation of reactants (sums of species) on pH and ionic strength at 298.15 K. This file of functions is available on MathSource (www.mathsource.com), and it can be used to calculate apparent equilibrium constants for enzyme-catalyzed reactions between the 131 reactants (sums of species) in the file of functions.(Alberty, 2001f). For the calculations described here, a program (calcappredpot) was written (see the Problems) to

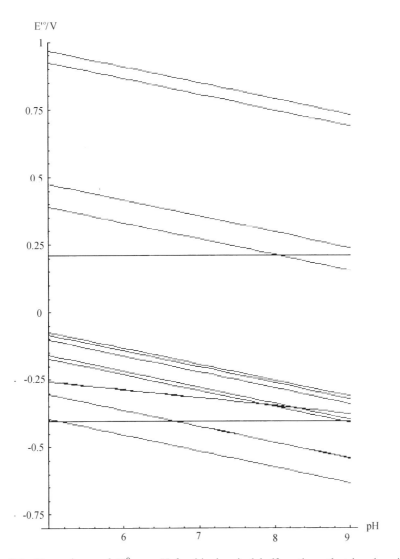

Figure 9.1 Dependence of E'^0 on pH for biochemical half rections that involve single species a 298.15 K and 0.25 M ionic strength. The values of E'^0 at pH 7 can be used to identify the number of the reaction in Table 9.2 if it is noted that te lines for reactions 6 and 7 and for reactions 14 and 15 overlap. [With permission from R. A. Alberty, *Arch. Biochem. Biophys.* 389, 94–109 (2001). Copyright Academic Press.]

calculate the standard apparent reduction potential for a biochemical half-reaction, given the typed half-reaction and the number of electrons transferred. All but two of the E'^0 values in Table 9.2 depend on both pH and ionic strength. The E'^0 values for cytochrome c and ferredoxin do not depend on the pH because there is no evidence of acid groups in the active sites of these proteins with pKs in the range 5 to 9. The pH dependencies of some of the half-reactions are the same because $-\Delta_r N_H / |\nu_e|$ is the same, but their ionic strength dependencies differ. The reduction potentials are rounded to ± 0.0001 volt because this corresponds with an error of ± 0.01 kJ mol^{-1} in the transformed Gibbs energy. The half-reactions are in the order of decreasing E'^0 at pH 7 and ionic strength 0.25 M.

Half-reactions higher in the table drive half-reactions lower in the table. Half-reactions with apparent reduction potentials at pH 7 and ionic strength 0.25 M that are more positive than 0.807 V at pH 7 and 0.25 M ionic strength tend to produce $O_2(g)$, and half reactions with apparent reduction potentials at pH 7 that are less than -0.422 V tend to produce $H_2(g)$. The pH dependencies of some of these half reactions are shown in Fig. 9.1.

Goldberg and Tewari (Goldberg et al., 1992; Goldberg and Tewari, 1994a, b, 1995a, b; Goldberg, 1999) have published critically evaluated data on experimental determinations of K' of 94 enzyme-catalyzed redox reactions. These measurements can all be used to calculate E'^0 for half-reactions and $\Delta_f G_j^0$ values for the species involved. Thus Table 9.2 can be extended considerably on the basis of experimental measurements. $\Delta_f G_j^0$ values of species can also be determined from enzyme-catalyzed reactions that are not redox reactions. When the $\Delta_f G_j^0$ value is unknown for either the oxidant or reductant, the oxidized form can arbitrarily be assigned $\Delta_f G_j^0 = 0$ at zero ionic strength.

■ 9.3   METHANE MONOOXYGENASE REACTION

The methane monooxygenase reaction (EC 1.14.13.25) is an especially interesting enzyme-catalyzed reaction for which all the reactants are single species. The chemical and biochemical forms of this reaction are given in the first two lines of Table 9.3. The methane monooxidase reaction is remarkable because it can be divided into three half-reactions. The chemical and biochemical equations for these three half-reactions are given in Table 9.3. These three half reactions are in a sense independent because they do not share reactants except for H_2O and H^+, which are everywhere. In other words, these half reactions could be catalyzed independently if there were appropriate sites on the enzyme for them to deposit and withdraw electrons at appropriate reduction potentials. The enzyme was able to couple the three half-reactions to give reaction 1 in Table 9.3. The fact that the methane monooxygenase reaction can be divided into three half-reactions is quite remarkable considering the statement at the beginning of this chapter that a redox reaction can be divided into two half reactions. The reason is that the methane monooxygenase reaction is the sum of the following two biochemical reactions,

Table 9.3   Standard Transformed Gibbs Energies (in kJ mol^{-1}) of Reactions and Standard Apparent Reduction Potentials (in volts) at 289.15 K, 1 bar, pH 7, and Ionic Strength 0.25 M for Reactions Involved in the Methane Monooxygenase Reaction

Chemical and Biochemical Reactions	$\Delta_r G'^0$	$|v_e|$	E'^0/V
1. $CH_4 + NADPH^{4-} + O_2 + H^+ = CH_3OH + NADP^{3-} + H_2O$			
2. methane $+ NADP_{red} + O_2 =$ methanol $+ NADP_{ox} + H_2O$	-374.13	4	0.969
3. $CH_3OH + 2H^+ + 2e^- = CH_4 + H_2O$			
4. methanol $+ 2e^- =$ methane $+ H_2O$	-14.68	2	0.076
5. $NADP^{3-} + H^+ + 2e^- = NADPH^{4-}$			
6. $NADP_{ox} + 2e^- = NADP_{red}$	61.09	2	-0.317
7 $O_2 + 4H^+ + 4e^- = 2H_2O$			
8. $O_2 + 4e^- = 2H_2O$	-327.72	4	0.849
9. $CH_3OH + NADP^{3-} + 3H^+ + 4e^- = CH_4(aq) + NADPH^{4-} + H_2O$			
10. methanol $+ NADP_{ox} + 4e^- =$ methane $+ NADP_{red} + H_2O$	46.41	4	-0.120
11. $CH_4 + (\frac{1}{2})O_2 = CH_3OH$			
12. methane $+ (\frac{1}{2})O_2 =$ methanol	-149.18	2	0.773
13. $(\frac{1}{2})O_2 + NADPH^{4-} + H^+ = NADP^{3-} + H_2O$			
14. $(\frac{1}{2})O_2 + NADP_{red} = NADP_{ox} + H_2O$	-224.95	2	1.166

Source: With permission from R. A. Alberty, *Arch. Biochem. Biophys.* 389, 94–109 (2001). Copyright Academic Press.
Note: All of these species are in aqueous solution, but the formal electron e$^-$ is not. The convention is that $\Delta_f G'^0(e^-) = 0$. The first two reactions and the last four reactions are whole reactions in contrast with the half-reactions. For the whole reactions the values given here can be used to calculate the apparent equilibrium constants under these conditions.

which could, in principle, be catalyzed separately:

$$\text{Methane} + (\tfrac{1}{2})O_2 = \text{methanol} \qquad E'^0 = 0.773 \text{ V} \qquad (9.3\text{-}1)$$

$$(\tfrac{1}{2})O_2 + NADP_{red} = NADP_{ox} + H_2O \qquad E'^0 = 1.166 \text{ V}$$

If these reactions were carried out in two galvanic cells in series, the electromotive force would be $0.773 + 1.166 = 1.939$ V for a two electron change, and the standard transformed Gibbs energy of the overall monooxygenase reaction would be $-2F(1.939) = -374.13$ kJ mol^{-1}, as expected.

The methane monooxygenase reaction can, in principle, be carried out in two other ways by the enzyme complex that catalyzes it: It can be carried out in three half-reactions at three catalytic sites as follows:

$$O_2 + 4e^- = 2H_2O \qquad (9.3\text{-}2)$$

$$\text{Methanol} + 2e^- = \text{methane} + H_2O$$

$$NADP_{ox} + 2e^- = NADP_{red}$$

Or it can be carried out in two half-reactions at two catalytic sites as follows:

$$O_2 + 4e^- = 2H_2O \qquad (9.3\text{-}3)$$

$$\text{Methanol} + NADP_{ox} + 4e^- = \text{methane} + NADP_{red} + H_2O$$

Table 9. 3 shows the reduction potentials for these half-reactions that would have to be somehow matched to the reduction potentials of sites in the enzyme. A good deal is known about the mechanism of this enzyme-catalyzed reaction (Gassner and Lippard, 1999).

■ 9.4 HALF-REACTIONS WITH REACTANTS INVOLVING MULTIPLE SPECIES AT SPECIFIED pH

The reason for a separate section on half reactions with reactants involving multiple species is that they cannot be represented by a single chemical equation. Acid dissociation reactions are also involved, and as a consequence the pH and ionic strength dependencies of standard apparent reduction potentials are more complicated than for the reactions in Tables 9.2 and 9.3. These biochemical half reactions and biochemical reactions considered involve reactants with pKs in the range pH 5 to 9. They include carbon dioxide (pK 6.2), malate (pK = 5.25), citrate(pK$_1$ = 6.39, pK$_2$ = 4.75) cysteine (pK 8.37), ammonia (pK = 9.25), and reduced glutathione (pK = 8.37), where the pKs are for 298.15 K and zero ionic strength. When carbon dioxide is a reactant in a biochemical reaction, the expression for the apparent equilibrium constant can be written in terms of $P(CO_2)$ or [CO$_2$tot], where [CO$_2$tot] is the sum of the concentrations in aqueous solution of CO_2, H_2CO_3, HCO_3^-, and CO_3^{2-}. The standard transformed Gibbs energies and enthalpies of CO_2tot have been calculated three different ways (Alberty, 1995b, 1997e, 1998b), which give the same results (see Section 8.7). When the apparent equilibrium constant of a biochemical reaction involving carbon dioxide is written in terms of $P(CO_2)$, the pK = 6.2 does not show up in the pH dependencies of E'^0 and K', but when the apparent equilibrium constant is written in terms of [CO$_2$tot] it does. The advantage of using [CO$_2$tot] is that it is more immediately relevant to the reactions inside of the living cell. Note that when CO_2(g) in a biochemical reaction is replaced by CO_2tot, H_2O has to be added to the other side of the biochemical reaction. The pH dependencies of some of the half reactions in Table 9.4 are shown in Fig. 9.2.

Table 9.4   Standard Apparent Reduction Potentials E'^0 (in volts) at 298.15 K of Half-reactions Involving Reactants with Multiple Species

	pH 5	pH 6	pH 7	pH 8	pH 9
	$HCO_3^- + 2H^+ + 2e^- = formate^- + H_2O$				
	$CO_2tot + 2e^- = formate + H_2O$				
$I = 0$ M	-0.3294	-0.3630	-0.4095	-0.4663	-0.5257
$I = 0.10$ M	-0.3297	-0.3653	-0.4148	-0.4726	-0.5330
$I = 0.25$ M	-0.3299	-0.3662	-0.4166	-0.4747	-0.5355
	$CO_2(g) + H^+ + 2e^- = formate^-$				
	$CO_2(g) + 2e^- = formate$				
$I = 0$ M	-0.3726	-0.4022	-0.4318	-0.4613	-0.4909
$I = 0.10$ M	-0.3726	-0.4022	-0.4318	-0.4613	-0.4909
$I = 0.25$ M	-0.3726	-0.4022	-0.4318	-0.4613	-0.4909
	acetyl-$CoA^{4-} + HCO^{3-} + 2H^+ + 2e^- =$ pyruvate- $+ CoA^{4-} + H_2O$				
	acetyl-$CoA + CO_2tot + 2e^- =$ pyruvate $+ CoA + H_2O$				
$I = 0$ M	-0.4205	-0.4541	-0.5005	-0.5573	-0.6168
$I = 0.10$ M	-0.4208	-0.4563	-0.5059	-0.5637	-0.6240
$I = 0.25$ M	-0.4209	-0.4572	-0.5077	-0.5657	-0.6266
	ketoglutarate$^{2-} + NH_4^+ + 2H^+ + 2e^- =$ glutamate$^{2-} + H_2O$				
	ketoglutarate $+$ ammonia $+ 2e^- =$ glutamate $+ H_2O$				
$I = 0$ M	0.0252	-0.0340	-0.0932	-0.1530	-0.2172
$I = 0.10$ M	0.0062	-0.0530	-0.1122	-0.1720	-0.2362
$I = 0.25$ M	0	-0.0592	-0.1184	-0.1782	-0.2423
	pyruvate$^- + NH_4^+ + 2H^+ + 2e^- =$ alanine $+ H_2O$				
	pyruvate $+$ ammonia $+ 2e^- =$ alanine $+ H_2O$				
$I = 0$ M	-0.00241	-0.0616	-0.1208	-0.1806	-0.2448
$I = 0.10$ M	-0.0151	-0.0743	-0.1335	-0.1933	-0.2575
$I = 0.25$ M	-0.0192	-0.0784	-0.1376	-0.1974	-0.2616
	$HCO_3^- +$ pyruvate$^- + 2H^+ + 2e^- =$ malate$^{-2} + H_2O$				
	$CO_2tot +$ pyruvate $+ 2e^- =$ malate $+ H_2O$				
$I = 0$ M	-0.2156	-0.2604	-0.3087	-0.3658	-0.4252
$I = 0.10$ M	-0.2163	-0.2576	-0.3079	-0.3658	-0.4261
$I = 0.25$ M	-0.2159	-0.2567	-0.3077	-0.3658	-0.4267
	$CO_2(g) +$ pyruvate$^- + H^+ + 2e^- =$ malate^{-2}				
	$CO_2(g) +$ pyruvate $+ 2e^- =$ malate				
$I = 0$ M	-0.2588	-0.2996	-0.3311	-0.3608	-0.3904
$I = 0.10$ M	-0.2591	-0.2945	-0.3249	-0.3545	-0.3841
$I = 0.25$ M	-0.2586	-0.2927	-0.3228	-0.3525	-0.3820
	cystine $+ 2H^+ + 2e^- =$ 2cysteine				
	cystine $+ 2e^- =$ 2cysteine				
$I = 0$ M	-0.2381	-0.2972	-0.3554	-0.4066	-0.4325
$I = 0.10$ M	-0.2445	-0.3035	-0.3611	-0.4085	-0.4281
$I = 0.25$ M	-0.2465	-0.3055	-0.3628	-0.4088	-0.4266
	citrate$^{3-} + CoA + 2e^- =$ malate$^{2-} +$ acetyl-$CoA + H_2O$				
	citrate $+ CoA + 2e^- =$ malate $+$ acetyl-$CoA + H_2O$				
$I = 0$ M	-0.1875	-0.2566	-0.3336	-0.4200	-0.5085
$I = 0.25$ M	-0.1986	-0.2727	-0.3570	-0.4452	-0.5338
$I = 0.25$ M	-0.2027	-0.2791	-0.3649	-0.4533	-0.5420

Table 9.4   Continued

	pH 5	pH 6	pH 7	pH 8	pH 9
	$GSSG^{2-} + 2H^+ + 2e^- = 2GS^-$				
	$glutathione_{ox} + 2e^- = 2glutathione_{red}$				
$I = 0$ M	-0.1565	-0.2156	-0.2737	-0.3243	-0.3491
$I = 0.25$ M	-0.1692	-0.2280	-0.2845	-0.3261	-0.3395
$I = 0.25$ M	-0.1732	-0.2320	-0.2876	-0.3254	-0.3359
	$TR_{ox} + 2e^- = TR_{red}HO^{2-}$				
	$Thioredoxin_{ox} + 2e^- = thioredoxin_{red}$				
$I = 0$ M	-0.1643	-0.2234	-0.2815	-0.3323	-0.3570
$I = 0.25$ M	-0.1707	-0.2296	-0.2871	-0.3325	-0.3474
$I = 0.25$ M	-0.1727	-0.2317	-0.2888	-0.3318	-0.3438

Source: With permission from R. A. Alberty, *Arch. Biochem. Biophys.* 389, 94–109 (2001). Copyright Academic Press.

■ 9.5   NITROGENASE REACTION

The nitrogenase reaction (EC 1.18.6.1) involves three biochemical reactions: (1) the fixation of molecular nitrogen, (2) the hydrogenase reaction when molecular nitrogen is absent (EC 1.18.99.1), and (3) the hydrolysis of ATP to ADP. About 15 moles of ATP are hydrolyzed per mole of nitrogen fixed (Burris, 1991), but this amount varies with the pH and temperature. This is why it is not based on conservation of atoms. It has been suggested (Alberty, 1994) that the role of the hydrolysis of ATP is to supply the hydrogen ions required in the fixation reaction so that the catalytic site does not become alkaline. The apparent reduction potentials in the nitrogenase reaction are of special interest because of the importance of nitrogen fixation and because of the extraordinarily large effect of pH on the apparent equilibrium constant for the fixation of nitrogen. The chemical reaction for the fixation reaction is

$$N_2(g \text{ or aq}) + 10H^+ + 8Fd_{red} = 2NH_4^+ + H_2(g \text{ or aq}) + 8Fd_{ox}^+ \quad (9.5\text{-}1)$$

where Fd_{red} and Fd_{ox}^+ represent the reactive site of the protein ferredoxin. Note that a mole of H_2 is produced for each mole of N_2 converted to ammonia. N_2 and H_2 can be in gaseous or aqueous states. Strombaugh et al. (1976) found that the standard apparent reduction potentials of eight ferredoxins at 298.15 K and pH 7 ranged from -0.377 V to -0.434 V, and so apparent equilibrium constants for biochemical reactions involving ferredoxin will depend on the ferredoxin used. The calculations here have been made with $E'^0 = -0.403$ V, which was obtained for *Claustridium pasteurianum*. Since other ferredoxins have different values of E'^0, different apparent equilibrium constants will be obtained. Since E'^0 for this ferredoxin is independent of pH in the range 6.1 to 7.4 (Tagawa and Arnon, 1968), it is assumed that there are no acid groups in the reactive site with pKs in the range considered here. Reaction 9.5-1 is referred to as a reference reaction, and it can be balanced with NH_3 rather than NH_4^+. When the pH is in the neighborhood of 9 and higher, it is necessary to include the acid dissociation of NH_4^+ (p$K = 9.25$) in calculating the equilibrium composition.

When the pH is specified, the fixation of molecular nitrogen is represented by the following biochemical reaction:

$$N_2(g \text{ or aq}) + 8 \text{ ferredoxin}_{red} = 2 \text{ ammonia} + H_2(g \text{ or aq}) + 8 \text{ ferredoxin}_{ox}$$

$$(9.5\text{-}2)$$

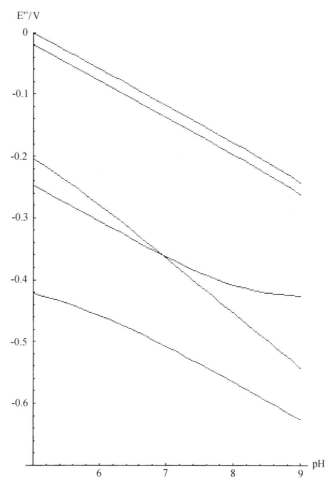

Figure 9.2 The pH dependence of the standard apparent reduction potentials at 298.15 K and 0.25 M ionic strength of the following biochemical half reactions, starting at the top of the ordinate [With permission from R. A. Alberty, *Biophys. Chem.* 389, 94–109 (2001). Copyright Academic Press.]:

> ketoglutarate + ammonia + $2e^-$ = glutamate + H_2O
> pyruvate + ammonia + $2e^-$ = alanine + H_2O
> CoA + citrate + $2e^-$ = acetyl-CoA + malate + H_2O
> cystine + $2e^-$ = 2 cysteine
> acetyl-CoA + CO_2tot = CoA + pyruvate + H_2O

Table 9.5 Standard Apparent Reduction Potentials E'^0 (in volts) at 298.15 K of Half-reactions in the Nitrogenase Reaction

	pH 5	pH 6	pH 7	pH 8	pH 9
	$N_2(g) + 10H^+ + 8e^- = 2NH_4^+ + H_2(g)$				
	$N_2(g) + 8e^- = 2$ ammonia $+ H_2(g)$				
$I = 0$ M	-0.1643	-0.2382	-0.3121	-0.3858	-0.4572
$I = 0.10$ M	-0.1706	-0.2445	-0.3185	-0.3921	-0.4635
I $= 0.25$ M	-0.1726	-0.2466	-0.3205	-0.3941	-0.4656
	$N_2(aq) + 10H^+ + 8e^- = 2NH^{4+} + H_2(aq)$				
	$N_2(aq) + 8e^- = 2$ ammonia $+ H_2(aq)$				
$I = 0$ M	-0.1628	-0.2368	-0.3107	-0.3843	-0.4558
$I = 0.10$ M	-0.1692	-0.2431	-0.3170	-0.3907	-0.4621
$I = 0.25$ M	-0.1712	-0.2452	-0.3191	-0.3927	-0.4642

Source: From R. A. Alberty, *Arch. Biochem. Biophys.* 389, 94–109 (2001). Copyright Academic Press.

Table 9.6   Apparent Equilibrium Constants K' for Nitrogenase Reactions at 298.15 K

pH 5	pH 6	pH 7	pH 8	pH 9
	$N_2(g) + 10H^+ + 8Fd_{red} = 2NH_4^+ + H_2(g) + 8Fd_{ox}^+$			
	$N_2(g) + 8\,ferredoxin_{red} = 2\,ammonia + H_2(g) + 8\,ferredoxin_{ox}$			
1.40×10^{31}	1.40×10^{21}	1.41×10^{11}	15.6	3.40×10^{-9}
	$N_2(aq) + 10H^+ + 8\,Fd_{red} = 2NH_4^+ + H_2(aq) + 8Fd_{ox}^+$			
	$N_2(aq) + 8\,ferredoxin_{red} = 2\,ammonia + H_2(aq) + 8\,ferredoxin_{ox}$			
2.18×10^{31}	2.18×10^{21}	2.20×10^{11}	24.3	5.30×10^{-9}
	$2H^+ + 2Fd_{red} = H_2(g) + 2Fd_{ox}^+$			
	$2\,ferredoxin_{red} = H_2(g) + 2\,ferredoxin_{ox}$			
2.180	21.8	0.218	2.18×10^{-3}	2.18×10^{-5}
	$2H^+ + 2Fd_{red} = H_2(aq) + 2Fd_{ox}^+$			
	$2\,ferredoxin_{red} = H_2(aq) + 2\,ferredoxin_{ox}$			
1.80	1.80×10^{-2}	1.80×10^{-4}	1.80×10^{-6}	1.80×10^{-8}

Source: From R. A. Alberty, *Arch. Biochem. Biophys.* 389, 94–109 (2001). Copyright Academic Press.
Note: The apparent equilibrium constants for these reactions do not depend on ionic strength because the equilibrium constants for the chemical reference reactions and acid dissociation do not depend on ionic strength.

where ammonia represents the sum of NH_4^+ and NH_3. Different abbreviations are used for the two forms of ferredoxin in chemical reactions and biochemical reactions to make it clear that the pH is held constant in the biochemical reaction. This distinction is especially important when there are acid groups in the reactive site of a protein, since the site at a specified pH consists of a sum of species (Alberty, 2000d).

In the absence of molecular nitrogen, nitrogenase produces molecular hydrogen. This hydrogenase reaction (EC 1.18.99.1) is represented by the following chemical reaction:

$$2H^+ + 2Fd_{red} = H_2(g \text{ or } aq) + 2Fd_{ox}^+ \tag{9.5-3}$$

The corresponding biochemical reaction at specified pH is

$$2\,ferredoxin_{red} = H_2(g \text{ or } aq) + 2\,ferredoxin_{ox} \tag{9.5-4}$$

It may seem strange not to show H^+ as a reactant, but $[H^+]$ is not involved in the expression for the apparent equilibrium constant for reaction 9.5-4 because it is held constant.

The effects of pH on the standard apparent reduction potentials of the half reactions involved in the nitrogenase reaction are shown in Table 9.5. The effects of pH on the apparent equilibrium constants of the reactions involved in the nitrogenase reaction as shown in Table 9.6.

The effect of pH on K' for the nitrogen fixation reaction (9.5-2) is so striking (a change of 10^{10} per pH unit) that it is shown in Fig. 9.3. Note that nitrogen cannot be fixed by this reaction above pH 8 when ferredoxin from *Claustridium pasteurianum* is used.

■ 9.6   CHANGES IN THE BINDING OF HYDROGEN IONS IN HALF-REACTIONS AT SPECIFIED pH

The change in binding of hydrogen ions in a biochemical reaction can be calculated from the derivative of the standard transformed Gibbs energy of

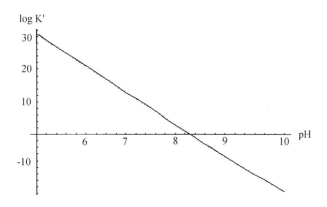

Figure 9.3 Plot of the base 10 logarithm of the apparent equilibrium constant K' for the nitrogen fixation reaction (see reaction 9.5-2) versus pH at 298.15 K (see Problem 9.2). [With permission from R. A. Alberty, *Arch. Biochem. Biophys.* **389**, 94–109 (2001). Copyright Academic Press.]

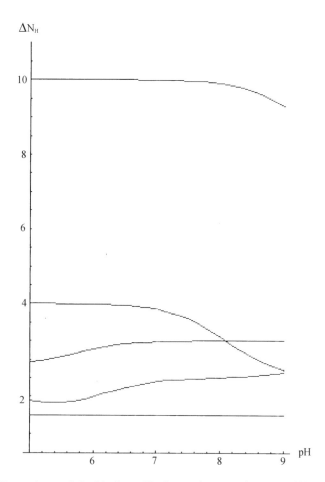

Figure 9.4 Dependence of the binding of hydrogen ions on the pH at 298.15 K and ionic strength 0.25 M for the following five biochemical half-reactions (starting at the top):

$$N_2(g) + 8e^- = 2 \text{ ammonia} + H_2(g)$$
$$\text{cystine} + 2e^- = 2 \text{ cysteine}$$
$$\text{citrate} + CoA + 2e^- = \text{malate} + \text{acetyl-CoA} + H_2O$$
$$CO_2\text{tot} + \text{pyruvate} + 2e^- = \text{malate} + H_2O$$
$$NAD_{ox} + 2e^- = NAD_{red}$$

(See Problem 9.4.) [With permission from R. A. Alberty, *Arch. Biochem. Biophys.* **389**, 94–109 (2001). Copyright Academic Press.]

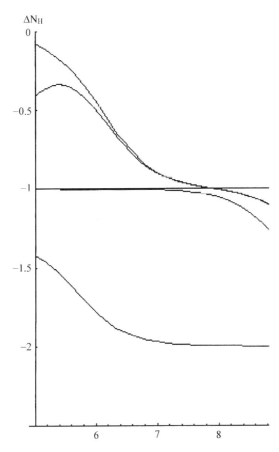

Figure 9.5 Change in the binding of H^+ in five biochemical reactions at 298.15 K, 1 bar, and ionic strength 0.25 M. Starting at the top the reactions are as follows:

NAD_{ox} + formate + H_2O = NAD_{red} + CO_2tot

NAD_{ox} + malate + H_2O = NAD_{red} + CO_2tot + pyruvate

NAD_{ox} + ethanol = NAD_{red} + acetaldehyde

NAD_{ox} + alanine + H_2O = NAD_{red} + pyruvate + ammonia

NAD_{ox} + malate + acetylcoA + H_2O = NAD_{red} + citrate + coA

(See Problem 9.5.) [With permission from R. A. Alberty, *Arch. Biochem. Biophys.* 389, 94–109 (2001). Copyright Academic Press.]

reaction with respect to the pH, as shown in equation 4.7-4. Since the standard transformed Gibbs energy of a redox reaction can be written as the difference between the standard transformed Gibbs energies of two half-reactions and the standard apparent reduction potentials of the half-reactions are proportional to the standard transformed Gibbs energies of the half reactions, equation 4.7-4 can also be written

$$\Delta_r N_H = -\frac{|v_e|F}{RT\ln(10)}\left(\frac{\partial E_R'^0}{\partial pH}\right)_{T,P} + \frac{|v_e|F}{RT\ln(10)}\left(\frac{\partial E_L'^0}{\partial pH}\right)_{T,P} \qquad (9.6\text{-}1)$$

$$= \Delta_r N_H(R) - \Delta_r N_H(L)$$

Thus the change in binding of hydrogen ions in a half-reaction is given by

$$\Delta_r N_H = -\frac{|v_e|F}{RT\ln(10)}\left(\frac{\partial E'^0}{\partial pH}\right)_{T,P} \qquad (9.6\text{-}2)$$

The changes in binding of hydrogen ions in five biochemical half reactions are shown in Fig. 9.4. Since $\partial E'^0/\partial pH$ is always negative for a half-reaction, $\Delta_r N_H$ is always positive. It is the magnitude of $\Delta_r N_H$ that determines the slope of the plot of E'^0 versus pH for a half reaction.

Figure 9.5 shows the change in binding of hydrogen ions in five biochemical reactions as a function of pH. The change in binding in one half-reaction may compensate for the change in binding in the other half reaction. The change in binding of H^+ may occur in only one of the two half reactions.

These last two plots have shown the importance of determinations of acid dissociation constants of biochemical reactants. In the next chapter we will see that it is also important to know the pKs of acid groups when heats of reaction are determined calorimetrically.

10

Calorimetry of Biochemical Reactions

The enthalpy H of a chemical reaction system is of special interest because when a reaction occurs at constant pressure, the change in enthalpy is equal to the heat absorbed; $\Delta H = q_P$. The standard enthalpy $\Delta_r H^0$ of a chemical reaction can be obtained by determining the heat of reaction calorimetrically or by measuring the dependence of the equilibrium constant K on temperature. The standard transformed enthalpy $\Delta_r H'^0$ of a biochemical reaction can be obtained by determining the heat of reaction at specified pH or by measuring the dependence of the apparent equilibrium constant K' on temperature. The determination of the transformed enthalpy of reaction of a biochemical reaction is complicated by the fact that the biochemical reaction may produce or consume hydrogen ions that react with the buffer to produce an additional heat effect in a calorimetric measurement. When the reactants bind metal ions, a similar effect will occur if the metal ion is bound by a reactant. If these effects are properly handled in the analysis of calorimetric data, the transformed enthalpies of reaction determined from equilibrium constants and from calorimetric measurements should agree. Enthalpies of reaction may be rather constant over narrow temperature ranges, but over wider temperature ranges, it is necessary to take into account the dependence of the heat of reaction on temperature. For a biochemical reaction the effect of temperature on the heat of reaction depends on the pH.

If the change in heat capacity in a chemical reaction is equal to zero, the enthalpy of the reaction is independent of temperature, and the equilibrium constant of the chemical reaction can be readily calculated over a range of temperature without making an integration, as described in Section 3.7. In general, the enthalpy of a chemical reaction is a function of temperature and ionic strength. When $\Delta_r G^0$ and $\Delta_r H^0$ are known, the standard reaction entropy $\Delta_r S^0$ can be calculated

The transformed enthalpy of a biochemical reaction is a function of temperature, pH, and ionic strength. Knowledge of $\Delta_f G^0$, $\Delta_f H^0$, and C_{Pm}^0 for all the species in a biochemical reaction makes it possible to calculate $\Delta_r G'^0$, $\Delta_r H'^0$, $\Delta_r C_P'^0$, and K' for the biochemical reaction at the desired T, pH, and ionic strength. Note that when ions are involved there is an electrostatic contribution that varies with temperature (see Section 3.7).

■ 10.1 CALORIMETRIC DETERMINATION OF THE STANDARD TRANSFORMED ENTHALPY OF REACTION

When a biochemical reaction that is affected by pH and pMg is carried out in a calorimeter in a buffer, the hydrogen ions and magnesium ions that are produced or consumed react with the buffer to produce a heat effect that is characteristic of the buffer, rather than the reaction being studied. Therefore this contribution should be calculated and should be used to correct the calorimetric heat effect to obtain the standard transformed enthalpy of the biochemical reaction $\Delta_r H'^0$. The analysis by Alberty and Goldberg (1993) shows that the enthalpy change in the calorimetric experiment $\Delta_r H(\text{cal})$ is given by

$$\Delta_r H(\text{cal}) = \Delta_r H'^0 + \Delta_r N_H \Delta_r H^0(\text{Buff}) + \Delta_r N_{Mg} \Delta_r H^0(\text{MgBuff}) \quad (10.1\text{-}1)$$

where the **enthalpy of dissociation** of hydrogen ions from the acidic form of the buffer is given by

$$\Delta_r H^0(\text{Buff}) = \Delta_f H^0(\text{H}^+) + \Delta_f H^0(\text{Buff}^-) - \Delta_f H^0(\text{HBuff}) \quad (10.1\text{-}2)$$

and the enthalpy of dissociation of magnesium ions from a complex ion is given by

$$\Delta_r H^0(\text{MgBuff}) = \Delta_f H^0(\text{Mg}^{2+}) + \Delta_f H^0(\text{Buff}^-) - \Delta_f H^0(\text{MgBuff}]) \quad (10.1\text{-}3)$$

Thus, if hydrogen ions or metal ions are produced or consumed, it is necessary to know all of the pK's and dissociation constants of magnesium complexes of the all of the species of all of the reactant in order to calculate $\Delta_r N_H$ and $\Delta_r N_{Mg}$. However, there is an experimental method for determining $\Delta_r N_H$ and $\Delta_r N_{Mg}$ that can be used with complicated reactants where it is not possible to determine all of the pK's and dissociation constants of magnesium complexes. This is based on determining the apparent equilibrium constant K' for the biochemical reaction as a function of pH and pMg. The changes in binding are calculated using

$$\Delta_r N_H = -\left(\frac{\partial \log K'}{\partial \text{pH}} \right)_{T,P,\text{pMg}} \quad (10.1\text{-}4)$$

$$\Delta_r N_{Mg} = -\left(\frac{\partial \log K'}{\partial \text{pMg}} \right)_{T,P,\text{pH}} \quad (10.1\text{-}5)$$

as shown in Section 4.5. The change in the binding of hydrogen ions $\Delta_r N_H$ can also be determined by use of a pHstat.

■ 10.2 CALCULATION OF STANDARD TRANSFORMED ENTHALPIES OF REACTIONS FROM THE STANDARD ENTHALPIES OF FORMATION OF SPECIES

If the enthalpies of formation of all the species involved are known, the **standard transformed enthalpy of a biochemical reaction** $\Delta_r H'^0$ at a specific T, P, pH, etc., can be calculated. The basic relation is (see equation 4.4-3)

$$\Delta_r H'^0 = \sum_{i=1}^{N'} v_i' \Delta_f H_i'^0 \qquad (10.2\text{-}1)$$

where the v_i' are the apparent stoichiometric numbers of the N' reactants (sums of species) in the biochemical reaction and the $\Delta_f H_i'^0$ are the standard transformed enthalpies of formation of the reactants under the desired conditions. In the following treatment it is assumed that Mg^{2+} is bound as well as H^+. In order to calculate $\Delta_f H_i'^0$ for a reactant, it is necessary to calculate the $\Delta_f H_j'^0$ for each species of the reactant at the desired conditions. For example, if species j contains $N_H(j)$ hydrogen atoms and $N_{Mg}(j)$ magnesium atoms, this adjustment is made with

$$\Delta_f H_j'^0 = \Delta_f H_j^0 - N_H(j)\Delta_f H^0(H^+) - N_{Mg}(j)\Delta_f H^0(Mg^{2+}) \qquad (10.2\text{-}2)$$

where $\Delta_f H_j^0$ is the standard enthalpy or formation of species j, $\Delta_f H^0(H^+)$ is the standard enthalpy of the hydrogen ion, and $\Delta_f H^0(Mg^{2+})$ is the standard enthalpy of the magnesium ion, all three of these quantities having been calculated at the desired ionic strength. When a reactant consists of two or more species, the standard transformed enthalpy of formation for the reactant is the mole fraction-weighted average of the standard transformed enthalpies of the N_{iso} species in the pseudoisomer group (see equation 4.5-3):

$$\Delta_f H_i'^0 = \sum_{j=1}^{N_{iso}} r_j \Delta_f H_j'^0 \qquad (10.2\text{-}3)$$

where r_j is the equilibrium mole fraction of species j, calculated using the standard transformed Gibbs energies of formation of the species under the desired conditions (see equation 4.3-7). Substituting equation 10.2-2 shows that the **standard transformed enthalpy of formation of a reactant** is made up of three contributions:

$$\Delta_f H_i'^0 = \langle \Delta_f H_j^0 \rangle - \bar{N}_H(i)\Delta_f H^0(H^+) - \bar{N}_{Mg}(i)\Delta_f H^0(Mg^{2+}) \qquad (10.2\text{-}4)$$

where $\langle \Delta_f H_j^0 \rangle$ is the **mole fraction-weighted enthalpy of formation** of the species in the pseudoisomer group. The average numbers of hydrogen atoms and magnesium atoms bound are given by

$$\bar{N}_H(i) = \sum_{j=1}^{N_{iso}} r_j N_H(j) \qquad (10.2\text{-}5)$$

and

$$\bar{N}_{Mg}(i) = \sum_{j=1}^{N_{iso}} r_j N_{Mg}(j) \qquad (10.2\text{-}6)$$

The convention is that $N_H(j)$ is the total number of hydrogen atoms in species j and $N_{Mg}(j)$ is the number of magnesium atoms in the species.

Substituting equation 10.2-4 for each reactant in equation 10.2-1 yields

$$\Delta_r H'^0 = \sum_{i=1}^{N'} v_i' \langle \Delta_f H_i^0 \rangle - \Delta_r N_H \Delta_f H^0(H^+) - \Delta_r N_{Mg}\Delta_f H^0(Mg^{2+}) \quad (10.2\text{-}7)$$

where the **changes in the binding** of H^+ and Mg^{2+} in the reaction are given by

$$\Delta_r N_H = \sum_{i=1}^{N'} v_i' \bar{N}_H(i) \qquad (10.2\text{-}8)$$

$$\Delta_r N_{Mg} = \sum_{i=1}^{N'} v_i' \bar{N}_{Mg}(i) \qquad (10.2\text{-}9)$$

Equation 10.2-7 shows that there are three contributions to the transformed enthalpy of reaction in this case: the effects of the standard enthalpies of formation of the species, the effect due to the change in binding of hydrogen ions, and the effect due to the change in binding of magnesium ions. In the hydrolysis of ATP to ADP and orthophosphate at 298.15 K, pH 7, pMg 4, and 0.25M ionic strength, $\Delta_r N_H = -0.62$, $\Delta_r N_{Mg} = -0.49$, and so

$$\Delta_r H'^0 = 177.74 - (-0.62)(0.41) - (-0.49)(465.36) = -30.76 \text{ kJ mol}^{-1}$$

(10.2-10)

Note that the effect of magnesium binding is larger than the first term and has the opposite sign in this case (Alberty and Goldberg, 1993).

■ 10.4 CALCULATION OF STANDARD TRANSFORMED ENTROPIES OF BIOCHEMICAL REACTIONS

The determination of standard transformed enthalpies of biochemical reactions at specified pH, either from temperature coefficients of apparent equilibrium constants or by calorimetric measurements, makes it possible to calculate the corresponding **standard transformed entropy of reaction** using

$$\Delta_r S'^0 = \frac{\Delta_r H'^0 - \Delta_r G'^0}{T}$$

(10.3-1)

Substituting $\Delta_r H'^0 = \Sigma v_i' \Delta_f H_i'^0$ and $\Delta_r G'^0 = \Sigma v_i' \Delta_f G_i'^0$ yields

$$\Delta_r S'^0 = \sum v_i' \Delta_f S_i'^0$$

(10.3-2)

where the **standard transformed entropy of formation** of pseudoisomer group i for a particular set of conditions is given by

$$\Delta_f S_i'^0 = \frac{\Delta_f H_i'^0 - \Delta_f G_i'^0}{T}$$

(10.3-3)

Thus tables given earlier can be used to calculate standard transformed entropies of formation. Standard transformed entropies of reactants have not been emphasized in previous chapters because the properties with the greatest practical interest are $\Delta_f G_i'^0$, which can be used to calculate K', and $\Delta_f H_i'^0$, which can be used to calculate the temperature coefficient of K' and the heat effect of a biochemical reaction. However, molar entropies $S_m^0(j)$ of species are sometimes considered more interpretable than $\Delta_f G_j^0$ and $\Delta_f H_j^0$. For example, $\Delta_r S^0(298.15 K) = -92.0 \text{ J K}^{-1} \text{ mol}^{-1}$ for the dissociation of acetic acid, whereas the $\Delta_r S^0$ values for gas dissociation reactions are always positive, corresponding with the increased degrees of freedom of the product gas molecules. The entropy of dissociation of acetic acid indicates that hydrogen ion and the acetate ion have fewer degrees of freedom than hydrated acetic acid molecules. The explanation of this contradiction is that the ions are more hydrated than the neutral acetic acid molecules because of the orienting effects of the electric fields in the neighborhood of an ion. Thus the freedom of H_2O molecules in an acetic acid solution is reduced by the dissociation, and consequently the entropy of the products is lower than the entropy of the acetic acid molecules. Therefore it is of interest to inquire more deeply into the relation between the molar entropies of species and the standard transformed entropy of formation of a pseudoisomer group at a specified pH.

The standard transformed enthalpy of formation of a pseudoisomer group is given by equation 10.2-3. The standard transformed Gibbs energy of formation is given by

$$\Delta_f G_i'^0 = \sum_{j=1}^{N_{iso}} r_j \Delta_f G_j'^0 + RT \sum_{j=1}^{N_{iso}} r_j \ln r_j$$

(10.3-4)

where N_{iso} is the number of species in the pseudoisomer group. Substituting equations 10.2-3 and 10.3-4 into equation 10.3-3 yields

$$\Delta_f S_i'^0 = \sum_{j=1}^{N_{iso}} r_j \Delta_f S_j'^0 - R \sum_{j=1}^{N_{iso}} r_j \ln r_j$$

(10.3-5)

since

$$\Delta_f S_j'^0 = \frac{\Delta_f H_j'^0 - \Delta_f G_j'^0}{T} \tag{10.3-6}$$

for a species.

Substituting (see Section 4.3)

$$\Delta_f H_j'^0 = \Delta_f H_j^0 - N_H(j)\Lambda_f H^0(H^+) - N_{Mg}(j)\Delta_f H^0(Mg^{2+}) \tag{10.3-7}$$

and

$$\Delta_f G_j'^0 = \Delta_f G_j^0 - N_H(j)[\Delta_f G_0(H^+) - RT\ln(10)pH]$$
$$- N_{Mg}(j)[\Delta_f G^0(Mg^{2+}) - RT\ln(10)pMg] \tag{10.3-8}$$

in equation 10.3-6 yields

$$\Delta_f S_j'^0 = \Delta_f S_j^0 - N_H(j)[\Delta_f S^0(H^+) + R\ln(10)pH]$$
$$- N_{Mg}(j)[\Delta_f S^0(Mg^{2+}) + R\ln(10)pMg] \tag{10.3-9}$$

Earlier the fundamental equation for G' (equation 4.1-22) was used to show that $S' = S - n_c(H)S_m(H^+)$ when hydrogen ions are bound, and that can be extended to

$$S' = S - n_c(H)S_m(H^+) - n_c(Mg)S_m(Mg^{2+}) \tag{10.3-10}$$

This can be written

$$\sum n_j S_m'(j) = \sum n_j S_m(j) - n_c(H)S_m(H^+) - n_c(Mg)S_m(Mg^{2+}) \tag{10.3-11}$$

Substituting $n_c(H) = \sum N_H(j)n_j$ and $n_c(Mg) = \sum N_{Mg}(j)n_j$ yields the expression for the **standard transformed molar entropy of a species**:

$$S_m'^0(j) = S_m^0(j) - N_H(j)[S_m^0(H^+) + R\ln(10)pH]$$
$$- N_{Mg}(j)[S_m^0(Mg^{2+}) + R\ln(10)pMg] \tag{10.3-12}$$

which can be compared with equation 10.3-9.

Equations 10.3-9 and 10.3-12 raise an issue about conventions for the hydrogen ion in thermodynamic tables. Since it is not possible to connect the standard thermodynamic properties of H^+ to those of molecular hydrogen, the convention is that $\Delta_f G^0(H^+) = 0$ and $\Delta_f H^0(H^+) = 0$ at each temperature. This indicates that the standard entropy of formation of a hydrogen ion $\Delta_f S^0(H^+)$ should be taken as zero at each temperature, but, for historical reasons, the convention adopted in current thermodynamic tables is $S_m^0(H^+) = 0$ at each temperature. In principle, the value of $S^0(H^+)$ should be calculated from $\Delta_f S^0(H^+)$ for the formation reaction for H^+. One way to write this reaction is

$$\tfrac{1}{2}H_2(g) = H^+(aq) + e^- \tag{10.3-13}$$

where e^- is the formal electron, not a hydrated electron in water. Both $\Delta_f S^0\{H^+(ao)\} = 0$ and $S_m^0(H^+) = 0$ can be satisfied if the formal electron is treated as a reactant and assigned a **standard molar entropy** of $S_m^0(e^-) = (\tfrac{1}{2})S_m^0(H_2, g)$ at each temperature. A different, but perhaps more logical convention would be to assign $\Delta_f G^0(e^-) = 0$, $\Delta_f H^0(e^-) = 0$, and $\Delta_f S^0(e^-) = 0$ so that equation 10.3-13 would lead to

$$\Delta_f S^0[H^+(ao)] = S_m^0(H^+) + S_m^0(e^-) - (\tfrac{1}{2})S_m(H_2) = S_m^0(H^+) + 0 - (\tfrac{1}{2})S_m^0(H_2, g) = 0 \tag{10.3-14}$$

In this case, $S_m^0(H^+) = (\tfrac{1}{2})S_m^0(H_2, g)$, rather than zero.

■ 10.4   EFFECT OF TEMPERATURE

In Chapter 3 equations and *Mathematica* programs were given for calculating $\Delta_f G^0$ and $\Delta_f H^0$ of species at temperatures other than 298.15 K on the basis of the assumption that $\Delta_f H^0$ at zero ionic strength is independent of temperature. More accurate calculations are possible when C_{Pm}^0 values are known for species and can be assumed to be independent of temperature. In this case,

$$\Delta_f H^0(T) = \Delta_f H^0(298.15 \text{ K}) + \Delta_f C_P^0 (T - 298.15 \text{ K}) \tag{10.4-1}$$

where $\Delta_f C_P^0$ is the difference between the standard molar heat capacity of the species and the sum of the molar heat capacities of the elements making it up. Therefore the standard enthalpy of a chemical reaction at temperature T is given by

$$\Delta_r H^0(T) = \Delta_r H^0(298.15 \text{ K}) + \Delta_r C_P^0 (T - 298.15 \text{ K}) \tag{10.4-2}$$

where

$$\Delta_r C_P^0 = \sum_{j=1}^{N_s} v_j \Delta_f C_P^0(j) = \sum_{j=1}^{N_s} v_j C_{Pm}^0(j) \tag{10.4-3}$$

C_{Pmj}^0 is the **standard heat capacity** of species j at constant pressure, and N_s is the number of species in the chemical reaction. Note that the standard molar heat capacities of the elements cancel in the summation. Equation 10.4-2 was given earlier as equation 3.2-19, and it was used to show that the pK of acetic acid goes through a maximum not far from room temperature. Equation 10.4-3 was given earlier as equation 3.2-16.

When the pH is specified, the standard transformed molar heat capacity of a species is given by (Alberty, 2001d)

$$C_{Pj}'^0(I) = C_{Pj}^0(I = 0) + RT \left[2 \left(\frac{\partial \alpha}{\partial T} \right)_P + T \left(\frac{\partial^2 \alpha}{\partial T^2} \right)_P \right] \frac{(z_j^2 - N_H(j))I^{1/2}}{1 + 1.6 I^{1/2}} \tag{10.4-4}$$

where z_j is the charge number for the ion and $N_H(j)$ is the number of hydrogen atoms it contains. Values of the coefficient of the ionic strength term are given as a function of temperature in Chapter 3. This equation is obtained by applying the Gibbs-Helmholtz equation to the expression for $\Delta_f H'^0(T)$.

Table 10.1   Standard Molar Heat Capacities of Species in Dilute Aqueous Solutions at 298.15 K

Species	C_{Pm}^0/J K^{-1} mol^{-1}
xylose	279 ± 20
ribose	276 ± 20
arabinose	276 ± 20
fructose	369 ± 14
galactose	319 ± 20
glucose	336 ± 7
mannose	342 ± 17
xylose	319 ± 28
lyxose	285 ± 20
ribose 5-phosphate^{2-}	10 ± 45
glucose 6-phosphate^{2-}	48 ± 17
mannose 6-phosphate^{2-}	57 ± 19
fructose 6-phosphate^{2-}	89 ± 17

Source: With permission from R. N. Goldberg and Y, Tewari, *J. Phys. Chem Ref. Data* 18, 809 (1989). Copyright American Institute of Physics.

Table 10.2   Standard Heat Capacities of Chemical Reactions in Dilute Aqueous
Solutions at 298.15 K

Chemical Reaction	$\Delta_r C_P^0 / J\ K^{-1}\ mol^{-1}$
glucose 6-phosphate^{2-} + H_2O = glucose + HPO_4^{2-}	-48 ± 20
mannose 6-phosphate^{2-} + H_2O = mannose + HPO_4^{2-}	-46 ± 10
fructose 6-phosphate^{2-} + H_2O = fructose + HPO_4^{2-}	-28 ± 40
ribose 5-phosphate^{2-} + H_2O = ribose + HPO_4^{2-}	-63 ± 40
ribulose 5-phosphate^{2-} + H_2O = ribulose + HPO_4^{2-}	-84 ± 30
glucose 6-phosphate^{2-} = fructose 6-phosphate^{2-}	44 ± 10
mannose 6-phosphate^{2-} = fructose 6-phosphate^{2-}	38 ± 30
glucose = fructose	76 ± 30
xylose = xylulose	40 ± 20
ATP4- + H_2O = ADP3- + HPO_4^{2-} + H^+	-237 ± 30

Source: With permission from R. N. Goldberg and Y, Tewari, *J. Phys. Chem Ref.
Data* 18, 809 (1989). Copyright American Institute of Physics.

In Chapter 4 the effects of temperature on $\Delta_f G'^0$ and $\Delta_f H'^0$ and on $\Delta_r G'^0$ and
$\Delta_r H'^0$ are discussed on the basis of the assumption that $\Delta_f H^0$ at zero ionic
strength is independent of temperature. Therefore the effects of heat capacities of
species were not treated. When a biochemical reactant contains two or more
species, the standard transformed molar heat capacity of the pseudoisomer group
is given by (Alberty, 1983a)

$$C_{mP}'^0(iso) = \sum_{j=1}^{N_{iso}} r_j C_{Pm}'^0(j) + \left(\frac{1}{RT}\right)\left\{\sum_{j=1}^{N_{iso}} r_j(\Delta_f H_j'^0)^2 - (\Delta_f H'^0(iso))^2\right\} \quad (10.4\text{-}5)$$

The second term is always positive because as the system is heated, the acid
dissociations shift in such a way as to absorb heat, as predicted by the Le
Chatelier principle.

Calorimetric measurements yield enthalpy changes directly, and they also
yield information on heat capacities, as indicated by equation 10.4-1. Heat
capacity calorimeters can be used to determine $C_{Pm}'^0$ directly. It is almost
impossible to determine $\Delta_r C_P'^0$ from measurements of apparent equilibrium
constants of biochemical reactions because the second derivative of $\ln K'$ is
required. Data on heat capacities of species in dilute aqueous solutions is quite
limited, although the NBS Tables give this information for most of their entries.
Goldberg and Tewari (1989) have summarized some of the literature on molar
heat capacities of species of biochemical interest in their survey on carbohydrates
and their monophosphates. Table 10.1 give some standard molar heat capacities
at 298.15 *K* and their uncertainties. The changes in heat capacities in some
chemical reactions are given in Table 10.2.

11

Use of Semigrand Partition Functions

The introduction of transformed thermodynamic properties to biochemical thermodynamics owes a lot to statistical mechanics because these calculations follow the pattern of calculations on reaction equilibria in systems of gaseous hydrocarbons at specified partial pressures of molecular hydrogen, ethylene, or acetylene. Alberty and Oppenheim (1989) used a semigrand partition function to describe the equilibrium distribution of alkyl benzenes at elevated temperatures as a function of the partial pressure of ethylene. The transformed Gibbs energy G' of a system at a specified pH can be calculated using a semigrand partition function (Alberty, 2001c). The further transformed Gibbs energy G'' of a system of biochemical reactions can be calculated using a semigrand partition function with the steady state concentrations of coenzymes as intensive variables (Alberty, 2001g, 2002a).

Statistical mechanics provides a bridge between the properties of atoms and molecules (microscopic view) and the thermodynmamic properties of bulk matter (macroscopic view). For example, the thermodynamic properties of ideal gases can be calculated from the atomic masses and vibrational frequencies, bond distances, and the like, of molecules. This is, in general, not possible for biochemical species in aqueous solution because these systems are very complicated from a molecular point of view. Nevertheless, statistical mechanmics does consider thermodynamic systems from a very broad point of view, that is, from the point of view of partition functions. A partition function contains all the thermodynamic information on a system. There is a different partition function for each choice of natural variables. Gibbs (1903) introduced the grand canonical partition function for a system containing a single species that is in contact with a large reservoir of that species through a permeable membrane. Semigrand partition functions were introduced later by statistical mechanicians to treat systems with several species, one or more of which are available from a large reservoir through a semipermeable membrane. The idea of holding the chemical potential of a species constant was often used in statistical mechanics long before it was used in thermodynamics.

In this chapter the usual convention in statistical mechanics of using numbers N_i of molecules (rather than amounts n_i), the Boltzmann constant k (rather than the gas constant R), and $\beta = 1/kT$ have been used, but the same symbols have been used for thermodynamic properties as in thermodynamics. Thus the proper interpretation of these latter symbols depends on context. Detailed information on various partition functions is provided by textbnooks on statistical mechanics (McQuarrie, 2000; Chandler, 1987; Greiner, Neise, and Stöcker, 1995; di Cera, 1995; Widom, 2002).

■ 11.1   INTRODUCTION TO SEMIGRAND PARTITION FUNCTIONS

Gibbs considered the statistical mechanics of a system containing one type of molecule in contact with a large reservoir of the same type of molecules through a permeable membrane. If the system has a specified volume and temperature and is in equilibrium with the resevoir, the chemical potential of the species in the system is determined by the chemical potential of the species in the reservoir. The natural variables of this system are T, V, and μ. We saw in equation 2.6-12 that the thermodynamic potential with these natural variables is $U[T, \mu]$ using Callen's nomenclature. The integration of the fundamental equation for $U[T, \mu]$ yields $-PV$ (see equation 2.6-20), and so $-PV$ can be considered a state function of the system under these conditions.

Statistical mechanics is based on the use of ensembles (collections of systems under specified conditions) that lead to partition functions. Partition functions are sums of exponential functions. Gibbs referred to the ensemble for a system at T and V in contact with a large reservoir of that species through a permeable membrane as the grand canonical ensemble. The corresponding **grand canonical partition function** is represented by $\Xi(T, V, \mu)$. Since thermodynamic potentials are given by $-kT$ times the natural logarithm of a partition function, the value of PV can be calculated using

$$PV = kT \ln \Xi(T, V, \mu) \tag{11.1-1}$$

where k is the Boltzmann constant (R/N_A) and N_A is the Avogadro constant.

If a system contains two types of species, but the membrane is permeable only to species number 1, the natural variables for the system are T, V, μ_1, and N_2, where N_2 is the number of molecules of type 2 in the system. The thermodynamic potential for this system containing two species is represented by $U[T, \mu_1]$. The corresponding ensemble is referred to as a semigrand ensemble, and the **semigrand partition function** can be represented by $\Psi(T, V, \mu_1, N_2)$. The thermodynamic potential of the system is related to the partition function by

$$U[T, \mu_1] = -kT \ln \Psi(T, V, \mu_1, N_2) \tag{11.1-2}$$

The Gibbs energy for a system at constant T and P containing a single species is given by

$$G(T, P, N) = -kT \ln \Delta(T, P, N) \tag{11.1-3}$$

where Δ is the **isothermal-isobaric partition function**.

When a system at specified T and P contains two species and one of them is in equilibrium with that species in a reservoir through a semipermeable membrane, the transformed Gibbs energy of the system is calculated from the **semigrand partition function** $\Gamma'(T, P, \mu_1, N_2)$ by use of

$$G'(T, P, \mu_1, N_2) = -kT \ln \Gamma'(T, P, \mu_1, N_2) \qquad (11.1\text{-}4)$$

Statistical mechanics is often thought of as a way to predict the thermodynamic properties of molecules from their microscopic properties, but statistical mechnics is more than that because it provides a complementary way of looking at thermodynamics. The transformed Gibbs energy G' for a biochemical reaction system at specified pH is given by

$$G' = -kT \ln \Gamma' \qquad (11.1\text{-}5)$$

The semigrand partition function Γ' corresponds with a system of enzyme-catalyzed reactions in contact with a reservoir of hydrogen ions at a specified pH. The semigrand partition function can be written for an aqueous solution of a biochemical reactant at specified pH or a system involving many biochemical reations. The other thermodynamic properties of the system can be calculated from Γ'.

The further transformed Gibbs energy G'' for a biochemical rection system at specified pH and specified concentrations of certain coenzymes is given by

$$G'' = -kT \ln \Gamma'' \qquad (11.1\text{-}6)$$

where Γ'' is the corresponding semigrand partition function. All the remaining thermodynamic properties of the system can be calculated by taking partial derivatives of G'' or Γ''.

This is not the place to discuss the partition functions Ξ, Ψ, Δ, and Γ in detail, but it is important to point out that Ξ and Ψ differ by the factor $\exp(\beta N_1 \mu_1)$ and Δ and Γ differ by the same factor. Thus holding the chemical potential of species 1 constant has the effect of introducing an exponential factor that depends on the number of molecules of species 1 and the chemical potential of species 1.

■ 11.2 TRANSFORMED GIBBS ENERGY FOR A SYSTEM CONTAINING A WEAK ACID AND ITS BASIC FORM AT A SPECIFIED pH

The forms of semigrand partition functions for biochemical rection systems can be illustrated, starting with an aqueous solution of a weak acid and its basic form at a specified pH (Alberty, 2001c). The semigrand partition function Γ' for this system is given by

$\Gamma'(T, P, \text{pH}, N'_{\text{iso}})$

$$= \{\exp(-\beta\mu_1)\exp(-N_{H1}\ln(10)\text{pH}) + \exp(-\beta\mu_2)\exp(-N_{H2}\ln(10)\text{pH}\}^{N'_{\text{iso}}}$$

$$= \{\exp(-\beta\mu'_1) + \exp(-\beta\mu'_2)\}^{N'_{\text{iso}}} \qquad (11.2\text{-}1)$$

where μ_1 and μ_2 are the chemical potentials of species 1 and 2 (i.e., A^- and HA) and $\mu'_i = \mu_i - N_{Hi}\mu(H^+)$. N_{H1} and N_{H2} are the numbers of hydrogen atoms in these two species N'_{iso} is the number of molecules in the pseudoisomer group, $N_1 + N_2$. Equation 11.2-1 has been given in the form of the partition function that applies at zero ionic strength in order to keep it simpler, but this complication can be taken into account by using

$$\mu'_j = \mu_j(I = 0) + N_H(j)RT\ln(10)\text{pH} - \frac{2.91482(z_j^2 - N_H(j))I^{1/2}}{1 + 1.6I^{1/2}} \qquad (11.2\text{-}2)$$

Note that the $\exp(-\beta\mu_i)$ terms for species are weighted according to the number of hydrogen atoms in the species. This sum of exponential terms for a pseudoisomer group may contain many terms and is really a Laplace transform (Greiner, Neise, and Stöcker, 1995).

Substituting equation 11.2-1 in 11.1-5 yields

$$G' = -N'_{iso}KT\ln\{\exp(-\beta\mu_1)\exp(-N_{H1}\ln(10)\text{pH})$$
$$+ \exp(-\beta\mu_2)\exp(-N_{H2}\ln(10)\text{pH})\}$$
$$= -N'_{iso}kT\ln\{\exp(-\beta\mu'_1) + \exp(-\beta\mu'_2)\} = N'_{iso}\mu'_{iso} \qquad (11.2\text{-}3)$$

The transformed chemical potential for the pseudoisomer group (i.e., A^- plus HA) is given by

$$\mu'_{iso} = -kT\ln\{\exp(-\beta\mu'_1) + \exp(-\beta\mu'_2)\} \qquad (11.2\text{-}4)$$

which can also be written as

$$\mu'_{iso} = \mu'^0_{iso} + kT\ln[A] \qquad (11.2\text{-}5)$$

where $[A] = [A^-] + [HA]$. Equation 11.2-4 leads to

$$\mu'^0_{iso} = -kT\ln\{\exp(-\beta\mu'^0_1) + \exp(-\beta\mu'^0_2)\} \qquad (11.2\text{-}6)$$

This can be demonstrated by substituting $\mu'_1 = \mu'^0_1 + kT\ln[HA]$ and $\mu'_2 = \mu'^0_2 + kT\ln[A^-]$ in equation 11.2-4 and using

$$r_i = \exp\left[\frac{\mu'^0_{iso} - \mu'^0_i}{kT}\right] \qquad (11.2\text{-}7)$$

where r_i is the mole fraction of species i.

The fundamental equation of thermodynamics for G' for a dilute aqueous solution containing a weak acid HA, and its basic form A^- when these species are at equilibrium at a specified pH is (see 4.3-1)

$$dG' = -S'dT + VdP + \mu'_{iso}dN'_{iso} + N_c(H)kT\ln(10)d\text{pH} \qquad (11.2\text{-}8)$$

where $N_c(H)$ is the number of atoms of hydrogen in the system. μ'_{iso} is the transformed chemical potential of the pseudoisomer group, and the number of molecules of the pseudoisomer group is given by $N'_{iso} = N_1 + N_2$. Equation 11.2-8 can be integrated at constant values of the intensive variables T, P, nd pH to obtain

$$G' = H'_{iso}\mu'_{iso} \qquad (11.2\text{-}9)$$

This is in agreement with statistical mechanical equation 11.2-3.

Equation 11.2-8 indicates that the other thermodynamic properties of the system can be obtained from

$$S' = -\left(\frac{\partial G'}{\partial T}\right)_{P,\text{pH},N'_{iso}} = kT\left(\frac{\partial \ln \Gamma'}{\partial T}\right)_{P,\text{pH},N'_{iso}} + k\ln\Gamma' \qquad (11.2\text{-}10)$$

$$V = \left(\frac{\partial G'}{\partial P}\right)_{T,\text{pH},N'_{iso}} = -kT\left(\frac{\partial \ln \Gamma'}{\partial P}\right)_{T,\text{pH},N'_{iso}} \qquad (11.2\text{-}11)$$

$$N_c(H) = \frac{1}{kT\ln(10)}\left(\frac{\partial G'}{\partial \text{pH}}\right)_{T,P,N'_{iso}} = -\frac{1}{\ln(10)}\left(\frac{\partial \ln \Gamma'}{\partial \text{pH}}\right)_{T,P,N'_{iso}} \qquad (11.2\text{-}12)$$

$$\mu'_{iso} = -kT\left(\frac{\partial \ln \Gamma'}{\partial N_{iso}}\right)_{T,P,\text{pH}} \qquad (11.2\text{-}13)$$

McQuarrie (2000) gives a nice table like this for several types of partition functions. The further transformed enthalpy H' of the system at specified pH can be calculated by use of the Gibbs-Helmholtz equation:

$$H' = -T^2 \left[\frac{\partial(G'/T)}{\partial T} \right]_{P,\text{pH}} = kT^2 \left[\frac{\partial \ln \Gamma'}{\partial T} \right]_{P,\text{pH}} \qquad (11.2\text{-}14)$$

Note that $G' = H' - TS'$. This shows that the semigrand partition function Γ' contains all the thermodynamic information about the system.

■ 11.3 SEMIGRAND PARTITION FUNCTION FOR A SYSTEM CONTAINING TWO PSEUDOISOMER GROUPS AT A SPECIFIED pH

For a system containing two pseudoisomer groups A and B at a specified pH, the semigrand partition function is given by

$$\Gamma'(T, P, \text{pH}, N'_{\text{isoA}}, N'_{\text{isoB}}) = \{\exp(-\beta\mu'_{\text{isoA}})\}^{N'_{\text{isoA}}} \{\exp(-\beta\mu'_{\text{isoB}})\}^{N'_{\text{isoB}}} \qquad (11.3\text{-}1)$$

Substituting this in $G' = -kT \ln \Gamma'$ yields

$$G' = -kT \ln[\{\exp(-\beta\mu'_{\text{isoA}})\}^{N'_{\text{isoA}}} \{\exp(-\beta\mu'_{\text{isoB}})\}^{N'_{\text{isoB}}}]$$
$$= N'_{\text{isoA}}\mu'_{\text{isoA}} + N'_{\text{isoB}}\mu'_{\text{isoB}} \qquad (11.3\text{-}2)$$

If pseudoisomer groups A and B are involved in a reaction

$$\text{A} = 2\text{B} \qquad (11.3\text{-}3)$$

and there are initially $(N_{\text{isoA}})_0$ molecules of A, the transformed Gibbs energy of the system at any time in the reaction is given by

$$G' = (N_{\text{isoA}})_0\mu'_{\text{isoA}} + (2\mu'_{\text{isoB}} - \mu'_{\text{isoA}})\xi' \qquad (11.3\text{-}4)$$

where ξ' is the extent of reaction. The transformed Gibbs energy is minimized at equilibrium, and so

$$\frac{\Delta G'}{\text{d}\xi'} = 2\mu'_{\text{isoB}} - \mu'_{\text{isoA}} = 0 \qquad (11.3\text{-}5)$$

which is the equilibrium condition.

■ 11.4 SEMIGRAND PARTITION FUNCTION FOR A BIOCHEMICAL REACTION SYSTEM AT SPECIFIED CONCENTRATIONS OF COENZYMES

At level 3 the semigrand partition function Γ'' for a sum of reactants that are pseudoisomers at specified concentrations of coenzymes is given by a sum of exponential terms raised to a power equal to the number of molecules in the pseudoisomer group (Alberty, 2001g):

$$\Gamma'' = \{\exp(-\beta\mu''_1) + \exp(-\beta\mu''_2) + \cdots\}^{N'_{\text{iso}}} \qquad (11.4\text{-}1)$$

Substituting this in $G'' = -kT \ln \Gamma''$ yields

$$G'' = -N''_{\text{iso}}kT \ln\{\exp(-\beta\mu''_1) + \exp(-\beta\mu''_2) + \cdots\} = N''_{\text{iso}}\mu''_{\text{iso}} \qquad (11.4\text{-}2)$$

The semigrand partition function can be written for a system of biochemical reactions at specified concentrations of coenzymes, and $G'' = -kT \ln \Gamma''$. The fundamental equation for G'' can be used to calculate the other thermodynamic properties of a system. The fundamental equation for G'' is

$$dG'' = -S'' \, dT + V \, dP + \Sigma \, \mu_i'' \, dN_i'' + kT \ln(10) N_c(H) dpH$$

$$- \Sigma N_c(coe_i) kT \, d\ln[coe_i] \qquad (11.4\text{-}3)$$

$$S'' = -\left(\frac{\partial G''}{\partial T}\right) = kT\left(\frac{\partial \ln \Gamma''}{\partial T}\right) + k \ln \Gamma'' \qquad (11.4\text{-}4)$$

$$V = \left(\frac{\partial G''}{\partial P}\right) = -kT\left(\frac{\partial \ln \Gamma''}{\partial P}\right) \qquad (11.4\text{-}5)$$

$$\mu_i'' = \left(\frac{\partial G''}{\partial N_i}\right) = -kT\left(\frac{\partial \ln \Gamma''}{\partial N_i}\right) \qquad (11.4\text{-}6)$$

$$N_c(H) = \frac{1}{kT \ln(10)}\left(\frac{\partial G''}{\partial pH}\right) = -\frac{1}{\ln(10)}\left(\frac{\partial \ln \Gamma''}{\partial pH}\right) \qquad (11.4\text{-}7)$$

$$N_c(coe_i) = \frac{1}{kT}\left(\frac{\partial G''}{\partial \ln[coe_i]}\right) = -\left(\frac{\partial \ln \Gamma''}{\partial \ln[coe_i]}\right) \qquad (11.4\text{-}8)$$

The subscripts on the partial derivatives have been omitted because they are complicated, as indicated by the fundamental equation. The change in "binding" of coenzymes in a reaction can be studied at constant concentrations of coenzymes, just as the change in binding of hydrogen ions in a reaction can be studied at constant pH. The further transformed enthalpy H'' of the system can be calculated by use of the Gibbs-Helmholtz equation or from $G'' = H'' - TS''$.

For glycolysis at specified T, P, pH, [ATP], [ADP], [P_i], [NAD_{ox}], and [NAD_{red}], the semigrand partition function is given by (see Section 6.6)

$$\Gamma'' = \{\exp(-\beta\mu_6'')\}^{N_6''}\{\exp(-\beta\mu_3'')\}^{N_3''} \qquad (11.4\text{-}9)$$

which leads to

$$G'' = N_6''\mu_6'' + N_3''\mu_3'' \qquad (11.4\text{-}10)$$

Note that these two further transformed chemical potentials each involve summations of exponential terms over the C_6 reactants and C_3 reactants. The standard further transformed Gibbs energies of formation of C_6 and C_3 can be calculated at the desired pH, ionic strength, and concentrations of coenzymes using equation 11.4-2. The apparent equilibrium contants K'' for $C_6 = 2C_3$ can be calculated. Solving a quadratic equation yields [C_6] and [C_3] as shown in Section 6.6. The distribution of reactants within these two pseudoisomer groups can then be calculated. The concentrations of species within the reactants can also be calculated. Thus no thermodynamic information is lost in going to the level 3 calculations. This method can be applied to larger systems by specifying the concentrations of more coenzymes, but the number of coenzymes that can be specified is $C'' - 1$ or less because at least one component must remain.

■ 11.5   DISCUSSION

The thermodynamics of biochemical rections at specified pH and specified concentrations of coenzymes can be represented by semigrand partition functions. Partition functions are always sums of exponential terms. For a single reactant at specified pH, Γ' has a term for each species that is weighted by a factor that is exponential in $N_H(i)pH$. For a mixture of reactants, the partition function Γ' for the system is a product of partition functions of reactants each raised to the power of the number of molecules of the reactant. Thus the partition function for a mixture of reactants is also a summation of exponential terms. For a sum of reactants at specified concentrations of coenzymes, for example, C_6, the partition function Γ'' is a sum of exponential terms, with a term for each reactant (e.g., glucose, G6P, F6P, and FBP) in the C_6 pseudoisomer group. Each term is weighted by a factor that gives the dependence on the concentrations of coenzymes for the pseudoisomer group being discussed. For a mixture of reactants, like C_6 and C_3, Γ'' is a product of two sums of exponentials, each raised to the power of the number of molecules in each pseudoisomer group.

This chapter is important because it shows that the thermodynamic properties of a biochemical reactant at a specified pH can be discussed in terms of a semigrand partition function in which terms for species are multiplied by $\exp(-N_H(j)pH)$, where $N_H(j)$ is the number of hydrogen atoms in the jth species. This partition function contains all the thermodynamic information about the reactant, and so it is of interest to note that the effect of the Legendre transform to make the pH an independent variable is to put the pH and the number of hydrogen atoms in a species into the exponent. Similarly the thermodynamic properties of a sum of reactants at specified concentrations of coenzymes can be discussed in terms of a semigrand partition function in which terms for reactants are multiplied by $\exp(-N_{ATP}(i)\Delta_f G'(ATP) + N_{ADP}(i)\Delta\Delta_f G'(ADP))$, where N_{ATP} molecules "contained in" reactant i. This partition function contains all the thermodynamic information about the sum of reactants, and so it is of interest to note that the effect of making a Legendre transform to make [ATP] and [ADP] independent variables is to put [ATP] and [ADP] into the exponent.

Glossary

Note: When primes are used on thermodynamic potentials, it is important to indicate in the context the intensive variables that have to be specified. This also applies when primes are used on equilibrium constants, amounts, or numbers like the number of components, degrees of freedom, and stoichiometric numbers. SI units are indicated in parentheses. When a physical quantity does not have units, no units are given. Dimensions of matrices are also indicated in parentheses.

a_i	activity of species i	D'	apparent number of variables needed to describe the extensive state of a system when the concentrations of one or more species have been specified
A	Helmholtz energy (J)		
A'	transformed Helmholtz energy (J)		
A	constant in the Debye-Hückel equation (0.510651 $L^{-1/2}$ $mol^{1/2}$ at 298.15 K)		
A_s	surface area (m^2)	D''	apparent number of variables needed to describe the extensive state of a system when the concentrations of one or more species and the concentrations of one or more reactants have been specified
$\boldsymbol{A}$	conservation matrix ($C \times N$)		
$\boldsymbol{A}'$	apparent conservation matrix when the concentrations of one or more species are held constant ($C' \times N'$)		
$\boldsymbol{A}''$	apparent conservation matrix when the concentrations of one or more species and one or more reactants are held constant ($C'' \times N''$)	E	electric field strength (V m^{-1})
		E	magnitude of the electric field strength (V m^{-1})
$\boldsymbol{B}$	magnetic flux density (magnetic field strength) (T)	E	electromotive force (electric potential difference) or reduction potential (V)
B	magnitude of the magnetic flux density (magnetic field strength) (T)	E^0	standard electromotive force of a cell or standard reduction potential (V)
B	empirical constant in the extended Debye-Hckel equation (1.6 $kg^{1/2}$ $mol^{-1/2}$)	E'	apparent electromotive force or apparent reduction potential at a specified pH (V)
c_i	concentration of species i (mol L^{-1})	E'^0	standard apparent electromotive force of a cell or standard apparent reduction potential (V)
c^0	standard concentration (1 mol L^{-1})		
C	number of components in a reaction system	f	force (N)
C'	apparent number of components in a reaction system when the concentrations of one or more species are held constant	F	Faraday constant (96,485 C mol^{-1})
		F	number of variables needed to describe the intensive state of a system
C''	apparent number of components when the concentrations of one or more species and one or more reactants are held constant	F'	apparent number of variables needed to describe the intensive state of a system after the concentrations of one or more species have been specified
$C^0_{Pm}(j)$	standard molar heat capacity of species j (J K^{-1} mol^{-1})		
$C'^0_{Pm}(i)$	standard transformed molar heat capacity of reactant i (J K^{-1} mol^{-1})	F''	apparent number of variables needed to describe the intensive state of a system when the concentrations of one or more species and the concentrations of one or more reactants have been specified
$\Delta_f C^0_P(j)$	standard heat capacity of formation of species j (J K^{-1} mol^{-1})		
$\Delta_r C^0_P$	standard heat capacity of reaction (J K^{-1} mol^{-1})	G	Gibbs energy of a system at specified T, P, and ionic strength (J)
$\Delta_r C'^0_P$	standard transformed heat capacity of reaction (J K^{-1} mol^{-1})		
D	number of variables needed to describe the extensive state of a system	G'	transformed Gibbs energy of a system at specified T, P, ionic strength, and concentrations of one or more species (J)

G''	further transformed Gibbs energy of a system at specified T, P, ionic strength, and concentrations of one or more species and one or more reactants (J)
$\Delta_f G_j^0$	standard Gibbs energy of formation of species j at specified T, P, and ionic strength (J mol^{-1})
$\Delta_f G_j$	Gibbs energy of formation of species j at specified T, P, and ionic strength (J mol^{-1})
$\Delta_r G$	Gibbs energy of chemical reaction (J mol^{-1})
$\Delta_r G^0$	standard Gibbs energy of chemical reaction (J mol^{-1})
$\Delta_f G_i'^0$	standard transformed Gibbs energy of formation of reactant i at specified T, P, ionic strength, and specified concentrations of one or more species (J mol^{-1})
$\Delta_f G_i''^0$	standard further transformed Gibbs energy of formation of a pseudoisomer group of reactants at specified T, P, ionic strength, and specified concentrations of one or more species and one or more reactants (J mol^{-1})
$\Delta_r G'$	transformed Gibbs energy of a biochemical reaction (J mol^{-1})
$\Delta_r G'^0$	standard transformed Gibbs energy of reaction at a specified concentration of a species (J mol^{-1})
$\Delta_r G''$	further transformed Gibbs energy of reaction at specified concentrations of one or more reactants (J mol^{-1})
$\Delta_r G''^0$	standard further transformed Gibbs energy of reaction at specified concentrations of one or more reactants (J mol^{-1})
$\Delta_f G_i''^0$	standard further transformed Gibbs energy of formation of reactant i at specified concentrations of one or more reactants (J mol^{-1})
h	Planck's constant
H	enthalpy of a system at specified T, P, and ionic strength (J)
H'	transformed enthalpy of a system at specified T, P, ionic strength, and concentrations of one or more species (J)
H''	further transformed enthalpy of a system at specified T, P, ionic strength, and concentrations of one or more species and one or more reactants (J)
$H_m(j)$	molar enthalpy of species j (J mol^{-1})
$\Delta_r H(cal)$	enthalpy change in a calorimetric experiment (J mol^{-1})
$\Delta_r H^0$	standard enthalpy of reaction (J mol^{-1})
$\Delta_f H_j^0$	standard enthalpy of formation of species i (J mol^{-1})
$H_m'(j)$	molar transformed enthalpy of species j (J mol^{-1})
$H_m'(i)$	molar transformed enthalpy of reactant i (J mol^{-1})
$\Delta_r H'^0$	standard transformed enthalpy of reaction at a specified concentration of a species (J mol^{-1})
$\Delta_r H'$	transformed enthalpy of reaction at a specified concentration of a species (J mol^{-1})
$\Delta_f H_i'^0$	standard transformed enthalpy of formation of reactant i at a specified concentration of a species (J mol^{-1})
$\Delta_f H_i'$	transformed enthalpy of formation of i at a specified concentration of a species (J mol^{-1})
$\langle \Delta_f H_i'^0 \rangle$	mole fraction-weighted standard transformed enthalpy of formation of reactant i at a specified concentration of a species (J mol^{-1})
I	ionic strength (mol L^{-1})

k	Boltzmann constant (J K^{-1})
K	equilibrium constant written in terms of concentrations of species at specified T, P, and ionic strength
K'	apparent equilibrium constant written in terms of concentrations of reactants (sums of species) at specified T, P, ionic strength and concentrations of one or more species
K''	apparent equilibrium constant written in terms of concentrations of pseudoisomer groups (sums of reactants) at specified T, P, ionic strength, and concentrations of one or more species and one or more reactants
K_H	Henry's law constant
K_H'	apparent Henry's law constant at a specified pH
K_a	acid dissociation constant
K_{Mg}	dissociation constant of a magnesium complex ion
L	elongation (m)
m	mass (kg)
$\boldsymbol{m}$	transformation matrix
$\boldsymbol{m}$	magnetic moment of a system (J T^{-1})
m	magnitude of the magnetic moment of a system (J T^{-1})
M_i	molar mass of species i (kg mol^{-1})
n	total amount in a system (mol)
n_{Bb}	amount of B bound in the system (mol)
n_i	amount of species i (mol)
$\{n_i\}$	set of amounts of species in a system (mol)
n_i'	amount of reactant i (sum of species) (mol)
n_i''	amount of pseudoisomer group i (sum of reactants) (mol)
n_{ci}	amount of component i (mol)
n_{ci}'	amount of apparent component i at specified concentrations of one or more species (mol)
n_{ci}''	amount of apparent component i at specified concentrations of one or more species and one or more reactants (mol)
$\boldsymbol{n}$	column vector of amounts of species ($N \times 1$) (mol)
$\boldsymbol{n}'$	column vector of amounts of reactants (sum of species) ($N' \times 1$) (mol)
$\boldsymbol{n}''$	column vector of amounts of pseudoisomer groups of reactants ($N'' \times 1$) (mol)
$\boldsymbol{n}_c$	column vector of amounts of components ($C \times 1$) (mol)
$\boldsymbol{n}_{nc}$	column vector of amounts of noncomponents $((N - C) \times 1)$ (mol)
$\boldsymbol{n}_c'$	column vector of amounts of apparent components at specified concentrations of one or more species ($C' \times 1$) (mol)
$\boldsymbol{n}_c''$	column vector of amounts of apparent components at specified concentrations of one or more species and one or more reactants ($C'' \times 1$) (mol)
N	number of molecules in a system
N_s	number of different kinds of species in a system
N_A	Avogadro constant (mol^{-1})
N'	number of different reactants (sums of species) in a system
N''	number of different pseudoisomer groups of reactants in a system
N_{iso}	number of isomers in an isomer group or pseudoisomers in a pseudoisomer group
$N_H(j)$	number of hydrogen atoms in species j

$N_{\text{ATP}}(i)$ — number of ATP in reactant i (can be positive or negative)

$N_{\text{Mg}}(j)$ — number of magnesium atoms in a molecule of j

$N_{\text{O}_2}(i)$ — number of oxygen molecules bound in i

N_c — matrix of numbers of specified components in the N_s species ($C \times N_s$)

$\bar{N}_H(i)$ — average number of hydrogen atoms bound by a molecule of i

$\bar{N}_{\text{Mg}}(i)$ — average number of magnesium atoms bound by a molecule of i

$\bar{N}_{\text{ATP}}(i)$ — rate of change of $n_c(\text{ATP})$ with respect to the amount of pseudoisomer group i

$\Delta_r N_H$ — change in binding of hydrogen ions in a biochemical reaction at specified T, P, ionic strength and concentrations of one or more species

$\Delta_r N_{\text{Mg}}$ — change in the binding of magnesium ions in a biochemical reaction at specified T, P, pH, and ionic strength

p — number of different kinds of phases in a system

P — pressure (bar)

P_i — partial pressure of i (bar)

P° — standard state pressure (1 bar)

P — binding polynomial (partition function)

P_k — intensive variable in Callen's nomenclature (varies)

$\boldsymbol{p}$ — dipole moment of the system (C m)

p — magnitude of the dipole moment of the system (C m)

pH_a — $-\log\{a(\text{H}^+)\}$

$\text{pH} = \text{pH}_c$ — $-\log[\text{H}^+]$ at specified T, P, and ionic strength

pMg — $-\log[\text{Mg}^{2+}]$ at specified T, P, and ionic strength

$\text{p}K$ — $-\log K$ for the dissociation of an acid at specified T, P, and ionic strength

q — heat flow into a system(J)

Q — amount of charge transferred across the center of a membrane (C)

Q — canonical ensemble partition function

Q_i — electric charge contributed to a phase by species i (C)

Q — reaction quotient at T and P

Q' — reaction quotient at specified T, P, pH, pMg, and ionic strength

r — distance from the axis of rotation (m)

r_i — mole fraction of isomer i within an isomer group or pseudoisomer i within a pseudoisomer group

R — gas constant (8.31451 J K^{-1} mol^{-1})

R — number of independent reactions in a system described in terms of species

R' — number of independent reactions in a system described in terms of reactants (sums of species)

R'' — number of independent reactions in a system described in terms of pseudoisomer groups of reactants

s — number of special constraints in the phase rule

S — entropy of a system at specified T, P, and ionic strength (J K^{-1})

S' — transformed entropy of a system at specified T, P, ionic strength, and concentrations of one or more species (J K^{-1})

S'' — further transformed entropy of a system at specified T, P, ionic strength, and concentrations of one

or more species and one or more reactants (J K^{-1})

ΔS — change in entropy in a change of state of a system (J K^{-1} mol^{-1})

$S_m(j)$ — molar entropy of species j (J K^{-1} mol^{-1})

$S'_m(i)$ — molar transformed entropy of reactant i (J K^{-1} mol^{-1})

$S^0_m(j)$ — standard molar entropy of species j (J K^{-1} mol^{-1})

$S'^0_m(i)$ — standard molar transformed entropy of reactant i (J K^{-1} mol^{-1})

$\Delta_r S$ — reaction entropy (J K^{-1} mol^{-1})

$\Delta_r S^0$ — standard reaction entropy (J K^{-1} mol^{-1})

$\Delta_f S^0_j$ — standard entropy of formation of species j (J K^{-1} mol^{-1})

$\Delta_r S'$ — transformed reaction entropy (J K^{-1} mol^{-1})

$\Delta_r S'^0$ — standard transformed reaction entropy (J K^{-1} mol^{-1})

$\Delta_f S'^0(j)$ — standard transformed entropy of formation of species j (J K^{-1} mol^{-1})

s'_i — stoichiometric number of step i

s' — pathway matrix ($R \times 1$)

T — temperature (K)

t — Celsius temperature ($^\circ$C)

U — internal energy (J)

U' — transformed internal energy (J)

$U[P_i]$ — Callen's nomenclature for the transformed internal energy that has intensive variable P_i as a natural variable (J)

V — volume (m^3)

V_m — molar volume (m^3 mol^{-1})

$\Delta_r V$ — reaction volume (m^3 mol^{-1})

w — work done on a system (J)

x_i — mole fraction of i

X_k — extensive variable in Callen's nomenclature (varies)

Y — fractional saturation

z_j — charge number of ion j

α — Debye-Hckel constant (1.17582 kg$^{1/2}$ mol$^{-1/2}$ at 298.15 K)

β — $1/kT$ in statistical mechanics (J)

γ_i — activity coefficient of species i

γ — surface tension (N m^{-1})

Δ — isothermal-isobaric partition function

Γ — semigrand partition function

Γ' — semigrand partition function at a specified pH

Γ'' — semigrand partition function at a specified pH and specified concentrations of coenzymes

μ_j — chemical potential of species j at specified T, P, and ionic strength (J mol^{-1})

$\{\mu_j\}$ — set of chemical potentials (J mol^{-1})

μ'_i — transformed chemical potential of reactant i at specified T, P, ionic strength, and concentrations of one or more species (J mol^{-1})

μ''_i — further transformed chemical potential of pseudoisomer group i at specified T, P, ionic strength, and concentrations of one or more species and one or more reactants (J mol^{-1})

μ^0_j — standard chemical potential of species i at specified T, P, and ionic strength (J mol^{-1})

$\mu^0_{j_0}$ — standard chemical potential of species i (J mol^{-1})

μ'_i — standard transformed chemical potential of reactant i (J mol^{-1})

$\mu_i''^0$ standard further transformed chemical potential of pseudoisomer group i at specified T, P, ionic strength, and concentrations of one or more species and one or more reactants (J mol^{-1})

$\boldsymbol{\mu}$ vector of chemical potentials of species at specified T, P, and ionic strength ($1 \times N$) (J mol^{-1})

$\boldsymbol{\mu}'$ vector of transformed chemical potentials of reactants at specified T, P, ionic strength, and concentrations of one or more species ($1 \times N'$) (J mol^{-1})

$\boldsymbol{\mu}''$ vector of further transformed chemical potentials of pseudoisomer groups of reactants at specified T, P, ionic strength, and concentrations of one or more species and one or more reactants ($1 \times N''$) (J mol^{-1})

μ_{ci} chemical potential of component i at specified T, P, and ionic strength (J mol^{-1})

μ_{ci}' transformed chemical potential of component i at specified T, P, ionic strength, and concentrations of one or more species (J mol^{-1})

μ_{ci}'' further transformed chemical potential of component i at specified T, P, ionic strength, and concentrations of one or more species and one or more reactants (J mol^{-1})

$\boldsymbol{\mu}_c$ vector of chemical potentials of components at specified T, P, and ionic strength ($1 \times C$) (J mol^{-1})

$\boldsymbol{\mu}_c'$ vector of transformed chemical potentials of components at specified T, P, ionic strength, and concentrations of one or more species ($1 \times C'$) (J mol^{-1})

$\boldsymbol{\mu}_{nc}'$ vector of chemical potentials of noncomponents at specified T, P, and ionic strength ($1 \times (N - C')$) (J mol^{-1})

$\boldsymbol{\mu}_c''$ vector of further transformed chemical potentials of components at specified T, P, ionic strength, and concentrations of one or more species and one or more reactants ($1 \times C''$) (J mol^{-1})

v_i stoichiometric number of species i in a chemical reaction

v_{ij} stoichiometric number of species i in reaction j

v_i' stoichiometric number of reactant i in a biochemical reaction

v_{ij}' stoichiometric number of reactant i in reaction j

v_{ij}'' stoichiometric number of pseudoisomer group i in reaction j

$\mathbf{v}$ stoichiometric number matrix in terms of species ($N \times R$)

$\mathbf{v}'$ stoichiometric number matrix in terms of reactants ($N' \times R'$)

$\mathbf{v}''$ stoichiometric number matrix in terms of pseudoisomer groups of reactants ($N'' \times R''$)

v_i stoichiometric number of species i

v_i' apparent stoichiometric number of reactant i (sum of species)

v_i'' apparent stoichiometric number of reactant i (sum of species) when the concentration of a reactant has been specified

$|v_e|$ number of electrons in a half-reaction

$\mathbf{v}$ stoichiometric number matrix ($N \times R$)

$\mathbf{v}'$ apparent stoichiometric number matrix ($N' \times R'$)

$\mathbf{v}'$ apparent stoichiometric number matrix for a net reaction ($N' \times 1$)

ξ extent of chemical reaction (mol)

ξ' extent of biochemical reaction (mol)

$\boldsymbol{\xi}$ extent of reaction column vector at specified T, P, and ionic strength ($R \times 1$) (mol)

$\boldsymbol{\xi}'$ extent of reaction column vector at specified T, P, ionic strength, and concentrations of one or more species ($R' \times 1$) (mol)

$\boldsymbol{\xi}''$ extent of reaction column vector at specified T, P, ionic strength, and concentrations of one or more species and one or more reactants ($R'' \times 1$) (mol)

π binding potential (J mol^{-1})

Ξ grand canonical partition function

τ shear stress (N m^{-2})

ϕ_i electric potential of the phase containing species i (V, J C^{-1})

Ψ semigrand partition function

ω angular velocity (s^{-1})

κ acid dissociation constant of an independent group

References

1873

J. W. Gibbs, *The Scientific Papers of J. Willard Gibbs, Vol. 1, Thermodynamics,* Dover, New York, 1961.

1948

J. Wyman, Heme proteins, *Adv. Protein Chem.* 4, 407–531 (1948).

1951

R. A. Alberty, R. H. Smith, and R. M. Bock, The apparent ionization constants of the adenosine phosphates and related compounds, *J. Biol. Chem.* 193, 425 (1951).

1953

K. Burton and H. A. Krebs, The free energy changes associated with the individual steps of the tricarboxylic acid cycle, glycolysis, alcoholic fermentation, and with the hydrolysis of the pyrophosphate group of adensosine triphosphate, *Biochem. J.* 54, 94–107 (1953).

1956

R. N. Smith and R. A. Alberty, The apparent stability constants of ionic complexes of various adenosine phosphates and divalent cations, *J. Am. Chem. Soc.,* 78, 2376–2380 (1956).

1957

H. A. Krebs and H. L. Kornberg, *Energy Transformations in Living Matter,* Springer-Verlag, Berlin (1957) with Appendix by K. Burton.

1958

J. T. Edsall and J. Wyman, *Biophysical Chemistry,* Academic Press, New York (1958).

1960

H. B. Callen, *Thermodynamics,* Wiley, New York, 1960.
P. W. Wigler and R. A. Alberty, The pH dependence of the competitive inhibition of fumarase, *J. Am. Chem. Soc.* 72, 5482–5488 (1960).

1961

W. M. Clark, *Oxidation-Reduction Potentials of Organic Systems,* Williams and Wilkins, Baltimore, 1961.
J. Hermans and H. Scheraga, Thermodynamic considerations of protein reactions. I. Modified reactivity of polar groups, *J. Am. Chem. Soc.* 83, 3284–3292 (1961).

1964

J. Wyman, Linked functions and reciprocal effects in hemoglobin: A second look, *Adv. Protein Chem.* 19, 223–286 (1964)..

1965

J. Wyman, The binding potential, a neglected concept, *J. Mol. Biol.* 11, 631–644 (1965).

1966

E. C. W. Clarke and D. N. Glew, Evaluation of thermodynamic functions from equilibrium constants, *Trans. Faraday. Soc.* 62, 539–547 (1966).
L. Tisza, *Generalized Thermodynamics,* MIT Press, Cambridge, 1966.

1967

E. A. Guggenheim, *Thermodynamics,* North-Holland, Amsterdam, 1967.

1968

R. A. Alberty, Effect of pH and metal ion concentrations on the equilibrium hydrolysis of ATP and ADP, *J. Biol. Chem.* 243, 1337–1342 (1968).
K. Tagawa and D. F. Arnon, Oxidation-reduction potentials and stoichiometry of electron transfer in ferredoxins, *Biochim. Biophys. Acta,* 153, 602–613 (1968).

1969

(a) Alberty, R. A., Gibbs free energy, enthalpy, and entropy changes as a function of pH and pMg for several reactions involving adenosine phosphates, *J. Biol. Chem.* 244, 3290–3302 (1969).
(b) R. A. Alberty, Maxell relations for thermodynamic quantities of biochemical reactions, *J. Am. Chem. Soc.,* 91, 3899–3903 (1969).
J. T. Edsall, CO_2: Chemical, biochemical, and physiological aspects, NASA SP-188, 1969.
R. C. Phillips, P. George, and R. J. Rutman, Thermodynamic data for the hydrolysis of adenosine triphosphate as a function of pH, pMg, and ionic strength, *J. Biol. Chem.* 244, 3330–3342 (1969).
R. C. Wilhoit, Thermodynamic properties of biochemical substances, in biochemical *Microcalorimetry,* H. D. Brown (ed.), Academic Press, New York, 1969.

1973

R. W. Guynn and R. L. Veech, The equilibrium constants of the adenosine triphosphate hydrolysis and the adenosine triphosphate-citrate lyase reactions, *J. Biol. Chem.* 248, 6966–6972 (1973).

1974

R. Parsons, Manual of symbols and terminology for physicochemical quantities and units, Appendix III, Electrochemical nomenclature, *Pure Appl. Chem.* 37, 499–516 (1974).
G. K. Akers and H. R. Halvorson, The linkage between oxygenation and subunit dissociation in human hemoglobin, *Proc. Nat. Acad. Sci. USA* 71, 4312 (1974)

1975

J. Wyman, A group of thermodynamic potentials applicable to ligand binding by a polyfunctional macromolecule, *Proc. Nat. Acad. Sci. USA* 72, 1464–1468 (1975).
J. A. Schellman, Macromolecular binding, *Biopolymers* 14, 999–1018 (1975).

190

1976

I. Wadsö, H. Gutfreund, P. Privalov, J. T. Edsall, W. P. Jencks, G. T. Armstrong, and R. L. Biltonen, Recommendations for measurement and presentation of biochemical equilibrium data, *J. Biol. Chem.* 251, 6879–6885; *Q. Rev. of Biophys.* 9, 439–456 (1976).

I. H. Segel, *Biochemical Calculations*, Wiley, New York, 1976.

J. A, Schellman, The effect of binding on the melting temperature of polymers, *Biopolymers* 15, 999–1000 (1976).

N. A. Strombaugh, J. E. Sundquist, R. H. Burris, and W. H. Orme-Johnson, Oxidation-Reduction Properties of Several Low Potential Iron-Sulfur Proteins of Methylviologen. *Biochem.*, 15, 2533–2641 (1976).

F. C. Mills, M. L. Johnson, and J. K. Akers, Oxygenation-linked subunit interactions in human hemoglobin: Experimental studies on the concentration dependence of oxygenation curves, *Biochemistry* 15, 5350 (1976).

1977

R. K. Thauer, K. Jungermann, and K. Decker, Energy conservation in chemotropic anerobic bacteria, *Bacteriol. Rev.* 41, 100–179 (1977).

1978

F. J. Krambeck, Presented at the 71st Annual Meeting of the AIChE, Miami Beach, FL, Nov. 16, 1978.

1979

J. A. Beattie and I. Oppenheim, *Principles of Thermodynamics*, Elsevier, Amsterdam, 1979.

R. L. Veech, J. W. Lawson, N. W. Cornell, and H. A. Krebs, Cystolic phosphorylation potential, *J. Biol. Chem.* 254: 6538–6547 (1979).

1980

H. Goldstein, *Classical Mechanics*, Addison-Wesley, Reading, MA, 1980.

E. C. W. Clarke, and D. N. Glew, Evaluation of Debye-Hückel limiting slopes for water between 0 and 50°C, *Chem. Soc.* 1, 76, 1911 (1980).

1982

W. R. Smith and R. W. Missen, *Chemical Reaction Equilibrium Analysis: Theory and Algorithms*, Wiley-Interscience, New York, 1982.

D. D. Wagman, W. H. Evans, V. B. Parker, R. H. Schumm, I. Halow, S. M. Bailey, K. L. Churney, and R. L. Nutall, The NBS Tables of Chemical Thermodynamic Properties, *J. Phys. Chem. Ref. Data*, 11, suppl. 2 (1982).

1983

(a) R. A. Alberty. Chemical thermodynamic properties of isomer groups, *I&EC Fund.* 22, 218–321 (1983).

(b) R. A. Alberty, Extrapolation of standard chemical thermodynamic properties of alkene isomer groups to higher carbon numbers, *J. Phys. Chem.* 87, 4999–5002 (1983).

1984

R. N. Goldberg, Compiled thermodynamic data sources for aqueous and biochemical systems: An annotated bibliography (1930–1983), National Bureau of Standards Special Publication 685, Government Printing Office, Washington, DC, 1984.

1985

H. B. Callen. *Thermodynamics and an Introduction to Thermostatics*, John Wiley, New York, 1985.

1987

D. Chandler, *Introduction to Modern Statistical Mechanics*, Oxford University Press, 1987.

1988

G. Strang, *Linear Algebra and Its Applications*, Harcourt Brace Jovanovich, San Diego, 1988.

R. A. Alberty and I. Oppenheim, Fundamental equation for systems in chemical equilibrium, *J. Chem. Phys.* 89, 3689–3693 (1988).

Y. B. Tewari and R. N. Goldberg; Thermodynamics of the conversion of penicillin G to 6-aminopenicillanic acid, *Biophys. Chem.* 29, 245–252 (1988).

E. Di Cera, S. J. Gill, and J. Wyman, Binding capacity: Cooperativity and buffering in biopolymers, *Proc. Natl. Acad. Sci. USA* 85, 449–452 (1988).

1989

J. D. Cox, D. D. Wagman, and V. A. Medvedev, *CODATA Key Values for Thermodynamics*, Hemisphere, Washington, DC, 1989.

R. N. Goldberg and Y. B. Tewari, Thermodynamic and transport properties of carbohydrates and their monophosphates: The pentoses and hexoses, *J. Phys. Chem. Ref. Data* 18, 809–880 (1989).

R. A. Alberty and I. Oppenheim, Use of semigrand ensembles in chemical equilibrium calculations on complex organic systems, *J. Chem. Phys.* 91, 1824–1828 (1989).

1990

S. L. Miller and D. Smith-Magowan, The thermodynamics of the Krebs cycle and related compounds, *J. Phys. Chem. Ref. Data* 19, 1049–1073 (1990).

J. Wyman, J. and S. J. Gill, *Binding and Linkage*, University Science Books, Mill Valley, CA. (1990).

1991

(a) R. A. Alberty, Equilibrium compositions of solutions of biochemical species and heats of biochemical reactions, *Proc. Nat. Acad. Sci.*, 88, 3268–3271 (1991).

(b) R. A. Alberty, Chemical equations are actually matrix equations, *J. Chem. Ed.* 68, 984 (1991).

(c) R. A. Alberty, in *Chemical Reactions in Complex Systems*, F. J. Krambeck, and A. M. Sapre (ed.) Van Nostrand Reinhold, New York, 1991.

R. H. Burris, Nitrogenases, *J. Biol. Chem.* 266, 9339–9342 (1991).

R. N. Goldberg and Y. B. Tewari, Thermodynamics of the disproportionation of adenosine 5'-diphosphate to adenosine 5'-triphosphate and adenosine 5'-monophosphate: I. Equilibrium model, *Biophys. Chem.* 40, 241–261 (1991).

F. J. Krambeck, in *Chemical Reactions in Complex Systems*, F. J. Krambeck and A. M. Sapre (eds.) Van Nostrand Reinhold, New York, 1991.

K. S. Pitzer, *Activity Coefficients in Electrolyte Solutions*, CRC Press, Boca Raton, FL, 1991.

1992

(a) R. A. Alberty, Equilibrium calculations on systems of biochemical reactions, *Biophys. Chem.* 42, 117–131 (1992).

(b) R. A. Alberty, Conversion of chemical equations to biochemical equations, *J. Chem. Ed.* 69, 493 (1992).

(c) R. A. Alberty, Calculation of transformed thermodynamic properties of biochemical reactants at specified pH and pMg, *Biophys. Chem.*, 43, 239–254 (1992).

(d) R. A. Alberty, Degrees of freedom in biochemical reaction systems at specified pH and pMg, *J. Phys. Chem.*, 96, 9614–9621 (1992).

R. A. Alberty, and I. Oppenheim, Use of the fundamental equation to derive expressions for the entropy and enthalpy of gaseous reaction systems at a specified partial pressure of a reactant, *J. Chem. Phys.*, 96, 9050–9054 (1992).

R. A. Alberty and R. N. Goldberg, Calculation of thermodynamic formation properties for the ATP series at specified pH and pMg, *Biochem.* 31, 10610–10615 (1992).

T. W. Larson, Y. B. Tewari, and R. N. Goldberg, Thermodynamics of the reactions between adenosine, adenosine 5'-monophosphate, inosine, and inosine 5'-monophosphate. The conversion of L-histidine to (urocanic acid and ammonia) *J. Chem. Thermodyn.* 25, 73–90 (1993).

E. C. Webb, *Enzyme Nomenclature*, Academic Press, San Diego, CA. (1992)

1993

(a) R. A. Alberty, The fundamental equation of thermodynamics for systems of biochemical reactions, *Pure Appl. Chem.* 65, 883–888 (1993).

(b) R. A. Alberty, Degrees of freedom in a gaseous reaction system at a specified partial pressure of a reactant, *J. Phys. Chem.*, 97, 6226–6232 (1993).

192 References

(c) R. A. Alberty, Levels of thermodynamic treatment of biochemical reaction systems, *Biophys. J.* 65, 1243–1254 (1993).

(d) R. A. Alberty, Applications of matrix methods to electrochemical reactions, *J. Electrochem. Soc.* 140, 3488–3492 (1993).

(e) R. A. Alberty, Thermodynamics of reactions of nicotinamide adenine dinucleotide and nicotinamide adenine dinucleotide phosphate, *Arch. Biochem. Biophys.* 307, 8–14 (1993).

R. A. Alberty and A. Cornish-Bowden, On the pH dependence of the apparent equilibrium constant K' of a biochemical reaction, *Trends Biochem. Sci.* 18, 288–291 (1993).

R. A. Alberty and R. N. Goldberg, Calorimetric dDetermination of the standard transformed enthalpy of a biochemical reaction at specified pH and pMg, *Biophys. Chem.* 47, 213–223 (1993).

(a) R. A. Alberty and I. Oppenheim, Thermodynamics of a reaction system at a specified partial pressure of a reactant, *J. Chem. Ed.*, 70, 729–735 (1993).

(b) R. A. Alberty and I. Oppenheim, New fundamental equations of thermodynamics for systems in chemical equilibrium at a specified partial pressure of a reactant and the standard transformed formation properties of reactants, *J. Chem. Phys.* 98, 8900–8904 (1993).

R. N. Goldberg, Y. B. Tewari, D. Bell, and K. Fazio, Thermodynamics of enzyme-catalyzed reactions: Part 1. Oxidoreductases, *J. Phys. Chem. Ref. Data* 22, 515 (1993).

Y. Guan, T. H. Lilly, and T. E. Treffry, Theory of phase equilibria for multicomponent aqueous solutions: Applications to aqueous polymer two-phase systems, *J. Chem. Soc., Faraday Trans.* 89 (24), 4283–4298 (1993).

J. W. Larson, Y. B. Tewari, and R. N. Goldberg, Thermochemistry of the reactions between adenosine, adenosine 5'-monophosphate, inosine, and inosine 5' monophosphate; the Conversion of L-histidine to (urocanic acid + ammonia), *J. Chem. Thermodyn.* 25, 73–90 (1993).

I. Mills, T. Cvitas, K. Homann, N. Kallay, and K. Kuchitsu, *Quantities, Units and Symbols in Physical Chemistry*, Blackwell Scientific, Oxford 1993.

1994

(a) R. A. Alberty, Thermodynamics of the nitrogenase reactions, *J. Biol. Chem.* 269, 7099–7102 (1994).

(b) R. A. Alberty, Constraints in biochemical reactions, *Biophys. Chem.*, 49, 251–261 (1994).

(c) R. A. Alberty, Biochemical thermodynamics (A review), *Biochem. Biophys. Acta* 1207, 1–11 (1994).

(d) R. A. Alberty, Legendre transforms in chemical thermodynamics, *Chem. Rev.* 94, 1457–1482 (1994).

R. A. Alberty, A. Cornish-Bowden,Q. H. Gibson,, R. N. Goldberg, G. G. Hammes, W. Jencks, K. F. Tipton, R. Veech, H. V. Westerhoff, and E. C. Webb, Recomendations for nomenclature and tables in biochemical thermodynamics, *Pure Appl. Chem.* 66, 1641–1666 (1994). Reprinted in *Europ. J. Biochem.* 240, 1–14 (1996). URL: *http://www.chem.gmw.as.uk/iabmb/thermod/*

M. Bailyn, *A Survey of Thermodynamics*, American Institute of Physics, New York, 1994.

(a) R. N. Goldberg and Y. B. Tewari, Thermodynamics of enzyme-catalyzed reactions: Part 2. Transferases, *J. Phys. Chem. Ref. Data* 23, 547–617 (1994).

(b) R. N. Goldberg and Y. B. Tewari, Thermodynamics of enzyme-catalyzed reactions: Part 3. Hydrolases, *J. Phys. Chem. Ref. Data* 23, 1035–1103 (1994).

1995

(a) R. A. Alberty, Chemical reactions in phases at different electric potentials, *J. Electrochem. Soc.* 142, 120–124 (1995).

(b) R. A. Alberty, Standard transformed Gibbs energy of formation of carbon dioxide in aqueous solution at specified pH, *J. Phys. Chem.* 99, 11028–11034 (1995).

(c) R. A. Alberty, Components in chemical thermodynamics, *J. Chem. Ed.*, 72, 820 (1995).

(d) R. A. Alberty, On the derivation of the Gibbs adsorption isotherm (Note), *Langmuir*, 11, 3598–3600 (1995).

(e) R. A. Alberty. Chemical reactions in phases of different electric potentials, *J. Electrochem. Soc.* 142, 120–124 (1995).

(a) R. N. Goldberg and Y. B. Tewari, Thermodynamics of enzyme-catalyzed reactions: Part 4. Lyases, *J. Phys. Chem. Ref. Data* 24, 1669–1698 (1995).

(b) R. N. Goldberg and Y. B. Tewari, Thermodynamics of enzyme-catalyzed reactions: Part 5. Isomerases and ligases, *J. Phys. Chem. Ref. Data* 24, 1765–1801 (1995).

W. Greiner, L. Neise, and H. Stocker, *Thermodynamics and Statistical Mechanics*, Springer, New York, 1995.

K. S. Pitzer, *Thermodynamics*, McGraw-Hill, New York, 1995.

E. Di Cera, *Thermodynamic Theory of Site-Specific Binding Processes in Biological Macromolecules*, Cambridge University Press, Camridge, (1995).

1996

(a) R. A. Alberty, Calculation of biochemical net reactions and pathways using matrix operations, *Biophys. J.* 71, 507–515 (1996).

(b) R. A. Alberty, Thermodynamics of the binding of ligands by macromolecules, *Biophys. Chem.*, 62, 141–159 (1996).

T. F. Weiss, *Cellular Biophysics*, Vol. 1, MIT Press: Cambridge, 1996,

1997

(a) R. A. Alberty, Determination of the seven apparent equilibrium constants for the binding of oxygen by hemoglobin from measured fractional saturations, *Biophys. Chem.*, 63, 119–132 (1997).

(b) R. A. Alberty, Legendre transforms in chemical thermodynamics (Rossini lecture), *Pure Appl. Chem.* 29, 501–516 (1997). Also published in *J. Chem. Thermo.* 69, 221–2230 (1997).

(c) R. A. Alberty, Constraints and missing reactions in the urea cycle, *Biophys. J.* 72, 2349–2356 (1997).

(d) R. A. Alberty, Thermodynamics of reactions involving phases at different electric potentials, *J. Phys. Chem. B*101, 7191–7196 (1997).

(e) R. A. Alberty, Apparent equilibrium constants and standard transformed Gibbs energies of biochemical reactions involving carbon dioxide, *Arch. Biochem. Biophys.* 348, 116–124 (1997).

I. M. Klotz, *Ligand Binding*, Wiley-Interscience, New York (1997).

1998

(a) R. A. Alberty, Change in binding of hydrogen ions and magnesium ions in the hydrolysis of ATP, *Biophys. Chem.*, 70, 109–119 (1998).

(b) R. A. Alberty, Calculation of standard transformed Gibbs energies and standard transformed enthalpies of biochemical reactants, *Arch. Biochem. Biophys.*, 353, 116–130 (1998).

(c) R. A. Alberty, Calculation of standard transformed entropies of formation of biochemical reactants and group contributions at specified pH, *J. Phys. Chem.*, 102, 8460–8466 (1998).

(d) R. A. Alberty, Calculation of standard transformed formation properties of biochemical reactants and standard apparent reduction potentials of half-reactions, *Arch. Biochem. Biophys.*, 358, 25–39 (1998).

1999

R. A. Alberty, Calculation of standard formation properties of species from standard transformed formation properties of reactants in biochemical reactions at specified pH, *J. Phys. Chem.*, 103, 261–265 (1999).

R. N. Goldberg, Thermodynamics of enzyme-catalyzed reactions: Part 6 — 1999 Update, *J. Phys. Chem. Ref. Data* 28, 931–965 (1999).

G. T. Gassner and S. J. Lippard, Component interactions in the soluble methane monooxygenase system, *Biochem.* 38, 12768–12785 (1999).

S. Wolfram, *The Mathematica Book*, 4th ed., Cambridge University Press, Cambridge, 1999.

2000

(a) R. A. Alberty, Calculation of apparent equilibrium constants of enzyme-catalyzed reactions at pH 7, *Biochem. Educ.* 28, 12–17 (2000).

(b) R. A. Alberty, Use of the matrix form of the fundamental equations for transformed Gibbs energies of biochemical reaction systems at three levels, *J. Phys. Chem. B*, 104, 650–657 (2000).

(c) R. A. Alberty, Calculation of equilibrium compositions of large systems of biochemical reactions, *J. Phys. Chem. B* 104, 4807–4814 (2000).

(d) R. A. Alberty, Effect of pH on protein-ligand equilibria, *J. Phys. Chem. B* 104, 9929–9934 (2000).

D. A. McQuarrie, *Statistical Mechanics*, University Science Books, Sausalito, CA, 2000.

2001

(a) R. A. Alberty, Calculation of equilibrium compositions of biochemical reaction systems involving water as a reactant, *J. Phys. Chem. B* 105, 1109–1114 (2001).

(b) R. A. Alberty, Standard apparent reduction potentials for biochemical half-reactions as a function of pH and ionic strength, *Arch. Biochem. Biophys.* 389, 94–109 (2001).

(c) R. A. Alberty, Biochemical reaction equilibria from the point of view of a semigrand partition function, *J. Chem. Phys.* 114, 8270–8274 (2001).

(d) R. A. Alberty, Effect of temperature on standard transformed Gibbs energies of formation of reactants at specified pH and ionic strength and apparent equilibrium constants of biochemical reactions, *J. Phys. Chem. B* 105, 7865–7870 (2001).

(e) R. A. Alberty, Thermodynamics in Biochemistry, *Encyclopedia of Life Sciences*, Macmillan, London, 2001.

(f) R. A. Alberty, BasicBiochemData, (2001). URL: http://www.mathsource.com/cgi-bin/msitem?0211-622.

(g) R. A. Alberty, Systems of biochemical reactions from the point of view of a semigrand partition function, *Biophys. Chem.*, 93, 1–10 (2001).

R. A. Alberty, J. M. G. Barthel, E. R. Cohen, M. B. Ewing, R. N. Goldberg, and E. Wilhelm, Use of Legendre transforms in chemical thermodynamics (an IUPAC technical report), *Pure Appl. Chem.* 73, No. 8 (2001).

D. L. Akers and R. N. Goldberg; BioEqCalc: A package for performing equilibrium calculations on biochemical reactions, *Mathematica J.*, 8, 86–113 (2001); URL: http://www.mathematica-journal.com/issue/v8i1/

R. J. Silbey and R. A. Alberty, *Physical Chemistry*, Wiley, New York, 2001.

R. J. Silbey and R. A. Alberty, *Solutions Manual for Physical Chemistry*, Wiley, New York, 2001.

J. Boerio-Goates, M. R. Francis, R. N. Goldberg, M. A. V. Ribeiro da Silva, M. D. M. C. Ribeiro da Silva, and Y. Tewari, Thermochemistry of adenosine, *J. Chem. Thermo.* 33, 929–947 (2001).

A. E. Martell, R. M. Smith, and R. J. Motekaitis, NIST critically selected stability constants of metal complexes database, *NIST Standard Reference Database* 46, Version 6.0, National Institute of Standards and Technology, Gaithersburg, Md, 2001.

C. B. Ould-Moulaye, C. G. Dussap, nd J. B. Gros, Purines and pyrimidines in the solid state and in aqueous solution, *Thermochimica Acta*, 385, 93–107 (2001).

2002

(a) R. A. Alberty, Thermodynamics of systems of bidochemical reaction systems, *J. Theoret. Biol.* 215, 491–501 (2002)

(b) R. A. Alberty, The role of water in the thermodynamics of dilute aqueous solutions, *Biophys. Chem.*, in press.

(c) R. A. Alberty, Inverse Legendre transform in biochemical thermodynamics; Applied to the last five reactions of glycolysis, *J. Phys. Chem.*, 106, 6594–6599 (2002).

(d) R. A. Alberty, BasicBiochemData2 (2002) URL: http://www.mathsource.com/cgi-bin/msitem?0211-622.

R. N. Goldberg, N. Kishore, and R. M. Lennen, Thermodynamic quantities for the ionization reactions of buffers, *J. Phys. Chem. Ref. Data* 31, 231 (2002).

R. N. Goldberg and Y. Tewari, Thermochemistry of the biochemical reaction: $Pyrophosphate(aq) + H_2O(l) = 2phosphate$ *J. Chem. Thermo.* 34, 827–839 (2002)

B. Widom, *Statistical Mechanics: Concise Introduction for Chemists*, Cambridge University Press, Cambridge, 2002.

C. B. Ould-Moulaye, C. G. Dussap, nd J. B. Gros, A consistent set of formation properties of nucleic acid compounds: nucleosides, nucleotides, and nucleotide phosphates in aqueous solution, *Thermochimica Ata*, 387, 1–15 (2002).

Second Part: Mathematica^R Solutions to Problems

Calculations on the thermodynamics of biochemical reactions are often very complicated because of the large numbers of independent variables that are involved: for example, T, pH, ionic strength, concentrations of free metal ions, and concentrations of coenzymes. Therefore, it is necessary to use a computer with a mathematical application installed. *Mathematica* is very convenient for these calculations, data storage, and making plots (Wolfram Research, 100 World Trade Center, Champaign, IL 61820-7237). This part of the book provides *Mathematica* solutions to problems and shows how to calculate figures and tables used in the book. Programs and all of the details involved in making these calculations are shown. These programs can be used to make calculations at other temperatures, pHs, ionic strengths, etc.

The basic data and most of the programs are available on the web in *MathSource* at

http://www.mathsource.com/cgi-bin/msitem?0211-622

BasicBiochemData2.nb is a *Mathematica* notebook that contains data, programs, explanations, and examples. It can be read using *MathReader*, which is free from Wolfram Research at

http://www.wolfram.com

This notebook is the first item in this part of the book. It can be downloaded into a personal computer with *Mathematica* installed and can be run by using Kernel/Evaluation/Evaluate Notebook. This brings in all the data and programs. *MathSource* also contains BasicBiochemData2.m, which is a package that contains data and programs. It can be loaded into a *Mathematica* notebook by simply typing <<BasicBiochemData2` after BasicBiochemData2.m has been downloaded into AddOns/ExtraPackages in your computer.

Many of the problems in this second part of the book require that the package has been loaded, but the command <<BasicBiochemData2` is not included in each problem. Although the package loads programs, the programs used in problems are usually repeated in the solutions shown here so that it is easier for the reader to see how the calculation is made.

A number of books have been written to help people get started with *Mathematica*. Examples are

J. H. Noggle, *Physical Chemistry Using Mathematica*, HarperCollins, New York, 1996.

K. R. Coombes, B. R. Hunt, R. L. Lipsman, J. E. Osborn, and G. J. Stuck, *The Mathematica Primer*, Cambridge University Press, 1998.

C-K. Cheung, G. E. Keough, C. Landraitis, and R. H. Gross, *Getting Started with Mathematica*, Wiley, New York, 1998.

H. F. W. Hoft and M. H. Hoft, *Computing with Mathematica*, Academic Press, San Diego, 1998.

W. H. Cropper, *Mathematica Computer Problems for Physical Chemistry*, Springer-Verlag, New York, 1998.

R. J. Silbey and R. A. Alberty, *Physical Chemistry-Solutions Manual*, Wiley, 2001.

It is not necessaary to be a programmer in order to use the programs and procedures illustrated in this collection of problems. Names, temperatures, pHs, and ionic strengths are readily changed in the solutions to problems.

Contents:

BasicBiochemData2

Data and Programs for Biochemical Thermodynamics

Robert A. Alberty
Department of Chemistry 6-215
Massachusetts Institute of Technology
Cambridge, MA 02139
alberty@mit.edu

Abstract: The objective of this package is to give the basic thermodynamic data on a large number of species involved in biochemical reactions at 298.15 K, 1 bar, and zero ionic strength and show how to use these data to calculate apparent equilibrium constants K' of biochemical reactions at desired pHs and ionic strengths. Biochemical reactions are written in terms of sums of species, which are referred to as reactants. The thermodynamic properties of reactants are referred to as transformed properties because the pH is specified. Programs are given for calculating the standard thermodynamic properties of reactants from the basic data on species. Four tables are included to show how the standard transformed Gibbs energies of formation and standard transformed enthalpies of formation of 131 reactants depend on pH and ionic strength. Programs are given for the calculation of apparent equilibrium constants and other thermodynamic properties of reactions by simply typing in the reaction in the computer. Examples are given of various uses of these tables, including how to plot the various properties versus pH.

■ 1.0 Introduction

The objective of this package is to give the basic thermodynamic data on a large number of species involved in biochemical reactions at 298.15 K, 1 bar, and zero ionic strength and show how to use these data to calculate apparent equilibrium constants K' of biochemical reactions at desired pHs and ionic strengths. Programs are given for making all of these calculations and more. The apparent equilibrium constant is related to the standard transformed Gibbs energy of reaction $\Delta_r G'^o$ and to the standard transformed Gibbs energies of the reactants $\Delta_f G_i'^o$ by (ref. 1-3)

$$\Delta_r G'^o = \sum v_i' \Delta_f G_i'^o = - RT \ln K' \tag{1}$$

The v_i' are the stoichiometric numbers of reactants in the biochemical equation (positive for reactants on the right side of the equation and negative for reactants on the left side). The prime is needed on the stoichiometric numbers to distinguish them from the stoichiomeric numbers in the underlying chemical reactions. The standard transformed enthalpy of reaction $\Delta_r H'^o$ (heat of reaction) is related to the standard transformed enthalpies of formation $\Delta_f H_i'^o$ of the reactants by

$$\Delta_r H'^o = \sum v_i' \Delta_f H_i'^o \tag{2}$$

These thermodynamic properties are functions of the pH and ionic strength, and they can be calculated from the standard Gibbs energies of formation $\Delta_f G^o$ and standard enthalpies of formation $\Delta_f H^o$ of the species involved. The $\Delta_f G_i'^o$ values of 131 reactants as functions of pH and ionic strength and the $\Delta_f H_i'^o$ values of 69 reactants are calculated using the *Mathematica* programs **calcdGmat** and **calcdHmat** (ref. 4). These functions make it possible to calculate values of these properties at 298.15 K and at pHs in the range 5 to 9 and ionic strengths in the range 0 to 0.35 M.

The following tables are given and can be printed out:

table1 gives standard transformed Gibbs energies of formation of 131 reactants at pH 7 and ionic strengths of 0, 0.10, and 0.25 M.

table2 gives standard transformed Gibbs energies of formation of 131 reactants at ionic strength 0.25 M and pH values of 5, 6, 7, 8, and 9.

table3 gives standard transformed enthalpies of formation of 69 reactants at pH 7 and ionic strengths of 0, 0.10, and 0.25 M.

table4 gives standard transformed enthalpies of formation of 69 reactants at ionic strength 0.25 M and pH values of 5, 6, 7, 8, and 9.

These tables can be used to calculate $\Delta_r G'^o$ and $\Delta_r H'^o$ at pH 7 and ionic strengths of 0, 0.10, and 0.25 M or at ionic strength 0.25 M and pHs of 5, 6, 7, 8, and 9 for any reaction for which all the reactants are in these tables. They can also be used to calculate standard apparent reduction potentials. The species data can be used to calculate average bindings of hydrogen ions by reactants. *Mathematica* programs for carrying out these calculations are provided.

The basic thermodynamic data comes from classical thermodynamic tables and from experimental measurements of K' and $\Delta_r H'^o$ at a particular pH and ionic strength together with measurements of acid dissociation constants. Some biochemical reactants consist of a single species, but others are sums of species; for example, ATP is made up of the species ATP^{4-}, $HATP^{3-}$, and $H_2 ATP^{2-}$ in the pH range 4 to 10 in the absence of metal ions that are bound reversibly. Therefore, the basic thermodynamic data on biochemical reactions includes the standard Gibbs energies of formation $\Delta_f G^o$ and the standard enthalpies of formation $\Delta_f H^o$ of species at zero ionic strength.

The basic data stored for each species is a list of { $\Delta_f G^o$, $\Delta_f H^o$, z_i , N_{Hi} }, where z_i is the charge number, and N_{Hi} is the number of hydrogen atoms in the species. When a reactant is made up of more than one species, the basic data is represented by a matrix with a row for each species. The values of the standard transformed Gibbs energies of formation $\Delta_f G'^o$ and standard transformed enthalpies of formation $\Delta_f H'^o$ of these species are functions of pH and ionic strength, where the effects of ionic strength are calculated using the extended Debye-Huckel equation and the effects of pH are calculated using the number of hydrogen atoms in the species. These functions for reactants are calculated using the *Mathematica* programs **calcdGmat** and **calcdHmat**.

This data base is set up in such a way that typing the name of a reactant, say atp (lower case letters are used because capital letters are used for *Mathematica* operations) yields the function of pH and ionic strength for $\Delta_f G'^o$ and typing the name atph yields the function of pH and ionic strength for $\Delta_f H'^o$. These functions can be evaluated at a specific pH and ionic strength by use of the replacement operator (/.), as illustrated by typing atp/.pH->7/.is->.1 or atp/.pH->{6,7,8}/.is->{0..1,.25}. In addition the average number of hydrogen atoms in a reactant at specified pH and ionic strength can be calculated by taking the derivative of $\Delta_f G'^\circ$ with respect to pH. For example, the number of hydrogen atoms bound by ATP at pH 7 and 0.10 M ionic strength is given by

(1/RTLog[10])*D[atp,pH]/.pH->7/.is->.1.

In writing chemical equations and biochemical equations it is important to be careful with names of reactants. Chemical reactions are written in terms of species. In chemical reaction equations, atoms of all elements and electric charges must balance. Biochemical reaction equations are written in terms of reactants, that is in terms of sums of species, H^+ is not included as a reactant and electric charges are not shown or balanced. In biochemical reaction equations, atoms of all elements other than hydrogen must balance. The names of the reactants that must be used in making calculations with this data base are given later.

The program **calctrGerx** can be used to calculate the standard transformed Gibbs energy of reaction $\Delta_r G'^\circ$ for a biochemical reaction in the form atp+h2o+de=adp+pi, where de is required for the *Mathematica* operation Solve. The desired pHs and ionic strengths can be specified. The program **calckprime** can be used to calculate the apparent equilibrium constant K' for a reaction at desired pHs and ionic strengths. The program **calctrGerx** can also be used to calculate $\Delta_f H'^\circ$ by typing in a biochemical reaction in the form atph+h2oh+de=adph+pih.

When oxidation and reduction are involved in an enzyme-catalyzed reaction, the standard apparent reduction potential for a half reaction can be calculated by typing the half reaction in **calcappredpot** and specifying the pHs and ionic strengths.

The mathematical functions for the standard transformed Gibbs energies of formation of biochemical reactants contain information about the average number of hydrogen atoms bound, as mentioned above. The change in binding of hydrogen atoms in a biochemical reaction can be calculated by taking the difference between products and reactants, but in using *Mathematica* there is an easier way and that is to take the derivative of $\Delta_r G'^\circ$ with respect to pH:

$$\Delta_r N_H = (1/RT\ln(10))(d\Delta_r G'^\circ /dpH) \tag{3}$$

The equilibrium composition for an enzyme-catalyzed reaction or a series of enzyme-catalyzed reactions can be calculated by using **equcalcc** or **equcalccrx**. The first of these programs requires a conservation matrix. The second requires a stoichiometric matrix. The second program is recommended, especially when water is involved as a reactant, because the convention that when dilute aqueous solutions are considered, the activity of water is taken to be unity, means that a second Legendre transform is necessary.

This version of the package provides eleven additional programs. One of the prgrams **calcdGHT** makes it possible to take the effect of temperature into account if enthalpy data are available (ref. 6). The uses of these programs are illustrated in the notebook.

Since the standard thermodynamic properties of adenosine have been determined (ref. 7), new values are given for the ATP series. These changes do not change the values of apparent equilibrium constants that are calculated between reactants in this series, but will be useful in investigating the production of adenosine..

The current table can be considerably extended by use of the compilations of Goldberg and Tewari of evaluated equilibrium data on biochemical reactions (ref. 8). Akers and Goldberg have published "BioEqCalc; A Package for Performing Equilibrium Calculations in Biohemical Reactions" (rcf. 9).

I am indebted to NIH 5-R01-GM48358 for support of the research that produced these tables and to Robert A. Goldberg and Ian Brooks for many helpful discussions.

References:
1. Alberty, R. A. Biophys. Chem. **1992** 42, 117; **1992** 43, 239.
2. Alberty, R. A.; Goldberg, R. N. Biochemistry **1992** 31, 10610.
3. Alberty, R. A. J. Phys. Chem. **1992** 96, 9614.
4. Alberty, R. A. Arch. Biochem. Biophys. **1998** 353, 116; **1998** 358, 25.
5. Alberty, R. A. J. Phys. Chem. B **2001** 105, 7865.
6. Boerio-Goates, J.; Francis, M. R.; Goldberg, R. N.; Ribeiro da Silva, M. A. V.; Ribeiro da Silva, M. D. M. C.; Tewari, Y. J. Chem. Thermo. **2001** 33, 929.
7. Goldberg, R. N. J. Phys. Chem. Ref. Data **1999** 28, 931 and earlier articles in this series.
8. Akers, D. L.; Goldberg, R. N. *Mathematica* J. **2001** 8, 1. (URL: http://www.mathematica-journal.com/issue/v8i1/)

```
Off[General::spell, General::spell1];

calcdGmat::usage =
  "calcdGmat[speciesmat_] produces the function of pH and ionic strength (is) that
     gives the standard transformed Gibbs energy of formation of a reactant (sum
     of species) at 298.15 K.  The input speciesmat is a matrix that gives the
     standard Gibbs energy of formation, the standard enthalpy of formation, the
     electric charge, and the number of hydrogen atoms in each species. There
     is a row in the matrix for each species of the reactant. gpfnsp is a list
     of the functions for the species.  Energies are expressed in kJ mol^-1.";
calcdG3I::usage = "calcdG3I[reactantname_] produces the standard transformed Gibbs
     energies of formation at 298 K, pH 7, and ionic strengths of 0, 0.10, and 0.25
     M.  The reactant name calls a function of pH and ionic strength or a constant.";
calcdG5pH::usage = "calcG5pH[reactantname_] produces the standard transformed
     Gibbs energies of formation at 298 K, ionic strength 0.25 M and pHs 5, 6,
     7, 8, and 9.  The reactant name calls a function of pH and ionic strength.";
calcdHmat::usage = "calcdHmat[speciesmat_] produces the function of pH and ionic
     strength that gives the standard transformed enthalpy of formation of a reactant (
     sum of species) at 298.15 K.  The input is a matrix that gives the standard
     Gibbs energy of formation,the standard enthalpy of formation, the electric
     charge,and the number of hydrogen atoms in the species in the reactant.  There
     is a row in the matrix for each species of the reactant.  dhfnsp is a list
     of the functions for the species.  Energies are expressed in kJ mol^-1.";
calcdH3I::usage = "calcdH3I[reactanth_] produces the standard transformed
     enthalpies of formation at 298 K,pH 7,and ionic strengths of 0,0.10,and 0.25 M.
     The reactanth name calls a function of pH and ionic strength or a constant.";
calcdH5pH::usage = "calcdHpH[reactanth_] produces the standard transformed enthalpies
     of formation at 298 K, ionic strength 0.25 M and pHs 5, 6, 7, 8, and 9.  The
     reactanth name with hf calls a function of pH and ionic strength or a constant.";
calctrGerx::usage = "calctrGerx[eq_,pHlist_,islist_] produces the standard
     transformed Gibbs energy of reaction in kJ mol^-1 at specified pHs
     and ionic strengths for a biochemical equation typed in the form atp+
     h2o+de==adp+pi.  The names of the reactants call the corresponding
     functions of pH and ionic strength. pHlist and islist can be lists.";
calckprime::usage = "calckprime[eq_,pHlist_,islist_] produces the apparent equilibrium
     constant K' at specified pHs and ionic strengths for a biochemical equation typed
     in the form atp+h2o+de==adp+pi.  The names of the reactants call the corresponding
     functions of pH and ionic strength. pHlist and islist can be lists.";
calcpK::usage = "calcpK[speciesmat_,no_,is_] calculates pKs of weak acids.";
calcdGHT::usage = "calcdGHT[speciesmat_] calculates the
     effect of temperature on transformed thermodynamic properties.";
calcGef1sp::usage = "calcGef1sp[equat_,pH_,ionstr_,z1_,nH1_]
     calculates the standard Gibbs energy of formation of
     the species of a reactant made up of a single species.";
calcGef2sp::usage = "calcGef2sp[equat_,pH_,ionstr_,z1_,nH1_,pK0_]
     calculates the standard Gibbs energies of formation of
     the two species of a reactant made up of a two species.";
calcGef3sp::usage = "calcGef3sp[equat_,pH_,ionstr_,z1_,nH1_,pK10_,
     pK20] calculates the standard Gibbs energies of formation of
     the three species of a reactant made up of a three species.";
calcNHrx::usage = "calcNHrx[eq_,pHlist_,islist_] calculates the change
     in binding of hydrogen ions in a biochemical reaction.";
rxthermotab::usage = "rxthermotab[eq_,pHlist_,islist_] calculates a table of
```

standard transformed reaction Gibbs energies, apparent equilibrium constants,
and changes in binding of hydrogen ions in a biochemical reaction.";
round::usage = "round[vec_,params_:{4,2}] rounds to the desired number of digits.";
mkeqm::usage = "mkeqm[c_List,s_List] types out a
biochemical reaction from its stoichiometric number vector.";
nameMatrix::usage = "nameMatrix[m_List,s_List] types out the biochemical reactions
in a system of biochemical reactions from the stoichiometric number matrix.";
calcappredpot::usage = "calcappredpot[eq_,nu_,pHlist_,islist_]
calculates the standard apparent reduction potential of
a half reaction at the desired pH and ionic strength.";
equcalcc::usage = "equcalcc[as_,lnk_,no_] calculates the equilibrium composition
for a biochemical reaction or a series of biochemical reactions,
given the conservation matrix, a list of transformed Gibbs energies
of formation, and a list of initial concentrations of reactants.";
equcalcrx::usage = "equcalcrx[nt_,lnkr_,no_] calculates the equilibrium composition
for a biochemical reaction or a series of biochemical reactions, given the
transposed stoichiometric number matrix, the apparent equilibrium constants
of the reactons, and a list of initial concentrations of reactants.";
listfnpHis::usage = "listfnpHis is a list of functions, like atp, that gives
the standard transformed Gibbs energies of formation of reactants.";
listspeciesdata::usage = "listspeciesdata is a list of names, like
atpsp, of data files on the species of a reactant.";
listdG3I::usage = "listdG3I is a list of vectors, like atpis,
that give the values of the standard transformed Gibbs
energies of formation at pH 7 and 3 ionic strengths.";
list::usage = "list is a list of names of reactants, each in quotation marks.";
listreactantspH::usage =
"listreactantspH is a list of vectors, like atppH,that give the values of the
standard transformed Gibbs energy of formation of a reactant at 5 pH values.";
listreactantsh::usage = "listreactantsh is a list of the names in
quotation marks for the reactants for which the standard
transformed enthalpy of formation is calculated.";
listhisreactants::usage = "listhisreactants is a vector of the values of standard
transformed enthalpies of formation of reactants at three ionic strengths.";
listfnpHish::usage = "listfnpHish is a function of pH and ionic strength that
gives the standard transformed enthalpy of formation of a reactant.";
listspeciesdatah::usage = "listspeciesdatah is a list of names, like atpsp, of
data files on the species of a reactant for which the enthalpy is known.";
listhpHreactants ::usage = "listhpHreactants is a list of names
of reactants, like atphpH, that give the values of the
standard transformed enthalpy of formation at 5 pH values.";
pH::usage = "pH is an independent variable.";
is::usage = "is is ionic strength, which is an independent variable.";
atp::usage = "atp is the name of a reactant that yields a function of pH and
ionic strength for the standard transformed Gibbs energy of formation.";
atpsp::usage = "atpsp yields the basic data on the species of the reactant.";
atpis::usage = "atpis yields the standard transformed
Gibbs energy of a reactant at 3 values of the ionic strength.";
atppH::usage = "atppH yields the standard transformed Gibbs energy
of a reactant at 5 values of the ionic strength.";
atph::usage = "atph yields the function of pH and ionic strength that
gives the standard transformed enthalpy of formation of atp.";
atphis::usage = "atphis yields the standard transformed
enthaply of atp at 3 ionic strengths.";
atphpH::usage = "atphis yields the standard transformed
enthaply of atp at 5 pH values.";
table1::usage = "table1 produces a table of standard transformed Gibbs energies of
formation of reactants at pH 7 and ionic strengths of 0, 0.10, and 0.25 M.";

```
table2::usage = "table2 produces a table of standard transformed Gibbs energies of
    formation of reactants at ionic strength 0.25 M and pHs of 5, 6, 7, 8, and 9.";
table3::usage = "table3 produces a table of standard transformed enthalpies of
    formation of reactants at pH 7 and ionic strengths of 0, 0.10, and 0.25 M.";
table4::usage = "table4 produces a table of standard transformed enthalpies of
    formation of reactants atonic strengths 0.25 M and pHs of 5, 6, 7, 8, and 9.";

Begin["BasicBiochemData2`Private`"]
```

■ 2.0 Basic data on the species that make up a reactant.

```
acetaldehydesp={{-139.,-212.23,0,4}};
acetatesp={{-369.31,-486.01,-1,3},{-396.45,-485.76,0,4}};
acetonesp={{-159.7,-221.71,0,6}};
acetylcoAsp={{-180.36,_,0,3}};
acetylphossp={{-1219.49,_,-2,3},{-1269.08,_,-1,4},{-1298.26,_,0,5}};
aconitatecissp={{-917.13,_,-3,3}};
adeninesp={{310.67, ,0,5},{286.7,_,1,6}};
adenosinesp={{-194.5,-621.3,0,13},{-214.28,637.7,1,14}};
adpsp={{-1906.13,-2626.54,-3,12},{-1947.1,-2620.94,-2,
        13},{-1971.98,-2638.54,-1,14}};
alaninesp={{-371.,-554.8,0,7}};
ammoniasp={{-26.5,-80.29,0,3},{-79.31,-132.51,1,4}};
ampsp={{-1040.45,-1635.37,-2,12},{-1078.86,-1629.97,-1,
        13},{-1101.63,-1648.07,0,14}};
arabinosesp={{-742.23,-1043.79,0,10}};
asparagineLsp={{-525.93,-766.09,0,8}};
aspartatesp={{-695.88,-943.41,-1,6}};
atpsp={{-2768.1,-3619.21,-4,12},{-2811.48,-3612.91,-3,
        13},{-2838.18,-3627.91,-2,14}};
bpgsp={{-2356.14,_,-4,4},{-2401.58,_,-3,5}};
butanolnsp={{-171.84,_,0,10}};
butyratesp={{-352.63,_,-1,7}};
citratesp={{-1162.69,-1515.11,-3,5},{-1199.18,-1518.46,-2,
        6},{-1226.33,-1520.88,-1,7}};
citrateisosp={{-1156.04,_,-3,5},{-1192.57,_,-2,6},{-1219.47,_,-1,7}};
coAsp={{0,_,-1,0},{-47.83,_,0,1}};
coAglutathionesp={{-35.85,_,-1,15}};
co2gsp={{-394.36,-393.5,0,0}};
co2totsp={{-527.81,-677.14,-2,0},{-586.77,-691.99,-1,1},{-623.11,-699.63,0,
        2}};
coaqsp={{-119.9,-120.96,0,0}};
cogsp={{-137.17,-110.53,0,0}};
creatinesp={{-259.2,_,0,9}};
creatininesp={{-23.14,_,0,7}};
cysteineLsp={{-291.,_,-1,6},{-338.82,_,0,7}};
cystineLsp={{-666.51,_,0,12}};
cytochromecoxsp={{0,_,3,0}};
cytochromecredsp={{-24.51,_,2,0}};
dihydroxyacetonephossp={{-1296.26,_,-2,5},{-1328.8,_,-1,6}};
ethanolsp={{-181.64,-288.3,0,6}};
ethylacetatesp={{-337.65,-482.,0,8}};
fadoxsp={{0,_,-2,31}};
fadredsp={{-38.88,_,-2,33}};
fadenzoxsp={{0,_,-2,31}};
fadenzredsp={{-88.6,_,-2,33}};
ferredoxinoxsp={{0,_,1,0}};
ferredoxinredsp={{38.07,_,0,0}};
fmnoxsp={{0,_,-2,19}};
fmnredsp={{-38.88,_,-2,21}};
formatesp={{-351.,-425.55,-1,1}};
fructosesp={{-915.51,-1259.38,0,12}};
fructose6phossp={{-1760.8,_,-2,11},{-1796.6,_,-1,12}};
fructose16phossp={{-2601.4,_,-4,10},{-2639.36,_,-3,11},{-2673.89,_,-2,12}};
fumaratesp={{-601.87,-777.39,-2,2},{-628.14,-774.46,-1,3},{-645.8,-774.88,0,
        4}};
galactosesp={{-908.93,-1255.2,0,12}};
galactose1phossp={{-1756.69,_,-2,11},{-1791.77,_,-1,12}};
glucosesp={{-915.9,-1262.19,0,12}};
glucose1phossp={{-1756.87,_,-2,11},{-1793.98,_,-1,12}};
```

```
glucose6phossp={{-1763.94,-2276.44,-2,11},{-1800.59,-2274.64,-1,12}};
glutamatesp={{-697.47,-979.89,-1,8}};
glutaminesp={{-528.02,-805.,0,10}};
glutathioneoxsp={{0,_,-2,30}};
glutathioneredsp={{34.17,_,-2,15},{-13.44,_,-1,16}};
glyceraldehydephossp={{-1288.6,_,-2,5},{-1321.14,_,-1,6}};
glycerolsp={{-497.48,-676.55,0,8}};
glycerol3phossp={{-1358.96,_,-2,7},{-1397.04,_,-1,8}};
glycinesp={{-379.91,-523.,0,5}};
glycolatesp={{-530.95,_,-1,3}};
glycylglycinesp={{-520.2,-734.25,0,8}};
glyoxylatesp={{-468.6,_,-1,1}};
h2aqsp={{17.6,-4.2,0,2}};
h2gsp={{0,0,0,2}};
h2osp={{-237.19,-285.83,0,2}};
h2o2aqsp={{-134.03,-191.17,0,2}};
hydroionsp={{0,0,1,1}};
hydroxypropionatebsp={{-518.4,_,-1,5}};
hypoxanthinesp={{89.5,_,0,4}};
indolesp={{223.8,97.5,0,7}};
ketoglutaratesp={{-793.41,_,-2,4}};
lactatesp={{-516.72,-686.64,-1,5}};
lactosesp={{-1567.33,-2233.08,0,22}};
leucineisoLsp={{-343.9,_,0,13}};
leucineLsp={{-352.25,-643.37,0,13}};
lyxosesp={{-749.14,_,0,10}};
malatesp={{-842.66,_,-2,4},{-872.68,_,-1,5}};
maltosesp={{-1574.69,-2238.06,0,22}};
mannitolDsp={{-942.61,_,0,14}};
mannosesp={{-910.,-1258.66,0,12}};
methanegsp={{-50.72,-74.81,0,4}};
methaneaqsp={{-34.33,-89.04,0,4}};
methanolsp={{-175.31,-245.93,0,4}};
methionineLsp={{-502.92,_,0,11}};
methylamineionsp={{-39.86,-124.93,1,6}};
n2aqsp={{18.7,-10.54,0,0}};
n2gsp={{0,0,0,0}};
nadoxsp={{0,0,-1,26}};
nadredsp={{22.65,-31.94,-2,27}};
nadpoxsp={{-835.18,0,-3,25}};
nadpredsp={{-809.19,-29.18,-4,26}};
o2aqsp={{16.4,-11.7,0,0}};
o2gsp={{0,0,0,0}};
oxalatesp={{-673.9,825.1,-2,0},{-698.33,_,-1,1}};
oxaloacetatesp={{-793.29,_,-2,2}};
oxalosuccinatesp={{-1138.88,_,-2,4}};
palmitatesp={{-259.4,_,-1,31}};
pepsp={{-1263.65,_,-3,2},{-1303.61,_,-2,3}};
pg2sp={{-1496.38,_,-3,4},{-1539.99,_,-2,5}};
pg3sp={{-1502.54,_,-3,4},{-1545.52,_,-2,5}};
phenylalanineLsp={{-207.1,_,0,11}};
pisp={{-1096.1,-1299.,-2,1},{-1137.3,-1302.6,-1,2}};
propanol2sp={{-185.23,-330.83,0,8}};
propanolnsp={{-175.81,_,0,8}};
ppisp={{-1919.86,-2293.47,-4,0},{-1973.86,-2294.87,-3,
        1},{-2012.21,-2295.37,-2,2},{-2025.11,-2290.37,-1,
        3},{-2029.85,-2281.17,0,4}};
pyruvatesp={{-472.27,-596.22,-1,3}};
retinalsp={{0,_,0,28}};
retinolsp={{-27.91,_,0,30}};
ribosesp={{-738.79,-1034.,0,10}};
ribose1phossp={{-1574.49,_,-2,9},{-1612.67,_,-1,10}};
ribose5phossp={{-1582.57,-2041.48,-2,9},{-1620.75,-2030.18,-1,10}};
ribulosesp={{-735.94,-1023.02,0,10}};
serineLsp={{-510.87,_,0,7}};
```

```
sorbosesp={{-911.95,-1263.3,0,12}};
succinatesp={{-690.44,-908.68,-2,4},{-722.62,-908.84,-1,5},{-746.64,-912.2,0,
        6}};
succinylcoAsp={{-509.72, _, -1, 4}, {-533.76, _, 0,
    5}};
sucrosesp={{-1564.7,-2199.87,0,22}};
thioredoxinoxsp={{0,_,0,0}};
thioredoxinredsp={{69.88,_,-2,0},{20.56,_,-1,1},{-25.37,_,0,2}};
tryptophaneLsp={{-114.7,-405.2,0,12}};
tyrosineLsp={{-370.7,_,0,11}};
ubiquinoneoxsp={{0,_,0,90}};
ubiquinoneredsp={{-89.92,_,0,92}};
uratesp={{-325.9,_,-1,3}};
ureasp={{-202.8,-317.65,0,4}};
uricacidsp={{-356.9,_,0,4}};
valineLsp={{-358.65,-611.99,0,11}};
xylosesp={{-750.49,-1045.94,0,10}};
xylulosesp={{-746.15,-1029.65,0,10}};
```

These thermodynamic values are based on the usual conventions of chemical thermodynamic tables that $\Delta_f G° = \Delta_f H° = 0$ for elements in defined reference states and for H^+ ($a=1$). Additional conventions are that $\Delta_f G° = \Delta_f H° = 0$ for coA^-, FAD_{ox}^{2-}, $FADenz^{2-}$, $cytochromec^{3+}$, $ferredoxin_{ox}^{-4}$, FMN^{2-}, $glutathiome_{ox}^{2-}$, NAD_{ox}^{-1}, $NADP_{ox}^{3-}$, $retinal^0$, $thioredoxin_{ox}^0$, and $ubiquinone_{ox}^0$.

■ 3.0 Calculation of the functions of pH and ionic strength for the standard transformed Gibbs energy of formation of reactants.

```
calcdGmat[speciesmat_] :=
Module[{dGzero, zi, nH, pHterm, isterm,gpfnsp},(*This program produces the function of
pH and ionic strength (is) that gives the standard transformed Gibbs energy of
formation of a reactant (sum of species) at 298.15 K.  The input speciesmat is a
matrix that gives the standard Gibbs energy of formation, the standard enthalpy of
formation, the electric charge, and the number of hydrogen atoms in each species.
There is a row in the matrix for each species of the reactant. gpfnsp is a list of the
functions for the species.  Energies are expressed in kJ mol^-1.*)
dGzero = speciesmat[[All,1]];
zi = speciesmat[[All,3]];
nH = speciesmat[[All,4]];
pHterm = nH*8.31451*.29815*Log[10^-pH];
isterm = 2.91482*((zi^2) - nH)*(is^.5)/(1 + 1.6*is^.5);
gpfnsp=dGzero - pHterm - isterm;
-8.31451*.29815*Log[Apply[Plus,Exp[-1*gpfnsp/(8.31451*.29815)]]]]]
```

The following is a list of names of the reactants. After the calculation using calcdGmat, typing one of these names yields the function of pH and ionic strength that gives the standard transformed Gibbs energy of formation of the reactant at 298.15 K in dilute aqueous solution.

```
listfnpHis = {acetaldehyde, acetate, acetone, acetylcoA, acetylphos, aconitatecis, adenine,
    adenosine, adp, alanine, ammonia, amp, arabinose, asparagineL, aspartate, atp, bpg,
    butanoln, butyrate, citrate, citrateiso, coA, coAglutathione, co2g, co2tot, coaq,
    cog, creatine, creatinine, cysteineL, cystineL, cytochromecox, cytochromecred,
    dihydroxyacetonephos, ethanol, ethylacetate, fadox, fadred, fadenzox, fadenzred,
    ferredoxinox, ferredoxinred, fmnox, fmnred, formate, fructose, fructose6phos,
    fructose16phos, fumarate, galactose, galactose1phos, glucose, glucose1phos,
    glucose6phos, glutamate, glutamine, glutathioneox, glutathionered, glyceraldehydephos,
    glycerol, glycerol3phos, glycine, glycolate, glycylglycine, glyoxylate, h2aq,
    h2g, h2o, h2o2aq, hydroxypropionateb, hypoxanthine, indole, ketoglutarate, lactate,
    lactose, leucineisoL, leucineL, lyxose, malate, maltose, mannitolD, mannose,
    methaneaq, methaneg, methanol, methionineL, methylamineion, n2aq, n2g, nadox,
    nadred, nadpox, nadpred, o2aq, o2g, oxalate, oxaloacetate, oxalosuccinate, palmitate,
    pep, pg2, pg3, phenylalanineL, pi, ppi, propanol2, propanoln, pyruvate, retinal,
    retinol, ribose, ribose1phos, ribose5phos, ribulose, serineL, sorbose, succinate,
    succinylcoA, sucrose, thioredoxinox, thioredoxinred, tryptophaneL, tyrosineL,
    ubiquinoneox, ubiquinonered, urate, urea, uricacid, valineL, xylose, xylulose};
```

The following is a list of names of the entries in the basic thermodynamic data on species making up a reactant.

```
listspeciesdata = {acetaldehydesp, acetatesp, acetonesp, acetylcoAsp, acetylphossp,
    aconitatecissp, adeninesp, adenosinesp, adpsp, alaninesp, ammoniasp, ampsp,
    arabinosesp, asparagineLsp, aspartatesp, atpsp, bpgsp, butanolnsp, butyratesp,
    citratesp, citrateisosp, coAsp, coAglutathionesp, co2gsp, co2totsp, coaqsp,
    cogsp, creatinesp, creatininesp, cysteineLsp, cystineLsp, cytochromecoxsp,
    cytochromecredsp, dihydroxyacetonephossp, ethanolsp, ethylacetatesp, fadoxsp,
    fadredsp, fadenzoxsp, fadenzredsp, ferredoxinoxsp, ferredoxinredsp, fmnoxsp,
    fmnredsp, formatesp, fructosesp, fructose6phossp, fructose16phossp, fumaratesp,
    galactosesp, galactose1phossp, glucosesp, glucose1phossp, glucose6phossp,
    glutamatesp, glutaminesp, glutathioneoxsp, glutathioneredsp, glyceraldehydephossp,
    glycerolsp, glycerol3phossp, glycinesp, glycolatesp, glycylglycinesp, glyoxylatesp,
    h2aqsp, h2gsp, h2osp, h2o2aqsp, hydroxypropionatebsp, hypoxanthinesp, indolesp,
    ketoglutaratesp, lactatesp, lactosesp, leucineisoLsp, leucineLsp, lyxosesp,
    malatesp, maltosesp, mannitolDsp, mannosesp, methaneaqsp, methanegsp, methanolsp,
    methionineLsp, methylamineionsp, n2aqsp, n2gsp, nadoxsp, nadredsp, nadpoxsp,
    nadpredsp, o2aqsp, o2gsp, oxalatesp, oxaloacetatesp, oxalosuccinatesp, palmitatesp,
    pepsp, pg2sp, pg3sp, phenylalanineLsp, pisp, ppisp, propanol2sp, propanolnsp,
    pyruvatesp, retinalsp, retinolsp, ribosesp, ribose1phossp, ribose5phossp,
    ribulosesp, serineLsp, sorbosesp, succinatesp, succinylcoAsp, sucrosesp,
    thioredoxinoxsp, thioredoxinredsp, tryptophaneLsp, tyrosineLsp, ubiquinoneoxsp,
    ubiquinoneredsp, uratesp, ureasp, uricacidsp, valineLsp, xylosesp, xylulosesp};
```

Now Map is used to apply calcdGmat to each of the matrices of species data.

```
Clear[acetaldehyde, acetate, acetone, acetylcoA, acetylphos, aconitatecis, adenine,
   adenosine, adp, alanine, ammonia, amp, arabinose, asparagineL, aspartate, atp, bpg,
   butanoln, butyrate, citrate, citrateiso, coA, coAglutathione, co2g, co2tot, coaq,
   cog, creatine, creatinine, cysteineL, cystineL, cytochromecox, cytochromecred,
   dihydroxyacetonephos, ethanol, ethylacetate, fadox, fadred, fadenzox, fadenzred,
   ferredoxinox, ferredoxinred, fmnox, fmnred, formate, fructose, fructose6phos,
   fructose16phos, fumarate, galactose, galactose1phos, glucose, glucose1phos,
   glucose6phos, glutamate, glutamine, glutathioneox, glutathionered, glyceraldehydephos,
   glycerol, glycerol3phos, glycine, glycolate, glycylglycine, glyoxylate, h2aq, h2g,
   h2o, h2o2aq, hydroxypropionateb, hypoxanthine, indole, ketoglutarate, lactate,
   lactose, leucineisoL, leucineL, lyxose, malate, maltose, mannitolD, mannose,
   methaneaq, methaneg, methanol, methionineL, methylamineion, n2aq, n2g, nadox,
   nadred, nadpox, nadpred, o2aq, o2g, oxalate, oxaloacetate, oxalosuccinate, palmitate,
   pep, pg2, pg3, phenylalanineL, pi, ppi, propanol2, propanoln, pyruvate, retinal,
   retinol, ribose, ribose1phos, ribose5phos, ribulose, serineL, sorbose, succinate,
   succinylcoA, sucrose, thioredoxinox, thioredoxinred, tryptophaneL, tyrosineL,
   ubiquinoneox, ubiquinonered, urate, urea, uricacid, valineL, xylose, xylulose];
```

```
Evaluate[listfnpHis] = Map[calcdGmat, listspeciesdata];
```

The following shows an example of a function of pH and ionic strength for a reactant that gives the standard transformed Gibbs energy of formation.

```
atp
```

$$-2.47897 \, \mathrm{Log}[E^{-0.403393 \, (-2838.18 + (29.1482 \, is^{0.5})/(1 + 1.6 \, is^{0.5}) - 34.7056 \, \mathrm{Log}[10^{-pH}])}$$
$$+ E^{-0.403393 \, (-2811.48 + (11.6593 \, is^{0.5})/(1 + 1.6 \, is^{0.5}) - 32.2266 \, \mathrm{Log}[10^{-pH}])}$$
$$+ E^{-0.403393 \, (-2768.1 - (11.6593 \, is^{0.5})/(1 + 1.6 \, is^{0.5}) - 29.7477 \, \mathrm{Log}[10^{-pH}])}]$$

The value of the standard transformed Gibbs energy of formation of ATP at pH 7 and ionic strength 0.25 M can be calculated in kJ/mol as follows:

```
atp/.pH->7/.is->.25
```

```
-2292.5
```

The values at pHs 6, 7, and 8 and at ionic strengths 0 and 0.25 M can also be calculated.

```
atp/.pH->{6,7,8}/.is->{0,.25}
```

```
{{-2366.43, -2363.76}, {-2292.61, -2292.5}, {-2220.96, -2223.44}}
```

■ 4.0 Calculation of a table of standard transformed Gibbs energies of formation of reactants at pH 7 and ionic strengths of 0, 0.10, and 0.25 M.

```
calcdG3I[reactantname_]:=Module[{out1,out2,out3},(*This program calculates the
standard transformed Gibbs energies of formation at 298 K, pH 7, and ionic strengths
of 0, 0.10, and 0.25 M.  The reactant name calls a function of pH and ionic strength
or a constant.*)
out1=reactantname/.{pH->7,is->0};
out2=reactantname/.{pH->7,is->.1};
out3=reactantname/.{pH->7,is->.25};
{out1,out2,out3}];
```

This is a list of the three values to be calculated for each of the reactants.

```
listdG3I = {acetaldehydeis, acetateis, acetoneis, acetylcoAis, acetylphosis,
    aconitatecisis, adenineis, adenosineis, adpis, alanineis, ammoniais, ampis,
    arabinoseis, asparagineLis, aspartateis, atpis, bpgis, butanolnis, butyrateis,
    citrateis, citrateisois, coAis, coAglutathioneis, co2gis, co2totis, coaqis,
    cogis, creatineis, creatinineis, cysteineLis, cystineLis, cytochromecoxis,
    cytochromecredis, dihydroxyacetonephosis, ethanolis, ethylacetateis, fadoxis,
    fadredis, fadenzoxis, fadenzredis, ferredoxinoxis, ferredoxinredis, fmnoxis,
    fmnredis, formateis, fructoseis, fructose6phosis, fumarate16phosis, fumarateis,
    galactoseis, galactose1phosis, glucoseis, glucose1phosis, glucose6phosis,
    glutamateis, glutamineis, glutathioneoxis, glutathioneredis, glyceraldehydephosis,
    glycerolis, glycerol3phosis, glycineis, glycolateis, glycylglycineis, glyoxylateis,
    h2aqis, h2gis, h2ois, h2o2aqis, hydroxypropionatebis, hypoxanthineis, indoleis,
    ketoglutarateis, lactateis, lactoseis, leucineisoLis, leucineLis, lyxoseis,
    malateis, maltoseis, mannitolDis, mannoseis, methaneaqis, methanegis, methanolis,
    methionineLis, methylamineionis, n2aqis, n2gis, nadoxis, nadredis, nadpoxis,
    nadpredis, o2aqis, o2gis, oxalateis, oxaloacetateis, oxalosuccinateis, palmitateis,
    pepis, pg2is, pg3is, phenylalanineLis, piis, ppiis, propanol2is, propanolnis,
    pyruvateis, retinalis, retinolis, riboseis, ribose1phosis, ribose5phosis,
    ribuloseis, serineLis, sorboseis, succinateis, succinylcoAis, sucroseis,
    thioredoxinoxis, thioredoxinredis, tryptophaneLis, tyrosineLis, ubiquinoneoxis,
    ubiquinoneredis, urateis, ureais, uricacidis, valineLis, xyloseis, xyluloseis};
```

Map is used to apply calcdG3I to each of the functions for reactants.

```
Clear[acetaldehydeis, acetateis, acetoneis, acetylcoAis, acetylphosis,
   aconitatecisis, adenineis, adenosineis, adpis, alanineis, ammoniais, ampis,
   arabinoseis, asparagineLis, aspartateis, atpis, bpgis, butanolnis, butyrateis,
   citrateis, citrateisois, coAis, coAglutathioneis, co2gis, co2totis, coaqis,
   cogis, creatineis, creatinineis, cysteineLis, cystineLis, cytochromecoxis,
   cytochromecredis, dihydroxyacetonephosis, ethanolis, ethylacetateis, fadoxis,
   fadredis, fadenzoxis, fadenzredis, ferredoxinoxis, ferredoxinredis, fmnoxis,
   fmnredis, formateis, fructoseis, fructose6phosis, fumarate16phosis, fumarateis,
   galactoseis, galactose1phosis, glucoseis, glucose1phosis, glucose6phosis,
   glutamateis, glutamineis, glutathioneoxis, glutathioneredis, glyceraldehydeis,
   glycerolis, glycerol3phosis, glycineis, glycolateis, glycylglycineis, glyoxylateis,
   h2aqis, h2gis, h2ois, h2o2aqis, hydroxypropionatebis, hypoxanthineis, indoleis,
   ketoglutarateis, lactateis, lactoseis, leucineisoLis, leucineLis, lyxoseis,
   malateis, maltoseis, mannitolDis, mannoseis, methaneaqis, methanegis, methanolis,
   methionineLis, methylamineionis, n2aqis, n2gis, nadoxis, nadredis, nadpoxis,
   nadpredis, o2aqis, o2gis, oxalateis, oxaloacetateis, oxalosuccinateis, palmitateis,
   pepis, pg2is, pg3is, phenylalanineLis, piis, ppiis, propanol2is, propanolnis,
   pyruvateis, retinalis, retinolis, riboseis, ribose1phosis, ribose5phosis,
   ribuloseis, serineLis, sorboseis, succinateis, succinylcoAis, sucroseis,
   thioredoxinoxis, thioredoxinredis, tryptophaneLis, tyrosineLis, ubiquinoneoxis,
   ubiquinoneredis, urateis, ureais, uricacidis, valineLis, xyloseis, xyluloseis];
```

```
Evaluate[listdG3I] = Map[calcdG3I, listfnpHis];
```

```
acetoneis
```

```
{80.0378, 83.7102, 84.8958}
```

This is a list of the names of reactants.

```
list = {"acetaldehyde", "acetate", "acetone", "acetylcoA", "acetylphos", "aconitatecis",
   "adenine", "adenosine", "adp", "alanine", "ammonia", "amp", "arabinose",
   "asparagineL", "aspartate", "atp", "bpg", "butanoln", "butyrate", "citrate",
   "citrateiso", "coA", "coAglutathione", "co2g", "co2tot", "coaq", "cog", "creatine",
   "creatinine", "cysteineL", "cystineL", "cytochromecox", "cytochromecred",
   "dihydroxyacetonephos", "ethanol", "ethylacetate", "fadox", "fadred", "fadenzox",
   "fadenzred", "ferredoxinox", "ferredoxinred", "fmnox", "fmnred", "formate",
   "fructose", "fructose6phos", "fructose16phos", "fumarate", "galactose",
   "galactose1phos", "glucose", "glucose1phos", "glucose6phos", "glutamate",
   "glutamine", "glutathioneox", "glutathionered", "glyceraldehydephos", "glycerol",
   "glycerol3phos", "glycine", "glycolate", "glycylglycine", "glyoxylate",
   "h2aq", "h2g", "h2o", "h2o2aq", "hydroxypropionateb", "hypoxanthine", "indole",
   "ketoglutarate", "lactate", "lactose", "leucineisoL", "leucineL", "lyxose",
   "malate", "maltose", "mannitolD", "mannose", "methaneaq", "methaneg", "methanol",
   "methionineL", "methylamineion", "n2aq", "n2g", "nadox", "nadred", "nadpox",
   "nadpred", "o2aq", "o2g", "oxalate", "oxaloacetate", "oxalosuccinate", "palmitate",
   "pep", "pg2", "pg3", "phenylalanineL", "pi", "ppi", "propanol2", "propanoln",
   "pyruvate", "retinal", "retinol", "ribose", "ribose1phos", "ribose5phos",
   "ribulose", "serineL", "sorbose", "succinate", "succinylcoA", "sucrose",
   "thioredoxinox", "thioredoxinred", "tryptophaneL", "tyrosineL", "ubiquinoneox",
   "ubiquinonered", "urate", "urea", "uricacid", "valineL", "xylose", "xylulose"};
```

Table 1 Standard Transformed Gibbs Energies of Formation in kJ mol^{-s} at 298.15 K, pH 7, and Ionic Strengths of 0, 0.10, and 0.25 M

```
table1=PaddedForm[TableForm[listdG3I,TableHeadings->{list,{"I = 0 M","I =
0.10 M","I = 0.25 M"}},TableSpacing->{.3,3}],{8,2}]
```

	I = 0 M	I = 0.10 M	I = 0.25 M
acetaldehyde	20.83	23.27	24.06
acetate	-249.46	-248.23	-247.83
acetone	80.04	83.71	84.90
acetylcoA	-60.49	-58.65	-58.06
acetylphos	-1109.34	-1107.57	-1107.02
aconitatecis	-797.26	-800.93	-802.12
adenine	510.45	513.51	514.50
adenosine	324.93	332.89	335.46
adp	-1428.93	-1425.55	-1424.70
alanine	-91.31	-87.02	-85.64
ammonia	80.50	82.34	82.93
amp	-562.04	-556.53	-554.83
arabinose	-342.67	-336.55	-334.57
asparagineL	-206.28	-201.38	-199.80
aspartate	-456.14	-453.08	-452.09
atp	-2292.61	-2292.16	-2292.50
bpg	-2202.06	-2205.69	-2207.30
butanoln	227.72	233.84	235.82
butyrate	-72.94	-69.26	-68.08
citrate	-963.46	-965.49	-966.23
citrateiso	-956.82	-958.84	-959.58
coA	-7.98	-7.43	-7.26
coAglutathione	563.49	572.06	574.83
co2g	-394.36	-394.36	-394.36
co2tot	-547.33	-547.15	-547.10
coaq	-119.90	-119.90	-119.90
cog	-137.17	-137.17	-137.17
creatine	100.41	105.92	107.69
creatinine	256.55	260.84	262.22
cysteineL	-59.23	-55.01	-53.65
cystineL	-187.03	-179.69	-177.32
cytochromecox	0.00	-5.51	-7.29
cytochromecred	-24.51	-26.96	-27.75
dihydroxyacetonephos	-1096.60	-1095.91	-1095.70
ethanol	58.10	61.77	62.96
ethylacetate	-18.00	-13.10	-11.52
fadox	1238.65	1255.17	1260.51
fadred	1279.68	1297.43	1303.16
fadenzox	1238.65	1255.17	1260.51
fadenzred	1229.96	1247.71	1253.44
ferredoxinox	0.00	-0.61	-0.81
ferredoxinred	38.07	38.07	38.07
fmnox	759.17	768.35	771.31
fmnred	800.20	810.61	813.97
formate	-311.04	-311.04	-311.04
fructose	-436.03	-428.69	-426.32
fructose6phos	-1321.71	-1317.16	-1315.74
fructose16phos	-2202.84	-2205.66	-2206.78
fumarate	-521.97	-523.19	-523.58
galactose	-429.45	-422.11	-419.74
galactose1phos	-1317.50	-1313.01	-1311.60
glucose	-436.42	-429.08	-426.71
glucose1phos	-1318.03	-1313.34	-1311.89
glucose6phos	-1325.00	-1320.37	-1318.92
glutamate	-377.82	-373.54	-372.15
glutamine	-128.46	-122.34	-120.36
glutathioneox	1198.69	1214.60	1219.74
glutathionered	625.75	634.76	637.62
glyceraldehydephos	-1088.94	-1088.25	-1088.04
glycerol	-177.83	-172.93	-171.35
glycerol3phos	-1080.22	-1077.83	-1077.13
glycine	-180.13	-177.07	-176.08
glycolate	-411.08	-409.86	-409.46
glycylglycine	-200.55	-195.65	-194.07
glyoxylate	-428.64	-428.64	-428.64
h2aq	97.51	98.74	99.13
h2g	79.91	81.14	81.53
h2o	-157.28	-156.05	-155.66
h2o2aq	-54.12	-52.89	-52.50
hydroxypropionateb	-318.62	-316.17	-315.38
hypoxanthine	249.33	251.77	252.56

indole	503.49	507.78	509.16
ketoglutarate	-633.58	-633.58	-633.58
lactate	-316.94	-314.49	-313.70
lactose	-688.29	-674.83	-670.48
leucineisoL	175.53	183.49	186.06
leucineL	167.18	175.14	177.71
lyxose	-349.58	-343.46	-341.48
malate	-682.88	-682.85	-682.85
maltose	-695.65	-682.19	-677.84
mannitolD	-383.22	-374.65	-371.89
mannose	-430.52	-423.18	-420.81
methaneaq	125.50	127.94	128.73
methaneg	109.11	111.55	112.34
methanol	-15.48	-13.04	-12.25
methionineL	-63.40	-56.67	-54.49
methylamineion	199.88	202.94	203.93
n2aq	18.70	18.70	18.70
n2g	0.00	0.00	0.00
nadox	1038.86	1054.17	1059.11
nadred	1101.47	1115.55	1120.09
nadpox	163.73	173.52	176.68
nadpred	229.67	235.79	237.77
o2aq	16.40	16.40	16.40
o2g	0.00	0.00	0.00
oxalate	-673.90	-676.35	-677.14
oxaloacetate	-713.38	-714.60	-715.00
oxalosuccinate	-979.05	-979.05	-979.05
palmitate	979.25	997.61	1003.54
pep	-1185.46	-1188.53	-1189.73
pg2	-1340.72	-1341.32	-1341.79
pg3	-1346.38	-1347.19	-1347.73
phenylalanineL	232.42	239.15	241.33
pi	-1058.56	-1059.17	-1059.49
ppi	-1934.95	-1939.13	-1940.66
propanol2	134.42	139.32	140.90
propanoln	143.84	148.74	150.32
pyruvate	-352.40	-351.18	-350.78
retinal	1118.78	1135.91	1141.45
retinol	1170.78	1189.14	1195.07
ribose	-339.23	-333.11	-331.13
ribose1phos	-1215.87	-1212.24	-1211.14
ribose5phos	-1223.95	-1220.32	-1219.22
ribulose	-336.38	-330.26	-328.28
serineL	-231.18	-226.89	-225.51
sorbose	-432.47	-425.13	-422.76
succinate	-530.72	-530.65	-530.64
succinylcoA	-349.90	-348.06	-347.47
sucrose	-685.66	-672.20	-667.85
thioredoxinox	0.00	0.00	0.00
thioredoxinred	54.32	55.41	55.74
tryptophaneL	364.78	372.12	374.49
tyrosineL	68.82	75.55	77.73
ubiquinoneox	3596.07	3651.15	3668.94
ubiquinonered	3586.06	3642.37	3660.55
urate	-206.03	-204.81	-204.41
urea	-42.97	-40.53	-39.74
uricacid	-197.07	-194.03	-193.84
valineL	80.87	87.60	89.78
xylose	-350.93	-344.81	-342.83
xylulose	-346.59	-340.47	-338.49

■ 5.0 Calculation of a table of standard transformed Gibbs energies of reactants at pH 5, 6, 7, 8, and 9 and ionic strength 0.25 M

```
calcdG5pH[reactantname_]:=Module[{out1,out2,out3,out4,out5},(*This program calculates
the standard transformed Gibbs energies of formation at 298 K, ionic strength 0.25 M
and pHs 5, 6, 7, 8, and 9.  The reactant name calls a function of pH and ionic
strength.*)
out1=reactantname/.{pH->5,is->.25};
out2=reactantname/.{pH->6,is->.25};
out3=reactantname/.{pH->7,is->.25};
out4=reactantname/.{pH->8,is->.25};
out5=reactantname/.{pH->9,is->.25};
{out1,out2,out3,out4,out5}];
```

This is a list of the values to be calculated for each of the reactants.

```
listreactantspH = {acetaldehydepH, acetatepH, acetonepH, acetylcoApH, acetylphospH,
    aconitatecispH, adeninepH, adenosinepH, adppH, alaninepH, ammoniapH, amppH,
    arabinosepH, asparagineLpH, aspartatepH, atppH, bpgpH, butanolnpH, butyratepH,
    citratepH, citrateisopH, coApH, coAglutathionepH, co2gpH, co2totpH, coaqpH,
    cogpH, creatinepH, creatininepH, cysteineLpH, cystineLpH, cytochromecoxpH,
    cytochromecredpH, dihydroxyacetonephospH, ethanolpH, ethylacetatepH, fadoxpH,
    fadredpH, fadenzoxpH, fadenzredpH, ferredoxinoxpH, ferredoxinredpH, fmnoxpH,
    fmnredpH, formatepH, fructosepH, fructose6phospH, fructose16phospH, fumaratepH,
    galactosepH, galactose1phospH, glucosepH, glucose1phospH, glucose6phospH,
    glutamatepH, glutaminepH, glutathioneoxpH, glutathioneredpH, glyceraldehydephospH,
    glycerolpH, glycerol3phospH, glycinepH, glycolatepH, glycylglycinepH, glyoxylatepH,
    h2aqpH, h2gpH, h2opH, h2o2aqpH, hydroxypropionatebpH, hypoxanthinepH, indolepH,
    ketoglutaratepH, lactatepH, lactosepH, leucineisoLpH, leucineLpH, lyxosepH,
    malatepH, maltosepH, mannitolDpH, mannosepH, methaneaqpH, methanegpH, methanolpH,
    methionineLpH, methylamineionpH, n2aqpH, n2gpH, nadoxpH, nadredpH, nadpoxpH,
    nadpredpH, o2aqpH, o2gpH, oxalatepH, oxaloacetatepH, oxalosuccinatepH, palmitatepH,
    peppH, pg2pH, pg3pH, phenylalanineLpH, pipH, ppipH, propanol2pH, propanolnpH,
    pyruvatepH, retinalpH, retinolpH, ribosepH, ribose1phospH, ribose5phospH,
    ribulosepH, serineLpH, sorbosepH, succinatepH, succinylcoApH, sucrosepH,
    thioredoxinoxpH, thioredoxinredpH, tryptophaneLpH, tyrosineLpH, ubiquinoneoxpH,
    ubiquinoneredpH, uratepH, ureapH, uricacidpH, valineLpH, xylosepH, xylulosepH};
```

Map is used to apply calcdG5pH to each of the functions for the reactants.

```
Clear[acetaldehydepH, acetatepH, acetonepH, acetylcoApH, acetylphospH,
    aconitatecispH, adeninepH, adenosinepH, adppH, alaninepH, ammoniapH, amppH,
    arabinosepH, asparagineLpH, aspartatepH, atppH, bpgpH, butanolnpH, butyratepH,
    citratepH, citrateisopH, coApH, coAglutathionepH, co2gpH, co2totpH, coaqpH,
    cogpH, creatinepH, creatininepH, cysteineLpH, cystineLpH, cytochromecoxpH,
    cytochromecredpH, dihydroxyacetonephospH, ethanolpH, ethylacetatepH, fadoxpH,
    fadredpH, fadenzoxpH, fadenzredpH, ferredoxinoxpH, ferredoxinredpH, fmnoxpH,
    fmnredpH, formatepH, fructosepH, fructose6phospH, fructose16phospH, fumaratepH,
    galactosepH, galactose1phospH, glucosepH, glucose1phospH, glucose6phospH,
    glutamatepH, glutaminepH, glutathioneoxpH, glutathioneredpH, glyceraldehydephospH,
    glycerolpH, glycerol3phospH, glycinepH, glycolatepH, glycylglycinepH, glyoxylatepH,
    h2aqpH, h2gpH, h2opH, h2o2aqpH, hydroxypropionatebpH, hypoxanthinepH, indolepH,
    ketoglutaratepH, lactatepH, lactosepH, leucineisoLpH, leucineLpH, lyxosepH,
    malatepH, maltosepH, mannitolDpH, mannosepH, methaneaqpH, methanegpH, methanolpH,
    methionineLpH, methylamineionpH, n2aqpH, n2gpH, nadoxpH, nadredpH, nadpoxpH,
    nadpredpH, o2aqpH, o2gpH, oxalatepH, oxaloacetatepH, oxalosuccinatepH, palmitatepH,
    peppH, pg2pH, pg3pH, phenylalanineLpH, pipH, ppipH, propanol2pH, propanolnpH,
    pyruvatepH, retinalpH, retinolpH, ribosepH, ribose1phospH, ribose5phospH,
    ribulosepH, serineLpH, sorbosepH, succinatepH, succinylcoApH, sucrosepH,
    thioredoxinoxpH, thioredoxinredpH, tryptophaneLpH, tyrosineLpH, ubiquinoneoxpH,
    ubiquinoneredpH, uratepH, ureapH, uricacidpH, valineLpH, xylosepH, xylulosepH];
```

```
Evaluate[listreactantspH] = Map[calcdG5pH, listfnpHis];
```

Table 2 Standard Transformed Gibbs Energies of Formation of Reactants at 298.15 K, Ionic Strength 0.25 M, and pHs of 5, 6, 7, 8, and 9

```
table2=PaddedForm[TableForm[listreactantspH,TableHeadings->{list,{"pH 5","pH 6","pH
7","pH 8","pH 9"}},TableSpacing->{.3,2}],{6,2}]
```

	pH 5	pH 6	pH 7	pH 8	pH 9
acetaldehyde	-21.60	1.23	24.06	46.90	69.73
acetate	-282.71	-265.02	-247.83	-230.70	-213.57
acetone	16.40	50.65	84.90	119.14	153.39
acetylcoA	-92.31	-75.19	-58.06	-40.94	23.81
acetylphos	-1153.77	-1129.84	-1107.02	-1085.39	-1066.49
aconitatecis	-836.37	-819.24	-802.12	-785.00	-767.87
adenine	457.06	485.92	514.50	543.04	571.58
adenosine	186.98	261.25	335.46	409.66	483.87
adp	-1569.05	-1495.55	-1424.70	-1355.78	-1287.24
alanine	-165.55	-125.59	-85.64	-45.68	-5.73
ammonia	37.28	60.11	82.93	105.64	127.51
amp	-698.40	-625.22	-554.83	-486.04	-417.51
arabinose	-448.73	-391.65	-334.57	-277.49	-220.41
asparagineL	-291.13	-245.47	-199.80	-154.14	-108.47
aspartate	-520.59	-486.34	-452.09	-417.85	-383.60
atp	-2437.46	-2363.76	-2292.50	-2223.44	-2154.88
bpg	-2262.15	-2233.92	-2207.30	-2183.36	-2160.38
butanoln	121.66	178.74	235.82	292.90	349.98
butyrate	-147.99	-108.03	-68.08	-28.12	11.83
citrate	-1027.23	-995.44	-966.23	-937.62	-909.07
citrateiso	-1020.58	-988.80	-959.58	-930.97	-902.42
coA	-18.48	-12.79	-7.26	-2.82	-1.10
coAglutathione	403.59	489.21	574.83	660.45	746.07
co2g	-394.36	-394.36	-394.36	-394.36	-394.36
co2tot	-564.61	-554.49	-547.10	-541.18	-535.80
coaq	-119.90	-119.90	-119.90	-119.90	-119.90
cog	-137.17	-137.17	-137.17	-137.17	-137.17
creatine	4.95	56.32	107.69	159.07	210.44
creatinine	182.31	222.27	262.22	302.18	342.13
cysteineL	-133.37	-93.43	-53.65	-14.97	20.99
cystineL	-314.31	-245.81	-177.32	-108.82	-40.33
cytochromecox	-7.29	-7.29	-7.29	-7.29	-7.29
cytochromecred	-27.75	-27.75	-27.75	-27.75	-27.75
dihydroxyacetonephos	-1154.88	-1124.53	-1095.70	-1067.13	-1038.59
ethanol	-5.54	28.71	62.96	97.20	131.45
ethylacetate	-102.85	-57.19	-11.52	34.14	79.81
fadox	906.61	1083.56	1260.51	1437.46	1614.40
fadred	926.43	1114.79	1303.16	1491.52	1679.89
fadenzox	906.61	1083.56	1260.51	1437.46	1614.40
fadenzred	876.71	1065.07	1253.44	1441.80	1630.17
ferredoxinox	-0.81	-0.81	-0.81	-0.81	-0.81
ferredoxinred	38.07	38.07	38.07	38.07	38.07
fmnox	554.41	662.86	771.31	879.77	988.22
fmnred	574.23	694.10	813.97	933.84	1053.70
formate	-322.46	-316.75	-311.04	-305.34	-299.63
fructose	-563.31	-494.81	-426.32	-357.82	-289.33
fructose6phos	-1445.66	-1379.42	-1315.74	-1252.84	-1190.04
fructose16phos	-2326.42	-2264.57	-2206.78	-2149.62	-2092.54
fumarate	-546.67	-535.02	-523.58	-512.16	-500.74
galactose	-556.73	-488.23	-419.74	-351.24	-282.75
galactose1phos	-1440.96	-1375.09	-1311.60	-1248.72	-1185.93
glucose	-563.70	495.20	426.71	358.21	289.72
glucose1phos	-1442.86	-1376.01	-1311.89	-1248.92	-1186.11
glucose6phos	-1449.53	-1382.88	-1318.92	-1255.98	-1193.18
glutamate	-463.48	-417.82	-372.15	-326.49	-280.82
glutamine	-234.52	-177.44	-120.36	-63.28	-6.20
glutathioneox	877.26	1048.50	1219.74	1390.98	1562.22
glutathionered	455.34	546.64	637.62	726.89	813.52
glyceraldehydephos	-1147.22	-1116.87	-1088.04	-1059.47	-1030.93
glycerol	-262.68	-217.02	-171.35	-125.69	-80.02
glycerol3phos	-1163.24	-1118.83	-1077.13	-1036.91	-996.93

glycine	-233.16	-204.62	-176.08	-147.54	-119.00
glycolate	-443.71	-426.59	-409.46	-392.34	-375.21
glycylglycine	-285.40	-239.74	-194.07	-148.41	-102.74
glyoxylate	-440.06	-434.35	-428.64	-422.94	-417.23
h2aq	76.30	87.72	99.13	110.55	121.96
h2g	58.70	70.12	81.53	92.95	104.36
h2o	-178.49	-167.07	-155.66	-144.24	-132.83
h2o2aq	-75.33	-63.91	-52.50	-41.08	-29.67
hydroxypropionateb	-372.46	-343.92	-315.38	-286.84	-258.30
hypoxanthine	206.90	229.73	252.56	275.40	298.23
indole	429.25	469.21	509.16	549.12	589.07
ketoglutarate	-679.25	-656.42	-633.58	-610.75	-587.92
lactate	-370.78	-342.24	-313.70	-285.16	-256.62
lactose	-921.63	-796.06	-670.48	-544.90	-419.32
leucineisoL	37.65	111.85	186.06	260.26	334.47
leucineL	29.30	103.50	177.71	251.91	326.12
lyxose	-455.64	-398.56	-341.48	-284.40	-227.32
malate	-729.49	-705.79	-682.85	-660.00	-637.17
maltose	-928.99	-803.42	-677.84	-552.26	-426.68
mannitolD	-531.71	-451.80	-371.89	-291.97	-212.06
mannose	-557.80	-489.30	-420.81	-352.31	-283.82
methaneaq	83.07	105.90	128.73	151.57	174.40
methaneg	66.68	89.51	112.34	135.18	158.01
methanol	-57.91	-35.08	-12.25	10.59	33.42
methionineL	-180.07	-117.28	-54.49	8.29	71.08
methylamineion	135.43	169.68	203.93	238.17	272.42
n2aq	18.70	18.70	18.70	18.70	18.70
n2g	0.00	0.00	0.00	0.00	0.00
nadox	762.29	910.70	1059.11	1207.51	1355.92
nadred	811.86	965.98	1120.09	1274.21	1428.33
nadpox	-108.72	33.98	176.68	319.38	462.08
nadpred	-59.05	89.36	237.77	386.18	534.59
o2aq	16.40	16.40	16.40	16.40	16.40
o2g	0.00	0.00	0.00	0.00	0.00
oxalate	-677.26	-677.15	-677.14	-677.14	-677.14
oxaloacetate	-737.83	-726.41	-715.00	-703.58	-692.16
oxalosuccinate	-1024.72	-1001.89	-979.05	-956.22	-933.39
palmitate	649.64	826.59	1003.54	1180.48	1357.43
pep	-1218.97	-1203.00	-1189.73	-1178.02	-1166.58
pg2	-1396.52	-1368.31	-1341.79	-1317.92	-1294.95
pg3	-1402.06	-1373.94	-1347.73	-1324.05	-1301.11
phenylalanineL	115.75	178.54	241.33	304.11	366.90
pi	-1079.46	-1068.49	-1059.49	-1052.97	-1047.17
ppi	-1957.07	-1947.46	-1940.66	-1935.64	-1933.29
propanol2	49.57	95.23	140.90	186.56	232.23
propanoln	58.99	104.65	150.32	195.98	241.65
pyruvate	-385.03	-367.91	-350.78	-333.66	-316.53
retinal	821.80	981.62	1141.45	1301.27	1461.10
retinol	852.59	1023.83	1195.07	1366.31	1537.55
ribose	-445.29	-388.21	-331.13	-274.05	-216.97
ribose1phos	-1320.16	-1264.30	-1211.14	-1159.50	-1108.09
ribose5phos	-1328.24	-1272.58	-1219.22	-1167.58	-1116.17
ribulose	-442.44	-385.36	-328.28	-271.20	-214.12
serineL	-305.42	-265.46	-225.51	-185.55	-145.60
sorbose	-559.75	-491.25	-422.76	-354.26	-285.77
succinate	-578.32	-553.72	-530.64	-507.79	-484.95
succinylcoA	-393.33	-370.32	-347.47	-324.63	-301.80
sucrose	-919.00	-793.43	-667.85	-542.27	-416.69
thioredoxinox	0.00	0.00	0.00	0.00	0.00
thioredoxinred	33.33	44.70	55.74	64.03	66.35
tryptophaneL	237.50	306.00	374.49	442.99	511.48
tyrosineL	-47.85	14.94	77.73	140.51	203.30
ubiquinoneox	2641.49	3155.21	3668.94	4182.66	4696.38
ubiquinonered	2610.27	3135.41	3660.55	4185.69	4710.83
urate	-238.66	-221.54	-204.41	-187.29	-170.16
urea	-85.40	-62.57	-39.74	-16.90	5.93
uricacid	-239.50	-216.67	-193.84	-171.00	-148.17
valineL	-35.80	26.99	89.78	152.56	215.35
xylose	-456.99	-399.91	-342.83	-285.75	-228.67
xylulose	-452.65	-395.57	-338.49	-281.41	-224.33

■ 6.0 Calculation of the functions of pH and ionic strength for the standard transformed enthalpies of formation of reactants

```
calcdHmat[speciesmat_] :=
 Module[{dHzero, zi, nH, dhfnsp, dGzero, pHterm, isenth, dgfnsp, dGreactant, ri},
  (*This program produces the function of ionic strength (is) that gives the standard
     transformed enthalpy of formation of a reactant (sum of species) at 298.15
     K.  The input is a matrix that gives the standard Gibbs energy of formation,
   the standard enthalpy of formation, the electric charge,
   and the number of hydrogen atoms in the species in the reactant.  There is
   a row in the matrix for each species of the reactant.  dhfnsp is a list
   of the functions for the species.  Energies are expressed in kJ mol^-1.*)
   dHzero = speciesmat[[All, 2]];
   zi = speciesmat[[All, 3]];
   nH = speciesmat[[All, 4]];
   isenth = 1.4775 * ((zi^2) - nH) * (is^.5) / (1 + 1.6 * is^.5);
   dhfnsp = dHzero + isenth;
  (*Now calculate the functions for
   the standard Gibbs energies of formation of the species.*)
  dGzero = speciesmat[[All, 1]];
  pHterm = nH * 8.31451 * .29815 * Log[10^-pH];
  gpfnsp = dGzero - pHterm - isenth * 2.91482 / 1.44775;
  (*Now calculate the standard
   transformed Gibbs energy of formation for the reactant.*)
  dGreactant = -8.31451 * .29815 * Log[Apply[Plus, Exp[-1 * gpfnsp / (8.31451 * .29815)]]];
  (*Now calculate the equilibrium mole fractions of the species
     in the reactant and the mole fraction-weighted average of the
     functions for the standard transformed enthalpies of the species.*)
  ri = Exp[(dGreactant - gpfnsp) / (8.31451 * .29815)];
  ri.dhfnsp]
```

The following is a list of functions for the reactants with which the standard transformed enthalpies of formation can be calculated. It is shorter than the list of reactants for the transformed Gibbs energies because these is less information about standard enthalpies of formation of species.

```
listreactantsh={"acetaldehyde","acetate","acetone","adenosine","adp","alanine","ammonia
","amp","arabinose","asparagineL","aspartate","atp","citrate","co2g","co2tot","coaq","c
og","ethanol","ethylacetate","formate","fructose","fumarate","galactose","glucose","glu
cose6phos","glutamate","glutamine","glycerol","glycine","glycylglycine","h2aq","h2g","h
2o","h2o2aq","indole","lactate","lactose","leucineL","maltose","mannose","methaneg","me
thaneaq","methanol","methylamineion","n2aq","n2g","nadox","nadpox","nadpred","nadred","
o2aq","o2g","pi","ppi","propanol2","pyruvate","ribose","ribose5phos","ribulose","sorbos
e", "succinate","sucrose","tryptophaneL","urea","valineL","xylose","xylulose"};
```

```
listfnpHish={acetaldehydeh,acetateh,acetoneh,adenosineh,adph,alanineh,ammoniah,amph,ara
binoseh,asparagineLh,aspartateh,
 atph,citrateh,co2gh, co2toth, coaqh, cogh,
   ethanolh, ethylacetateh,
   formateh, fructoseh, fumarateh, galactoseh,  glucoseh,
 glucose6phosh,glutamateh, glutamineh,
 glycerolh,  glycineh, glycylglycineh,  h2aqh, h2gh, h2oh,
 h2o2aqh,indoleh, lactateh, lactoseh,leucineLh,
   maltoseh, mannoseh, methanegh,
 methaneaqh, methanolh,methylamineionh, n2aqh,
 n2gh, nadoxh, nadpoxh, nadpredh, nadredh, o2aqh,
 o2gh, pih,ppih, propanol2h,
 pyruvateh, riboseh,  ribose5phosh,ribuloseh,
   sorboseh, succinateh, sucroseh, tryptophaneLh,  ureah,  valineLh, xyloseh,
 xyluloseh};
```

```
listspeciesdatah={acetaldehydesp,acetatesp,acetonesp,adenosinesp,adpsp,alaninesp,ammoni
asp,ampsp,arabinosesp,asparagineLsp,aspartatesp,
 atpsp,citratesp,co2gsp, co2totsp, coaqsp, cogsp,
   ethanolsp, ethylacetatesp,
   formatesp, fructosesp, fumaratesp, galactosesp,  glucosesp,
 glucose6phossp,glutamatesp, glutaminesp,
 glycerolsp,  glycinesp, glycylglycinesp,  h2aqsp, h2gsp, h2osp,
 h2o2aqsp,indolesp, lactatesp, lactosesp,leucineLsp,
   maltosesp, mannosesp, methanegsp,
 methaneaqsp, methanolsp,methylamineionsp, n2aqsp,
 n2gsp, nadoxsp, nadpoxsp, nadpredsp, nadredsp, o2aqsp,
 o2gsp,pisp,ppisp, propanol2sp,
 pyruvatesp, ribosesp,  ribose5phossp,ribulosesp,
   sorbosesp, succinatesp, sucrosesp, tryptophaneLsp,  ureasp,  valineLsp, xylosesp,
 xylulosesp};
```

Map is used to apply calcdHmat to each of the matrices of species data called up by listspeciesdatah.

```
Clear[acetaldehydeh,acetateh,acetoneh,adenosineh,adph,alanineh,ammoniah,amph,arabinoseh
,asparagineLh,aspartateh,
 atph,citrateh,co2gh, co2toth, coaqh, cogh,
   ethanolh, ethylacetateh,
   formateh, fructoseh, fumarateh, galactoseh,  glucoseh,
 glucose6phosh,glutamateh, glutamineh,
 glycerolh,  glycineh, glycylglycineh,  h2aqh, h2gh, h2oh,
 h2o2aqh,indoleh, lactateh, lactoseh,leucineLh,
   maltoseh, mannoseh, methanegh,
 methaneaqh, methanolh,methylamineionh, n2aqh,
 n2gh, nadoxh, nadpoxh, nadpredh, nadredh, o2aqh,
 o2gh, pih,ppih, propanol2h,
 pyruvateh, riboseh,  ribose5phosh,ribuloseh,
   sorboseh, succinateh, sucroseh, tryptophaneLh,  ureah,  valineLh, xyloseh,
 xyluloseh];
```

```
Evaluate[listfnpHish] = Map[calcdHmat, listspeciesdatah];
```

■ 7.0 Calculation of a table of standard transformed enthalpies of formation of reactants at pH 7 and ionic strengths of 0, 0.10, and 0.25 M

```
listhisreactants={acetaldehydehis,acetatehis,acetonehis,adenosinehis,adphis,alaninehis,
ammoniahis,amphis,arabinosehis,asparagineLhis,aspartatehis,
  atphis,citratehis,co2ghis, co2tothis, coaqhis, coghis,
    ethanolhis, ethylacetatehis,
    formatehis, fructosehis, fumaratehis, galactosehis,  glucosehis,
  glucose6phoshis,glutamatehis, glutaminehis,
  glycerolhis,  glycinehis, glycylglycinehis,  h2aqhis, h2ghis, h2ohis,
  h2o2aqhis,indolehis, lactatehis, lactosehis,leucineLhis,
   maltosehis, mannosehis, methaneghis,
  methaneaqhis, methanolhis,methylamineionhis, n2aqhis,
  n2ghis, nadoxhis, nadpoxhis, nadpredhis, nadredhis, o2aqhis,
  o2ghis, pihis,ppihis, propanol2his,
  pyruvatehis, ribosehis,  ribose5phoshis,ribulosehis,
   sorbosehis, succinatehis, sucrosehis, tryptophaneLhis,  ureahis,  valineLhis,
xylosehis,
  xylulosehis};

calcdH3I[reactanth_]:=Module[{out1,out2,out3},(*This program calculates the standard
transformed enthalpies of formation at 298 K, pH 7, and ionic strengths of 0, 0.10,
and 0.25 M.  The reactanth name calls a function of pH and ionic strength or a
constant.*)
out1=reactanth/.{pH->7,is->0};
out2=reactanth/.{pH->7,is->.1};
out3=reactanth/.{pH->7,is->.25};
{out1,out2,out3}];

Clear[acetaldehydehis,acetatehis,acetonehis,adenosinehis,adphis,alaninehis,ammoniahis,a
mphis,arabinosehis,asparagineLhis,aspartatehis,
  atphis,citratehis,co2ghis, co2tothis, coaqhis, coghis,
    ethanolhis, ethylacetatehis,
    formatehis, fructosehis, fumaratehis, galactosehis,  glucosehis,
  glucose6phoshis,glutamatehis, glutaminehis,
  glycerolhis,  glycinehis, glycylglycinehis,  h2aqhis, h2ghis, h2ohis,
  h2o2aqhis,indolehis, lactatehis, lactosehis,leucineLhis,
   maltosehis, mannosehis, methaneghis,
  methaneaqhis, methanolhis,methylamineionhis, n2aqhis,
  n2ghis, nadoxhis, nadpoxhis, nadpredhis, nadredhis, o2aqhis,
  o2ghis, pihis,ppihis, propanol2his,
  pyruvatehis, ribosehis,  ribose5phoshis,ribulosehis,
   sorbosehis, succinatehis, sucrosehis, tryptophaneLhis,  ureahis,  valineLhis,
xylosehis,
  xylulosehis];
```

Map is used to apply calcdH3I to each of the functions for reactants.

```
Evaluate[listhisreactants] = Map[calcdH3I, listfnpHish];
```

Table 3 Standard Transformed Enthalpies of Formation (in kJ mol^{-1}) of Biochemical Reactants at pH 7 and Ionic Strengths of 0, 0.10, and 0.25 M.

```
table3=PaddedForm[TableForm[listhisreactants,TableHeadings->{listreactantsh,{"I = 0
M","I = 0.10 M","I= 0.25 M"}},TableSpacing->{.3,2}],{8,2}]
```

	I = 0 M	I = 0.10 M	I= 0.25 M
acetaldehyde	-212.23	-213.47	-213.87
acetate	-486.01	-486.63	-486.83
acetone	-221.71	-223.57	-224.17
adenosine	-620.93	-624.97	-626.27
adp	-2623.20	-2626.54	-2627.24
alanine	-554.80	-556.97	-557.67
ammonia	-132.22	-133.15	-133.45
amp	-1633.49	-1637.17	-1638.19
arabinose	-1043.79	-1046.89	-1047.89
asparagineL	-766.09	-768.57	-769.37
aspartate	-943.41	-944.96	-945.46
atp	-3614.23	-3616.65	-3616.92
citrate	-1515.78	-1514.14	-1513.66
co2g	-393.50	-393.50	-393.50
co2tot	-693.43	-692.99	-692.86
coaq	-120.96	-120.96	-120.96
cog	-110.53	-110.53	-110.53
ethanol	-288.30	-290.16	-290.76
ethylacetate	-482.00	-484.48	-485.28
formate	-425.55	-425.55	-425.55
fructose	-1259.38	-1263.10	-1264.31
fumarate	-777.38	-776.77	-776.57
galactose	-1255.20	-1258.92	-1260.13
glucose	-1262.19	-1265.91	-1267.12
glucose6phos	-2276.06	-2278.56	-2279.30
glutamate	-979.89	-982.06	-982.76
glutamine	-805.00	-808.10	-809.10
glycerol	-676.55	-679.03	-679.83
glycine	-523.00	-524.55	-525.05
glycylglycine	-734.25	-736.73	-737.53
h2aq	-4.20	-4.82	-5.02
h2g	0.00	-0.62	-0.82
h2o	-285.83	-286.45	-286.65
h2o2aq	-191.17	-191.79	-191.99
indole	97.50	95.33	94.63
lactate	-686.64	-687.88	-688.28
lactose	-2233.08	-2239.91	-2242.11
leucineL	-643.37	-647.40	-648.71
maltose	-2238.06	-2244.89	-2247.09
mannose	-1258.66	-1262.38	-1263.59
methaneg	-74.81	-76.05	-76.45
methaneaq	-89.04	-90.28	-90.68
methanol	-245.93	-247.17	-247.57
methylamineion	-124.93	-126.48	-126.98
n2aq	-10.54	-10.54	-10.54
n2g	0.00	0.00	0.00
nadox	0.00	-7.76	-10.26
nadpox	0.00	-4.96	-6.57
nadpred	-29.18	-32.28	-33.28
nadred	-31.94	-39.08	-41.38
o2aq	-11.70	-11.70	-11.70
o2g	0.00	0.00	0.00
pi	-1301.24	-1299.89	-1299.36
ppi	-2295.04	-2292.54	-2291.57
propanol2	-330.83	-333.31	-334.11
pyruvate	-596.22	-596.84	-597.04
ribose	-1034.00	-1037.10	-1038.10
ribose5phos	-2037.77	-2041.51	-2042.43
ribulose	-1023.02	-1026.12	-1027.12
sorbose	-1263.30	-1267.02	-1268.23
succinate	-908.69	-908.70	-908.70
sucrose	-2199.87	-2206.70	-2208.90
tryptophaneL	-405.20	-408.92	-410.13
urea	-317.65	-318.89	-319.29
valineL	-611.99	-615.40	-616.50
xylose	-1045.94	-1049.04	-1050.04
xylulose	-1029.65	-1032.75	-1033.75

■ 8.0 Calculation of the standard transformed enthalpies of formation of reactants at pH 5, 6, 7, 8, and 9 and ionic strength 0.25 M

```
listhpHreactants={acetaldehydehpH,acetatehpH,acetonehpH,adenosinehpH,adphpH,alaninehpH,
ammoniahpH,amphpH,arabinosehpH,asparagineLhpH,aspartatehpH,
 atphpH,citratehpH,co2ghpH, co2tothpH, coaqhpH, coghpH,
   ethanolhpH, ethylacetatehpH,
   formatehpH, fructosehpH, fumaratehpH, galactosehpH, glucosehpH,
   glucose6phoshpH,glutamatehpH, glutaminehpH,
   glycerolhpH, glycinehpH, glycylglycinehpH, h2aqhpH, h2ghpH, h2ohpH,
   h2o2aqhpH,indolehpH, lactatehpH, lactosehpH,leucineLhpH,
   maltosehpH, mannosehpH, methaneghpH,
   methaneaqhpH, methanolhpH,methylamineionhpH, n2aqhpH,
   n2ghpH, nadoxhpH, nadpoxhpH, nadpredhpH, nadredhpH, o2aqhpH,
   o2ghpH, pihpH,ppihpH, propanol2hpH,
   pyruvatehpH, ribosehpH, ribose5phoshpH,ribulosehpH,
   sorbosehpH, succinatehpH, sucrosehpH, tryptophaneLhpH, ureahpH, valineLhpH,
xylosehpH,
   xylulosehpH};

Clear[acetaldehydehpH,acetatehpH,acetonehpH,adenosinehpH,adphpH,alaninehpH,ammoniahpH,a
mphpH,arabinosehpH,asparagineLhpH,aspartatehpH,
 atphpH,citratehpH,co2ghpH, co2tothpH, coaqhpH, coghpH,
   ethanolhpH, ethylacetatehpH,
   formatehpH, fructosehpH, fumaratehpH, galactosehpH, glucosehpH,
   glucose6phoshpH,glutamatehpH, glutaminehpH,
   glycerolhpH, glycinehpH, glycylglycinehpH, h2aqhpH, h2ghpH, h2ohpH,
   h2o2aqhpH,indolehpH, lactatehpH, lactosehpH,leucineLhpH,
   maltosehpH, mannosehpH, methaneghpH,
   methaneaqhpH, methanolhpH,methylamineionhpH, n2aqhpH,
   n2ghpH, nadoxhpH, nadpoxhpH, nadpredhpH, nadredhpH, o2aqhpH,
   o2ghpH, pihpH,ppihpH, propanol2hpH,
   pyruvatehpH, ribosehpH, ribose5phoshpH,ribulosehpH,
   sorbosehpH, succinatehpH, sucrosehpH, tryptophaneLhpH, ureahpH, valineLhpH,
xylosehpH,
   xylulosehpH];

calcdH5pH[reactanth_]:=Module[{out1,out2,out3,out4,out5},(*This program calculates the
standard transformed enthalpies of formation at 298 K, ionic strength 0.25 M and pHs
5, 6, 7, 8, and 9.  The reactanth name calls a function of pH and ionic strength or a
constant.*)
out1=reactanth/.{pH->5,is->.25};
out2=reactanth/.{pH->6,is->.25};
out3=reactanth/.{pH->7,is->.25};
out4=reactanth/.{pH->8,is->.25};
out5=reactanth/.{pH->9,is->.25};
{out1,out2,out3,out4,out5}];
```

Map is used to apply calcdH5pH to each of the functions for the reactants.

```
Evaluate[listhpHreactants] = Map[calcdH5pH, listfnpHish];
```

Table 4 Standard Transformed Enthalpies of Formation (in kJ mol^{-s}) of Biochemical Reactants at I = 0.25 M and pHs of 5, 6, 7, 8, and 9.

```
table4=PaddedForm[TableForm[listhpHreactants,TableHeadings->{listreactantsh,{"pH
5","pH 6 ","pH 7", "pH 8","pH 9"}},TableSpacing->{.3,2}],{6,2}]
```

	pH 5	pH 6	pH 7	pH 8	pH 9
acetaldehyde	-213.87	-213.87	-213.87	-213.87	-213.87
acetate	-486.96	-486.85	-486.83	-486.83	-486.83
acetone	-224.17	-224.17	-224.17	-224.17	-224.17
adenosine	-590.92	-622.97	-626.27	-626.60	-626.63
adp	-2625.82	-2625.74	-2627.24	-2627.71	-2627.76
alanine	-557.67	-557.67	-557.67	-557.67	-557.67
ammonia	-133.74	-133.71	-133.45	-130.97	-115.00
amp	-1635.98	-1636.50	-1638.19	-1638.60	-1638.65
arabinose	-1047.89	-1047.89	-1047.89	-1047.89	-1047.89
asparagineL	-769.37	-769.37	-769.37	-769.37	-769.37
aspartate	-945.46	-945.46	-945.46	-945.46	-945.46
atp	-3615.67	-3615.43	-3616.92	-3617.49	-3617.56
citrate	-1518.50	-1514.96	-1513.66	-1513.49	-1513.47
co2g	-393.50	-393.50	-393.50	-393.50	-393.50
co2tot	-699.80	-696.59	-692.86	-691.80	-689.51
coaq	-120.96	-120.96	-120.96	-120.96	-120.96
cog	-110.53	-110.53	-110.53	-110.53	-110.53
ethanol	-290.76	-290.76	-290.76	-290.76	-290.76
ethylacetate	-485.28	-485.28	-485.28	-485.28	-485.28
formate	-425.55	-425.55	-425.55	-425.55	-425.55
fructose	-1264.31	-1264.31	-1264.31	-1264.31	-1264.31
fumarate	-776.45	-776.56	-776.57	-776.57	-776.57
galactose	-1260.13	-1260.13	-1260.13	-1260.13	-1260.13
glucose	-1267.12	-1267.12	-1267.12	-1267.12	-1267.12
glucose6phos	-2279.17	-2279.25	-2279.30	-2279.31	-2279.31
glutamate	-982.76	-982.76	-982.76	-982.76	-982.76
glutamine	-809.10	-809.10	-809.10	-809.10	-809.10
glycerol	-679.83	-679.83	-679.83	-679.83	-679.83
glycine	-525.05	-525.05	-525.05	-525.05	-525.05
glycylglycine	-737.53	-737.53	-737.53	-737.53	-737.53
h2aq	-5.02	-5.02	-5.02	-5.02	-5.02
h2g	-0.82	-0.82	-0.82	-0.82	-0.82
h2o	-286.65	-286.65	-286.65	-286.65	-286.65
h2o2aq	-191.99	-191.99	-191.99	-191.99	-191.99
indole	94.63	94.63	94.63	94.63	94.63
lactate	-688.28	-688.28	-688.28	-688.28	-688.28
lactose	-2242.11	-2242.11	-2242.11	-2242.11	-2242.11
leucineL	-648.71	-648.71	-648.71	-648.71	-648.71
maltose	-2247.09	-2247.09	-2247.09	-2247.09	-2247.09
mannose	-1263.59	-1263.59	-1263.59	-1263.59	-1263.59
methaneg	-76.45	-76.45	-76.45	-76.45	-76.45
methaneaq	-90.68	-90.68	-90.68	-90.68	-90.68
methanol	-247.57	-247.57	-247.57	-247.57	-247.57
methylamineion	-126.98	-126.98	-126.98	-126.98	-126.98
n2aq	-10.54	-10.54	-10.54	-10.54	-10.54
n2g	0.00	0.00	0.00	0.00	0.00
nadox	-10.26	-10.26	-10.26	-10.26	-10.26
nadpox	-6.57	-6.57	-6.57	-6.57	-6.57
nadpred	-33.28	-33.28	-33.28	-33.28	-33.28
nadred	-41.38	-41.38	-41.38	-41.38	-41.38
o2aq	-11.70	-11.70	-11.70	-11.70	-11.70
o2g	0.00	0.00	0.00	0.00	0.00
pi	-1302.89	-1302.03	-1299.36	-1297.99	-1297.79
ppi	-2294.18	-2292.80	-2291.57	-2290.05	-2287.69
propanol2	-334.11	-334.11	-334.11	-334.11	-334.11
pyruvate	-597.04	-597.04	-597.04	-597.04	-597.04
ribose	-1038.10	-1038.10	-1038.10	-1038.10	-1038.10
ribose5phos	-2034.57	-2038.10	-2042.43	-2043.41	-2043.52
ribulose	-1027.12	-1027.12	-1027.12	-1027.12	-1027.12
sorbose	-1268.23	-1268.23	-1268.23	-1268.23	-1268.23
succinate	-909.85	-908.87	-908.70	-908.68	-908.68
sucrose	-2208.90	-2208.90	-2208.90	-2208.90	-2208.90
tryptophaneL	-410.13	-410.13	-410.13	-410.13	-410.13
urea	-319.29	-319.29	-319.29	-319.29	-319.29
valineL	-616.50	-616.50	-616.50	-616.50	-616.50
xylose	-1050.04	-1050.04	-1050.04	-1050.04	-1050.04
xylulose	-1033.75	-1033.75	-1033.75	-1033.75	-1033.75

These tables have been given to 0.01 kJ mol^{-1}. In general this overemphasizes the accuracy with which these formation properties are known. However for some reactants for which species are in classical tables, this accuracy is warranted. An error of 0.01 kJ mol^{-1} in the standard transformed Gibbs energy of a reaction at 298 K corresponds with an error of about 1% in the value of the apparent equilibrium constant. It is important to understand that the large number of digits in these tables is required because the thermodynamic information is in differences between entries.

These tables can be used to calculate changes in standard transformed Gibbs energies of reaction and standard transformed enthalpies of reaction for any biochemical reaction for which all the reactants are in the tables. Standard transfomed entropies of formation and of reaction can also be calculated. The advantage of having this notebook is that these propertiec can be calculated at any desired pH in the range 5 to 9 and any ionic strength in the range 0 to 0.35 M.

■ 9.0 Examples of calculations using the database on species

■ 9.1 Calculation of pKs of weak acids

When weak acids have pKs in the range between 4 to 10, the standard Gibbs energies of formation at 298.15 K and zero ionic strength of the various species are given in the database on species. In using the program calcpK, it is necessary to give the number associated with the pK. pKs are numbered 1, 2, 3,... from the highest to the lowest in the pH range approximately 4 to 10.

```
calcpK[speciesmat_, no_, is_] :=
  Module[{lnkzero, sigmanuzsq, lnK},(*Calculates pKs for a weak acid at 298.15 K at
specified ionic strengths (is) when the number no of the pK is specified.  pKs are
numbered 1,2, 3,... from the highest pK to the lowest pK, but the highest pK for a
weak acid may be omitted if it is outside of the range 5 t0 9.  For h3PO4,
pK1=calc[pisp,1,{0}] = 7.22.*)
    lnkzero = (speciesmat[[no + 1,1]] - speciesmat[[no,1]])/
      (8.31451*0.29815); sigmanuzsq = speciesmat[[no,3]]^2 -
      speciesmat[[no + 1,3]]^2 + 1;
    lnK = lnkzero + (1.17582*is^0.5*sigmanuzsq)/
      (1 + 1.6*is^0.5); N[-(lnK/Log[10])]]]

calcpK[atpsp,1,.25]

6.46502

calcpK[atpsp,2,.25]

3.82652
```

■ 9.2 Calculation of changes in thermodynamic properties of reactions

```
calctrGerx[eq_, pHlist_, islist_] :=
  Module[{energy}, (*Calculates the standard transformed
      Gibbs energy of reaction in kJ mol^-1 at specified pHs and ionic
        strengths for a biochemical reaction typed in the form atp+h2o+de=
    adp+pi.  The names of reactants call the appropriate functions of
      pH and ionic strength.  pHlist and is list can be lists.*)
    energy = Solve[eq, de]; energy[[1, 1, 2]] /. pH → pHlist /. is → islist]
```

This program can be used to calculate standard transformed Gibbs energies of reaction or standard transformed enthalpies of reaction in kJ mol^{-1}. To calculate the changes in standard transformed enthalpy, an h is appended to the name of the reactant.

```
dGrxatp = calctrGerx[atp + h2o + de == adp + pi, {5, 6, 7, 8, 9}, 0.25]
```

{-32.5633, -33.2166, -36.0353, -41.0742, -46.7021}

```
dHrxatp = calctrGerx[atph + h2oh + de == adph + pih, {5, 6, 7, 8, 9}, 0.25]
```

{-26.386, -25.6892, -23.0327, -21.558, -21.3455}

The corresponding standard transformed entropies of reaction are given by

$$\frac{\text{dHrxatp} - \text{dGrxatp}}{0.29815}$$

{20.7188, 25.2473, 43.611, 65.4575, 85.0466}

where these values are in J K^{-1} mol^{-1}. The standard transformed entropy of reaction increases rapidly above pH 7 because of the production of hydrogen ions. There is an increase in the standard transformed entropy because of the increase in the number of species in the reaction.

This program can also be used to produce plots of standard transformed properties of reaction versus pH or ionic strength.

```
Plot[calctrGerx[atp+h2o+de==adp+pi,pH,.25],{pH,5,9},AxesLabel->{"pH","ΔG'°/(kJ/mol)"}];
```

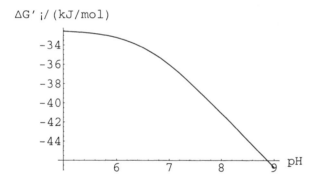

■ 9.3 Calculation of apparent equilibrium constants

```
calckprime[eq_, pHlist_, islist_] := Module[{energy, dG},(*Calculates the apparent
equilibrium constant at specified pHs and ionic strengths for a biochemical reaction
typed in the form atp+h2o+de==adp+pi.  The names of reactants call the appropriate
functions of pH and ionic strength.  pHlist and is list can be lists.*)
    energy = Solve[eq, de];
    dG = energy[[1,1,2]] /. pH -> pHlist /. is -> islist;
    E^(-(dG/(8.31451*0.29815)))]

calckprime[atp + h2o + de == adp + pi, {5, 6, 7, 8, 9}, 0.25]
```

{506774., 659585., $2.05626\ 10^{6}$, $1.5698\ 10^{7}$, $1.51989\ 10^{8}$}

This program can be also used to pepare plots versus pH or ionic strength.

```
Plot[Evaluate[calckprime[atp+h2o+de==adp+pi,pH,0.25]],{pH,6,8},AxesLabel->{"pH","K'"}];
```

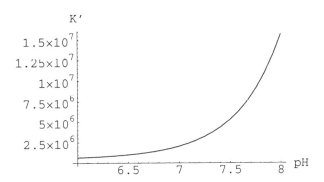

■ 9.4 Calculations of changes in binding of hydrogen ions in biochemical reactions

Biochemical reactions are different from chemical reactions in that they may produce or consume non-integer amounts of hydrogen ions. This is discussed in terms of the change in binding of hydrogen ions $\Delta_f N_H$ in a biochemical reaction, which is given by

$$\Delta_f N_H = -(d\log K'/dpH) = -(1/\ln(10))(d\ln K'/dpH)$$

The change in the binding of hydrogen ions in the hydrolysis of ATP can also be plotted as follows:

```
Plot[Evaluate[-
(1/Log[10])*D[Log[calckprime[atp+h2o+de==adp+pi,pH,0.25]],pH]],{pH,5,9},AxesOrigin-
>{5,-1},AxesLabel->{"pH","\!\(N\_H\)"}];
```

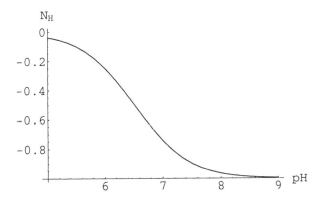

Since there is a decrease in binding of hydrogen ions in this reaction, hydrogen ions are produced at constant pH.

■ 9.5 Calculation of standard transformed reaction Gibbs energies at other temperatures

The program calcdGHT can be used to produce the function of T, pH, and ionic strength that will give the standard transformed Gibbs energies of formation at other temperatures, provided the standard enthalpies of formation of all the species are known.

```
calcdGHT[speciesmat_] := Module[{dGzero, dGzeroT, dHzero,
    zi, nH, gibbscoeff, pHterm, isterm, gpfnsp, dGfn, dHfn},(*This program produces
the function of T (in Kelvin), pH, and ionic strength (is) that gives the stndard
transformed Gibbs energy of formation of a reactant (sum of species) and the standard
transformed enthalpy.  The input speciesmat is a matrix that gives the standard Gibbs
energy of formation at 198.15 K, the standard enthalpy of formation at 298.15 K, the
electric charge, and the number of hydrogen atoms in each species.  There is a row in
the matrix for each species of the reactant.  gpfnsp is a list of the functions for
the transformed Gibbs energies of the species.  The output is in the form {dGfn,dHfn},
and the energies are expressed in kJ mol^-1.  The values of the standard transformed
Gibbs energy of formation and the standard transformed enthalpy of formation can be
calculated at any temperature in the range 273.15 K to 313.15 K, any pH in the range 5
to 9, and any ionic strength in the range 0 to 0.35 M by use of the assignment
operator (/.).*)
    {dGzero, dHzero, zi, nH} = Transpose[speciesmat];
    gibbscoeff = (9.20483*t)/10^3 - (1.284668*t^2)/10^5 +
      (4.95199*t^3)/10^8; dGzeroT = (dGzero*t)/298.15 +
      dHzero*(1 - t/298.15); pHterm =
    (nH*8.31451*t*Log[10^(-pH)])/1000;
    istermG = (gibbscoeff*(zi^2 - nH)*is^0.5)/
      (1 + 1.6*is^0.5); gpfnsp = dGzeroT - pHterm - istermG;
    dGfn = -((8.31451*t*Log[Plus @@
         (E^(-(gpfnsp/((8.31451*t)/1000)))))])/1000);
    dHfn = -(t^2*D[dGfn/t, t]); {dGfn, dHfn}]
```

Calculate the functions that yield the standard transformed Gibbs energies of the reactants in the reaction atp + h2o = adp + pi at 313.15 K.

```
atp313 = calcdGHT[atpsp] [[1]] /. t → 313.15;

h2o313 = calcdGHT[h2osp] [[1]] /. t → 313.15;

adp313 = calcdGHT[adpsp] [[1]] /. t → 313.15;

pi313 = calcdGHT[pisp] [[1]] /. t → 313.15;
```

Calculate the standard transformed reaction Gibbs energies for the hydrolysis of ATP at 313.15 K, pHs 5, 6, 7, 8, and 9 and ionic strengths of 0, 0.10, and 0.25 M.

```
dGerx313 = calctrGerx[atp313 + h2o313 + de == adp313 + pi313, {5, 6, 7, 8, 9}, {0, 0.1, 0.25}];

TableForm[Transpose[dGerx313], TableHeadings →
  {{"I = 0 M", "I = 0.10 M", "I = 0.25 M"}, {"pH 5", "pH 6", "pH 7", "pH 8", "pH 9"}}]
```

	pH 5	pH 6	pH 7	pH 8	pH 9
I = 0 M	-35.9046	-36.4876	-38.272	-43.5265	-49.6803
I = 0.10 M	-33.6656	-34.2845	-37.1598	-42.482	-48.4002
I = 0.25 M	-32.8677	-33.5895	-36.6852	-42.0518	-47.9736

■ 9.6 Calculation of the standard transformed Gibbs energies of formation of the species of a reactant from the apparent equilibrium constant of a reaction

To do this all the reactants have to be in the database, except for one or two, The following three programs, which make this calculation in one step, are based on the concept of the inverse Legendre transform. The program to be used depends on the number of species in the reactant. The programs produce entries for the database on species.

```
calcGef1sp[equat_, pHc_, ionstr_, z1_, nH1_] :=
    Module[{energy, trGereactant},(*This program uses ∑viΔfGi'°=-RTlnK' to calculate the
standard Gibbs energy of formation of the species of a reactant that does not have a
pK in the range 4 to 10. The equation is of the form pyruvate+atp-x-
adp==-8.31451*.29815*Log[K'], where K' is the apparent equilibrium constant at 298.15
K, pHc, and ionic strength is.  The reactant has charge number z1 and hydrogen atom
number nH1.  The output is the species vector without the standard enthalpy of
formation.*)
    energy = Solve[equat, x] /. pH -> pHc /. is -> ionstr;
     trGereactant = energy[[1,1,2]];
     gef1 = trGereactant - nH1*8.31451*0.29815*Log[10]*pHc +
        (2.91482*(z1^2 - nH1)*ionstr^0.5)/
        (1 + 1.6*ionstr^0.5); {{gef1, _, z1, nH1}}]

calcGef2sp[equat_, pHc_, ionstr_, z1_, nH1_, pK0_] :=
    Module[{energy, trGereactant, pKe, trgefpHis,gef1, gef2},(*This program uses ∑viΔ
fGi'°=-RTlnK' to calculate the standard Gibbs energies of formation of the two species
of a reactant for which the pK at zero ionic strength is pK0. The equation is of the
form pyruvate+atp-x-adp==-8.31451*.29815*Log[K'], where K' is the apparent equilibrium
constant at 298.15 K, pHc, and ionic strength is.  The more basic form of the reactant
has charge number z1 and hydrogen atom number nH1.  The output is the species matrix
without the standard enthalpies of formation.*)
    energy = Solve[equat, x] /. pH -> pHc /. is -> ionstr;
     trGereactant = energy[[1,1,2]];
     pKe = pK0 + (0.510651*ionstr^0.5*2*z1)/
        (1 + 1.6*ionstr^0.5); trgefpHis =
      trGereactant + 8.31451*0.29815*Log[1 + 10^(pKe - pHc)];
     gef1 = trgefpHis - nH1*8.31451*0.29815*Log[10]*pHc +
        (2.91482*(z1^2 - nH1)*ionstr^0.5)/
        (1 + 1.6*ionstr^0.5);
     gef2 = gef1 + 8.31451*0.29815*Log[10^(-pK0)];
     {{gef1, _, z1, nH1}, {gef2, _, z1 + 1, nH1 + 1}}]

calcGef3sp[equat_, pHc_, ionstr_, z1_, nH1_, pK10_,
    pK20_] := Module[{energy, trGereactant, pKe, trgefpHis,
    gef1, gef2, gef3, pK1e, pK2e},(*This program uses ∑viΔfGi'°=-RTlnK' to calculate
the standard Gibbs energies of formation of the three species of a reactant for which
the pKs at zero ionic strength is pK10 and pK20. The equation is of the form
pyruvate+atp-x-adp==-8.31451*.29815*Log[K'], where K' is the apparent equilibrium
constant at 298.15 K, pHc, and ionic strength is.  The more basic form of the reactant
has charge number z1 and hydrogen atom number nH1.  The output is the species matrix
without the standard enthalpies of formation of the three species.*)
    energy = Solve[equat, x] /. pH -> pHc /. is -> ionstr;
     trGereactant = energy[[1,1,2]];
     pK1e = pK10 + (0.510651*ionstr^0.5*2*z1)/
        (1 + 1.6*ionstr^0.5);
     pK2e = pK20 + (0.510651*ionstr^0.5*(2*z1 + 2))/
        (1 + 1.6*ionstr^0.5); trgefpHis =
      trGereactant + 8.31451*0.29815*
        Log[1 + 10^(pK1e - pHc) + 10^(pK1e + pK2e - 2*pHc)];
     gef1 = trgefpHis - nH1*8.31451*0.29815*Log[10]*pHc +
        (2.91482*(z1^2 - nH1)*ionstr^0.5)/
        (1 + 1.6*ionstr^0.5);
     gef2 = gef1 + 8.31451*0.29815*Log[10^(-pK10)];
     gef3 = gef2 + 8.31451*0.29815*Log[10^(-pK20)];
     {{gef1, _, z1, nH1}, {gef2, _, z1 + 1, nH1 + 1},
      {gef3, _, z1 + 2, nH1 + 2}}]
```

The following example is concerned with biochemical reaction EC 1.1.1.37. If the standard Gibbs energies of formation of both coA and acetyl coA are unknown, the convention can be adopted that the standard Gibbs energy of formation of RS^- is zero. The standard Gibbs energy of formatin of RSH can be calculated using the pK at zero ionic strength.

```
coA2sp = {{0, _, -1, 0}, {-47.83, _, 0, 1}};
```

```
coA2 = calcdGmat[coA2sp];
```

Now the data entry for acetylcoA can be calculated from the apparent equilibrium constant (10.8) of this reaction at pH 7.12 and ionic strength 0.05 M.

```
acetylcoAsp2 = calcGef1sp[
  citrate + coA2 + nadred - malate - x - nadox - h2o == -8.31451 0.29815 Log[10.8], 7.12, 0.05, 0,
{{-188.523, _, 0, 3}}

acetylcoA2 = calcdGmat[acetylcoAsp2];
```

This calculation can be verified by using the data on coA and acetylcoA to calculate the apparent equilibrium constant for this reaction at the experimental conditions.

```
calckprime[malate + acetylcoA2 + nadox + h2o + de == citrate + coA2 + nadred, 7.12, 0.05]

10.8
```

■ 9.7 Calculation of the thermodynamic properties of a biochemical reaction

It is convenient to be able to calculate the standard transformed reaction Gibbs energy, apparent equilibrium constant, and change in binding of hydrogen ions in a biochemical reaction at a series of pHs and ionic strengths. The pH dependencies of the standard transformed reaction Gibbs energies and K' can be regarded as a consequence of the change in binding of hydrogen ions.

```
calcNHrx[eq_, pHlist_, islist_] := Module[{energy},(*This program calculates the
change in the binding of hydrogen ions in a biochemical reaaction at specified pHs and
ionic strengths.*)
   energy = Solve[eq, de];
    D[energy[[1,1,2]], pH]/(8.31451*0.29815*Log[10]) /.
     pH -> pHlist /. is -> islist]

rxthermotab[eq_, pHlist_, islist_] := Module[{energy, tg, tk, tn},
   (*This program uses three other programs to make a thermodynamic table of
     standard transformed reaction Gibbs energies, apparent equilibrium constants,
     and changes in the number of hydrogen ions bound in a biochemical reaction.*)
   tg = calctrGerx[eq, pHlist, islist]; tk = calckprime[eq, pHlist, islist];
   tn = calcNHrx[eq, pHlist, islist]; TableForm[Join[{tg, tk, tn}]]]

rxthermotab[pep + adp + de == pyruvate + atp, {5, 6, 7, 8, 9}, {0, 0.1, 0.25}]
```

```
-33.4613      -32.7661      -30.6224      -25.1722      -19.2526
-34.1841      -33.1449      -29.2617      -23.6955      -17.993
-34.4724      -33.1063      -28.8451      -23.2908      -17.5968

728017.       549974.       231625.       25700.8       2359.9
974465.       640772.       133784.       14166.3       1419.79
           6
1.09465 10    630867.       113085.       12032.5       1210.04

0.150557      0.151752      0.700597      1.06297       1.01353
0.0719355     0.38045       0.915575      0.99698       0.999809
0.0826948     0.480624      0.927296      0.993667      0.999384
```

The first row gives the standard transformed reaction Gibbs energies in kJ mol^{-1} at zero ionic strength, the second row at 0.10 M ionic strength, and the third at 0.25 M ionic strength. The second part of the tsable is made up of apparent equilibrium constants, and the third is made up of changes in the binding of hydrogen ions. TableHeadings could be added, but this program allows you to use various numbers of pHs and various numbers of ionic strengths.

A vector can be rounded by use of round.

```
round[vec_, params_:{4, 2}] :=
  Flatten[Map[NumberForm[#1, params] & , {vec}, {2}]]

round[calctrGerx[glucose + atp + de == glucose6phos + adp,
  {5, 6, 7, 8, 9}, 0.25], {4, 2}]

{-17.41, -19.47, -24.42, -30.11, -35.82}
```

■ 9.8 Printing out biochemical reactions that correspond with a stoichiometric matrix

A biochemical reaction can be represented by a vector of its stoichiometric numbers. A system of biochemical reactions is represented by a stoichiometric number matrix. This stoichiometric number matrix can be used to print out the reactions. The programs that can be used to print out the biochemical reactions are mkeqm and nameMatrix.

```
mkeqm[c_List,s_List]:=(*c_List is the list of stoichiometric numbers for a reaction.
s_List is a list of the names of species or reactants.  These names have to be put in
quotation marks.*)Map[Max[#,0]&,-c].s->Map[Max[#,0]&,c].s

nameMatrix[m_List,s_List]:=(*m_List is the transposed stoichiometric number matrix for
the system of reactions. s_List is a list of the names of species or reactants.  These
names have to be put in quotation marks.*)Map[mkeqm[#,s]&,m]
```

The first three reactions of glycolysis are

ATP + glucose = ADP + G6P

G6P = F6P

ATP + F6P = ADP + F16P

This system of reactions is represented by the following stoichiometric number matrix.

```
nu = {{-1, 0, 0}, {-1, 0, -1}, {1, -1, 0}, {1, 0, 1}, {0, 1, -1}, {0, 0, 1}};

TableForm[nu]

-1   0    0
-1   0    -1
1    -1   0
1    0    1
0    1    -1
0    0    1

names3 = {"Glc", "ATP", "G6P", "ADP", "F6P", "F16BP"};
```

```
TableForm[nu, TableHeadings -> {names3, {"rx 13", "rx 14", "rx 15"}}]
```

	rx 13	rx 14	rx 15
Glc	-1	0	0
ATP	-1	0	-1
G6P	1	-1	0
ADP	1	0	1
F6P	0	1	-1
F16BP	0	0	1

```
mkeqm[{-1, -1, 1, 1, 0, 0}, names3]
```

ATP + Glc -> ADP + G6P

```
nameMatrix[Transpose[nu], names3]
```

{ATP + Glc -> ADP + G6P, G6P -> F6P, ATP + F6P -> ADP + F16BP}

■ 9.9 Calculation of standard apparent reduction potentials for half reactions

The standard apparent reduction potential in volts at 298.15 K for a half reaction can be calculated using calcappredpots. As an example of a half reaction consider $NAD_{ox} + 2\ e^- = NAD_{red}$:

```
calcappredpot[eq_, nu_, pHlist_, islist_] :=
  Module[{energy},(*Calculates the standard apparent reduction potential of a half
reaction at specified pHs and ionic strengths for a biochemical half reaction typed in
the form nadox+de==nadred.  The names of the reactants call the corresponding
functions of pH and ionic strength.  nu is the number of electrons involved.  pHlist
and islist can be lists.*)
  energy = Solve[eq, de];
    -(energy[[1,1,2]]/(nu*96.485)) /. pH -> pHlist /.
    is -> islist]
```

```
TableForm[Transpose[calcappredpot[nadox + de == nadred, 2, {5, 6, 7, 8, 9}, {0, 0.1, 0.25}]],
  TableHeadings -> {{"I=0 M", "I=0.10 M", "I=0.25 M"},
  {"pH 5", "pH 6", "pH 7", "pH 8", "pH 9"}}, TableSpacing -> {1, 1}]
```

	pH 5	pH 6	pH 7	pH 8	pH 9
I=0 M	-0.265275	-0.294855	-0.324435	-0.354015	-0.383595
I=0.10 M	-0.258932	-0.288512	-0.318092	-0.347672	-0.377252
I=0.25 M	-0.256884	-0.286464	-0.316044	-0.345624	-0.375204

```
Plot[calcappredpot[nadox+de==nadred,2,pH,.25],{pH,5,9},AxesLabel->{"pH","E'/V"}];
```

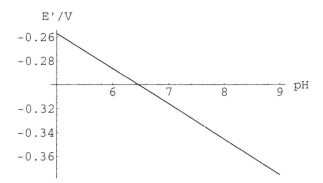

∎ 9.10 Calculation of equilibrium compositions of a single biochemical reaction or a system of biochemical reactions at specified pH

Equilibrium compositions of systems of biochemical reactions can be calculated using the following two programs. The first was written by Fred Krambeck (Mobil Research and Development) and the second was written by Krambeck and Alberty. The Newton-Raphson method is used to iterate to the composition with the lowest possible Gibbs energy or transformed Gibbs energy.

```
equcalcc[as_,lnk_,no_]:=Module[{l,x,b,ac,m,n,e,k},
(*  as=conservation matrix
lnk=-(1/RT)(Gibbs energy of formation vector at T)
no=initial composition vector *)
(*Setup*)
{m,n}=Dimensions[as];
b=as.no;
ac=as;
(*Initialize*)
l=LinearSolve[ as.Transpose[as],-as.(lnk+Log[n]) ];
(*Solve*)
Do[ e=b-ac.(x=E^(lnk+l.as) );
If[(10^-10)>Max[ Abs[e] ], Break[] ];
l=l+LinearSolve[ac.Transpose[as*Table[x,{m}]],e],
{k,100}];
If[ k=100,Return["Algorithm Failed"] ];
Return[x]
]

equcalcrx[nt_,lnkr_,no_]:=Module[{as,lnk},
(*nt=transposed stoichiometric number matrix
lnkr=ln of equilibrium constants of rxs (vector)
no=initial composition vector*)
(*Setup*)
lnk=LinearSolve[nt,lnkr];
as=NullSpace[nt];
equcalcc[as,lnk,no]
]
```

The reaction considered here is the hydrolysis of glucose 6-phosphate to glucose and inorganic hosphate at 298.15 K, pH 7, and 0.25 M ionic strength. The objective is to calculate the equilibrium concentrations when the enzyme is added to a 0.10 M solution of glucose 6-phosphate.

G6P + H_2O = Glu + P_i

This is a single reaction, and so the equilibrium comosition can be readily calculated without a computer program. However, it involves an important issue that often arises in enzyme-catalyzed reactions, and that is the problem of water as a reactant in dilute aqueous solutions. The problem is that the activity of water is taken as unity by convention, and so its concentration does not appear in the expression for the apparent equilibrium constant. The conservation matrix at constant pH is

		G6P	H_2O	Glu	P_i
A' =	C	6	0	6	0
	O	9	1	6	4
	P	1	0	0	1

Since the activity of water is taken as unity in dilute aqueous solutions independent of the extent of reaction, the H_2O column and the oxygen row must be deleted from the conservation matrix:

		G6P	Glu	P_i
A' =	C	6	6	0
	P	1	0	1

Therefore the conservation matrix is given by

```
as = {{6, 6, 0}, {1, 0, 1}};
```

The standard transformed Gibbs energies of the reactants at 298.15 K, pH 7, and 0.25 M ionic strength are

```
glucose6phos /. pH → 7 /. is → 0.25
```

-1318.92

```
glucose /. pH → 7 /. is → 0.25
```

-426.708

```
pi /. pH → 7 /. is → 0.25
```

-1059.49

```
h2o /. pH → 7 /. is → 0.25
```

-155.658

The standard further transformed Gibbs energies of formation in kJ mol^{-1} are

```
glucose6phosft = -1318.92 - 9 (-155.66)
```

82.02

```
glucoseft = -426.71 - 6 (-155.66)
```

507.25

```
pift = -1059.49 - 4 (-155.66)
```

-436.85

The equilibrium composition is calculated using

$$equcalcc\left[as, -\frac{\{82.02, 507.25, -436.85\}}{8.31451 \, 0.29815}, \{0.1, 0, 0\}\right]$$

{0.0000919341, 0.0999081, 0.0999081}

It is more convenient to use equcalcrx because it takes a stoichiometric number matrix and a vector of the apparent equilibrium constants for a set of independent reactions in the system. Note that this program calculates a consevation matrix that is consistent with the stoichiometric number matrix, and uses it in equcalc. The transposed stoichiometric number matrix nt for the reactin without H_2O is given by

```
nt = {{-1, 1, 1}};
```

The apparent equilibrium constant is given by

```
calckprime[glucose6phos + h2o + de == glucose + pi, 7, 0.25]
```

```
108.375
```

The equilibrium composition is calculated using

```
equcalcrx[nt, {Log[108.4]}, {0.1, 0, 0}]
```

```
{0.0000920811, 0.0999079, 0.0999079}
```

The equilibrium compostion can be verified by calculating the apparent equilibrium constant with the calculated composition

```
E^(nt . Log[{9.208/10^5, 9.991/10^2, 9.991/10^2}])
```

```
{108.406}
```

A single enzyme-catalyzed reaction has been used in this example, but a system of reactions can be handled in the same way. The advantage of using matrices and vectors in a computer program is that the same program can be used for large systems of reactions.

```
End[];
```

```
EndPackage[];
```

Chapter 1 Apparent Equilibrium Constants

1.1 Plot the fractions r_i of ATP in the forms ATP^{4-}, $HATP^{3-}$, and ATP^{2-} versus pH at 298.15 K and 0.25 M ionic strength.

1.2 Plot the average binding $\overline{N}_H$ of hydrogen ions by ATP at 298.15 K and 0.25 M ionic strength as a function of pH. Show that plotting equation 1.3-13 yields the same result as plotting 1.3-7.

1.3 Plot the average binding $\overline{N}_H$ of hydrogen ions by ATP at 298.15 K and 0.25 M ionic strength as a function of pH at pMg = 2, 3 4, 5, and 6.

1.4 Plot the average binding $\overline{N}_{Mg}$ of magnesium ions by ATP at 298.15 K and 0.25 M ionic strength as a function of pMg at pH = 3 4, 5, 6,7, 8, and 9.

1.5 Plot the average binding $\overline{N}_H$ of hydrogen ions by ATP at 298.15 K and 0.25 M ionic strength versus pH and pMg. Also plot the rate of change of $\overline{N}_H$ with pMg for comparison with Problem 1.6 to verify the reciprocity relation.

1.6 Plot the average binding $\overline{N}_{Mg}$ of magnesium ions by ATP at 298.15 K and 0.25 M ionic strength versus pH and pMg. Also plot the rate of change of $\overline{N}_{Mg}$ with pH for comparison with Problem 1.5 to verify the reciprocity relation..

1.7 Plot (a) the base 10 logarithm of the apparent equilibrium constant K' and (b) -RTlnK' in kJ mol^{-1} for ATP + H_2O = ADP + P_i versus pH and pMg at 298.15 K and 0.25 M ionic strength.

1.8 Plot the change in the binding of hydrogen ions $\Delta_r N(H^+)$ in ATP + H_2O = ADP + P_i versus pH and pMg at 298.15 K and 0.25 M ionic strength.

1.9 Plot the change in the binding of magnesium ions $\Delta_r N(Mg^{2+})$ in ATP + H_2O = ADP + P_i versus pH and pMg at 298.15 K and 0.25 M ionic strength.

1.10 Calculate the acid pKs at 298.15 K and ionic strengths of 0, 0.05, 0.10, 0.15, 0.20, and 0.25 M for all of the acids for which data are given in the table BasicBiochemData.

1.11 Plot the pKs of acetate, ammonia, atp, and pyrophosphate versus ionic strength from I = 0 to I = 0.3 M at 298.15 K. Biochemists are usually only concerned with the pHs in the pH 5 to 9 range.

1.12 (a) Calculate the acid titration curve for ATP at 298.15 K and 0.25 M ionic strength from the binding polynomial P. (b) Integrate the calculated binding curve to obtain ln P plus a constant of integration. The needed equations are
N_H = (-1/ln(10))(dlnP/dpH)
$-\ln(10) \int N_H dpH = \ln P + const$

1.13 (a) Test the differentiation of ln P to obtain the equation for the binding N_H of hydrogen ions by ATP at 298.15 K and ionic strength 0.25 M in the region pH 2 to 10. (b) Test the integration of the equation for N_H to obtain the equation for ln P. (c) Plot ln p versus pH and N_H versus pH. The equations involved are
$N_H = [H^+] \frac{dlnp}{d[H^+]}$
$\int \frac{N_H}{[H^+]} d[H^+] = \ln P + const$

1.1 Plot the equilibrium mole fractions r_i of ATP in the forms ATP^{4-}, $HATP^{3-}$, and ATP^{2-} versus pH at 298.15 K and 0.25 M ionic strength.

```
k1ATP = 3.43 * 10 ^ -7;

k2ATP = .000148;
```

Type in the binding polynomial.

```
p = 1 + (10 ^ -pH) / k1ATP + (10 ^ (-2 * pH)) / (k1ATP * k2ATP)
```

$$1 + \frac{1.9699 \ 10^{10}}{10^{2\ pH}} + \frac{2.91545 \ 10^{6}}{10^{pH}}$$

```
Plot[{1 / p, (10 ^ -pH) / (k1ATP * p), (10 ^ (-2 * pH)) / ((k1ATP * k2ATP) * p)},
  {pH, 3, 9}, AxesLabel → {"pH", "r_i"}];
```

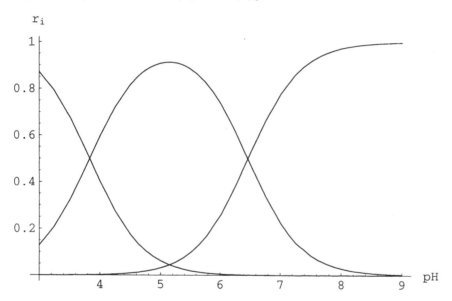

1.2 Plot the average binding $\overline{N}_H$ of hydrogen ions by ATP at 298.15 K and 0.25 M ionic strength as a function of pH. Show that plotting equation 1.3-13 yields the same result as plotting 1.3-7.

```
k1ATP = 3.43 * 10 ^ -7;
k2ATP = 0.000148;

nH = ((10 ^ -pH) / k1ATP + 2 * (10 ^ (-2 * pH)) / (k1ATP * k2ATP)) /
  (1 + (10 ^ -pH) / k1ATP + (10 ^ (-2 * pH)) / (k1ATP * k2ATP))
```

$$\frac{\dfrac{3.9398 \ 10^{10}}{10^{2\ pH}} + \dfrac{2.91545 \ 10^{6}}{10^{pH}}}{1 + \dfrac{1.9699 \ 10^{10}}{10^{2\ pH}} + \dfrac{2.91545 \ 10^{6}}{10^{pH}}}$$

```
Plot[nH, {pH, 3, 9}, AxesLabel → {"pH", "N_H"}];
```

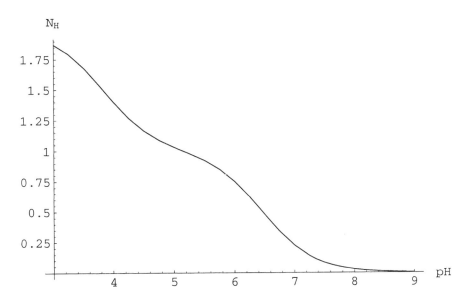

Type in the binding polynomial.

p = 1 + (10 ^ -pH) / k1ATP + 2 * (10 ^ (-2 * pH)) / (k1ATP * k2ATP)

$$1 + \frac{3.9398 \ 10^{10}}{10^{2 \ pH}} + \frac{2.91545 \ 10^{6}}{10^{pH}}$$

Use equation 1.3-7.

nHp = -D[Log[p], pH] / Log[10]

$$-\left(\frac{\dfrac{-9.07172 \ 10^{10} \ 2^{1 \ - \ 2 \ pH}}{5^{2 \ pH}} - \dfrac{6.71308 \ 10^{6}}{10^{pH}}}{\left(1 + \dfrac{3.9398 \ 10^{10}}{10^{2 \ pH}} + \dfrac{2.91545 \ 10^{6}}{10^{pH}}\right) \ Log[10]} \right)$$

Plot[nHp, {pH, 3, 9}, AxesLabel → {"pH", "N$_H$"}];

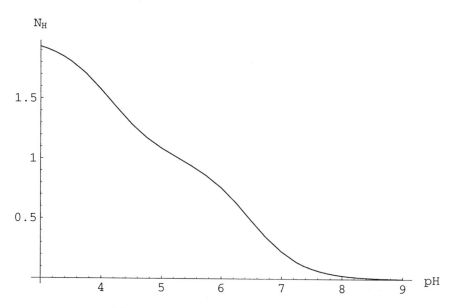

1.3 Plot the average binding $\overline{N}_H$ of hydrogen ions by ATP at 298.15 K and 0.25 M ionic strength as a function of pH at pMg = 2, 3 4, 5, and 6.

```
k1ATP = 3.43 * 10 ^ -7;
k2ATP = 0.000148;
k3ATP = 0.000122;
k4ATP = 0.0118;
k5ATP = 0.0279;
```

The binding polynomial is given by

```
p = 1 + (10 ^ -pH) / k1ATP + (10 ^ (-2 * pH)) / (k1ATP * k2ATP) + (10 ^ -pMg) / k3ATP +
    ((10 ^ -pH) * (10 ^ -pMg)) / (k1ATP * k4ATP) + (10 ^ (-2 * pMg)) / (k3ATP * k5ATP)
```

$$1 + \frac{1.9699 \; 10^{10}}{10^{2 \; pH}} + \frac{2.91545 \; 10^{6}}{10^{pH}} + 2.47072 \; 10^{8} \; 10^{-pH \; - \; pMg} + \frac{293789.}{10^{2 \; pMg}} + \frac{8196.72}{10^{pMg}}$$

The average binding of hydrogen ions is given by

```
nH = -D[Log[p], pH] / Log[10]
```

$$-\left(\frac{\dfrac{-4.53586 \; 10^{10} \; 2^{1 \; - \; 2 \; pH}}{5^{2 \; pH}} - \dfrac{6.71308 \; 10^{6}}{10^{pH}} - 5.68905 \; 10^{8} \; 10^{-pH \; - \; pMg}}{\left(1 + \dfrac{1.9699 \; 10^{10}}{10^{2 \; pH}} + \dfrac{2.91545 \; 10^{6}}{10^{pH}} + 2.47072 \; 10^{8} \; 10^{-pH \; - \; pMg} + \dfrac{293789.}{10^{2 \; pMg}} + \dfrac{8196.72}{10^{pMg}}\right) \; Log[10]} \right)$$

```
Plot[Evaluate[nH /. {pMg → {2, 3, 4, 5, 6}}], {pH, 3, 9}, AxesLabel → {"pH", "N_H"}];
```

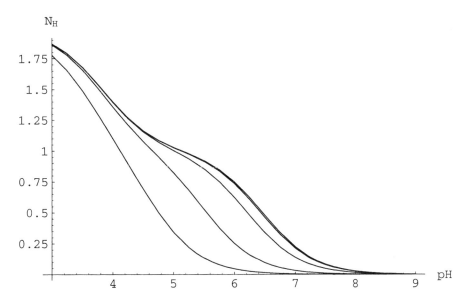

1.4 Plot the average binding $\overline{N}_{Mg}$ of magnesium ions by ATP at 298.15 K and 0.25 M ionic strength as a function of pMg at pH = 3, 4, 5, 6,7, 8, and 9.

```
k1ATP = 3.43 * 10 ^ -7;
k2ATP = 0.000148;
k3ATP = 0.000122;
k4ATP = 0.0118;
k5ATP = 0.0279;
```

The binding polynomial is given by

```
p = 1 + (10 ^ -pH) / k1ATP + (10 ^ (-2 * pH)) / (k1ATP * k2ATP) + (10 ^ -pMg) / k3ATP +
    ((10 ^ -pH) * (10 ^ -pMg)) / (k1ATP * k4ATP) + (10 ^ (-2 * pMg)) / (k3ATP * k5ATP)
```

$$1 + \frac{1.9699\ 10^{10}}{10^{2\ pH}} + \frac{2.91545\ 10^{6}}{10^{pH}} + 2.47072\ 10^{8}\ 10^{-pH\ -\ pMg} +$$

$$\frac{293789.}{10^{2\ pMg}} + \frac{8196.72}{10^{pMg}}$$

The binding of magnesium ions is given by

```
nMg = -D[Log[p], pMg] / Log[10]
```

$$-\left(\left(\frac{-676475.\ 2^{1-2\,pMg}}{5^{2\,pMg}} - 5.68905\ 10^{8}\ 10^{-pH-pMg} - \frac{18873.6}{10^{pMg}}\right)\ /\right.$$

$$\left(\left(1 + \frac{1.9699\ 10^{10}}{10^{2\,pH}} + \frac{2.91545\ 10^{6}}{10^{pH}} + \right.\right.$$

$$\left.\left.\left. 2.47072\ 10^{8}\ 10^{-pH-pMg} + \frac{293789.}{10^{2\,pMg}} + \frac{8196.72}{10^{pMg}}\right)\ \mathrm{Log[10]}\right.\right.$$

$$\left.\left.\right)\right)$$

```
Plot[Evaluate[nMg /. {pH → {3, 4, 5, 6, 7, 8, 9}}],
  {pMg, 2, 7}, AxesLabel → {"pMg", "N_Mg"}, PlotRange → {0, 1.5}];
```

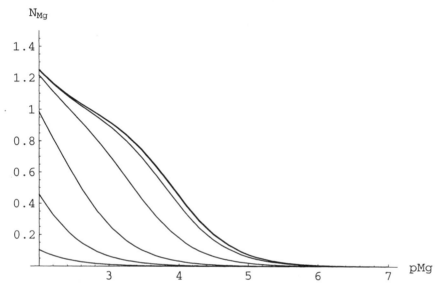

1.5 Plot the average binding $\overline{N}_H$ of hydrogen ions by ATP at 298.15 K and 0.25 M ionic strength versus pH and pMg. Also plot the rate of change of $\overline{N}_H$ with pMg for comparison with Problem 1.6 to verify the reciprocity relation.

```
k1ATP = 3.43 * 10 ^ -7;
k2ATP = 0.000148;
k3ATP = 0.000122;
k4ATP = 0.0118;
k5ATP = 0.0279;
```

The binding polynomial is given by

```
p = 1 + (10 ^ -pH) / k1ATP + (10 ^ (-2 * pH)) / (k1ATP * k2ATP) + (10 ^ -pMg) / k3ATP +
  ((10 ^ -pH) * (10 ^ -pMg)) / (k1ATP * k4ATP) + (10 ^ (-2 * pMg)) / (k3ATP * k5ATP)
```

$$1 + \frac{1.9699\ 10^{10}}{10^{2\,pH}} + \frac{2.91545\ 10^{6}}{10^{pH}} + 2.47072\ 10^{8}\ 10^{-pH-pMg} +$$

$$\frac{293789.}{10^{2\,pMg}} + \frac{8196.72}{10^{pMg}}$$

The binding of hydrogen ions is given by

```
nH = -D[Log[p], pH] / Log[10]
```

$$-((\frac{-4.53586 \ 10^{10} \ 2^{1 \ - \ 2 \ pH}}{5^{2 \ pH}} - \frac{6.71308 \ 10^6}{10^{pH}} -$$

$$5.68905 \ 10^8 \ 10^{-pH \ - \ pMg}) \ /$$

$$((1 + \frac{1.9699 \ 10^{10}}{10^{2 \ pH}} + \frac{2.91545 \ 10^6}{10^{pH}} +$$

$$2.47072 \ 10^8 \ 10^{-pH \ - \ pMg} + \frac{293789.}{10^{2 \ pMg}} + \frac{8196.72}{10^{pMg}}) \ Log[10]$$

$$))$$

```
Plot3D[nH, {pH, 3, 9}, {pMg, 1, 6}, AxesLabel → {"pH", "pMg", "N_H"}];
```

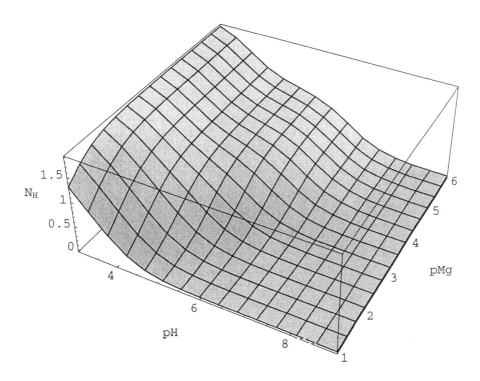

```
nHMg = -D[nH, pMg] / Log[10];
```

```
Plot3D[nHMg, {pH, 3, 9}, {pMg, 1, 6}, AxesLabel → {"pH", "pMg", "N_HMg"}];
```

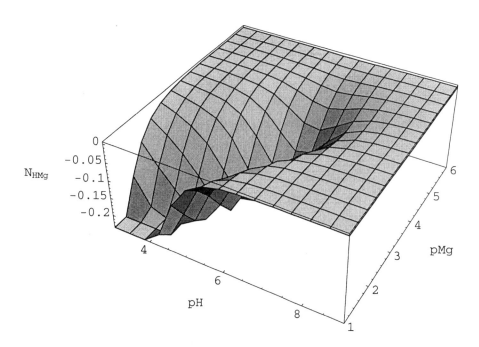

1.6 Plot the average binding $\overline{N}_{Mg}$ of magnesium ions by ATP at 298.15 K and 0.25 M ionic strength versus pH and pMg. Also plot the rate of change of $\overline{N}_{Mg}$ with pH for comparison with Problem 1.5 to verify the reciprocity relation.

```
k1ATP = 3.43 * 10 ^ -7;
k2ATP = 0.000148;
k3ATP = 0.000122;
k4ATP = 0.0118;
k5ATP = 0.0279;
```

The binding polynomial is given by

```
p = 1 + (10 ^ -pH) / k1ATP + (10 ^ (-2 * pH)) / (k1ATP * k2ATP) + (10 ^ -pMg) / k3ATP +
   ((10 ^ -pH) * (10 ^ -pMg)) / (k1ATP * k4ATP) + (10 ^ (-2 * pMg)) / (k3ATP * k5ATP)
```

$$1 + \frac{1.9699 \; 10^{10}}{10^{2\,pH}} + \frac{2.91545 \; 10^{6}}{10^{pH}} + 2.47072 \; 10^{8} \; 10^{-pH \, - \, pMg} \, +$$

$$\frac{293789.}{10^{2\,pMg}} + \frac{8196.72}{10^{pMg}}$$

The binding of magnesium ions is given by

```
nMg = -D[Log[p], pMg] / Log[10]
```

$$-\left(\left(\frac{-676475.\ 2^{1\ -\ 2\ pMg}}{5^{2\ pMg}}\ -\ 5.68905\ 10^8\ 10^{-pH\ -\ pMg}\ -\right.\right.$$

$$\left.\frac{18873.6}{10^{pMg}}\right)\ /$$

$$\left(\left(1\ +\ \frac{1.9699\ 10^{10}}{10^{2\ pH}}\ +\ \frac{2.91545\ 10^6}{10^{pH}}\ +\right.\right.$$

$$\left.\left.2.47072\ 10^8\ 10^{-pH\ -\ pMg}\ +\ \frac{293789.}{10^{2\ pMg}}\ +\ \frac{8196.72}{10^{pMg}}\right)\ Log[10]\right)$$

$$\left.\right)$$

```
Plot3D[nMg, {pH, 3, 9}, {pMg, 1, 6}, AxesLabel → {"pH", "pMg", "N_Mg"}];
```

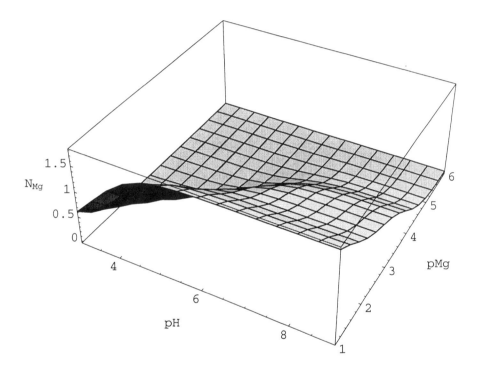

```
nMgH = -D[nMg, pH] / Log[10];

Plot3D[nMgH, {pH, 3, 9}, {pMg, 1, 6}, AxesLabel → {"pH", "pMg", "N_MgH    "}];
```

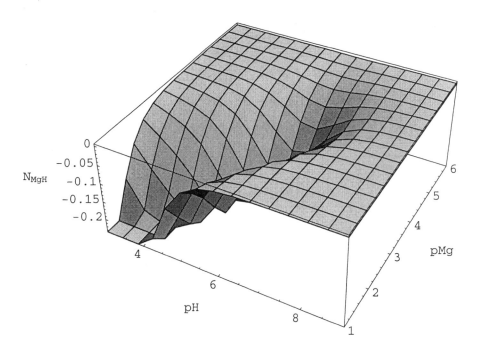

1.7 Plot (a) the base 10 logarithm of the apparent equilibrium constant K' and (b) -RTlnK' in kJ mol⁻¹ for ATP + H₂O = ADP + P_i versus pH and pMg at 298.15 K and 0.25 M ionic strength.

1.7 Plot (a) the base 10 logarithm of the apparent equilibrium constant K' and (b) $-RT\ln K'$ in kJ mol^{-1} for ATP + H_2O = ADP + P_i versus pH and pMg at 298.15 K and 0.25 M ionic strength.

```
k1ATP = 3.43 * 10 ^ -7;
k2ATP = .000148;
k3ATP = .000123;
k4ATP = .0118;
k5ATP = .0279;
k1ADP = 4.70 * 10 ^ -7;
k2ADP = .000161;
k3ADP = .00113;
k4ADP = .0431;
k1P = 2.24 * 10 ^ -7;
k2P = .0266;
kref = .222;

pATP = 1 + (10 ^ -pH) / k1ATP + (10 ^ (-2 * pH)) / (k1ATP * k2ATP) + (10 ^ -pMg) / k3ATP +
    ((10 ^ -pH) * (10 ^ -pMg)) / (k1ATP * k4ATP) + (10 ^ (-2 * pMg)) / (k3ATP * k5ATP);
pADP = 1 + (10 ^ -pH) / k1ADP + (10 ^ (-2 * pH)) / (k1ADP * k2ADP) +
    (10 ^ -pMg) / k3ADP + ((10 ^ -pH) * (10 ^ -pMg)) / (k1ADP * k4ADP);
pP = 1 + (10 ^ -pH) / k1P + (10 ^ -pMg) / k2P;
kapp = kref * pADP * pP / ((10 ^ -pH) * pATP);

Plot3D[Log[10, kapp], {pH, 3, 9}, {pMg, 1, 6}, AxesLabel → {"pH", "pMg", "logK'"}];
```

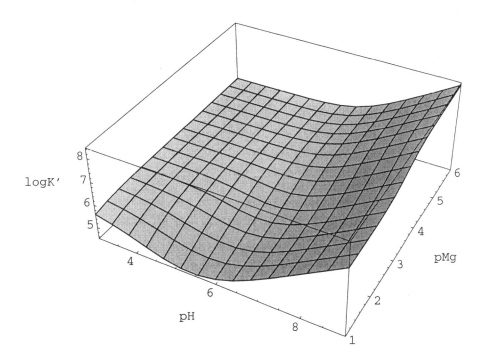

```
Plot3D[-8.31451 * .29815 * Log[kapp], {pH, 3, 9},
  {pMg, 1, 6}, AxesLabel → {"pH", "pMg", "-RTlnK'  "}];
```

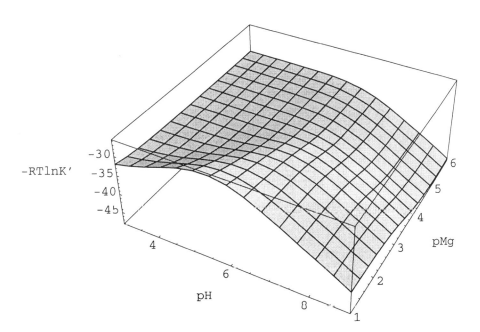

1.8 Plot the change in the binding of hydrogen ions $\Delta_r N(H^+)$ in ATP + H$_2$O = ADP + P$_i$ versus pH and pMg at 298.15 K and 0.25 M ionic strength.

```
k1ATP = 3.43 * 10 ^ -7;
k2ATP = .000148;
k3ATP = .000123;
k4ATP = .0118;
k5ATP = .0279;
k1ADP = 4.70 * 10 ^ -7;
k2ADP = .000161;
k3ADP = .00113;
k4ADP = .0431;
k1P = 2.24 * 10 ^ -7;
k2P = .0266;
kref = .222;

pATP = 1 + (10^-pH) / k1ATP + (10 ^ (-2 * pH)) / (k1ATP * k2ATP) + (10 ^ -pMg) / k3ATP +
    ((10 ^ -pH) * (10 ^ -pMg)) / (k1ATP * k4ATP) + (10 ^ (-2 * pMg)) / (k3ATP * k5ATP);
pADP = 1 + (10^-pH) / k1ADP + (10 ^ (-2 * pH)) / (k1ADP * k2ADP) +
    (10 ^ -pMg) / k3ADP + ((10 ^ -pH) * (10 ^ -pMg)) / (k1ADP * k4ADP);
pP = 1 + (10^-pH) / k1P + (10^-pMg) / k2P;
kapp = kref * pADP * pP / ((10^-pH) * pATP);

nHrx = -D[Log[kapp], pH] / Log[10];

Plot3D[nHrx, {pH, 3, 9}, {pMg, 1, 6}, AxesLabel → {"pH", "pMg", "Δ_r N ( H^+ )"}];
```

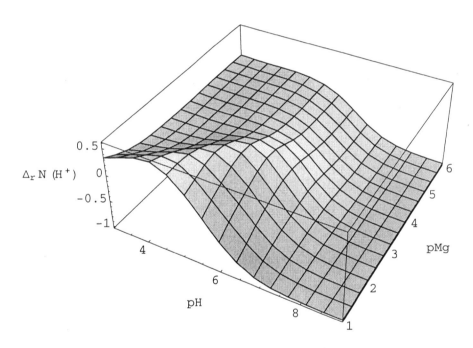

1.9 Plot the change in the binding of magnesium ions $\Delta_r N(Mg^{2+})$ in $ATP + H_2O = ADP + P_i$ versus pH and pMg at 298.15 K and 0.25 M ionic strength.

```
k1ATP = 3.43 * 10 ^ -7;
k2ATP = .000148;
k3ATP = .000123;
k4ATP = .0118;
k5ATP = .0279;
k1ADP = 4.70 * 10 ^ -7;
k2ADP = .000161;
k3ADP = .00113;
k4ADP = .0431;
k1P = 2.24 * 10 ^ -7;
k2P = .0266;
kref = .222;

pATP = 1 + (10 ^ -pH) / k1ATP + (10 ^ (-2 * pH)) / (k1ATP * k2ATP) + (10 ^ -pMg) / k3ATP +
    ((10 ^ -pH) * (10 ^ -pMg)) / (k1ATP * k4ATP) + (10 ^ (-2 * pMg)) / (k3ATP * k5ATP);
pADP = 1 + (10 ^ -pH) / k1ADP + (10 ^ (-2 * pH)) / (k1ADP * k2ADP) +
    (10 ^ -pMg) / k3ADP + ((10 ^ -pH) * (10 ^ -pMg)) / (k1ADP * k4ADP);
pP = 1 + (10 ^ -pH) / k1P + (10 ^ -pMg) / k2P;
kapp = kref * pADP * pP / ((10 ^ -pH) * pATP);

nMgrx = -D[Log[kapp], pMg] / Log[10];

Plot3D[nMgrx, {pH, 3, 9}, {pMg, 1, 6}, AxesLabel → {"pH", "pMg", "Δr N ( Mg 2+ )  "}];
```

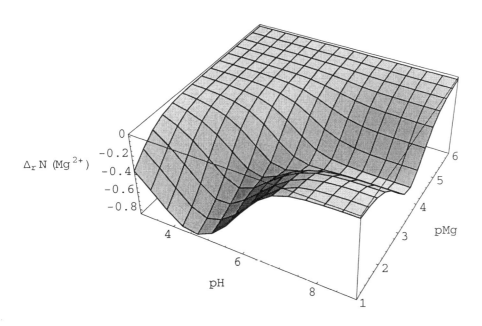

1.10 Calculate the acid pKs at 298.15 K and ionic strengths of 0, 0.05, 0.10, 0.15, 0.20, and 0.25 M for all the acids for which data are given in the package BasicBiochemData2.

(BasicBiochemData2 needs to be loaded.)

```
calcpK[speciesmat_, no_, is_] := Module[{lnkzero, sigmanuzsq, lnK},
  (*Calculates pKa for a weak acid at 298.15 K at specified ionic strengths
    (is) when the number no of the pK is specified.  pKs are numbered 1,
    2, 3,... from the highest pK to the lowest pK,
    but the highest pK for a weak acid may be omitted if it is outside of the
    physiological range pH 5 to 9.  For H3PO4, pK1=calcpK[pisp,1,{0}] = 7.22*)
  lnkzero = (speciesmat[[no + 1, 1]] - speciesmat[[no, 1]]) / (8.31451 * .29815);
  sigmanuzsq = speciesmat[[no, 3]] ^ 2 - speciesmat[[no + 1, 3]] ^ 2 + 1;
  lnK = lnkzero + (1.17582 * is ^ .5) * sigmanuzsq / (1 + 1.6 * is ^ .5);
  N[-lnK / Log[10]]
  ]
```

As a check on this program, pK1(I=0) for ATP is 7.60. sigmanusq is 16+1-9=8. pK1(I=0.25) can be calculated outside the program using

```
-(Log[10^-7.60]+(1.17582*is^.5)*8/(1+1.6*is^.5))/Log[10]/.is->.25

6.46522
```

```
names = {"acetatepK1", "acetylphospK1", "acetylphospK2", "adeninepK1", "adenosinepK1",
    "adppK1", "adppK2", "ammoniapK1", "amppK1", "amppK2", "atppK1", "atppK2", "bpgpK1",
    "citratepK1", "citratepK2", "citrateisopK1", "citrateisopK2", "carbondioxidepK1",
    "carbondioxidepK2", "coApK1", "cysteineLpK1", "dihydroxyacetonephospK1",
    "fructose6phospK1", "fructose16phospK1", "fructose16phospK2", "fumaratepK1",
    "fumaratepK2", "galactose1phospK1", "glucose6phospK1", "glucose1phospK1",
    "glutathioneredpK1", "glyceraldehydephospK1", "glycerol3phospK1",
    "malatepK1", "oxalatepK1", "phosphoenolpyruvateK1", "phosphoglycerate2pK1",
    "phosphoglycerate3pK1", "phosphatepK1", "pyrophosphatepK1", "pyrophosphatepK2",
    "pyrophosphatepK3", "ribose1phospK1", "ribose5phospK1", "succinatepK1",
    "succinatepK2", "succinylcoApK1", "thioredoxinredpK1", "thioredoxinredpK2"};
```

```
acetatepK1 = NumberForm[calcpK[acetatesp, 1, {0, .05, .1, .15, .2, .25}], 3];
```

```
acetylphospK1=NumberForm[calcpK[acetylphossp,1,{0,.05,.1,.15,.2,.25}],3];
```

```
acetylphospK2=NumberForm[calcpK[acetylphossp,2,{0,.05,.1,.15,.2,.25}],3];
```

```
adeninepK1=NumberForm[calcpK[adeninesp,1,{0,.05,.1,.15,.2,.25}],3];
```

```
adenosinepK1 = NumberForm[calcpK[adenosinesp, 1, {0, .05, .1, .15, .2, .25}], 3];
```

```
adppK1 = NumberForm[calcpK[adpsp, 1, {0, .05, .1, .15, .2, .25}], 3];
```

```
adppK2 = NumberForm[calcpK[adpsp, 2, {0, .05, .1, .15, .2, .25}], 3];
```

```
ammoniapK1 = NumberForm[calcpK[ammoniasp, 1, {0, .05, .1, .15, .2, .25}], 3];
```

```
amppK1 = NumberForm[calcpK[ampsp, 1, {0, .05, .1, .15, .2, .25}], 3];
```

```
amppK2 = NumberForm[calcpK[ampsp, 2, {0, .05, .1, .15, .2, .25}], 3];
```

```
atppK1 = NumberForm[calcpK[atpsp, 1, {0, .05, .1, .15, .2, .25}], 3];
```

```
atppK2 = NumberForm[calcpK[atpsp, 2, {0, .05, .1, .15, .2, .25}], 3];
```

```
bpgpK1=NumberForm[calcpK[bpgsp,1,{0,.05,.1,.15,.2,.25}],3];
```

```
citratepK1 = NumberForm[calcpK[citratesp, 1, {0, .05, .1, .15, .2, .25}], 3];

citratepK2 = NumberForm[calcpK[citratesp, 2, {0, .05, .1, .15, .2, .25}], 3];

citrateisopK1 = NumberForm[calcpK[citrateisosp, 1, {0, .05, .1, .15, .2, .25}], 3];

citrateisopK2 = NumberForm[calcpK[citratesp, 2, {0, .05, .1, .15, .2, .25}], 3];

carbondioxidepK1 = NumberForm[calcpK[co2totsp, 1, {0, .05, .1, .15, .2, .25}], 3];

carbondioxidepK2 = NumberForm[calcpK[co2totsp, 2, {0, .05, .1, .15, .2, .25}], 3];

coApK1=NumberForm[calcpK[coAsp,1,{0,.05,.1,.15,.2,.25}],3];

cysteineLpK1 = NumberForm[calcpK[cysteineLsp, 1, {0, .05, .1, .15, .2, .25}], 3];

dihydroxyacetonephospK1=NumberForm[calcpK[dihydroxyacetonephossp,1,{0,.05,.1,.15,.2,.25
}],3];

fructose6phospK1 = NumberForm[calcpK[fructose6phossp, 1, {0, .05, .1, .15, .2, .25}], 3];

fructose16phospK1=NumberForm[calcpK[fructose16phossp,1,{0,.05,.1,.15,.2,.25}],3];

fructose16phospK2=NumberForm[calcpK[fructose16phossp,2,{0,.05,.1,.15,.2,.25}],3];

fumaratepK1=NumberForm[calcpK[fumaratesp,1,{0,.05,.1,.15,.2,.25}],3];

fumaratepK2=NumberForm[calcpK[fumaratesp,2,{0,.05,.1,.15,.2,.25}],3];

galactose1phospK1 = NumberForm[calcpK[galactose1phossp, 1, {0, .05, .1, .15, .2, .25}], 3];

glucose6phospK1 = NumberForm[calcpK[glucose6phossp, 1, {0, .05, .1, .15, .2, .25}], 3];

glucose1phospK1 = NumberForm[calcpK[glucose1phossp, 1, {0, .05, .1, .15, .2, .25}], 3];

glutathioneredpK1 = NumberForm[calcpK[glutathioneredsp, 1, {0, .05, .1, .15, .2, .25}], 3];

glyceraldehydephospK1=NumberForm[calcpK[glyceraldehydephossp,1,{0,.05,.1,.15,.2,.25}],3
];

glycerol3phospK1=NumberForm[calcpK[glycerol3phossp,1,{0,.05,.1,.15,.2,.25}],3];

malatepK1 = NumberForm[calcpK[malatesp, 1, {0, .05, .1, .15, .2, .25}], 3];

oxalatepK1 = NumberForm[calcpK[oxalatesp, 1, {0, .05, .1, .15, .2, .25}], 3];

phosphoenolpyruvatepK1=NumberForm[calcpK[pepsp,1,{0,.05,.1,.15,.2,.25}],3];

phosphoglycerate2pK1=NumberForm[calcpK[pg2sp,1,{0,.05,.1,.15,.2,.25}],3];

phosphoglycerate3pK1=NumberForm[calcpK[pg3sp,1,{0,.05,.1,.15,.2,.25}],3];

phosphatepK1 = NumberForm[calcpK[pisp, 1, {0, .05, .1, .15, .2, .25}], 3];

pyrophosphatepK1 = NumberForm[calcpK[ppisp, 1, {0, .05, .1, .15, .2, .25}], 3];

pyrophosphatepK2 = NumberForm[calcpK[ppisp, 2, {0, .05, .1, .15, .2, .25}], 3];

pyrophosphatepK3 = NumberForm[calcpK[ppisp, 3, {0, .05, .1, .15, .2, .25}], 3];
```

```
ribose1phospK1 = NumberForm[calcpK[ribose1phossp, 1, {0, .05, .1, .15, .2, .25}], 3];

ribose5phospK1 = NumberForm[calcpK[ribose5phossp, 1, {0, .05, .1, .15, .2, .25}], 3];

succinatepK1=NumberForm[calcpK[succinatesp,1,{0,.05,.1,.15,.2,.25}],3];

succinatepK2=NumberForm[calcpK[succinatesp,2,{0,.05,.1,.15,.2,.25}],3];

succinylcoApK1=NumberForm[calcpK[succinylcoAsp,1,{0,.05,.1,.15,.2,.25}],3];

thioredoxinredpK1 = NumberForm[calcpK[thioredoxinredsp, 1, {0, .05, .1, .15, .2, .25}], 3];

thioredoxinredpK2 = NumberForm[calcpK[thioredoxinredsp, 2, {0, .05, .1, .15, .2, .25}], 3];

TableForm[{acetatepK1, acetylphospK1, acetylphospK2, adeninepK1, adenosinepK1, adppK1,
    adppK2, ammoniapK1, amppK1, amppK2, atppK1, atppK2, bpgpK1, citratepK1, citratepK2,
    citrateisopK1, citrateisopK2, carbondioxidepK1, carbondioxidepK2, coApK1, cysteineLpK1,
    dihydroxyacetonephospK1, fructose6phospK1, fructose16phospK1, fructose16phospK2,
    fumaratepK1, fumaratepK2, galactose1phospK1, glucose6phospK1, glucose1phospK1,
    glutathioneredpK1, glyceraldehydephospK1, glycerol3phospK1, malatepK1, oxalatepK1,
    phosphoenolpyruvatepK1, phosphoglycerate2pK1, phosphoglycerate3pK1, phosphatepK1,
    pyrophosphatepK1, pyrophosphatepK2, pyrophosphatepK3, ribose1phospK1, ribose5phospK1,
    succinatepK1, succinatepK2, succinylcoApK1, thioredoxinredpK1, thioredoxinredpK2},
  TableHeadings → {names, {"I=0", "I=0.05", "I=0.10", "I=0.15", "I=0.20", "I=0.25"}}]
```

	I=0	I=0.05	I=0.10	I=0.15	I=0.20	I=0.25
acetatepK1	{4.75,	4.59,	4.54,	4.51,	4.49,	4.47}
acetylphospK1	{8.69,	8.35,	8.26,	8.2,	8.16,	8.12}
acetylphospK2	{5.11,	4.94,	4.9,	4.87,	4.85,	4.83}
adeninepK1	{4.2,	4.2,	4.2,	4.2,	4.2,	4.2}
adenosinepK1	{3.47,	3.47,	3.47,	3.47,	3.47,	3.47}
adppK1	{7.18,	6.67,	6.53,	6.44,	6.38,	6.33}
adppK2	{4.36,	4.02,	3.93,	3.87,	3.83,	3.79}
ammoniapK1	{9.25,	9.25,	9.25,	9.25,	9.25,	9.25}
amppK1	{6.73,	6.39,	6.3,	6.24,	6.2,	6.16}
amppK2	{3.99,	3.82,	3.77,	3.74,	3.72,	3.71}
atppK1	{7.6,	6.93,	6.74,	6.62,	6.53,	6.47}
atppK2	{4.68,	4.17,	4.03,	3.94,	3.88,	3.83}
bpgpK1	{7.96,	7.29,	7.1,	6.98,	6.9,	6.83}
citratepK1	{6.39,	5.89,	5.75,	5.66,	5.59,	5.54}
citratepK2	{4.76,	4.42,	4.33,	4.27,	4.22,	4.19}
citrateisopK1	{6.4,	5.9,	5.76,	5.67,	5.6,	5.55}
citrateisopK2	{4.76,	4.42,	4.33,	4.27,	4.22,	4.19}
carbondioxidepK1	{10.3,	9.99,	9.9,	9.84,	9.8,	9.76}
carbondioxidepK2	{6.37,	6.2,	6.15,	6.12,	6.1,	6.08}
coApK1	{8.38,	8.21,	8.16,	8.14,	8.11,	8.1}

cysteineLpK1	{8.38, 8.21, 8.16, 8.13, 8.11, 8.09}
dihydroxyacetonephospK1	{5.7, 5.36, 5.27, 5.21, 5.17, 5.13}
fructose6phospK1	{6.27, 5.94, 5.84, 5.78, 5.74, 5.7}
fructose16phospK1	{6.65, 5.98, 5.79, 5.67, 5.59, 5.52}
fructose16phospK2	{6.05, 5.54, 5.41, 5.32, 5.25, 5.2}
fumaratepK1	{4.6, 4.27, 4.17, 4.11, 4.07, 4.03}
fumaratepK2	{3.09, 2.93, 2.88, 2.85, 2.83, 2.81}
galactose1phospK1	{6.15, 5.81, 5.72, 5.66, 5.61, 5.58}
glucose6phospK1	{6.42, 6.08, 5.99, 5.93, 5.89, 5.85}
glucose1phospK1	{6.5, 6.16, 6.07, 6.01, 5.97, 5.93}
glutathioneredpK1	{8.34, 8., 7.91, 7.85, 7.81, 7.77}
glyceraldehydephospK1	{5.7, 5.36, 5.27, 5.21, 5.17, 5.13}
glycerol3phospK1	{6.67, 6.33, 6.24, 6.18, 6.14, 6.1}
malatepK1	{5.26, 4.92, 4.83, 4.77, 4.73, 4.69}
oxalatepK1	{4.28, 3.94, 3.85, 3.79, 3.75, 3.71}
phosphoenolpyruvateK1	{7., 6.5, 6.36, 6.27, 6.2, 6.15}
phosphoglycerate2pK1	{7.64, 7.14, 7., 6.91, 6.84, 6.79}
phosphoglycerate3pK1	{7.53, 7.03, 6.89, 6.8, 6.73, 6.68}
phosphatepK1	{7.22, 6.88, 6.79, 6.73, 6.69, 6.65}
pyrophosphatepK1	{9.46, 8.79, 8.6, 8.48, 8.4, 8.33}
pyrophosphatepK2	{6.72, 6.21, 6.08, 5.99, 5.92, 5.87}
pyrophosphatepK3	{2.26, 1.92, 1.83, 1.77, 1.73, 1.69}
ribose1phospK1	{6.69, 6.35, 6.26, 6.2, 6.16, 6.12}
ribose5phospK1	{6.69, 6.35, 6.26, 6.2, 6.16, 6.12}
succinatepK1	{5.64, 5.3, 5.21, 5.15, 5.11, 5.07}
succinatepK2	{4.21, 4.04, 3.99, 3.96, 3.94, 3.92}
succinylcoApK1	{4.21, 4.04, 4., 3.97, 3.95, 3.93}
thioredoxinredpK1	{8.64, 8.3, 8.21, 8.15, 8.11, 8.07}
thioredoxinredpK2	{8.05, 7.88, 7.83, 7.8, 7.78, 7.76}

1.11 Plot the pKs of acetate, ammonia, atp, and pyrophosphate versus ionic strength from I = 0 to I = 0.3 M at 298.15 K. Biochemists are usually only concerned with the pHs in the pH 5 to 9 range.

(BasicBiochemData2 needs to be loaded)

```
calcpK[speciesmat_, no_, is_] := Module[{lnkzero, sigmanuzsq, lnK},
   (*Calculates pKa for a weak acid at 298.15 K at specified ionic strengths
     (is) when the number no of the pK is specified.  pKs are numbered 1,
    2, 3,... from the highest pK to the lowest pK,
     but the highest pK for a weak acid may be omitted if it is outside of the
      physiological range pH 5 to 9.   For H3PO4, pK1=calcpK[pisp,1,{0}] = 7.22*)
  lnkzero = (speciesmat[[no + 1, 1]] - speciesmat[[no, 1]]) / (8.31451 * .29815);
  sigmanuzsq = speciesmat[[no, 3]]^2 - speciesmat[[no + 1, 3]]^2 + 1;
  lnK = lnkzero + (1.17582 * is^.5) * sigmanuzsq / (1 + 1.6 * is^.5);
  N[-lnK / Log[10]]
  ]
```

acetatepK1=calcpK[acetatesp,1,is]

$$0.434294 \left(10.9481 - \frac{2.35164 \; is^{0.5}}{1. + 1.6 \; is^{0.5}}\right)$$

plotacetate=Plot[acetatepK1,{is,0,.3},AxesOrigin->{0,4.45},AxesLabel->{"I/M","pK"}];

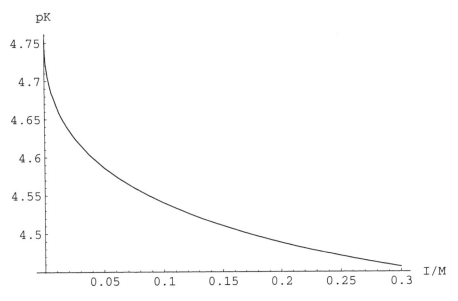

ammoniapK1=calcpK[ammoniasp,1,is];

plotammonia=Plot[ammoniapK1,{is,0,.3},AxesLabel->{"I/M","pK"}];

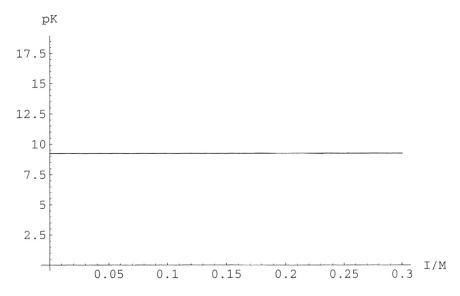

```
atppK1 = calcpK[atpsp, 1, is];

plotatp1=Plot[atppK1,{is,0,.3},AxesOrigin->{0,6.4},AxesLabel->{"I/M","\!\(pK\_1\)"}];
```

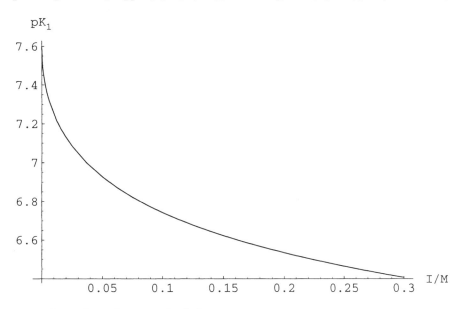

```
atppK2 = calcpK[atpsp, 2, is];

plotatp2=Plot[atppK2,{is,0,.3},AxesOrigin->{0,3.8},AxesLabel->{"I/M","\!\(pK\_2\)"}];
```

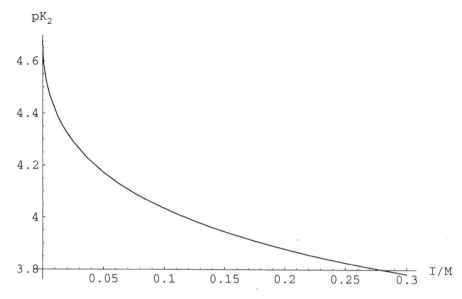

```
pyrophosphatepK1 = calcpK[ppisp, 1, is];

plotpyro1=Plot[pyrophosphatepK1,{is,0,.3},AxesOrigin->{0,8.2},AxesLabel-
>{"I/M","\!\(pK\_1\)"}];
```

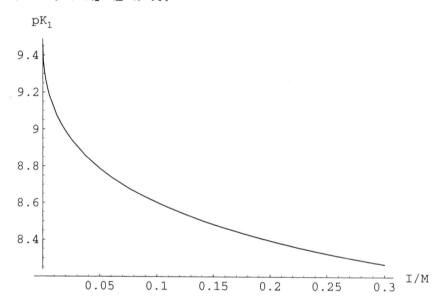

```
pyrophosphatepK2 = calcpK[ppisp, 2, is];

plotpyro2=Plot[pyrophosphatepK2,{is,0,.3},AxesOrigin->{0,5.8},AxesLabel-
>{"I/M","\!\(pK\_2\)"}];
```

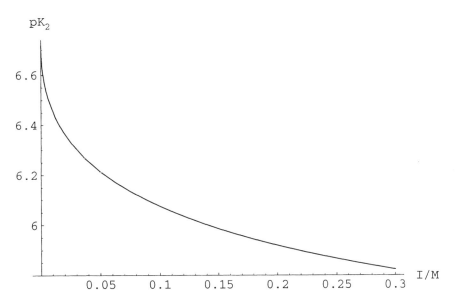

```
Show[plotacetate,plotammonia,plotatp1,plotatp2,plotpyro1,plotpyro2,AxesOrigin-
>{0,3},PlotRange->{3,10},AxesLabel->{"I/M","pK"}];
```

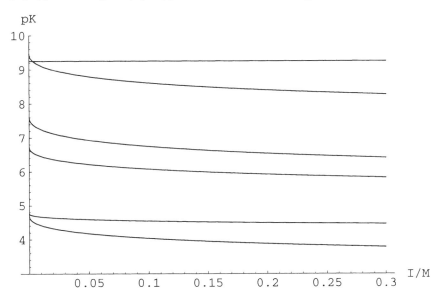

1.12 (a) Calculate the acid titration curve for ATP at 298.15 K and 0.25 M ionic strength from the binding polynomial P.
(b) Integrate the calculated binding curve to obtain ln P plus a constant of integration. The needed equations are

N_H = (-1/ln(10))(dlnP/dpH)

$-\ln$ (10) $\int N_H dpH = \ln P + const$

(a) The titration curve can be calculated by differentiating the logarithm of the binding potential P.

```
p=1+(10^-pH)/(10^-6.47)+((10^-pH)^2)/((10^-6.47)*(10^-3.83))
```

$$1 + \frac{1.99526 \ 10^{10}}{10^{2 \ pH}} + \frac{2.95121 \ 10^{6}}{10^{pH}}$$

`Plot[p,{pH,2,10},AxesLabel->{"pH","p"}];`

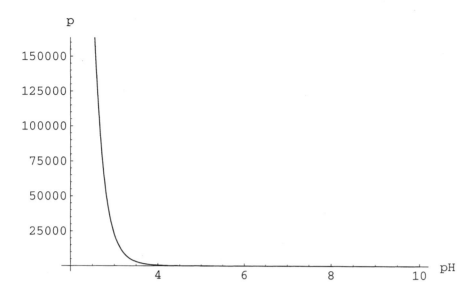

`Plot[Log[p],{pH,2,10},AxesLabel->{"pH","ln p"}];`

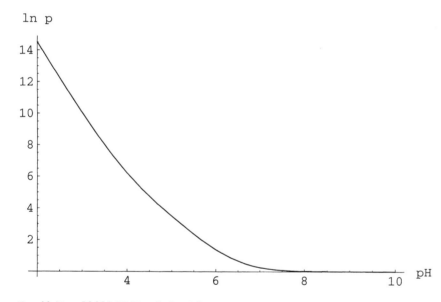

`nH=-(1/Log[10])*D[Log[p],pH]`

$$-\left(\frac{\dfrac{-4.59426 \; 10^{10} \; 2^{1-2\,pH}}{5^{2\,pH}} - \dfrac{6.79541 \; 10^{6}}{10^{pH}}}{\left(1 + \dfrac{1.99526 \; 10^{10}}{10^{2\,pH}} + \dfrac{2.95121 \; 10^{6}}{10^{pH}}\right) \; Log[10]}\right)$$

`Plot[nH,{pH,2,10},AxesLabel->{"pH","\!\(N\_H\)"}];`

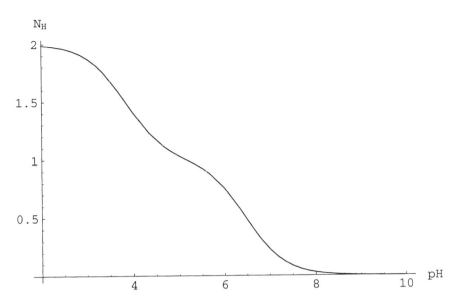

(b) Integration of the binding curve gives lnP plus a constant.

```
lnpcalc=-Log[10]*Integrate[nH,pH]
```

$$-4.60517 \text{ pH} + 1. \text{ Log}[6776.39 + 10^{\text{pH}}] + 1. \text{ Log}[2.94443 \; 10^{6} + 10^{\text{pH}}]$$

```
(lnpcalc-Log[p])/.pH->{5,6,7}
```

$$\{0., \; 3.55271 \; 10^{-15}, \; 7.10543 \; 10^{-15}\}$$

Thus the integration constant is zero within machine accuracy.

```
Plot[lnpcalc,{pH,2,10},AxesLabel->{"pH","ln p"}];
```

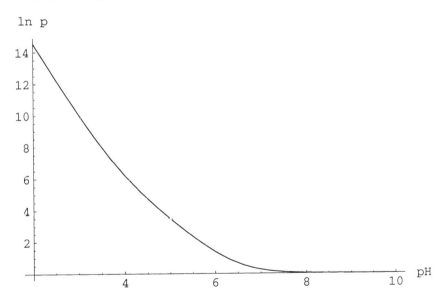

1.13 (a) Test the differentiation of ln P to obtain the equation for the binding N_H of hydrogen ions by ATP at 298.15 K and ionic strength 0.25 M in the region pH 2 to 10. (b) Test the integration of the equation for N_H to obtain the equation for ln P. (c) Plot ln p versus pH and N_H versus pH. The equations involved are

$N_H = [H^+]\frac{d\ln p}{d[H^+]}$

$\int \frac{N_H}{[H^+]} d[H^+] = \ln P + \text{const}$

(a) First we express the binding potential P as a function of the hydrogen ion concentration, represented by h..

p=1+h/10^-6.47+h^2/((10^-6.47)*(10^-3.83))

$1 + 2.95121\ 10^6\ h + 1.99526\ 10^{10}\ h^2$

Plot[p,{h,10^-2,10^-10},AxesLabel->{"h","p"}];

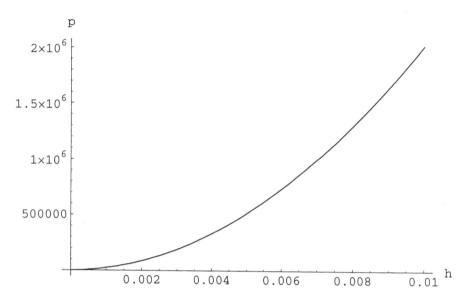

Plot[Log[p],{h,10^-2,10^-10},AxesLabel->{"h","ln p"}];

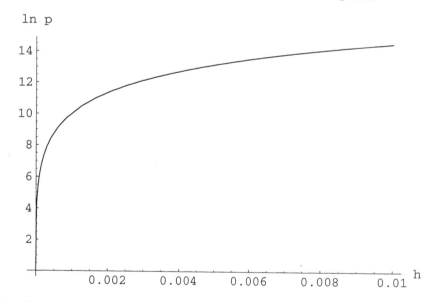

The equation for the titration curve is obtained by use of the following differentiation:

nH=h*D[Log[p],h]

$$\frac{h\ (2.95121\ 10^6 + 3.99052\ 10^{10}\ h)}{1 + 2.95121\ 10^6\ h + 1.99526\ 10^{10}\ h^2}$$

(b) This equation can be integrated to obtain lnp:

Integrate[(nH/h),h]

1. Log[3.39624 10^{-7} + h] + 1. Log[0.000147571 + h]

This should be lnp, which is

Log[p]

Log[1 + 2.95121 10^6 h + 1.99526 10^{10} h^2]

except for an integration constant.

We can compare the two expressions for ln p by calculating numerical values at pH 2, 2.5, 3, 3.5, ...10:

ph=Table[n,{n,2,10,.5}]

{2, 2.5, 3., 3.5, 4., 4.5, 5., 5.5, 6., 6.5, 7., 7.5, 8., 8.5, 9., 9.5, 10.}

hh=10^-ph

{$\frac{1}{100}$, 0.00316228, 0.001, 0.000316228, 0.0001, 0.0000316228, 0.00001, 3.16228 10^{-6}, 1. 1(

3.16228 10^{-7}, 1. 10^{-7}, 3.16228 10^{-8}, 1. 10^{-8}, 3.16228 10^{-9}, 1. 10^{-9}, 3.16228 10^{-10}, 1.

Log[p]/.h->hh

{14.521, 12.2494, 10.0391, 7.98259, 6.20586, 4.73863, 3.48147, 2.35442, 1.37906, 0.6602:

0.0892422, 0.0290869, 0.00928946, 0.00294688, 0.000932821, 0.000295078}

These are the values of ln p calculated from pK1 and pK2.

Integrate[(nH/h),h]/.h->hh

{-9.19566, -11.4672, -13.6775, -15.734, -17.5108, -18.978, -20.2352, -21.3622, -22.3376,

-23.4579, -23.6274, -23.6875, -23.7073, -23.7137, -23.7157, -23.7163}

These are the values of ln p calculated by integration of the binding curve (that is, nH versus h)

(Integrate[(nH/h),h]/.h->hh)-(Log[p]/.h->hh)

{-23.7166, -23.7166, -23.7166, -23.7166, -23.7166, -23.7166, -23.7166, -23.7166, -23.71(

-23.7166, -23.7166, -23.7166, 23.7166, -23.7166, -23.7166, -23.7166}

This is the integration constant. Now calculate values of p at 0.5 pHs.

Exp[(Integrate[(nH/h),h]/.h->hh)+23.7166]

{2.02472 10^6, 208854., 22904.2, 2929.44, 495.634, 114.275, 32.5065, 10.5318, 3.97106, 1.

1.29529, 1.09332, 1.02949, 1.00931, 1.00292, 1.00091, 1.00027}

This can be compared with the values calculated directly from ln p:

p/.h->hh

{2.02478 10^6, 208860., 22904.8, 2929.52, 495.647, 114.278, 32.5074, 10.5321, 3.97116, 1.
   1.29532, 1.09335, 1.02951, 1.00933, 1.00295, 1.00093, 1.0003}

The agreement is quite good as expected.

We can also plot the value of P calculated by the successive differentiation and integration to see whether it is the same as the plot in part b.

Plot[Evaluate[Exp[Integrate[(nH/h),h]+23.7166]],{h,10^-2,10^-10},AxesLabel->{"h","p"}];

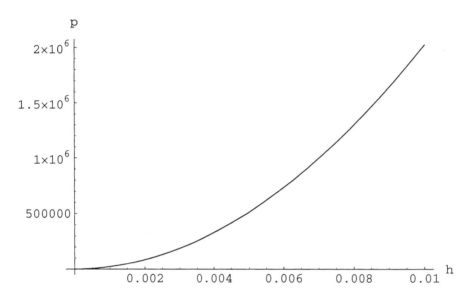

(c) If we introduce the relation between h and pH, the expressions for p and nH will be converted to functions of pH.

h=10^-pH

10^{-pH}

p

$$1 + \frac{1.99526\ 10^{10}}{10^{2\ pH}} + \frac{2.95121\ 10^6}{10^{pH}}$$

Plot[Log[p],{pH,2,10},AxesLabel->{"pH","ln p"}];

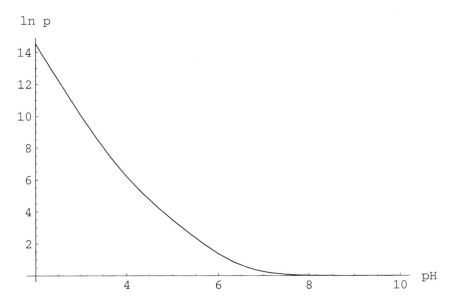

Since the computer will substitute h=10^-pH, we obtain the equation for the binding cuve by simply typing nH.

nH

$$\frac{2.95121\ 10^6 + \dfrac{3.99052\ 10^{10}}{10^{pH}}}{10^{pH}\ \left(1 + \dfrac{1.99526\ 10^{10}}{10^{2\ pH}} + \dfrac{2.95121\ 10^6}{10^{pH}}\right)}$$

```
Plot[nH,{pH,2,10},AxesLabel->{"pH","\!\(N\_H\)"}];
```

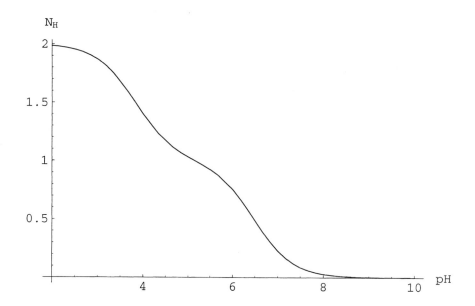

Chapter 3 Chemical Equilibrium in One Phase Systems

3.1 Use the table of basic data to calculate the acid dissociation constants of ATP, ADP, and P_i and the equilibrium constant for the reference reaction $ATP^{4-} + H_2O = ADP^{3-} + HPO_4^{2-} + H^+$ at 298.15 K and ionic strengths of 0, 05, 0.10, 0.15, 0.20, and 0.25 M.

3.2 For the solution reaction A = B, assume that the standard Gibbs energy of formation of A is 20 kJ mol^{-1} and of B is 18 kJ mol^{-1} at 298.15 K. (a) For a reaction starting with a mole of A at a concentration of 1 M, plot the Gibbs energy of the mixture versus extent of reaction from zero to unity and identify the approximate extent of reaction at equilibrium. (b) Identify the equilibrium extent of reaction more precisely by plotting the derivative of the Gibbs energy of the mixture with respect to extent of reaction. (c) Calculate the equilibrium constant and verify the equilibrium extent of reaction.

3.3 Calculate the standard Gibbs energy changes and equilibrium constants in terms of species for the following reactions at 298.15 K and ionic strengths of 0, 0.10, and 0.25 M. Summarize the calculations in two tables.
(a) $NAD^- + H_2(g) = NADH^{2-} + H^+$
(b) $NADP^{3-} + H_2(g) = NADPH^{4-} + H^+$
(c) $NAD^- + NADPH^{4-} = NADH^{2-} + NADP^{3-}$
(d) $CH_3 CH_2 OH + NAD^- = CH_3 CHO + NADH^{2-} + H^+$
(e) $CH_3 CHO + NAD^- + H_2O = CH_3 CO_2^- + NADH^{2-} + 2 H^+$
(f) $C_3 H_7 NO_2 + NAD^- + H_2O = C_3 H_3 O^- + NADH^{2-} + NH_4^+ + H^+$
The last reaction involves L-alanine and pyruvate.

3.4 Plot the acid dissociation constant of acetic acid from 0 °C to 50 °C given that at 298.15 K, $\Delta_f G° = 27.14$ kJ mol^{-s}, $\Delta_f H° = -0.39$ kJ mol^{-s}, and $\Delta_f C_P° = -155$ J K^{-1}mol^{-s}. Assume zero ionic strength.

3.5 (a) Calculate the function of T that gives the values of the Debye-Huckel constant α at the temperatures in Table 3.1 in Section 3.7. Plot the data and the function. (b) Calculate the function of T for $RT\alpha$. (c) Calculate the function of T for $RT^2(\partial\alpha/\partial T)_P$. (d) Use these functions to calculate these coefficients at 0, 10, 20, 25, 30, and 40 °C.

3.6 Calculate the standard Gibbs energies of formation of the three species of ATP at temperatures of 283.15 K, 298.15 K, 298.15 K and ionic strengths of 0, 0.10, and 0.25 M.

3.7 Calculate the adjustments to be subtracted from pH_a obtained with a pH meter to obtain $pH_c = -\log[H^+]$ at 0 °C to 40 °C and ionic strengths of 0, 0.10, and 0.25 M.

3.8 Calculate the standard enthalpies of formation of the three species of ATP at 283.15 K, 298.15 K, and 313.15 K at ionic strengths of 0, 0.10, and 0.25 M

3.9 There are two ways to obtain values for the enthalpy coefficient in equation 3.6-5 as a function of temperature: (a) Calculate the derivative with respect to T of the Gibbs energy coefficient divided by T. (b) Fit the enthalpy coefficients of Clarke and Glew to $AT^2 + BT^3$. Use both of these methods and make plots to compare these functions with the values in Table 3.1

3.10 Plot the activity coefficients of ions with charges 1, 2, 3, and 4 versus the ionic strength at 0 °C. Repeat these calculations at 25 °C and 40 °C

3.1 Use the table of basic data to calculate the acid dissociation constants of ATP, ADP, and P_i and the equilibrium constant for the reference reaction $ATP^{4-} + H_2O = ADP^{3-} + HPO_4{}^{2-} + H^+$ at 298.15 K and ionic strengths of 0, 05, 0.10, 0.15, 0.20, and 0.25 M.

(BasicBiochemData2 has to be loaded)

The following program calculate the function of ionic strength that gives the standard Gibbs energy of formation of a species.

```
calcdGis[species_] := Module[{dGzero, zi, isterm},
    (*This program calculates the function of ionic strength (is) that gives the
      standard Gibbs energy of formation of a species at298.15 K.  The input is
      a list for the species that gives the standard Gibbs energy of formation,
      the standard enthalpy of formation,the electric charge, and the number of
      hydrogen atoms in the species.  Energies are expressed in kJ mol^-1.*)
        dGzero = species[[1]];
        zi = species[[3]];
        isterm = 2.91482 * (zi^2) * (is^.5) / (1 + 1.6*is^.5);
        dGzero - isterm]
```

The basic data on the hydrogen ion species is

```
hydroionsp={0,0,1,1}
```

$\{0, 0, 1, 1\}$

```
hydroion=calcdGis[hydroionsp]
```

$$-\frac{2.91482 \; is^{0.5}}{1 + 1.6 \; is^{0.5}}$$

Now we need to produce a list for each species from the entries in BasicBiochemData2.

```
atpH0=calcdGis[atpsp[[1]]]
```

$$-2768.1 - \frac{46.6371 \; is^{0.5}}{1 + 1.6 \; is^{0.5}}$$

```
atpH1=calcdGis[atpsp[[2]]];

atpH2=calcdGis[atpsp[[3]]];

adpH0=calcdGis[adpsp[[1]]];

adpH1=calcdGis[adpsp[[2]]];

adpH2=calcdGis[adpsp[[3]]];

phosphateH1=calcdGis[pisp[[1]]];

phosphateH2=calcdGis[pisp[[2]]];

h2o=calcdGis[h2osp[[1]]];
```

```
calckrx[eq_,islist_]:=Module[{energy,dG},(*Calculates the equilibrium constant K for a
chemical equation typed in the form atpH1+de==hydroion+atpH0.*)
energy=Solve[eq,de];
dG=energy[[1,1,2]]/.is->islist;
Exp[-dG/(8.31451*.29815)]]

atpk1=calckrx[atpH1+de==hydroion+atpH0,{0,.1,.25}]
```

$\{2.51302 \ 10^{-8}, \ 1.81144 \ 10^{-7}, \ 3.4275 \ 10^{-7}\}$

```
atpk2=calckrx[atpH2+de==hydroion+atpH1,{0,.1,.25}];

adpk1=calckrx[adpH1+de==hydroion+adpH0,{0,.1,.25}];

adpk2=calckrx[adpH2+de==hydroion+adpH1,{0,.1,.25}];

pik1=calckrx[phosphateH2+de==hydroion+phosphateH1,{0,.1,.25}];

kref=calckrx[atpH0+h2o+de==adpH0+phosphateH1+hydroion,{0,.1,.25}];

TableForm[{atpk1,atpk2,adpk1,adpk2,pik1,kref},TableHeadings-
>{{"atpk1","atpk2","adpk1","adpk2","pik1","kref"},{0,.1,.25}}]
```

	0	0.1	0.25
atpk1	$2.51302 \ 10^{-8}$	$1.81144 \ 10^{-7}$	$3.4275 \ 10^{-7}$
atpk2	0.0000210082	0.0000924186	0.000149099
adpk1	$6.64366 \ 10^{-8}$	$2.92266 \ 10^{-7}$	$4.71512 \ 10^{-7}$
adpk2	0.0000437761	0.000117531	0.000161669
pik1	$6.05499 \ 10^{-8}$	$1.62565 \ 10^{-7}$	$2.23617 \ 10^{-7}$
kref	0.291014	0.177606	0.151432

3.2 For the solution reaction A = B, assume that the standard Gibbs energy of formation of A is 20 kJ mol^{-1} and of B is 18 kJ mol^{-1} at 298.15 K. (a) For a reaction starting with a mole of A at a concentration of 1 M, plot the Gibbs energy of the mixture versus extent of reaction from zero to unity and identify the approximate extent of reaction at equilibrium. (b) Identify the equilibrium extent of reaction more precisely by plotting the derivative of the Gibbs energy of the mixture with respect to extent of reaction. (c) Calculate the equilibrium constant and verify the equilibrium extent of reaction.

(a) The Gibbs energy of the reaction mixture is given by

```
Clear[x]

g = (1 - x) * 20 + 18 * x + (8.31451 * 10^-3) * 298.15 * ((1 - x) * Log[1 - x] + x * Log[x])

20 (1 - x) + 18 x + 2.47897 ((1 - x) Log[1 - x] + x Log[x])

Plot[g, {x, 0, 1}, AxesOrigin -> {0, 15},
  AxesLabel -> {"£/mol", "G/kJ mol⁻¹"}, PlotRange -> {15, 21}];
```

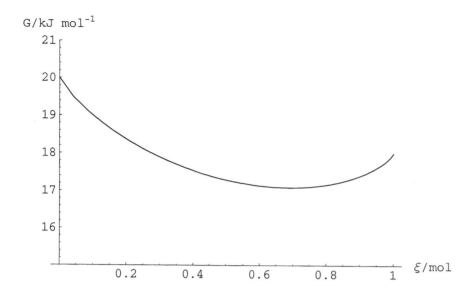

The approximate extent of reaction at equilibrium 298 K and 1 bar is 0.7.

(b) The derivative of G with respect to x is given by

dvt = D[g, x]

-2 + 2.47897 (-Log[1 - x] + Log[x])

Plot[dvt, {x, 0, 1}, PlotRange → {-10, 10}, AxesLabel → {"ξ/mol", "dG/dξ"}];

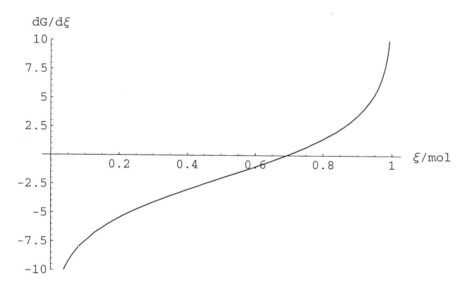

(c) The equilibrium constant is given by k = x/(1-x), and so the equilibrium extent of reaction is given by x=k/(1+k).

k = Exp[(18 - 20) / ((8.3145 ∗ 10^-3) ∗ 298.15)]

0.44629

k / (1 + k)

```
0.308576
```

This agrees with the preceding plot.

3.3 Calculate the standard Gibbs energy changes and equilibrium constants in terms of species for the following reactions at 298.15 K and ionic strengths of 0, 0.10, and 0.25 M. Summarize the calculations in two tables.

(a) $\quad NAD^- + H_2(g) = NADH^{2-} + H^+$

(b) $\quad NADP^{3-} + H_2(g) = NADPH^{4-} + H^+$

(c) $\quad NAD^- + NADPH^{4-} = NADH^{2-} + NADP^{3-}$

(d) $\quad CH_3\,CH_2\,OH + NAD^- = CH_3\,CHO + NADH^{2-} + H^+$

(e) $\quad CH_3\,CHO + NAD^- + H_2O = CH_3\,CO_2^- + NADH^{2-} + 2\,H^+$

(f) $\quad C_3\,H_7\,NO_2 + NAD^- + H_2O = C_3\,H_3\,O^- + NADH^{2-} + NH_4^+ + H^+$

The last reaction involves L-alanine and pyruvate.

(BasicBiochemData2 has to be loaded)

Since these reactants are all single species, calcdGis in the preceding problem can be used to calculate the function that gives the standard Gibbs energy of formation. The values at these three ionic strengths can be calculated using the assignment operation (/.). We can add the values for the products and subtract the values for the reactants to obtain the values of the standard Gibbs energy for the reaction. Finally, we can calculate the corresponding equilibrium constants and put them in a table.

```
calcdGis[species_] := Module[{dGzero, zi, isterm},
   (*This program calculates the function of ionic strength (is) that gives the
     standard Gibbs energy of formation of a species at298 .15 K.  The input is
     a list for the species that gives the standard Gibbs energy of formation,
     the standard enthalpy of formation,the electric charge, and the number of
     hydrogen atoms in the species.  Energies are expressed in kJ mol^-1.*)
      dGzero = species[[1]];
      zi = species[[3]];
      isterm = 2.91482 * (zi^2) * (is^.5) / (1 + 1.6 * is^.5);
      dGzero - isterm]
```

```
hydroionis=calcdGis[hydroionsp]/.is->{0,.1,.25}
```

```
{0, -0.612064, -0.809672}
```

```
nadoxis=calcdGis[Flatten[nadoxsp]]/.is->{0,.1,.25}
```

```
{0, -0.612064, -0.809672}
```

The Flatten is needed to remove the outer curly brackets in the BasicBiochemData2.

```
h2gis={0,0,0}
```

```
{0, 0, 0}
```

```
nadredis=calcdGis[Flatten[nadredsp]]/.is->{0,.1,.25};
```

```
nadpoxis=calcdGis[Flatten[nadpoxsp]]/.is->{0,.1,.25};
```

```
nadpredis=calcdGis[Flatten[nadpredsp]]/.is->{0,.1,.25};
```

```
calcdGis[Flatten[ethanolsp]]/.is->{0,.1,.25}
```

```
-181.64

ethanolis={-181.64,-181.64,-181.64}

{-181.64, -181.64, -181.64}
```

This had to be put in by hand because ethanol is not an ion.

```
calcdGis[Flatten[acetaldehydesp]]/.is->{0,.1,.25}

-139.

acetaldehydeis={-139.,-139.,-139.};

calcdGis[Flatten[h2osp]]/.is->{0,.1,.25}

-237.19

h2ois={-237.19,-237.19,-237.19};

calcdGis[Flatten[alaninesp]]/.is->{0,.1,.25}

-371.

alanineis={-371.,-371.,-371.};


acetateis=calcdGis[Flatten[acetatesp]]/.is->{0,.1,.25};

pyruvateis=calcdGis[Flatten[pyruvatesp]]/.is->{0,.1,.25};

nh4is=calcdGis[Flatten[ammoniasp[[2]]]]/.is->{0,.1,.25};

dGrxa=nadredis+hydroionis-nadoxis-h2gis;

krxa=Exp[-dGrxa/(8.31451*.29815)];

dGrxb=nadpredis+hydroionis-nadpoxis-h2gis;

krxb=Exp[-dGrxb/(8.31451*.29815)];

dGrxc=nadredis+nadpoxis-nadoxis-nadpredis;

krxc=Exp[-dGrxc/(8.31451*.29815)];

dGrxd=acetaldehydeis+nadredis+hydroionis-ethanolis-nadoxis;

krxd=Exp[-dGrxd/(8.31451*.29815)];

dGrxe=acetateis+nadredis+2*hydroionis-acetaldehydeis-nadoxis-h2ois;

krxe=Exp[-dGrxe/(8.31451*.29815)];

dGrxf=pyruvateis+nadredis+nh4is+hydroionis-alanineis-nadoxis-h2ois;

krxf=Exp[-dGrxf/(8.31451*.29815)];
```

Table 1 Standard Reaction Gibbs Energies in kJ mol^{-1} at 298.15 K

```
TableForm[{dGrxa,dGrxb,dGrxc,dGrxd,dGrxe,dGrxf},TableHeadings-
>{{"dGrxa","dGrxb","dGrxc","dGrxd","dGrxe","dGrxf"},{0,0.10,0.25}}]
```

	0	0.1	0.25
dGrxa	22.65	20.2017	19.4113
dGrxb	25.99	21.0935	19.5126
dGrxc	-3.34	-0.891743	-0.101311
dGrxd	65.29	62.8417	62.0513
dGrxe	29.53	25.8576	24.672
dGrxf	79.26	75.5876	74.402

Table 2 Equilibrium constants at 298.15 K

```
TableForm[{krxa,krxb,krxc,krxd,krxe,krxf},TableHeadings-
>{{"krxa","krxb","krxc","krxd","krxe","krxf"},{0,0.10,0.25}}]
```

	0	0.1	0.25
krxa	0.000107625	0.000288953	0.00039747
krxb	0.0000279753	0.000201652	0.000381554
krxc	3.84715	1.43293	1.04171
krxd	$3.64546\ 10^{-12}$	$9.78736\ 10^{-12}$	$1.3463\ 10^{-11}$
krxe	$6.70807\ 10^{-6}$	0.0000295099	0.0000476083
krxf	$1.30115\ 10^{-14}$	$5.72398\ 10^{-14}$	$9.23451\ 10^{-14}$

These values agree with R. A. Alberty, Arch. Biochem. Biophys. 307, 8-14 (1993).

3.4 Plot the acid dissociation constant of acetic acid from 0 °C to 50 °C given that at 298.15 K, $\Delta_f G° = 27.14$ kJ mol^{-s}, $\Delta_f H° = -0.39$ kJ mol^{-s}, and $\Delta_f C_P° = -155$ J K^{-1} mol^{-s}. Assume zero ionic strength.

$\Delta_f S° = (\Delta_f H° - \Delta_f G°)/T$

```
deltaS=(-390-27140)/298.15
```

-92.3361

```
lnk=390/(8.31451*t)-92.34/8.31451+(155/8.31451)*(1-298.15/t-Log[t/298.15])
```

$-11.1059 + \dfrac{46.906}{t} + 18.6421\ (1 - \dfrac{298.15}{t} - \mathrm{Log}[0.00335402\ t])$

```
Plot[lnk,{t,273.15,323.15},AxesOrigin->{270,-11.02},AxesLabel->{"T/K","ΔG"}];
```

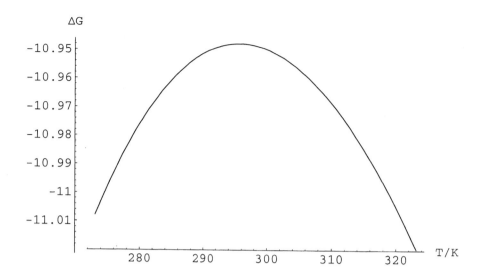

Plot[Exp[lnk],{t,273.15,323.15},AxesOrigin->{270,.0000164},AxesLabel->{"T/K","K"}];

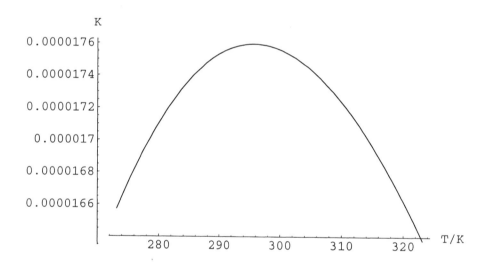

3.5 (a) Calculate the function of T that gives the values of the Debye-Huckel constant α at the temperatures in Table 3.1 in Section 3.7. Plot the data and the function. (b) Calculate the function of T for RTα. (c) Calculate the function of T for RT$^2 (\partial\alpha/\partial T)_P$. (d) Use these functions to calculate these coefficients at 0, 10, 20, 25, 30, and 40 °C.

(a) The Debye-Huckel constant α as a function of temperature:

**data={{273.15,1.12938},{283.15,1.14717},{293.15,1.16598},{298.15,1.17582},{303.15,1.185
99},{313.15,1.20732}}**

{{273.15, 1.12938}, {283.15, 1.14717}, {293.15, 1.16598}, {298.15, 1.17582}, {303.15, 1.
    {313.15, 1.20732}}

TableForm[data]

```
273.15    1.12938
283.15    1.14717
293.15    1.16598
298.15    1.17582
303.15    1.18599
313.15    1.20732
```

alpha=Fit[data,{1,t,t^2},t]

$$1.10708 - 0.00154508 \ t + 5.95584 \ 10^{-6} \ t^2$$

plot1=Plot[alpha,{t,270,314},DisplayFunction->Identity];

plot2=ListPlot[data,AxesOrigin->{270,1.12},Prolog->AbsolutePointSize[4],AxesLabel->{"T/K","α"},DisplayFunction->Identity];

Show[plot2,plot1,DisplayFunction->$DisplayFunction];

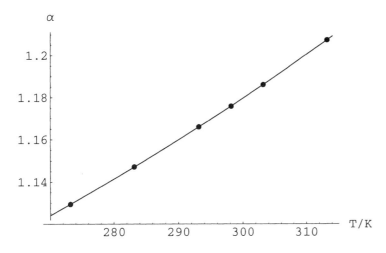

(b) $RT\alpha$ as a function of temperature:

   rtalpha=Expand[8.31451*10^-3*t*alpha]

$$0.00920485 \ t - 0.0000128466 \ t^2 + 4.95199 \ 10^{-8} \ t^3$$

The factor 10^-3 converts the value of R from J to kJ.

(c) $RT^2 \ (\partial \alpha / \partial T)_P$ as a function of temperature:

   rt2Dalpha=Expand[8.31451*10^-3*t^2*D[alpha,t]]

$$-0.0000128466 \ t^2 + 9.90399 \ 10^{-8} \ t^3$$

Take the derivative of RTalpha

(d) Make a table

   col2=alpha/.t->{0,10,20,25,30,40}+273.15

   {1.12942, 1.1471, 1.16597, 1.17585, 1.18603, 1.20729}

```
col3=rtalpha/.t->{0,10,20,25,30,40}+273.15
```

{2.56503, 2.70056, 2.84193, 2.91491, 2.98945, 3.14341}

```
col4=rt2Dalpha/.t->{0,10,20,25,30,40}+273.15
```

{1.05994, 1.21837, 1.39106, 1.48294, 1.5786, 1.78158}

```
tab={{0,10,20,25,30,40}+273.15,col2,col3,col4}
```

{{273.15, 283.15, 293.15, 298.15, 303.15, 313.15},
 {1.12942, 1.1471, 1.16597, 1.17585, 1.18603, 1.20729},
 {2.56503, 2.70056, 2.84193, 2.91491, 2.98945, 3.14341},
 {1.05994, 1.21837, 1.39106, 1.48294, 1.5786, 1.78158}}

```
TableForm[Transpose[tab],TableHeadings->{None,{"T/K","α","RTα","RT^2(∂α/∂T)"}}]
```

T/K	α	RTα	RT^2($\partial\alpha/\partial T$)
273.15	1.12942	2.56503	1.05994
283.15	1.1471	2.70056	1.21837
293.15	1.16597	2.84193	1.39106
298.15	1.17585	2.91491	1.48294
303.15	1.18603	2.98945	1.5786
313.15	1.20729	3.14341	1.78158

These values can be compared with the values in Table 3.1 in the text.

3.6 Calculate the standard Gibbs energies of formation of the three species of ATP at temperatures of 283.15 K, 298.15 K, 313.15 K and ionic strengths of 0, 0.10, and 0.25 M.

(BasicBiochemData2 has to be loaded)

```
calcdGTsp[speciesmat_,temp_,ionstr_] :=
Module[{dGzero, dGzeroT,dHzero,zi, nH, gibbscoeff, istermG,gfnsp},(*This program
calculates the functions of T and ionic strength for the standard Gibbs energy of
formation for all of the species in a reactant. The temperature, temp in K,can be
specified in approximately the range 273.15 K to 313.15 K, and the ionic strength,
ionstr in M, can be specified in the range 0 to 0.35 M. Lists of temperatures and
ionic strengths can also be used. The standard Gibbs energies of formation in the
output are in kJ mol^-1.
    The input speciesmat is a matrix that gives the standard Gibbs energy of formation
at 298.15 K, the standard enthalpy of formation at 298.15 K, the electric charge, and
the numbers of hydrogen atoms in each species. There is a row in the matrix for each
species of the reactant. gfnsp is alist of the functions for the species.*)
{dGzero,dHzero,zi,nH} = Transpose[speciesmat];
gibbscoeff=9.20483*10^-3*t-1.284668*10^-5*t^2+4.95199*10^-8*t^3;
dGzeroT=dGzero*t/298.15+dHzero*(1-t/298.15);
istermG = gibbscoeff*(zi^2)*(is^.5)/(1 + 1.6*is^.5);
gfnsp=dGzeroT - istermG;
gfnsp/.t->temp/.is->ionstr]
```

```
gatp=calcdGTsp[atpsp,298.15,0]
```

{-2768.1, -2811.48, -2838.18}

```
atpsp
```

{{-2768.1, -3619.21, -4, 12}, {-2811.48, -3612.91, -3, 13}, {-2838.18, -3627.91, -2, 14}

This program is run three times at I = 0, 0.10, and 0.25 M.

```
TableForm[calcdGTsp[atpsp,{283.15,298.15,313.15},0],TableHeadings-
>{{"ATP-4","HATP-3","H2ATP-2"},{"283.15 K","298.15 K","313.15 K"}}]
```

	283.15 K	298.15 K	313.15 K
ATP-4	-2810.92	-2768.1	-2725.28
HATP-3	-2851.8	-2811.48	-2771.16
H2ATP-2	-2877.91	-2838.18	-2798.45

```
TableForm[calcdGTsp[atpsp,{283.15,298.15,313.15},.1],TableHeadings-
>{{"ATP-4","HATP-3","H2ATP-2"},{"283.15 K","298.15 K","313.15 K"}}]
```

	283.15 K	298.15 K	313.15 K
ATP-4	-2819.99	-2777.89	-2735.84
HATP-3	-2856.9	-2816.99	-2777.1
H2ATP-2	-2880.18	-2840.63	-2801.09

```
TableForm[calcdGTsp[atpsp,{283.15,298.15,313.15},.25],TableHeadings-
>{{"ATP-4","HATP-3","H2ATP-2"},{"283.15 K","298.15 K","313.15 K"}}]
```

	283.15 K	298.15 K	313.15 K
ATP-4	-2822.92	-2781.06	-2739.25
HATP-3	-2858.55	-2818.77	-2779.02
H2ATP-2	-2880.91	-2841.42	-2801.94

3.7 Calculate the adjustments to be subtracted from pH_a obtained with a pH meter to obtain $pH_c = -\log[H^+]$ at 0 °C to 40 °C and ionic strengths of 0, 0.10, and 0.25 M.

The values of α in the Debye-Huckel equation are given by

```
data={{273.15,1.12938},{283.15,1.14717},{293.15,1.16598},{298.15,1.17582},{303.15,1.185
99},{313.15,1.20732}}
```

{{273.15, 1.12938}, {283.15, 1.14717}, {293.15, 1.16598}, {298.15, 1.17582},
  {303.15, 1.18599}, {313.15, 1.20732}}

TableForm[data]

273.15	1.12938
283.15	1.14717
293.15	1.16598
298.15	1.17582
303.15	1.18599
313.15	1.20732

alpha=Fit[data,{1,t,t^2},t]

$1.10708 - 0.00154508\ t + 5.95584\ 10^{-6}\ t^2$

alpha

$1.10708 - 0.00154508\ t + 5.95584\ 10^{-6}\ t^2$

Equation 3.7-6 yields the following table: These are the adjustments to be subtracted from pH_a to give $pH_c = -\log[H^+]$.

```
TableForm[Transpose[alpha*is^.5/(Log[10]*(1+1.6*is^.5))/.t->{283.15,298.15,313.15}/.is-
>{0,.05,.1,.15,.2,.25}],TableHeadings->{{"0",".05",".10",".15",".20",".25"},{"10
C","25 C","40 C"}}]
```

	10 C	25 C	40 C
0	0	0	0
.05	0.0820433	0.0841	0.0863484
.10	0.104609	0.107232	0.110098
.15	0.119125	0.122111	0.125376
.20	0.129867	0.133122	0.136681
.25	0.138383	0.141852	0.145644

3.8 Calculate the standard enthalpies of formation of the three species of ATP at 283.15 K, 298.15 K, and 313.15 K at ionic strengths of 0, 0.10, and 0.25 M.

(BasicBiochemData2 has to be loaded)

Since we have a program to calculate standard Gibbs energies of formation, we can calculate the standard enthalpies of formation by use of the Gibbs Helmholtz equation.

```
calcdHTsp[speciesmat_,temp_,ionstr_] :=
Module[{dGzero, dGzeroT,dHzero,zi, nH, gibbscoeff, istermG,gfnsp,hfnsp},(*This program
first calculates the functions of T and ionic strength for the standard Gibbs energy
of formation for all of the species in a reactant, and then uses the Gibbs Duhem
equation to calculate the functions of T and ionic strength for the standard enthalpy
of formation. The temperature, temp in K,can be specified in approximately the range
273.15 K to 313.15 K, and the ionic strength, ionstr in M, can be specified in the
range 0 to 0.35 M.  Lists of temperatures and ionic strengths can also be used.  The
standard enthalpies of formation in the output are in kJ mol^-1.
   The input speciesmat is a matrix that gives the standard Gibbs energy of formation
at 298.15 K, the standard enthalpy of formation at 298.15 K, the electric charge, and
the number of hydrogen atoms in each species. There is a row in the matrix for each
species of the reactant. gfnsp is alist of the functions for the species.  hfnsp is a
list of the functions for the enthalpies of the species.*)
{dGzero,dHzero,zi,nH} = Transpose[speciesmat];
gibbscoeff=9.20483*10^-3*t-1.284668*10^-5*t^2+4.95199*10^-8*t^3;
dGzeroT=dGzero*t/298.15+dHzero*(1-t/298.15);
istermG = gibbscoeff*(zi^2)*(is^.5)/(1 + 1.6*is^.5);
gfnsp=dGzeroT - istermG;
hfnsp=-t^2*D[gfnsp/t,t];
hfnsp/.t->temp/.is->ionstr]

calcdHTsp[atpsp,298.15,0]

{-3619.21, -3612.91, -3627.91}
```

These are the expected values for these three species at zero ionic strength.

```
atpsp

{{-2768.1, -3619.21, -4, 12}, {-2811.48, -3612.91, -3, 13},
  {-2838.18, -3627.91, -2, 14}}
```

```
TableForm[calcdHTsp[atpsp,{283.15,298.15,313.15},0],TableHeadings-
>{{"ATP-4","HATP-3","H2ATP-2"},{"283.15 K","298.15 K","313.15 K"}}]
```

	283.15 K	298.15 K	313.15 K
ATP-4	-3619.21	-3619.21	-3619.21
HATP-3	-3612.91	-3612.91	-3612.91
H2ATP-2	-3627.91	-3627.91	-3627.91

This calculation is based on the assumption that the standard enthalpies of formation of the three species are independent of temperature, and so this is the expected result.

```
TableForm[calcdHTsp[atpsp,{283.15,298.15,313.15},.1],TableHeadings-
>{{"ATP-4","HATP-3","H2ATP-2"},{"283.15 K","298.15 K","313.15 K"}}]
```

	283.15 K	298.15 K	313.15 K
ATP-4	-3615.12	-3614.23	-3613.22
HATP-3	-3610.61	-3610.11	-3609.54
H2ATP-2	-3626.89	-3626.66	-3626.41

Even though the standard enthalpies of formation are independent of temperature at zero ionic strength, they are not independent of temperature at a finite ionic strength. The largest effect is for the -4 ion.

```
TableForm[calcdHTsp[atpsp,{283.15,298.15,313.15},.25],TableHeadings-
>{{"ATP-4","HATP-3","H2ATP-2"},{"283.15 K","298.15 K","313.15 K"}}]
```

	283.15 K	298.15 K	313.15 K
ATP-4	-3613.8	-3612.62	-3611.29
HATP-3	-3609.86	-3609.2	-3608.46
H2ATP-2	-3626.56	-3626.26	-3625.93

3.9 There are two ways to obtain values for the enthalpy coefficient in equation 3.6-5 as a function of temperature: (a) Calculate the derivative of the Gibbs energy coefficient divided by T. (b) Fit the enthalpy coefficients of Clarke and Glew to $AT^2 + BT^3$. Use both of these methods and make plots to compare these functions with the values in Table 3.1

(a) According to Problem 3.5, $RT\alpha$ is given by

```
rtalpha=9.20485*10^-3*t-1.28466*10^-5*t^2+4.95199*10^-8*t^3
```

$$0.00920485\ t - 0.0000128466\ t^2 + 4.95199\ 10^{-8}\ t^3$$

Using the Gibbs-Helmholtz equation

```
rt2Dalpha=Expand[t^2*D[rtalpha/t,t]]
```

$$0.\ t - 0.0000128466\ t^2 + 9.90398\ 10^{-8}\ t^3$$

```
plot1=Plot[rt2Dalpha,{t,270,314},DisplayFunction->Identity];
```

The following are the enthalpy coefficients given by Clarke and Glew.

```
datah={{273.15,1.075},{283.15,1.213},{293.15,1.3845},{298.15,1.4775},{303.15,1.5775},{3
13.15,1.800}}
```

{{273.15, 1.075}, {283.15, 1.213}, {293.15, 1.3845}, {298.15, 1.4775},
   {303.15, 1.5775}, {313.15, 1.8}}

TableForm[datah]

273.15   1.075
283.15   1.213
293.15   1.3845
298.15   1.4775
303.15   1.5775
313.15   1.8

plot2=ListPlot[datah,AxesOrigin->{270,1},Prolog->AbsolutePointSize[4],AxesLabel->{"T/K","coefh"},DisplayFunction->Identity];

Show[plot2,plot1,DisplayFunction->$DisplayFunction];

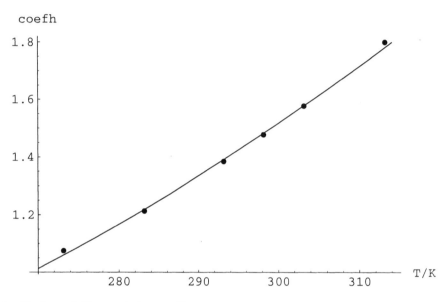

(b) Fit the Clarke and Glew enthalpy coefficients

coefh=Fit[datah,{t^2,t^3},t]

$-0.0000132345 \ t^2 + 1.0045 \ 10^{-7} \ t^3$

Note -1.28466*10^-5+9.90398*10^-8t^3 in (a),

coefh/.t->{273.15,283.15,293.15,298.15,303.15,313.15}

{1.05973, 1.21928, 1.39324, 1.48582, 1.58223, 1.78684}

These values can be compared with the table in (a).

plot3=Plot[coefh,{t,270,314},DisplayFunction->Identity];

plot4=ListPlot[datah,AxesOrigin->{270,1},Prolog->AbsolutePointSize[4],AxesLabel->{"T/K","coefh"},DisplayFunction->Identity];

Show[plot4,plot3,DisplayFunction->$DisplayFunction];

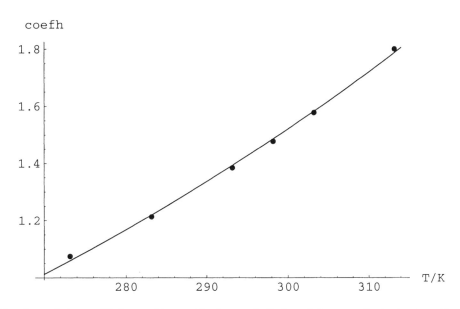

3.10 Plot the activity coefficients of ions with charges 1, 2, 3, and 4 versus the ionic strength at 0 °C. Repeat these calculations at 25 °C and 40 °C.

At 0 °C

```
gamma0[z_,is_]:=Module[{},(*Calculates the activity coefficient of an ion of charge z
at 0 C as a function of ionic strength at 0 C.*)
Exp[-1.12942*(z^2)*(is^.5)/(1+1.6*is^.5)]]

plot01=Plot[gamma0[1,is],{is,0,.25},PlotRange->{0,1},AxesOrigin->{0,0},AxesLabel-
>{"I/M","γ"},DisplayFunction->Identity];

plot02=Plot[gamma0[2,is],{is,0,.25},PlotRange->{0,1},AxesOrigin->{0,0},AxesLabel-
>{"I/M","γ"},DisplayFunction->Identity];

plot03=Plot[gamma0[3,is],{is,0,.25},PlotRange->{0,1},AxesOrigin->{0,0},AxesLabel-
>{"I/M","γ"},DisplayFunction->Identity];

plot04=Plot[gamma0[4,is],{is,0,.25},PlotRange->{0,1},AxesOrigin->{0,0},AxesLabel-
>{"I/M","γ"},DisplayFunction->Identity];

Show[plot01,plot02,plot03,plot04,DisplayFunction->$DisplayFunction];
```

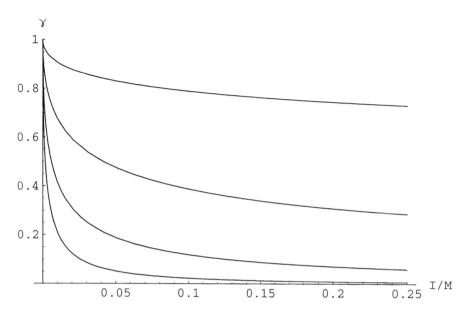

At 25 °C

```
gamma25[z_,is_]:=Module[{},(*Calculates the activity coefficient of an ion of charge z
at 0 C as a function of ionic strength at 0 C.*)
Exp[-1.17585*(z^2)*(is^.5)/(1+1.6*is^.5)]]

plot251=Plot[gamma25[1,is],{is,0,.25},PlotRange->{0,1},AxesOrigin->{0,0},AxesLabel-
>{"I/M","γ"},DisplayFunction->Identity];

plot252=Plot[gamma25[2,is],{is,0,.25},PlotRange->{0,1},AxesOrigin->{0,0},AxesLabel-
>{"I/M","γ"},DisplayFunction->Identity];

plot253=Plot[gamma25[3,is],{is,0,.25},PlotRange->{0,1},AxesOrigin->{0,0},AxesLabel-
>{"I/M","γ"},DisplayFunction->Identity];

plot254=Plot[gamma25[4,is],{is,0,.25},PlotRange->{0,1},AxesOrigin->{0,0},AxesLabel-
>{"I/M","γ"},DisplayFunction->Identity];

Show[plot251,plot252,plot253,plot254,DisplayFunction->$DisplayFunction];
```

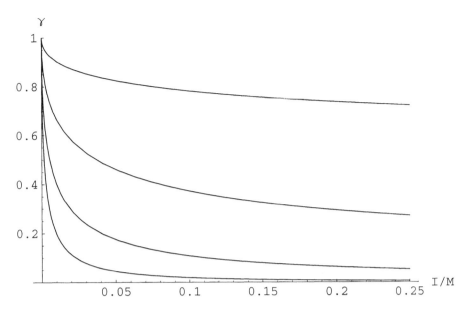

At 40 °C

```
gamma40[z_,is_]:=Module[{},(*Calculates the activity coefficient of an ion of charge z
at O C as a function of ionic strength at 0 C.*)
Exp[-1.20729*(z^2)*(is^.5)/(1+1.6*is^.5)]]

plot401=Plot[gamma40[1,is],{is,0,.25},PlotRange->{0,1},AxesOrigin->{0,0},AxesLabel-
>{"I/M","γ"},DisplayFunction->Identity];

plot402=Plot[gamma40[2,is],{is,0,.25},PlotRange->{0,1},AxesOrigin->{0,0},AxesLabel-
>{"I/M","γ"},DisplayFunction->Identity];

plot403=Plot[gamma40[3,is],{is,0,.25},PlotRange->{0,1},AxesOrigin->{0,0},AxesLabel-
>{"I/M","γ"},DisplayFunction->Identity];

plot404=Plot[gamma40[4,is],{is,0,.25},PlotRange->{0,1},AxesOrigin->{0,0},AxesLabel-
>{"I/M","γ"},DisplayFunction->Identity];

Show[plot401,plot402,plot403,plot404,DisplayFunction->$DisplayFunction];
```

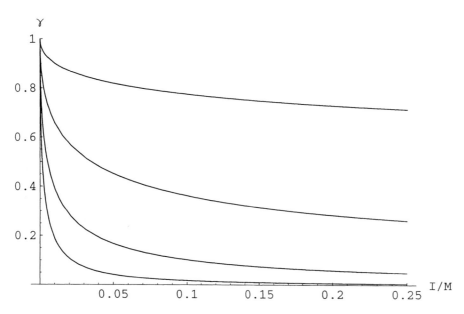

At 0 C and 40 C.

```
Show[plot01,plot02,plot03,plot04,plot401,plot402,plot403,plot404,DisplayFunction-
>$DisplayFunction];
```

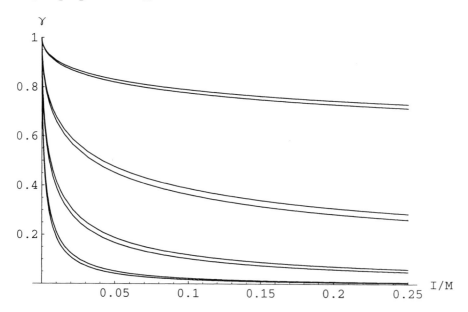

Chapter 4 Thermodynamics of Biochemical Reactions at Specified pH

4.1 (a) Calculate $\Delta_f G'^\circ$ for the species of ATP at 298.15 K, pH 7, and I = 0.25. (b) Calculate $\Delta_f G'^\circ$ for ATP at 298.15 K, pH 7, and I = 0.25 M. (c) Plot $\Delta_f G'^\circ$ for ATP at 298.15 K, and I = 0.25 M versus pH. (d) Plot N_H for ATP at I = 0.25 M versus pH.

4.2 (a) Calculate $\Delta_f H'^\circ$ for ATP at 298.15 K, pH 7, and I = 0.25 M. (b) Plot $\Delta_f H'^\circ$ for ATP at 298.15 K, and I = 0.25 M versus pH. (c) Calculate the standard transformed enthalpy of formation at pH 7 and 0.25 M ionic strength at several temperatures in the range 273 K to 313 K.

4.3 (a) Calculate $\Delta_r G'^\circ$ in kJ mol^{-1} at 298.15 K, pH 7, and I = 0.25 M for ATP + H$_2$O = ADP + P$_i$. (b) Calculate the corresponding $\Delta_r H'^\circ$. (c) Calculate the corresponding $\Delta_r S'^\circ$ in J K^{-1} mol^{-1}. (d) Calculate logK'. (e) Plot the values of each of these properties versus pH at I = 0.25 M on the assumption that the standard enthalpies of formation of ions are independent of temperature.

4.4 (a) For ATP + H$_2$O = ADP + P$_i$ plot K' versus pH at ionic strengths of 0, 0.10, and 0.25 M. (b) Plot logK' versus pH at ionic strengths of 0, 0.10, and 0.25 M. (c) Plot K' versus ionic strength at pHs 5, 7, and 9.

4.5 For ATP + H$_2$O = ADP + P$_i$ plot $\Delta_r N_H$ versus pH at ionic strengths of 0, 0.10, and 0.25 M.

4.6 Calculate the standard transformed Gibbs energies of reaction for ATP + H$_2$O = ADP + P$_i$ at temperatures of 283.15 K, 298.15 K, and 313.15 K, at pHs 5, 6, 7, 8, 9, and ionic strengths of 0, 0.10, and 0.25 M.

4.7 Calculate the standard transformed Gibbs energies of reaction at 298.15 K and the experimental pH and ionic strength for the reactions in the Goldberg and Tewari series of six critical reviews for which all the reactants are in BasicBiochem-Data2. Compare the calculated standard transformed Gibbs energies of reaction with the experimental values. At present there is not enough information to calculate the effects of temperature and metal ions, and so these effects are ignored. The three steps in this process are: (a) Make a table of the calculated standard transformed Gibbs energies of reaction. (b) Make a table of the relevant data in the Goldberg and Tewari Tables. (c) Make a table of the differences between the values of the standard transformed Gibbs energies of reaction calculated in part (a) and the experimental values in the Goldberg and Tewari tables.

4.8 Calculate the standard transformed reaction Gibbs energies, apparent equilibrium constants, and changes in the binding of hydrogen ions for the ten reactions of glycolysis at 298.15 K, pHs 5, 6, 7, 8, and 9, and ionic strengths 0, 0.10, and 0.25 M. Also calculate these properties for the net reaction.

4.9 Calculate the standard transformed reaction Gibbs energies, apparent equilibrium constants, and changes in the binding of hydrogen ions for the four reactions of gluconeogenesis that are different from those in glycolysis at 298.15 K, pHs 5, 6, 7, 8, and 9 and ionic strengths 0, 0.10, and 0.25 M. Also calculate the properties of the net of pyruvate carboxylase and phosphoenolpyruvate carboxykinase reactions and the net reaction of gluconeogenesis.

4.10 Calculate the standard transformed reaction Gibbs energies, apparent equilibrium constants, and changes in the binding of hydrogen ions for the pyruvate dehydrogenase reaction and the nine reactions of the citric acid cycle at 298.15 K, pHs 5, 6, 7, 8, and 9 and ionic strengths 0, 0.10, and 0.25 M. Also calculate these properties for the net reaction of the citric acid cycle, the net reaction for pyruvate dehydrogenase plus the citric acid cycle, and the net reaction for glycolysis, pyruvate dehydrogenase, and the citric acid cycle.

4.1 (a) Calculate $\Delta_f G'^\circ$ for the species of ATP at 298.15 K, pH 7, and I = 0.25. (b) Calculate $\Delta_f G'^\circ$ for ATP at 298.15 K, pH 7, and I = 0.25 M. (c) Plot $\Delta_f G'^\circ$ for ATP at 298.15 K, and I = 0.25 M versus pH. (d) Plot N_H for ATP at I = 0.25 M versus pH.

(BasicBiochemData2 has to be loaded)

(a) Use equation 4.4-10

```
dgsp-nH*8.31451*.29815*Log[10^-pH]-2.91482*(zi^2-nH)*(is^.5)/(1+1.6*is^.5)
```

$$dgsp - \frac{2.91482 \ is^{0.5} \ (-nH + zi^2)}{1 + 1.6 \ is^{0.5}} - 2.47897 \ nH \ Log[10^{-pH}]$$

The data on the three species of ATP are in \the package.

The basic data for species of ATP are given by

```
atpsp
```

```
{{-2768.1, -3619.21, -4, 12}, {-2811.48, -3612.91, -3, 13},
  {-2838.18, -3627.91, -2, 14}}
```

```
TableForm[atpsp]
```

```
-2768.1    -3619.21   -4   12
-2811.48   -3612.91   -3   13
-2838.18   -3627.91   -2   14
```

The first row gives the standard Gibbs energy of formation, the standard enthalpy of formation, the charge number, and the number of hydrogen atoms of the species with the fewest hydrogen atoms. The standard transformed Gibbs energy of this species can be calculated by using the function in the first line of this problem.

```
atpsp[[2,1]]
```

```
-2811.48
```

```
g1=atpsp[[1,1]]-12*8.31451*.29815*Log[10^-7]-2.91482*((-4)^2-12)*(.25^.5)/(1+1.6*.25^.5
)
```

```
-2291.86
```

```
g2=atpsp[[2,1]]-13*8.31451*.29815*Log[10^-7]-2.91482*((-3)^2-13)*(.25^.5)/(1+1.6*(.25^.
5))
```

```
-2288.81
```

```
g3=atpsp[[3,1]]-14*8.31451*.29815*Log[10^-7]-2.91482*((-2)^2-14)*(.25^.5)/(1+1.6*(.25^.
5))
```

```
-2270.7
```

(b) Use equation 4.5-1 to calculate $\Delta_f G'^\circ$ for the pseudoisomer group

```
-8.31451*.29815*Log[Exp[-g1/(8.31451*.29815)]+Exp[-g2/(8.31451*.29815)]+Exp[-
g3/(8.31451*.29815)]]
```

-2292.5

calcdGmat[atpsp]/.pH->7/.is->.25

-2292.5

This illustrates the convenience of the program calcdGmat..

(c) When the package BasicBiochemData2 is loaded, typing atp yields a function of pH and ionic strength as we have
seen in (b).

Plot[atp/.is->.25,{pH,5,9},AxesLabel->{"pH","\!\(TraditionalForm\`\(Δ_r\) G'\^o\)"}];

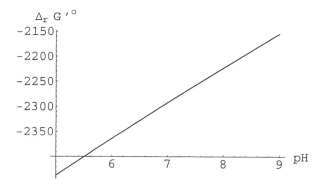

(d) Equation 4.7-3 shows that the average number of hydrogen atoms in a reactant is given by $(1/RT\ln(10))$ times the
derivative of the standard transformed Gibbs energy of the reactant. First we calculate the function of pH at ionic strength
0.25 M

fpH=atp/.is->.25

$$-2.47897 \; \text{Log}[E^{-0.403393 \; (-2830.08 \; - \; 34.7056 \; \text{Log}[10^{-pH}])} \; +$$

$$E^{-0.403393 \; (-2808.24 \; - \; 32.2266 \; \text{Log}[10^{-pH}])} \; +$$

$$E^{-0.403393 \; (-2771.34 \; - \; 29.7477 \; \text{Log}[10^{-pH}])} \;]$$

dvt=D[fpH,pH]

$$
(-2.47897\ \left(\cfrac{-32.2362}{E^{0.403393\ (-2830.08\ -\ 34.7056\ Log[10^{-pH}])}}\ -\right.
$$

$$
\cfrac{29.9336}{E^{0.403393\ (-2808.24\ -\ 32.2266\ Log[10^{-pH}])}}\ -
$$

$$
\left.\cfrac{27.631}{E^{0.403393\ (-2771.34\ -\ 29.7477\ Log[10^{-pH}])}}\right))\ /
$$

$$
(E^{-0.403393\ (-2830.08\ -\ 34.7056\ Log[10^{-pH}])}\ +
$$

$$
E^{-0.403393\ (-2808.24\ -\ 32.2266\ Log[10^{-pH}])}\ +
$$

$$
E^{-0.403393\ (-2771.34\ -\ 29.7477\ Log[10^{-pH}])})
$$

```
Plot[Evaluate[(1/(8.31451*.29815*Log[10]))*D[fpH,pH]],{pH,5,9},AxesLabel-
>{"pH","\!\(N\_H\)"}];
```

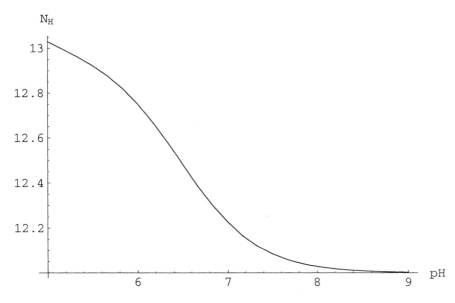

4.2 (a) Calculate $\Delta_f H'^\circ$ for ATP at 298.15 K, pH 7, and I = 0.25 M. (b) Plot $\Delta_f H'^\circ$ for ATP at 298.15 K, and I = 0.25 M versus pH. (c) Calculate the standard transformed enthalpy of formation at pH 7 and 0.25 M ionic strength at several temperatures in the range 273 K to 313 K on the assumption that the standard enthalpies of formation of ions are independent of pH..

(BasicBiochemData2 has to be loaded)

(a) There are two different kinds of programs to make this calculation. The program calcdHmat calculates the equilibrium mole fractions of the species and the standard transformed enthalpies of formation of the species and calculates the mole fraction weighted average. The program calcdHTgp calculates the standard transformed Gibbs energy of formation of the reactant as a function of pH, ionic strength, and temperature and uses the Gibbs-Helmholtz equation to calculate the standard transfromed enthalpy of formation of the reactant.

```
calcdHmat[speciesmat_] :=
 Module[{dHzero, zi, nH, dhfnsp, dGzero, pHterm, isenth, dgfnsp, dGreactant, ri},
  (*This program produces the function of ionic strength (is) that gives
    the standard transformed enthalpies of formation of the specie at
    298.15 K.  It then calculates the standard transformed Gibbs energy for
    the reactant and the equilibrium mole fractions of the species.  The
    function of pH and ionic strength for the standard transformed enthalpy
    of formation of the reactant is calculated by a dot product.  The
    input is a matrix that gives the standard Gibbs energy of formation,
   the standard enthalpy of formation, the electric charge,and the number of
    hydrogen atoms in the species in the reactant.  There is a row in the matrix
    for each species of the reactant.  Energies are expressed in kJ mol^-1.*)
    {dGzero, dHzero, zi, nH} = Transpose[speciesmat];
    isenth = 1.4775 * ((zi ^ 2) - nH) * (is ^ .5) / (1 + 1.6 * is ^ .5);
    dhfnsp = dHzero + isenth;
  (*Now calculate the functions for
    the standard Gibbs energies of formation of the species.*)
  pHterm = nH * 8.31451 * .29815 * Log[10 ^ -pH];
  gpfnsp = dGzero - pHterm - isenth * (2.91482 / 1.4775);
   (*Now calculate the standard
    transformed Gibbs energy of formation for the reactant.*)
  dGreactant = -8.31451 * .29815 * Log[Apply[Plus, Exp[-1 * gpfnsp / (8.31451 * .29815)]]];
   (*Now calculate the equilibrium mole fractions of the species
      in the reactant and the mole fraction-weighted average of the
      functions for the standard transformed enthalpies of the species.*)
  ri = Exp[(dGreactant - gpfnsp) / (8.31451 * .29815)];
  ri.dhfnsp]

fnh=calcdHmat[atpsp]/.is->.25/.pH->7

-3616.89


calcdHTgp[speciesmat_] :=
Module[{dGzero, dGzeroT,dHzero,zi, nH, gibbscoeff,pHterm, isterm,gpfnsp,dGfn},(*This
program first produces the function of T (in Kelvin), pH and ionic strength (is) that
gives the standard transformed Gibbs energy of formation of a reactant (sum of
species).  It then uses the Gibbs-Helmholtz equation to calculate the function for the
standard transformed enthalpy of formation of the pseudoisomer group.  The input
speciesmat is a matrix that gives the standard Gibbs energy of formation at 298.15 K,
the standard enthalpy of formation at 298.15 K, the electric charge, and the number of
hydrogen atoms in each species. There is a row in the matrix for each species of the
reactant. gpfnsp is a list of the functions for the transformed Gibbs energies of the
species.  Energies are expressed in kJ mol^-1.  The value of the standard transformed
enthalpy of formation can be calculated at any temperature in the approximate range
273.15 K to 313.15 K, any pH in the range 5 to 9, and any ionic strength in tho range
0 to 0.35 m by use of the assignment operator(/.).*)
{dGzero,dHzero,zi,nH}=Transpose[speciesmat];
gibbscoeff=9.20483*10^-3*t-1.284668*10^-5*t^2+4.95199*10^-8*t^3;
dGzeroT=dGzero*t/298.15+dHzero*(1-t/298.15);
pHterm = nH*8.31451*(t/1000)*Log[10^-pH];
istermG = gibbscoeff*((zi^2) - nH)*(is^.5)/(1 + 1.6*is^.5);
gpfnsp=dGzeroT - pHterm - istermG;
dGfn=-8.31451*(t/1000)*Log[Apply[Plus,Exp[-1*gpfnsp/(8.31451*(t/1000))]]];
-t^2*D[dGfn/t,t]]

calcdHTgp[atpsp]/.pH->7/.is->.25/.t->298.15

-3616.89
```

This second program has the advantage that the standard transformed enthalpy of formation can be calculated at other temperatures.

(b) This second program can be used to construct a plot of the standard transformed enthalpy of formation as a function of pH.

```
Plot[calcdHTgp[atpsp]/.is->.25/.t->298.15,{pH,5,9},AxesOrigin->{5,-3618},AxesLabel-
>{"pH","\!\(TraditionalForm\`\(\(\Delta\_f\) H'\^o\)\)"},PlotRange->{-3618,-3615}];
```

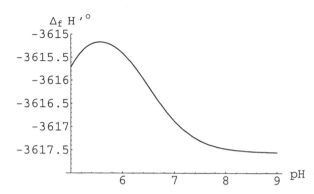

(c) This second program can be used to calculate the standard transformed enthalpy of formation at several temperature.

```
calcdHTgp[atpsp]/.pH->7/.is->.25/.t->{270,298,313}
```

{-3617.28, -3616.89, -3616.68}

This calculation is based on the assumption that the standard enthalpies of formation of the ions at zero ionic strength are independent of temperature. However, at finite ionic strength there is a small electrostatic effect on the enthalpy.

4.3 (a) Calculate $\Delta_r G'^\circ$ in kJ mol^{-1} at 298.15 K, pH 7, and I = 0.25 M for ATP + H$_2$O = ADP + P$_i$. (b) Calculate the corresponding $\Delta_r H'^\circ$. (c) Calculate the corresponding $\Delta_r S'^\circ$ in J K^{-1}mol^{-1}. (d) Calculate logK'. (e) Plot the values of each of these properties versus pH at I = 0.25 M.

(BasicBiochemData2 must be loaded)

```
calctrGerx[eq_,pHlist_,islist_]:=Module[{energy},(*Calculates the standard transformed
Gibbs energy of reaction in kJ mol^-1 at specified pHs and ionic strengths for a
biochemical equation typed in the form atp+h2o+de==adp+pi.  The names of the reactants
call the appropriate functions of pH and ionic strength. pHlist and islist can be
lists. This program can be used to calculate the standard transformed enthalpy of
reaction by appending an h to the name of each reactant.*)
energy=Solve[eq,de];
energy[[1,1,2]]/.pH->pHlist/.is->islist]
```

(a) Calculate the standard transformed reaction Gibbs energy

```
dg298pH7is25=calctrGerx[atp+h2o+de==adp+pi,7,.25]
```

-36.0353

(b) Append h to the name of each reactaant to obtain its standard transformed reaction enthalpy

```
calctrGerx[atph+h2oh+de==adph+pih,7,.25]
```

-23.0327

(c) The standard transformed entropy of reaction at pH 7 and I = 0.25 M is given by

(-23.03+36.07)*1000/298.15

43.7364

in $J\,K^{-1}\,mol^{-1}$.

(d) The base 10 logarithm of the apparent equilibrium constant is given by

Log[10,Exp[-dg298pH7is25/(8.31451*.29815)]]

6.31308

(e) Plots

Plot[calctrGerx[atp+h2o+de==adp+pi,pH,.25],{pH,5,9},AxesLabel->{"pH","\!\(TraditionalForm\`\(Δ_r\) G'\^o\)"}];

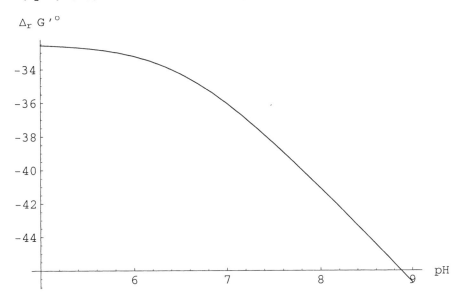

Plot[calctrGerx[atph+h2oh+de==adph+pih,pH,.25],{pH,5,9},AxesLabel->{"pH","\!\(TraditionalForm\`\(Δ_r\) H'\^o\)"},AxesOrigin->{5,-27}];

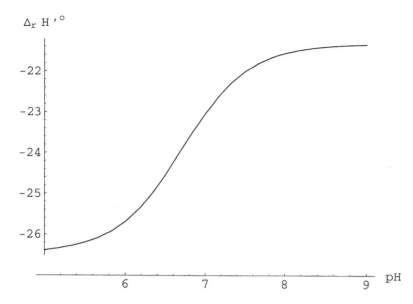

```
Plot[(calctrGerx[atph+h2oh+de==adph+pih,pH,.25]-
calctrGerx[atp+h2o+de==adp+pi,pH,.25])/.29815,{pH,5,9},AxesLabel-
>{"pH","\!\(TraditionalForm\`\(Δ\_r\) S'\^o\)"}];
```

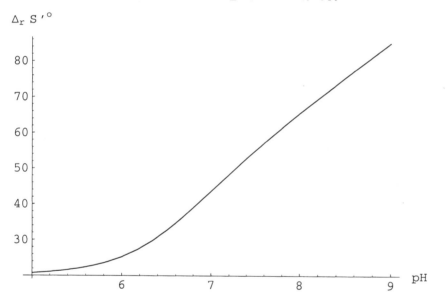

As the pH increases, the change in the transformed enthalpy becomes less favorable for hydrolysis, but the effect is small. On the other hand the transformed entropy increases rapidly with pH and causes the equilibrium to move further to the right.

```
calckprime[eq_,pHlist_,islist_]:=Module[{energy,dG},(*Calculates the apparent
equilibrium constant K' at specified pHs and ionic strengths for a biochemical
equation typed in the form atp+h2o+de==adp+pi.  The names of the reactants call the
appropriate functions of pH and ionic strength. pHlist and islist can be entered as
lists.*)
energy=Solve[eq,de];
dG=energy[[1,1,2]]/.pH->pHlist/.is->islist;
Exp[-dG/(8.31451*.29815)]]

Plot[Log[10,calckprime[atp+h2o+de==adp+pi,pH,.25]],{pH,5,9},AxesOrigin-
>{5,5.5},AxesLabel->{"pH","LogK'"}];
```

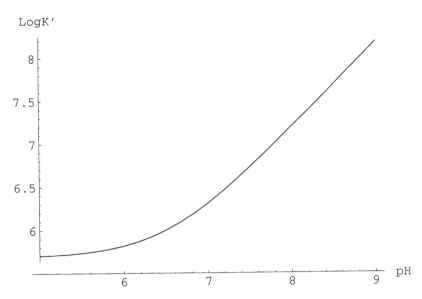

4.4 (a) For ATP + H_2O = ADP + P_i plot K' versus pH at ionic strengths of 0, 0.10, and 0.25 M. (b) Plot logK' versus pH at ionic strengths of 0, 0.10, and 0.25 M. (c) Plot K' versus ionic strength at pHs 5, 7, and 9.

(BasicBiochemData2 has to be loaded)

```
calckprime[eq_,pHlist_,islist_]:=Module[{energy,dG},(*Calculates the apparent
equilibrium constant K' at specified pHs and ionic strengths for a biochemical
equation typed in the form atp+h2o+de==adp+pi.  The names of the reactants call the
appropriate functions of pH and ionic strength. pHlist and islist can be entered as
lists.*)
energy=Solve[eq,de];
dG=energy[[1,1,2]]/.pH->pHlist/.is->islist;
Exp[-dG/(8.31451*.29815)]]
```

(a)

```
kprime00=calckprime[atp+h2o+de==adp+pi,pH,0];

kprime10=calckprime[atp+h2o+de==adp+pi,pH,.10];

kprime25=calckprime[atp+h2o+de==adp+pi,pH,.25];

plotKp00=Plot[kprime00,{pH,5,9},DisplayFunction->Identity];

plotKp10=Plot[kprime10,{pH,5,9},DisplayFunction->Identity];

plotKp25=Plot[kprime25,{pH,5,9},DisplayFunction->Identity];

Show[plotKp00,plotKp10,plotKp25,AxesOrigin->{5,10^5},PlotRange->{10^5,10^7},AxesLabel-
>{"pH","K'"},DisplayFunction->$DisplayFunction];
```

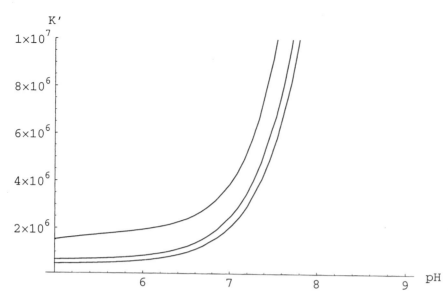

The upper curve is for zero ionic strength.

(b)

```
plotlogKp00=Plot[Log[10,kprime00],{pH,5,9},DisplayFunction->Identity];

plotlogKp10=Plot[Log[10,kprime10],{pH,5,9},DisplayFunction->Identity];

plotlogKp25=Plot[Log[10,kprime25],{pH,5,9},DisplayFunction->Identity];

Show[plotlogKp00,plotlogKp10,plotlogKp25,PlotRange->{5,8.5},AxesOrigin-
>{5,5},AxesLabel->{"pH","logK'"},DisplayFunction->$DisplayFunction];
```

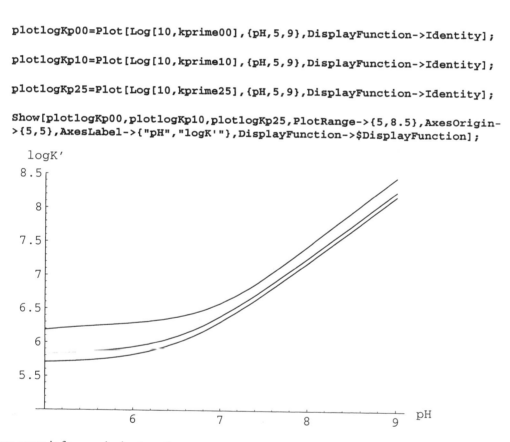

The upper curve is for zero ionic strength.

(c) In order to compare the ionic strength effects, the calculated apparent equilibrium constants are divided by the value at zero ionic strength at that pH.

```
kprime5=calckprime[atp+h2o+de==adp+pi,5,is]/calckprime[atp+h2o+de==adp+pi,5,0];

kprime7=calckprime[atp+h2o+de==adp+pi,7,is]/calckprime[atp+h2o+de==adp+pi,7,0];

kprime9=calckprime[atp+h2o+de==adp+pi,9,is]/calckprime[atp+h2o+de==adp+pi,9,0];

plotK5is=Plot[kprime5,{is,0,.3},DisplayFunction->Identity];

plotK7is=Plot[kprime7,{is,0,.3},DisplayFunction->Identity];

plotK9is=Plot[kprime9,{is,0,.3},DisplayFunction->Identity];

Show[plotK5is,plotK7is,plotK9is,AxesOrigin->{0,.2},AxesLabel-
>{"I/M","K'(I)/K'(I=0)"},PlotRange->{.2,1},DisplayFunction->$DisplayFunction];
```

The lowest curve is for pH 5.

4.5 For ATP + H$_2$O = ADP + P$_i$ plot $\Delta_r N_H$ versus pH at ionic strengths of 0, 0.10, and 0.25 M.

(BasicBiochemData has to be loaded)

```
calckprime[eq_,pHlist_,islist_]:=Module[{energy,dG},(*Calculates the apparent
equilibrium constant K' at specified pHs and ionic strengths for a biochemical
equation typed in the form atp+h2o+de==adp+pi.  The names of the reactants call the
appropriate functions of pH and ionic strength. pHlist and islist can be entered as
lists.*)
energy=Solve[eq,de];
dG=energy[[1,1,2]]/.pH->pHlist/.is->islist;
Exp[-dG/(8.31451*.29815)]]

kprime00=calckprime[atp+h2o+de==adp+pi,pH,0];

nH00=(-1/Log[10])*D[Log[kprime00],pH];

plotnH00=Plot[nH00,{pH,5,9},DisplayFunction->Identity];

kprime10=calckprime[atp+h2o+de==adp+pi,pH,.10];
```

```
nH10=(-1/Log[10])*D[Log[kprime10],pH];

plotnH10=Plot[nH10,{pH,5,9},DisplayFunction->Identity];

kprime25=calckprime[atp+h2o+de==adp+pi,pH,.25];

nH25=(-1/Log[10])*D[Log[kprime25],pH];

plotnH25=Plot[nH25,{pH,5,9},DisplayFunction->Identity];

Show[plotnH00,plotnH10,plotnH25,AxesLabel->{"pH"," ΔrNH "},AxesOrigin-
>{5,-1},DisplayFunction->$DisplayFunction];
```

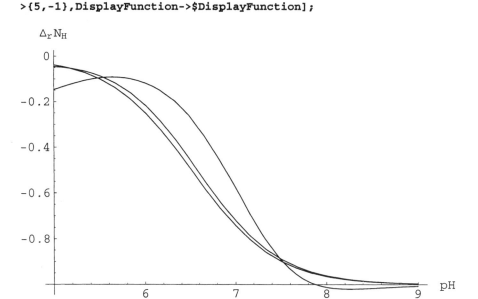

4.6 Calculate the standard transformed Gibbs energies of reaction for ATP + H_2O = ADP + P_i at temperatures of (a) 298.15 K, (b) 283.15 K, and (c) 313.15 K, at pHs 5, 6, 7, 8, 9, and ionic strengths of 0, 0.10, and 0.25 M.

(BasicBiochemData2 has to be loaded)

(a) Calculations at 298.15 K

```
calctrGerx[eq_,pHlist_,islist_]:=Module[{energy},(*Calculates the standard transformed
Gibbs energy of reaction in kJ mol^-1 at specified pHs and ionic strengths for a
biochemical equation typed in the form atp+h2o+de==adp+pi. The names of the reactants
call the appropriate functions of pH and ionic strength. pHlist and islist can be
lists. This program can be used to calculate the standard transformed enthalpy of
reaction by appending an h to the name of each reactant.*)
energy=Solve[eq,de];
energy[[1,1,2]]/.pH->pHlist/.is->islist]

dGerx298=calctrGerx[atp+h2o+de==adp+pi,{5,6,7,8,9},{0,.1,.25}];

TableForm[Transpose[dGerx298],TableHeadings->{{"I = 0 M","I = 0.10 M","I = 0.25
M"},{"pH 5","pH 6","pH 7","pH 8","pH 9"}}]
```

	pH 5	pH 6	pH 7	pH 8	pH 9
I = 0 M	-35.299	-35.9137	-37.6048	-42.5001	-48.2933
I = 0.10 M	-33.2951	-33.8717	-36.5	-41.4784	-47.0983
I = 0.25 M	-32.5633	-33.2166	-36.0353	-41.0742	-46.7021

To make the calculation at the other temperatures, we can use calcdGHT to produce the function of T, pH and ionic strength that will give ΔG'° and ΔH'° at other temperatures.

```
calcdGHT[speciesmat_] :=
Module[{dGzero, dGzeroT,dHzero,zi, nH, gibbscoeff,pHterm,
isterm,gpfnsp,dGfn,dHfn},(*This program produces the function of T (in Kelvin), pH and
ionic strength (is) that gives the standard transformed Gibbs energy of formation of a
reactant (sum of species) and the standard transformed enthalpy.  The input speciesmat
is a matrix that gives the standard Gibbs energy of formation at 298.15 K, the
standard enthalpy of formation at 298.15 K, the electric charge, and the number of
hydrogen atoms in each species. There is a row in the matrix for each species of the
reactant. gpfnsp is a list of the functions for the transformed Gibbs energies of the
species.  The output is in the form {dGfn,dHfn}, and energies are expressed in kJ
mol^-1.  The values of the standard transformed Gibbs energy of formation and the
standard transformed enthalpy of formation can be calculated at any temperature in the
range 273.15 K to 313.15 K, any pH in the range 5 to 9, and any ionic strength in the
range 0 to 0.35 m by use of the assignment operator(/.).*)
{dGzero,dHzero,zi,nH}=Transpose[speciesmat];
gibbscoeff=9.20483*10^-3*t-1.284668*10^-5*t^2+4.95199*10^-8*t^3;
dGzeroT=dGzero*t/298.15+dHzero*(1-t/298.15);
pHterm = nH*8.31451*(t/1000)*Log[10^-pH];
istermG = gibbscoeff*((zi^2) - nH)*(is^.5)/(1 + 1.6*is^.5);
gpfnsp=dGzeroT - pHterm - istermG;
dGfn=-8.31451*(t/1000)*Log[Apply[Plus,Exp[-1*gpfnsp/(8.31451*(t/1000))]]];
dHfn=-t^2*D[dGfn/t,t];
{dGfn,dHfn}]
```

(b) Calculate the standard transformed Gibbs energies of of the reactants and the reaction at 283.15 K.

```
atp283=calcdGHT[atpsp][[1]]/.t->283.15;
```

```
h2o283=calcdGHT[h2osp][[1]]/.t->283.15;
```

```
adp283=calcdGHT[adpsp][[1]]/.t->283.15;
```

```
pi283=calcdGHT[pisp][[1]]/.t->283.15;
```

```
dGerx283=calctrGerx[atp283+h2o283+de==adp283+pi283,{5,6,7,8,9},{0,.1,.25}]
```

```
{{-34.6913, -32.9147, -32.2468}, {-35.3381, -33.4522, -32.8351}, {-36.9415, -35.8386, -3
  {-41.4708, -40.4698, -40.0894}, {-46.9055, -45.7905, -45.4228}}
```

```
TableForm[Transpose[dGerx283],TableHeadings->{{"I = 0 M","I = 0.10 M","I = 0.25
M"},{"pH 5","pH 6","pH 7","pH 8","pH 9"}}]
```

	pH 5	pH 6	pH 7	pH 8	pH 9
I = 0 M	-34.6913	-35.3381	-36.9415	-41.4708	-46.9055
I = 0.10 M	-32.9147	-33.4522	-35.8386	-40.4698	-45.7905
I = 0.25 M	-32.2468	-32.8351	-35.3814	-40.0894	-45.4228

(c) Calculate the standard transformed Gibbs energies of the reactants and the reaction at 313.15 K.

```
atp313=calcdGHT[atpsp][[1]]/.t->313.15;
```

```
h2o313=calcdGHT[h2osp][[1]]/.t->313.15;
```

```
adp313=calcdGHT[adpsp][[1]]/.t->313.15;
```

```
pi313=calcdGHT[pisp][[1]]/.t->313.15;
```

```
dGerx313=calctrGerx[atp313+h2o313+de==adp313+pi313,{5,6,7,8,9},{0,.1,.25}]
```

{{-35.9046, -33.6656, -32.8677}, {-36.4876, -34.2845, -33.5895}, {-38.272, -37.1598, -36
  {-43.5265, -42.482, -42.0518}, {-49.6803, -48.4002, -47.9736}}

```
TableForm[Transpose[dGerx313],TableHeadings->{{"I = 0 M","I = 0.10 M","I = 0.25
M"},{"pH 5","pH 6","pH 7","pH 8","pH 9"}}]
```

	pH 5	pH 6	pH 7	pH 8	pH 9
I = 0 M	-35.9046	-36.4876	-38.272	-43.5265	-49.6803
I = 0.10 M	-33.6656	-34.2845	-37.1598	-42.482	-48.4002
I = 0.25 M	-32.8677	-33.5895	-36.6852	-42.0518	-47.9736

Since the hydrolysis of ATP evolves heat, Le Chatelier's principle says raising the temperature will cause the reaction to go less far to the right. But at 313 K the transformed Gibbs energy of reaction is more negative. To apply Le Chatelier we have to look at the apparent equilibrium constants. At pH 7 and ionic strength 0, we obtain

```
k283=Exp[36.98/(8.31451*.28315)]
```

$6.63414 \ 10^6$

```
k298=Exp[37.64/(8.31451*.29815)]
```

$3.9283 \ 10^6$

```
k313=Exp[38.31/(8.31451*.31315)]
```

$2.45525 \ 10^6$

These apparent equilibrium constants are in accord with what we expect.

4.7 Calculate the standard transformed Gibbs energies of reaction at 298.15 K and the experimental pH and ionic strength for the reactions in the Goldberg and Tewari series of six critical reviews for which all the reactants are in BasicBiochemData2. Compare the calculated standard transformed Gibbs energies of reaction with the experimental values. At present there is not enough information to calculate the effects of temperature and metal ions, and so these effects are ignored. The three steps in this process are: (a) Make a table of the calculated standard transformed Gibbs energies of reaction. (b) Make a table of the relevant data in the Goldberg and Tewari Tables. (c) Make a table of the differences between the values of the standard transformed Gibbs energies of reaction calculated in part (a) and the experimental values inthe Goldberg and Tewari Tables.

(BasicBiochemData has to be loaded)

(a) Make a table of the calculated standard transformed Gibbs energies of reaction.

```
calcrow[eq_, pHlist_, islist_, rx_] :=
 Module[{energy, dG, gvector}, (*Calculates the standard transformed Gibbs energy
      of reaction for a specified reaction at specified pHs and a single
      ionic strength. It then prepends the reaction without the +de=.
    This second form of the reaction must be in quotation marks.*)
  energy = Solve[eq, de];
  dG = energy[[1, 1, 2]] /. pH -> pHlist /. is -> islist;
  gvector = Flatten[Map[NumberForm[#, {4, 2}] &, {dG}, {2}]];
  Prepend[gvector, rx]]
```

```
g1x1x1x1 = calcrow[ethanol + nadox + de == acetaldehyde + nadred,
    7, .25, "ethanol+nadox==acetaldehyde+nadred"];

g1x1x1x26 = calcrow[glycolate + nadox + de == glyoxylate + nadred,
    7.8, .03, "glycolate+nadox=gloxylate+nadred"];

g1x1x1x27b = calcrow[lactate + nadox + de == pyruvate + nadred,
    6.8, .05, "lactate+nadox=pyruvate+nadred"];

g1x1x1x37=
    calcrow[malate+nadox+de==oxaloacetate+nadred,7.1,.25,
      "malate+nadox=oxaloacetate+nadred"];

g1x1x1x39=
    calcrow[malate+nadox+h2o+de==pyruvate+nadred+co2tot,6.5,.05,
      "malate+nadox+h2o=pyruvate+nadred+co2tot"];

g1x1x1x40=
    calcrow[malate+nadpox+h2o+de==pyruvate+nadpred+co2tot,7,.25,
      "malate+nadpox+h2o=pyruvate+nadpred+co2tot"];

g1x1x1x42=
    calcrow[citrateiso+nadpox+h2o+de==ketoglutarate+nadpred+co2tot,6.9,.25,
      "citrateiso+nadpox+h2o=ketoglutarate+nadpred+co2tot"];

g1x1x1x67=
    calcrow[mannitolD+nadox+de==fructose+nadred,8.05,.05,
      "mannitolD+nadox=fructose+nadred"];

g1x1x1x79=
    calcrow[glycolate+nadpox+de==glyoxylate+nadpred,7.92,.1,
      "glycolate+nadpox=glyoxylate+nadpred"];

g1x1x99x7=
    calcrow[lactate+oxaloacetate+de==malate+pyruvate,7.8,.1,
      "lactate+oxaloacetate=malate+pyruvate"];

g1x2x1x2=
    calcrow[formate+nadox+h2o+de==co2tot+nadred,6.2,.07,
      "formate+nadox+h2o=co2tot+nadred"];

g1x2x1x10=
    calcrow[acetaldehyde+coA+nadox+de==acetylcoA+nadred,7.3,.03,
      "acetaldehyde+coA+nadox=acetylcoA+nadred"];

g1x2x1x43=
    calcrow[formate+nadpox+h2o+de==co2tot+nadpred,7.5,.1,
      "formate+nadpox+h2o=co2tot+nadpred"];

g1x4x1x1 = calcrow[alanine + nadox + h2o + de == pyruvate + nadred + ammonia,
    7.97, .1, "alanine+nadox+h2o=pyruvate+nadred+ammonia"];

g1x4x1x2 = calcrow[glutamate + nadox + h2o + de == ketoglutarate + nadred + ammonia,
    6.9, .24, "glutamate+nadox+h2o=ketoglutarate+nadred+ammonia"];

g1x4x1x3 = calcrow[glutamate + nadpox + h2o + de == ketoglutarate + nadpred + ammonia,
    7, .47, "glutamate+nadpox+h2o=ketoglutarate+nadpred+ammonia"];

g1x4x1x10 = calcrow[glycine + nadox + h2o + de == glyoxylate + nadred + ammonia,
    6.4, .08, "glycine+nadox+h2o=glyoxylate+nadred+ammonia"];
```

```
g1x6x1x1=
  calcrow[nadox+nadpred+de==nadred+nadpox,7.5,.1,
    "nadox+nadpred=nadred+nadpox"];

g1x6x4x2=
  calcrow[nadpox+2*glutathionered+de==nadpred+glutathioneox,6.9,.25,
    "nadpox+2*glutathionered=nadpred+glutathioneox"];

g1x6x4x5=
  calcrow[nadpox+thioredoxinred+de==nadpred+thioredoxinox,7,.1,
    "nadpox+2*thioredoxinred=nadpred+thioredoxinox"];

g2x3x1x54=
  calcrow[acetylcoA+formate+de==coA+pyruvate,7.2,.05,
    "acetylcoA+formate=coA+pyruvate"];

g2x4x1x7=
  calcrow[sucrose+pi+de==glucose1phos+fructose,8.25,.1,
    "sucrose+pi=glucose1phos+fructose"];

g2x4x1x8=
  calcrow[maltose+pi+de==glucose+glucose1phos,7,.1,
    "maltose+pi=glucose+glucose1phos"];

g2x4x2x1=
  calcrow[adenosine+pi+de==adenine+ribose1phos,7,.1,
    "adenosine+pi=adenine+ribose1phos"];

g2x6x1x1=
  calcrow[aspartate+ketoglutarate+de==oxaloacetate+glutamate,7.4,.1,
    "aspartate+ketoglutarate=oxaloacetate+glutamate"];

g2x6x1x2=
  calcrow[alanine+ketoglutarate+de==pyruvate+glutamate,7.4,.5,
    "alanine+ketoglutarate=pyruvate+glutamate"];

g2x6x1x35=
  calcrow[glycine+oxaloacetate+de==glyoxylate+aspartate,7.1,.1,
    "glycine+oxaloacetate=glyoxylate+aspartate"];

g2x7x1x1=
  calcrow[atp+glucose+de==adp+glucose6phos,6.07,.1,
    "atp+glucose=adp+glucose6phos"];

g2x7x1x23=
  calcrow[atp+nadox+de==adp+nadpox,6.1,.1,
    "atp+nadox=adp+nadpox"];

g2x7x4x3=
  calcrow[2*adp+de==amp+atp,7.41,.04,
    "2*adp=amp+atp"];

g3x1x3x1=
  calcrow[amp+h2o+de==adenosine+pi,8.86,.3,
    "amp+h2o=adenosine+pi"];

g3x1x3x1b=
  calcrow[fructose6phos+h2o+de==fructose+pi,8.4,.29,
    "fructose6phos+h2o=fructose+pi"];

g3x1x3x1c=
  calcrow[glucose6phos+h2o+de==glucose+pi,6.99,.25,
    "glucose6phos+h2o=glucose+pi"];
```

```
g3x1x3x1d=
  calcrow[glycerol3phos+h2o+de==glycerol+pi,7,.1,
    "glycerol3phos+h2o=glycerol+pi"];

g3x1x3x1e=
  calcrow[ppi+h2o+de==2*pi,7,.1,
    "ppi+h2o=2*pi"];

g3x2x1x3=
  calcrow[maltose+h2o+de==2*glucose,5,.01,
    "maltose+h2o=2*glucose"];

g3x2x1x23=
  calcrow[lactose+h2o+de==glucose+galactose,5.65,.1,
    "lactose+h2o=glucose+galactose"];

g3x2x2x4=
  calcrow[amp+h2o+de==adenine+ribose5phos,8,.1,
    "amp+h2o=adenine+ribose5phos"];

g3x2x2x7=
  calcrow[adenosine+h2o+de==adenine+ribose,7,.1,
    "adenosine+h2o=adenine+ribose"];

g3x5x1x2=
  calcrow[glutamine+h2o+de==glutamate+ammonia,5.5,.1,
    "glutamine+h2o=glutamate+ammonia"];

g4x1x3x1 = calcrow[citrateiso + de == succinate + glyoxylate,
    7.7, .1, "citrateiso=succinate+glyoxylate"];

g4x1x3x6 =
  calcrow[citrate + de == acetate + oxaloacetate, 7, .2, "citrate=acetate+oxaloacetate"];

g4x1x3x7 = calcrow[oxaloacetate + acetylcoA + h2o + de == citrate + coA,
    7.05, .25, "oxaloacetate+acetylcoA=citrate+coA"];

g4x1x99x1 = calcrow[tryptophaneL + h2o + de == indole + pyruvate + ammonia,
    7.97, .1, "tryptophaneL+h2o=indole+pyruvate+ammonia"];

g4x2x1x2 = calcrow[fumarate + h2o + de == malate, 7.3, .1, "fumarate+h2o=malate"];

g4x2x1x3 =
  calcrow[citrate + de == aconitatecis + h2o, 7.4, .1, "citrate=aconitatecis+h2o"];

g4x3x1x1 =
  calcrow[aspartate + de == fumarate + ammonia, 7.37, .1, "aspartate=fumarate+ammonia"];

g5x3x1x5 = calcrow[glucose + de == fructose, 7.5, .1, "glucose=fructose"];

g5x3x1x5b = calcrow[xylose + de == xylulose, 8.7, .1, "xylose=xylulose"];

g5x3x1x7 = calcrow[mannose + de == fructose, 7, .05, "mannose=fructose"];

g5x3x1x9 =
  calcrow[glucose6phos + de == fructose6phos, 8.7, .3, "glucose6phos=fructose6phos"];

g5x3x1x15 = calcrow[lyxose + de == xylulose, 7, .06, "lyxose=xylulose"];

g5x3x1x20 = calcrow[ribose + de == ribulose, 7.4, .04, "ribose=ribulose"];
```

```
g5x3x1xa = calcrow[mannose + de == fructose, 7.6, .02, "mannose=fructose"];

g5x4x2x2 =
  calcrow[glucose1phos + de == glucose6phos, 8.48, .1, "glucose1phos=glucose6phos"];

g5x4x2x7 = calcrow[ribose1phos + de == ribose5phos, 7, .25, "ribose1phos=ribose5phos"];

g6x2x1x1 = calcrow[atp + acetate + coA + de == amp + ppi + acetylcoA,
    7, .25, "atp+acetate+coA=amp+ppi+acetylcoA"];

g6x3x1x2 = calcrow[atp + glutamate + ammonia + de == adp + pi + glutamine,
    7, .3, "atp+glutamate+ammonia=adp+pi+glutamine"];

g6x4x1x1 = calcrow[atp + pyruvate + co2tot + de == adp + pi + oxaloacetate,
    7.4, .02, "atp+pyruvate+co2tot=adp+pi+oxaloacetate"];

TableForm[{g1x1x1x1, g1x1x1x26, g1x1x1x27b, g1x1x1x37, g1x1x1x39, g1x1x1x40, g1x1x1x42,
    g1x1x1x67, g1x1x1x79, g1x1x99x7, g1x2x1x2, g1x2x1x10, g1x2x1x43, g1x4x1x1,
    g1x4x1x2, g1x4x1x3, g1x4x1x10, g1x6x1x1, g1x6x4x2, g1x6x4x5, g2x3x1x54, g2x4x1x7,
    g2x4x1x8, g2x4x2x1, g2x6x1x1, g2x6x1x2, g2x6x1x35, g2x7x1x1, g2x7x1x23, g2x7x4x3,
    g3x1x3x1, g3x1x3x1b, g3x1x3x1c, g3x1x3x1d, g3x1x3x1e, g3x2x1x3, g3x2x1x23,
    g3x2x2x4, g3x2x2x7, g3x5x1x2, g4x1x3x1, g4x1x3x6, g4x1x3x7, g4x1x99x1, g4x2x1x2,
    g4x2x1x3, g4x3x1x1, g5x3x1x5, g5x3x1x5b, g5x3x1x7, g5x3x1x9, g5x3x1x15,
    g5x3x1x20, g5x3x1xa, g5x4x2x2, g5x4x2x7, g6x2x1x1, g6x3x1x2, g6x4x1x1},
  TableHeadings -> {{"1.1.1.1", "1.1.1.26", "1.1.1.27b", "1.1.1.37", "1.1.1.39",
    "1.1.1.40", "1.1.1.42", "1.1.1.67", "1.1.1.79", "1.1.99.7", "1.2.1.2",
    "1.2.1.10", "1.2.1.43", "1.4.1.1", "1.4.1.2", "1.4.1.3", "1.4.1.10", "1.6.1.1",
    "1.6.4.2", "1.6.4.5", "2.3.1.54", "2.4.1.7", "2.4.1.8", "2.4.2.1", "2.6.1.1",
    "2.6.1.2", "2.6.1.35", "2.7.1.1", "2.7.1.23", "2.7.4.3", "3.1.3.1", "3.1.3.1b",
    "3.1.3.1c", "3.1.3.1d", "3.1.3.1e", "3.2.1.3", "3.2.1.23", "3.2.2.4",
    "3.2.2.7", "3.5.1.2", "4.1.3.1", "4.1.3.6", "4.1.3.7", "4.1.99.1", "4.2.1.2",
    "4.2.1.3", "4.3.1.1", "5.3.1.5", "5.3.1.5b", "5.3.1.7", "5.3.1.9", "5.3.1.15",
    "5.3.1.20", "5.3.1.a", "5.4.2.2", "5.4.2.7", "6.2.1.1", "6.3.1.2", "6.4.1.1"},
  {"Reaction", "Calc G'"}}, TableSpacing -> {1, 1}]
```

	Reaction	Calc G'
1.1.1.1	ethanol+nadox==acetaldehyde+nadred	22.095
1.1.1.26	glycolate+nadox=gloxylate+nadred	38.896
1.1.1.27b	lactate+nadox=pyruvate+nadred	26.3652
1.1.1.37	malate+nadox=oxaloacetate+nadred	28.2639
1.1.1.39	malate+nadox+h2o=pyruvate+nadred+co2tot	4.45812
1.1.1.40	malate+nadpox+h2o=pyruvate+nadpred+co2tot	1.71088
1.1.1.42	citrateiso+nadpox+h2o=ketoglutarate+nadpred+co2tot	-3.83041
1.1.1.67	mannitolD+nadox=fructose+nadred	1.88013
1.1.1.79	glycolate+nadpox=glyoxylate+nadpred	38.2358
1.1.99.7	lactate+oxaloacetate=malate+pyruvate	-4.92266
1.2.1.2	formate+nadox+h2o=co2tot+nadred	-15.1398
1.2.1.10	acetaldehyde+coA+nadox=acetylcoA+nadred	-13.861
1.2.1.43	formate+nadpox+h2o=co2tot+nadpred	-20.4151
1.4.1.1	alanine+nadox+h2o=pyruvate+nadred+ammonia	29.9682

1.4.1.2	glutamate+nadox+h2o=ketoglutarate+nadred+ammonia	38.7895
1.4.1.3	glutamate+nadpox+h2o=ketoglutarate+nadpred+ammonia	36.5242
1.4.1.10	glycine+nadox+h2o=glyoxylate+nadred+ammonia	51.8996
1.6.1.1	nadox+nadpred=nadred+nadpox	-0.891743
1.6.4.2	nadpox+2*glutathionered=nadpred+glutathioneox	6.01048
1.6.4.5	nadpox+2*thioredoxinred=nadpred+thioredoxinox	6.86709
2.3.1.54	acetylcoA+formate=coA+pyruvate	11.0295
2.4.1.7	sucrose+pi=glucose1phos+fructose	-11.5121
2.4.1.8	maltose+pi=glucose+glucose1phos	-1.06837
2.4.2.1	adenosine+pi=adenine+ribose1phos	27.5509
2.6.1.1	aspartate+ketoglutarate=oxaloacetate+glutamate	-1.47
2.6.1.2	alanine+ketoglutarate=pyruvate+glutamate	-3.39595
2.6.1.35	glycine+oxaloacetate=glyoxylate+aspartate	9.94413
2.7.1.1	atp+glucose=adp+glucose6phos	-20.0611
2.7.1.23	atp+nadox=adp+nadpox	-8.32968
2.7.4.3	2*adp=amp+atp	2.7092
3.1.3.1	amp+h2o=adenosine+pi	-12.9696
3.1.3.1b	fructose6phos+h2o=fructose+pi	-13.6567
3.1.3.1c	glucose6phos+h2o=glucose+pi	-11.6294
3.1.3.1d	glycerol3phos+h2o=glycerol+pi	1.78075
3.1.3.1e	ppi+h2o=2*pi	-23.15
3.2.1.3	maltose+h2o=2*glucose	-19.92
3.2.1.23	lactose+h2o=glucose+galactose	-20.31
3.2.2.4	amp+h2o=adenine+ribose5phos	5.74391
3.2.2.7	adenosine+h2o=adenine+ribose	3.5668
3.5.1.2	glutamine+h2o=glutamate+ammonia	-12.7946
4.1.3.1	citrateiso=succinate+glyoxylate	-0.531653
4.1.3.6	citrate=acetate+oxaloacetate	3.2174
4.1.3.7	oxaloacetate+acetylcoA=citrate+coA	-45.0692
4.1.99.1	tryptophaneL+h2o=indole+pyruvate+ammonia	22.7596
4.2.1.2	fumarate+h2o=malate	-3.60654
4.2.1.3	citrate=aconitatecis+h2o	8.42485
4.3.1.1	aspartate=fumarate+ammonia	12.2178
5.3.1.5	glucose=fructose	0.39
5.3.1.5b	xylose=xylulose	4.34
5.3.1.7	mannose=fructose	-5.51

5.3.1.9	glucose6phos=fructose6phos	3.14096
5.3.1.15	lyxose=xylulose	2.99
5.3.1.20	ribose=ribulose	2.85
5.3.1.a	mannose=fructose	-5.51
5.4.2.2	glucose1phos=glucose6phos	-7.06836
5.4.2.7	ribose1phos=ribose5phos	-8.08
6.2.1.1	atp+acetate+coA=amp+ppi+acetylcoA	-5.97683
6.3.1.2	atp+glutamate+ammonia=adp+pi+glutamine	-22.6469
6.4.1.1	atp+pyruvate+co2tot=adp+pi+oxaloacetate	-10.7687

(b) Make a table of the relevant data in the Goldberg and Tewari Tables. First make a vector of the experimental apparent equilibrium constants.

```
vector={9.25*10^-5,1.03*10^-7,1.7*10^-5,1.2*10^-5,1.12*10^-3,3.4*10^-2,.72,5.6*10^-2,9.
0*10^-11,4.3,420,2*10^3,6.5*10^2,6*10^-6,1.5*10^-6,1.2*10^-6,5.7*10^-5,1.41,1.56*10^-2,
3.6*10^-2,23,31.5,.23,5.41*10^-3,.148,.658,.016,294,29.3,.365,189,262,110,68,2*10^4,200
,35.2,170,53,896,.0023,.877,1.14*10^6,2.2*10^-4,4.87,.032,4.71*10^-3,.87,.172,3.0,.299,
.23,.391,1.5,17.1,26,11.1,270,6.55};
```

```
gradevector={A,B,A,A,C,A,A,B,B,B,D,B,C,B,A,A,C,B,A,C,C,A,C,B,A,A,B,A,B,A,A,A,A,B,B,C,A,
B,B,B,B,A,A,A,A,A,A,A,A,B,A,B,A,B,A,A,A,A,A};
```

```
pHvector={7.0,7.8,6.8,7.1,6.5,7.0,6.9,8.05,7.92,7.8,6.2,7.3,7.5,7.97,6.9,7.0,6.4,7.5,6.
9,7.0,7.2,8.25,7.0,7.0,7.4,7.4,7.1,6.07,6.1,7.41,8.86,8.4,6.99,7.0,7.0,5.0,5.65,8.0,7.0
,5.5,7.7,7.0,7.05,7.97,7.3,7.4,7.37,7.5,8.7,7.0,8.7,7.0,7.4,7.6,8.48,7.0,7.0,7.0,7.4};
```

```
isvector={.25,.03,.05,.25,.05,.25,.25,.05,.1,.1,.07,.03,.1,.1,.24,.47,.08,.1,.25,.1,.05
,.1,.1,.1,.1,.5,.1,.1,.1,.04,.3,.24,.25,.1,.1,.01,.1,.1,.1,.1,.1,.02,.25,.1,.1,.1,.1,.1
,.1,.05,.3,.06,.04,.02,.1,.25,.25,.3,.02}
```

```
{0.25, 0.03, 0.05, 0.25, 0.05, 0.25, 0.25, 0.05, 0.1, 0.1, 0.07, 0.03, 0.1, 0.1, 0.24, (
  0.1, 0.1, 0.1, 0.5, 0.1, 0.1, 0.1, 0.04, 0.3, 0.24, 0.25, 0.1, 0.1, 0.01, 0.1, 0.1, 0.
  0.1, 0.1, 0.1, 0.05, 0.3, 0.06, 0.04, 0.02, 0.1, 0.25, 0.25, 0.3, 0.02}
```

```
rxvector={g1x1x1x1,g1x1x1x26,g1x1x1x27b,g1x1x1x37,g1x1x1x39,g1x1x1x40,g1x1x1x42,g1x1x1x
67,g1x1x1x79,g1x1x99x7,g1x2x1x2,g1x2x1x10,g1x2x1x43,g1x4x1x1,g1x4x1x2,g1x4x1x3,g1x4x1x1
0,g1x6x1x1,g1x6x4x2,g1x6x4x5,g2x3x1x54,g2x4x1x7,g2x4x1x8,g2x4x2x1,g2x6x1x1,g2x6x1x2,g2x
6x1x35,g2x7x1x1,g2x7x1x23,g2x7x4x3,g3x1x3x1,g3x1x3x1b,g3x1x3x1c,g3x1x3x1d,g3x1x3x1e,g3x
2x1x3,g3x2x1x23,g3x2x2x4,g3x2x2x7,g3x5x1x2,g4x1x3x1,g4x1x3x6,g4x1x3x7,g4x1x99x1,g4x2x1x
2,g4x2x1x3,g4x3x1x1,g5x3x1x5,g5x3x1x5b,g5x3x1x7,g5x3x1x9,g5x3x1x15,g5x3x1x20,g5x3x1xa,g
5x4x2x2,g5x4x2x7,g6x2x1x1,g6x3x1x2,g6x4x1x1}[[All,1]];
```

```
calcgvector={g1x1x1x1,g1x1x1x26,g1x1x1x27b,g1x1x1x37,g1x1x1x39,g1x1x1x40,g1x1x1x42,g1x1
x1x67,g1x1x1x79,g1x1x99x7,g1x2x1x2,g1x2x1x10,g1x2x1x43,g1x4x1x1,g1x4x1x2,g1x4x1x3,g1x4x
1x10,g1x6x1x1,g1x6x4x2,g1x6x4x5,g2x3x1x54,g2x4x1x7,g2x4x1x8,g2x4x2x1,g2x6x1x1,g2x6x1x2,
g2x6x1x35,g2x7x1x1,g2x7x1x23,g2x7x4x3,g3x1x3x1,g3x1x3x1b,g3x1x3x1c,g3x1x3x1d,g3x1x3x1e,
g3x2x1x3,g3x2x1x23,g3x2x2x4,g3x2x2x7,g3x5x1x2,g4x1x3x1,g4x1x3x6,g4x1x3x7,g4x1x99x1,g4x2
x1x2,g4x2x1x3,g4x3x1x1,g5x3x1x5,g5x3x1x5b,g5x3x1x7,g5x3x1x9,g5x3x1x15,g5x3x1x20,g5x3x1x
a,g5x4x2x2,g5x4x2x7,g6x2x1x1,g6x3x1x2,g6x4x1x1}[[All,2]]
```

```
{22.095, 38.896, 26.3652, 28.2639, 4.45812, 1.71088, -3.83041, 1.88013, 38.2358, -4.922€
  38.7895, 36.5242, 51.8996, -0.891743, 6.01048, 6.86709, 11.0295, -11.5121, -1.06837, 2
  -20.0611, -8.32968, 2.7092, -12.9696, -13.6567, -11.6294, 1.78075, -23.15, -19.92, -2(
  -0.531653, 3.2174, -45.0692, 22.7596, -3.60654, 8.42485, 12.2178, 0.39, 4.34, -5.51, ;
  -8.08, -5.97683, -22.6469, -10.7687}
```

```
round[vec_,params_:{4,2}]:=(*When a list of numbers ver has more digits to the right
of the decimal point than you want, say 5, you can request 2 by using
round[vec,{5,2}].*)
Flatten[Map[NumberForm[#,params]&,{vec},{2}]]
```

```
calcgvec=round[calcgvector,{5,2}]
```

{22.1, 38.9, 26.37, 28.26, 4.46, 1.71, -3.83, 1.88, 38.24, -4.92, -15.14, -13.86, -20.4Z

  6.87, 11.03, -11.51, -1.07, 27.55, -1.47, -3.4, 9.94, -20.06, -8.33, 2.71, -12.97, -1:

  -20.31, 5.74, 3.57, -12.79, -0.53, 3.22, -45.07, 22.76, -3.61, 8.42, 12.22, 0.39, 4.3₄

  -8.08, -5.98, -22.65, -10.77}

```
ecnos={"1.1.1.1","1.1.1.26","1.1.1.27b","1.1.1.37","1.1.1.39","1.1.1.40","1.1.1.42",\
"1.1.1.67","1.1.1.79","1.1.99.7","1.2.1.2","1.2.1.10","1.2.1.43","1.4.1.1","1.\
4.1.2","1.4.1.3","1.4.1.10","1.6.1.1","1.6.4.2","1.6.4.5","2.3.1.54","2.4.1.\
7","2.4.1.8","2.4.2.1","2.6.1.1","2.6.1.2","2.6.1.35","2.7.1.1","2.7.1.23","2.\
7.4.3","3.1.3.1","3.1.3.1b","3.1.3.1c","3.1.3.1d","3.1.3.1e","3.2.1.3","3.2.1.\
23","3.2.2.4","3.2.2.7","3.5.1.2","4.1.3.1","4.1.3.6","4.1.3.7","4.1.99.1","4.\
2.1.2","4.2.1.3","4.3.1.1","5.3.1.5","5.3.1.5b","5.3.1.7","5.3.1.9","5.3.1.\
15","5.3.1.20","5.3.1.a","5.4.2.2","5.4.2.7","6.2.1.1","6.3.1.2","6.4.1.1"};
```

```
TableForm[Transpose[{ecnos,rxvector,vector,pHvector,isvector,gradevector}],TableHeading
s->{None,{"EC Nos.","Reaction","K'","pH","I","Gr"}},TableSpacing->{1,1}]
```

EC Nos.	Reaction	K'	pH	I	Gr
1.1.1.1	ethanol+nadox⇌acetaldehyde+nadred	0.0000925	7.	0.25	A
1.1.1.26	glycolate+nadox=gloxylate+nadred	$1.03\ 10^{-7}$	7.8	0.03	B
1.1.1.27b	lactate+nadox=pyruvate+nadred	0.000017	6.8	0.05	A
1.1.1.37	malate+nadox=oxaloacetate+nadred	0.000012	7.1	0.25	A
1.1.1.39	malate+nadox+h2o=pyruvate+nadred+co2tot	0.00112	6.5	0.05	C
1.1.1.40	malate+nadpox+h2o=pyruvate+nadpred+co2tot	0.034	7.	0.25	A
1.1.1.42	citrateiso+nadpox+h2o=ketoglutarate+nadpred+co2tot	0.72	6.9	0.25	A
1.1.1.67	mannitolD+nadox=fructose+nadred	0.056	8.05	0.05	B
1.1.1.79	glycolate+nadpox=glyoxylate+nadpred	$9.\ 10^{-11}$	7.92	0.1	B
1.1.99.7	lactate+oxaloacetate=malate+pyruvate	4.3	7.8	0.1	B
1.2.1.2	formate+nadox+h2o=co2tot+nadred	420	6.2	0.07	D
1.2.1.10	acetaldehyde+coA+nadox=acetylcoA+nadred	2000	7.3	0.03	B
1.2.1.43	formate+nadpox+h2o=co2tot+nadpred	650.	7.5	0.1	C
1.4.1.1	alanine+nadox+h2o=pyruvate+nadred+ammonia	500000	7.97	0.1	B
1.4.1.2	glutamate+nadox+h2o=ketoglutarate+nadred+ammonia	$1.5\ 10^{-6}$	6.9	0.24	A
1.4.1.3	glutamate+nadpox+h2o=ketoglutarate+nadpred+ammonia	$1.2\ 10^{-6}$	7.	0.47	A
1.4.1.10	glycine+nadox+h2o=glyoxylate+nadred+ammonia	0.000057	6.4	0.08	C
1.6.1.1	nadox+nadpred=nadred+nadpox	1.41	7.5	0.1	B
1.6.4.2	nadpox+2*glutathionered=nadpred+glutathioneox	0.0156	6.9	0.25	A

The "3" over a line appears to the right near the 1.2.1.43 row (reading as part of K' formatting: $\frac{3}{\ }$).

1.6.4.5	nadpox+2*thioredoxinred=nadpred+thioredoxinox	0.036	7.	0.1	C
2.3.1.54	acetylcoA+formate=coA+pyruvate	23	7.2	0.05	C
2.4.1.7	sucrose+pi=glucose1phos+fructose	31.5	8.25	0.1	A
2.4.1.8	maltose+pi=glucose+glucose1phos	0.23	7.	0.1	C
2.4.2.1	adenosine+pi=adenine+ribose1phos	0.00541	7.	0.1	B
2.6.1.1	aspartate+ketoglutarate=oxaloacetate+glutamate	0.148	7.4	0.1	A
2.6.1.2	alanine+ketoglutarate=pyruvate+glutamate	0.658	7.4	0.5	A
2.6.1.35	glycine+oxaloacetate=glyoxylate+aspartate	0.016	7.1	0.1	B
2.7.1.1	atp+glucose=adp+glucose6phos	294	6.07	0.1	A
2.7.1.23	atp+nadox=adp+nadpox	29.3	6.1	0.1	B
2.7.4.3	2*adp=amp+atp	0.365	7.41	0.04	A
3.1.3.1	amp+h2o=adenosine+pi	189	8.86	0.3	A
3.1.3.1b	fructose6phos+h2o=fructose+pi	262	8.4	0.24	A
3.1.3.1c	glucose6phos+h2o=glucose+pi	110	6.99	0.25	A
3.1.3.1d	glycerol3phos+h2o=glycerol+pi	68	7.	0.1	B
3.1.3.1e	ppi+h2o=2*pi	20000	7.	0.1	B
3.2.1.3	maltose+h2o=2*glucose	200	5.	0.01	C
3.2.1.23	lactose+h2o=glucose+galactose	35.2	5.65	0.1	A
3.2.2.4	amp+h2o=adenine+ribose5phos	170	8.	0.1	B
3.2.2.7	adenosine+h2o=adenine+ribose	53	7.	0.1	B
3.5.1.2	glutamine+h2o=glutamate+ammonia	896	5.5	0.1	B
4.1.3.1	citrateiso=succinate+glyoxylate	0.0023	7.7	0.1	B
4.1.3.6	citrate=acetate+oxaloacetate	0.877	7.	0.02	A
4.1.3.7	oxaloacetate+acetylcoA=citrate+coA	$1.14\ 10^6$	7.05	0.25	A
4.1.99.1	tryptophaneL+h2o=indole+pyruvate+ammonia	0.00022	7.97	0.1	A
4.2.1.2	fumarate+h2o=malate	4.87	7.3	0.1	A
4.2.1.3	citrate=aconitatecis+h2o	0.032	7.4	0.1	A
4.3.1.1	aspartate=fumarate+ammonia	0.00471	7.37	0.1	A
5.3.1.5	glucose=fructose	0.87	7.5	0.1	A
5.3.1.5b	xylose=xylulose	0.172	8.7	0.1	A
5.3.1.7	mannose=fructose	3.	7.	0.05	B
5.3.1.9	glucose6phos=fructose6phos	0.299	8.7	0.3	A
5.3.1.15	lyxose=xylulose	0.23	7.	0.06	B
5.3.1.20	ribose=ribulose	0.391	7.4	0.04	A
5.3.1.a	mannose=fructose	1.5	7.6	0.02	B

```
5.4.2.2    glucose1phos=glucose6phos                    17.1    8.48 0.1  A

5.4.2.7    ribose1phos=ribose5phos                       26      7.  0.25 A

6.2.1.1    atp+acetate+coA=amp+ppi+acetylcoA            11.1     7.  0.25 A

6.3.1.2    atp+glutamate+ammonia=adp+pi+glutamine       270      7.  0.3  A

6.4.1.1    atp+pyruvate+co2tot=adp+pi+oxaloacetate      6.55    7.4 0.02 A
```

The references are given in Goldberg and Tewari.

(c) Make a table of the differences between the values of the standard transformed Gibbs energies of reaction calculated in part (a) at pH 7 and the experimental values in Goldberg and Tewari.
Calculate the experimental standard transformed Gibbs energies of reaction.

gprime=-8.31451*.29815*Log[vector]

{23.0254, 39.883, 27.2248, 28.0882, 16.8432, 8.38238, 0.814352, 7.1454, 57.3416, -3.6158
  33.2431, 33.7963, 24.2256, -0.851749, 10.3137, 8.24069, -7.7728, -8.55242, 3.64328, 12
  -8.37294, 2.49845, -12.9941, -13.8038, -11.6524, -10.46, -24.5505, -13.1344, -8.82773,
  0.325361, -34.5731, 20.8776, -3.92444, 8.53267, 13.2825, 0.345227, 4.36364, -2.72343,
  -7.03799, -8.07673, -5.96675, -13.8783, -4.65914}

gprimer=round[gprime,{5,2}]

{23.03, 39.88, 27.22, 28.09, 16.84, 8.38, 0.81, 7.15, 57.34, -3.62, -14.97, -18.84, -16.
  10.31, 8.24, -7.77, -8.55, 3.64, 12.94, 4.74, 1.04, 10.25, -14.09, -8.37, 2.5, -12.99,
  -8.83, -12.73, -9.84, -16.85, 15.06, 0.33, -34.57, 20.88, -3.92, 8.53, 13.28, 0.35, 4.
  -7.04, -8.08, -5.97, -13.88, -4.66}

Calculate differences before rounding.

diff=calcgvector-gprime

{-0.930416, -0.986987, -0.859612, 0.175639, -12.3851, -6.6715, -4.64476, -5.26526, -19.1
  -4.35884, 0.161716, 5.54638, 2.72787, 27.6739, -0.039994, -4.30324, -1.3736, 18.8023,
  -4.43352, -0.30683, -5.97168, 0.0432646, 0.210751, 0.0244942, 0.147033, 0.0229674, 12.
  18.4754, 13.409, 4.05733, -15.591, 2.89204, -10.4961, 1.882, 0.317903, -0.107815, -1.0
  0.148067, -0.653284, 0.522128, -4.50486, -0.0303697, -0.00327266, -0.0100851, -8.76853

dif=round[diff,{6,2}]

{-0.93, -0.99, -0.86, 0.18, -12.39, -6.67, -4.64, -5.27, -19.11, -1.31, -0.17, 4.98, -4.
  -1.37, 18.8, -2.96, -4.71, 14.61, -6.21, -4.43, -0.31, -5.97, 0.04, 0.21, 0.02, 0.15,
  13.41, 4.06, -15.59, 2.89, -10.5, 1.88, 0.32, -0.11, -1.06, 0.04, -0.02, -2.79, 0.15,
  -8.77, 6.11}

Calculate the differences.and make a table.

TableForm[Transpose[{ecnos,rxvector,calcgvec,gprimer,dif}],TableHeadings->{None,{"EC Nos.","Reaction","Calc","Exptl","Calc-Exptl"}},TableSpacing->{1,1}]

EC Nos.	Reaction	Calc	Exptl	Calc-Exptl
1.1.1.1	ethanol+nadox=acetaldehyde+nadred	22.1	23.03	-0.93
1.1.1.26	glycolate+nadox=gloxylate+nadred	38.9	39.88	-0.99
1.1.1.27b	lactate+nadox=pyruvate+nadred	26.37	27.22	-0.86

1.1.1.37	malate+nadox=oxaloacetate+nadred	28.26	28.09	0.18
1.1.1.39	malate+nadox+h2o=pyruvate+nadred+co2tot	4.46	16.84	-12.39
1.1.1.40	malate+nadpox+h2o=pyruvate+nadpred+co2tot	1.71	8.38	-6.67
1.1.1.42	citrateiso+nadpox+h2o=ketoglutarate+nadpred+co2tot	-3.83	0.81	-4.64
1.1.1.67	mannitolD+nadox=fructose+nadred	1.88	7.15	-5.27
1.1.1.79	glycolate+nadpox=glyoxylate+nadpred	38.24	57.34	-19.11
1.1.99.7	lactate+oxaloacetate=malate+pyruvate	-4.92	-3.62	-1.31
1.2.1.2	formate+nadox+h2o=co2tot+nadred	-15.14	-14.97	-0.17
1.2.1.10	acetaldehyde+coA+nadox=acetylcoA+nadred	-13.86	-18.84	4.98
1.2.1.43	formate+nadpox+h2o=co2tot+nadpred	-20.42	-16.06	-4.36
1.4.1.1	alanine+nadox+h2o=pyruvate+nadred+ammonia	29.97	29.81	0.16
1.4.1.2	glutamate+nadox+h2o=ketoglutarate+nadred+ammonia	38.79	33.24	5.55
1.4.1.3	glutamate+nadpox+h2o=ketoglutarate+nadpred+ammonia	36.52	33.8	2.73
1.4.1.10	glycine+nadox+h2o=glyoxylate+nadred+ammonia	51.9	24.23	27.67
1.6.1.1	nadox+nadpred=nadred+nadpox	-0.89	-0.85	-0.04
1.6.4.2	nadpox+2*glutathionered=nadpred+glutathioneox	6.01	10.31	-4.3
1.6.4.5	nadpox+2*thioredoxinred=nadpred+thioredoxinox	6.87	8.24	-1.37
2.3.1.54	acetylcoA+formate=coA+pyruvate	11.03	-7.77	18.8
2.4.1.7	sucrose+pi=glucose1phos+fructose	-11.51	-8.55	-2.96
2.4.1.8	maltose+pi=glucose+glucose1phos	-1.07	3.64	-4.71
2.4.2.1	adenosine+pi=adenine+ribose1phos	27.55	12.94	14.61
2.6.1.1	aspartate+ketoglutarate=oxaloacetate+glutamate	-1.47	4.74	-6.21
2.6.1.2	alanine+ketoglutarate=pyruvate+glutamate	-3.4	1.04	-4.43
2.6.1.35	glycine+oxaloacetate=glyoxylate+aspartate	9.94	10.25	-0.31
2.7.1.1	atp+glucose=adp+glucose6phos	-20.06	-14.09	-5.97
2.7.1.23	atp+nadox=adp+nadpox	-8.33	-8.37	0.04
2.7.4.3	2*adp=amp+atp	2.71	2.5	0.21
3.1.3.1	amp+h2o=adenosine+pi	-12.97	-12.99	0.02
3.1.3.1b	fructose6phos+h2o=fructose+pi	-13.66	-13.8	0.15
3.1.3.1c	glucose6phos+h2o=glucose+pi	-11.63	-11.65	0.02
3.1.3.1d	glycerol3phos+h2o=glycerol+pi	1.78	-10.46	12.24
3.1.3.1e	ppi+h2o=2*pi	-23.15	-24.55	1.4
3.2.1.3	maltose+h2o=2*glucose	-19.92	-13.13	-6.79
3.2.1.23	lactose+h2o=glucose+galactose	-20.31	-8.83	-11.48
3.2.2.4	amp+h2o=adenine+ribose5phos	5.74	-12.73	18.48
3.2.2.7	adenosine+h2o=adenine+ribose	3.57	-9.84	13.41

3.5.1.2	glutamine+h2o=glutamate+ammonia	-12.79 -16.85 4.06
4.1.3.1	citrateiso=succinate+glyoxylate	-0.53 15.06 -15.59
4.1.3.6	citrate=acetate+oxaloacetate	3.22 0.33 2.89
4.1.3.7	oxaloacetate+acetylcoA=citrate+coA	-45.07 -34.57 -10.5
4.1.99.1	tryptophaneL+h2o=indole+pyruvate+ammonia	22.76 20.88 1.88
4.2.1.2	fumarate+h2o=malate	-3.61 -3.92 0.32
4.2.1.3	citrate=aconitatecis+h2o	8.42 8.53 -0.11
4.3.1.1	aspartate=fumarate+ammonia	12.22 13.28 -1.06
5.3.1.5	glucose=fructose	0.39 0.35 0.04
5.3.1.5b	xylose=xylulose	4.34 4.36 -0.02
5.3.1.7	mannose=fructose	-5.51 -2.72 -2.79
5.3.1.9	glucose6phos=fructose6phos	3.14 2.99 0.15
5.3.1.15	lyxose=xylulose	2.99 3.64 -0.65
5.3.1.20	ribose=ribulose	2.85 2.33 0.52
5.3.1.a	mannose=fructose	-5.51 -1.01 -4.5
5.4.2.2	glucose1phos=glucose6phos	-7.07 -7.04 -0.03
5.4.2.7	ribose1phos=ribose5phos	-8.08 -8.08 -0.00
6.2.1.1	atp+acetate+coA=amp+ppi+acetylcoA	-5.98 -5.97 -0.01
6.3.1.2	atp+glutamate+ammonia=adp+pi+glutamine	-22.65 -13.88 -8.77
6.4.1.1	atp+pyruvate+co2tot=adp+pi+oxaloacetate	-10.77 -4.66 -6.11

There are problems with adenine, glyolate, glyoxylate,,and glycerol3phos, which indicates that their values should be recalculated or should be calculated from other reactions. Since most of the species properties in BasicBiochemData2 can be obtained by several different paths, it is possible to find and correct errors.

4,8 Calculate the standard transformed reaction Gibbs energies, appparent equilibrium constants, and changes in the binding of hydrogen ions for the ten reactions of glycolysis at 298,15 K, pHs 5, 6, 7, 8, and 9 and ionic strengths of .0, 0.10, and 0.25 M. Also calculate these properties for the net reaction..

(BasicBiochemData2 has to be loaded)

```
calcNHrx[eq_, pHlist_, islist_] := Module[{energy},(*This program calculates the
change in the binding of hydrogen ions in a biochemical reaaction at specified pHs and
ionic strengths.*)
    energy = Solve[eq, de];
      D[energy[[1,1,2]], pH]/(8.31451*0.29815*Log[10]) /.
      pH -> pHlist /. is -> islist]

rxthermotab[eq_, pHlist_, islist_] := Module[{energy, tg, tk, tn},
    (*This program uses three other programs to make a thermodynamic table of
      standard transformed reaction Gibbs energies, apparent equilibrium constants,
      and changes in the number of hydrogen ions bound in a biochemical reaction.*)
  tg = calctrGerx[eq, pHlist, islist];
  tk = calckprime[eq, pHlist, islist];
  tn = calcNHrx[eq, pHlist, islist]; TableForm[Join[{tg, tk, tn}]]]]
```

Reaction 1

rxthermotab[glucose+atp+de==glucose6phos+adp,{5,6,7,8,9},{0,.1,.25}]

-19.9564	-21.1455	-24.8972	-31.3157	-37.3892
-18.0757	-19.82	-24.6736	-30.4846	-36.2155
-17.4137	-19.4683	-24.4199	-30.1133	-35.8228

3134.61	5064.24	23002.6	306367.	$3.55032 \ 10^{6}$
1467.94	2966.78	21018.9	219099.	$2.21129 \ 10^{6}$
1123.89	2574.39	18973.7	188625.	$1.88735 \ 10^{6}$

-0.177235	-0.336077	-0.992429	-1.12833	-1.02082
-0.122329	-0.579942	-1.01164	-1.00952	-1.0011
-0.139449	-0.649491	-0.984363	-1.0005	-1.00008

This program produces three tables. The top table gives the standard transformed reaction Gibbs energies at three ionic strengths and five pH values. The second table gives the corresponding equilibrium constants, and the third table gives the changes in the binding of hydrogen ions.

Reaction 2

rxthermotab[glucose6phos+de==fructose6phos,{5,6,7,8,9},{0,.1,.25}]

3.9532	3.72567	3.2948	3.15854	3.14189
3.89797	3.5248	3.20516	3.147	3.14071
3.86839	3.45934	3.18839	3.1451	3.14051

0.202971	0.222482	0.264715	0.279673	0.281558
0.207543	0.24126	0.274463	0.280978	0.281693
0.210035	0.247715	0.276326	0.281193	0.281714

-0.0141959	-0.0733161	-0.0509985	-0.00731544	-0.000761426
-0.0330731	-0.0847485	-0.0242531	-0.00280164	-0.000284407
-0.0419873	-0.080239	-0.0183996	-0.00204602	-0.000206853

Reaction 3

rxthermotab[fructose6phos+atp+de==fructose16phos+adp,{5,6,7,8,9},{0,.1,.25}]

-12.5514	-13.7161	-17.4527	-23.8749	-29.9491
-12.258	-15.6976	-21.884	-27.9151	-33.6694
-12.3548	-16.9525	-23.2451	-29.1287	-34.858

158.085	252.901	1141.72	15229.1	176534.
140.444	562.446	6821.56	77710.5	791752.
146.033	933.1	11812.6	126794.	$1.27886 \ 10^{6}$

-0.175363	-0.330035	-0.993053	-1.12864	-1.02085
-0.260983	-0.95943	-1.10486	-1.01996	-1.00216
-0.410005	-1.08946	-1.06644	-1.00934	-1.00097

Reaction 4

rxthermotab[fructose16phos+de==dihydroxyacetonephos+glyceraldehydephos,{5,6,7,8,9},{0,.1,.25}]

23.2659	20.3466	17.2974	16.6248	16.5486
23.9815	22.0132	21.4975	21.4426	21.4371
24.3294	23.1829	23.0316	23.0188	23.0175

0.0000839493	0.000272551	0.0009325	0.00122317	0.00126134
0.0000629003	0.000139146	0.000171327	0.000175161	0.00017555
0.000054664	0.0000868079	0.0000922697	0.0000927491	0.0000927959

-0.246757	-0.714037	-0.269856	-0.0337308	-0.00345608
-0.340222	-0.211319	-0.024558	-0.00246096	-0.000246092
-0.289279	-0.0732304	-0.00589343	-0.000561605	-0.0000558653

Reaction 5

rxthermotab[dihydroxyacetonephos+de==glyceraldehydephos,{5,6,7,8,9},{0,.1,.25}]

7.66	7.66	7.66	7.66	7.66
7.66	7.66	7.66	7.66	7.66
7.66	7.66	7.66	7.66	7.66

0.0455023	0.0455023	0.0455023	0.0455023	0.0455023
0.0455023	0.0455023	0.0455023	0.0455023	0.0455023
0.0455023	0.0455023	0.0455023	0.0455023	0.0455023

$2.36514\ 10^{-14}$	$2.48962\ 10^{-14}$	$1.86721\ 10^{-15}$	$1.24481\ 10^{-15}$	$6.22405\ 10^{-16}$
$5.35268\ 10^{-14}$	$6.22405\ 10^{-15}$	$3.11202\ 10^{-15}$	$1.24481\ 10^{-15}$	$1.24481\ 10^{-15}$
$5.66389\ 10^{-14}$	$1.24481\ 10^{-14}$	$1.24481\ 10^{-15}$	$-6.22405\ 10^{-16}$	$-6.22405\ 10^{-16}$

Reaction 6

rxthermotab[glyceraldehydephos+pi+nadox+de==bpg+nadred,{5,6,7,8,9},{0,.1,.25}]

22.8921	13.849	8.05124	4.32833	-0.337223
16.1673	8.4353	3.1141	-1.94228	-7.5227
14.0984	6.71564	1.21649	-4.21929	-9.88359

0.0000976101	0.00374781	0.038859	0.174467	1.14572
0.00147106	0.0332818	0.284732	2.18912	20.7928
0.00338903	0.0666004	0.612183	5.48509	53.8916

-1.82897	-1.28797	-0.769332	-0.669347	-0.933047
-1.64339	-1.09084	-0.840272	-0.947342	-0.993787
-1.569	-1.06683	-0.921309	-0.981391	-0.997935

Reaction 7

rxthermotab[bpg+adp+de==pg3+atp,{5,6,7,8,9},{0,.1,.25}]

-8.77058	-8.3272	-8.00276	-7.96743	-8.29556
-8.35908	-8.21794	-8.11648	-8.30786	-8.36297
-8.31749	-8.22167	-8.22407	-8.34411	-8.36722

34.3977	28.7643	25.2357	24.8786	28.3995
29.1366	27.5241	26.4203	28.5408	29.1824
28.6519	27.5655	27.5922	28.9611	29.2325

0.13882	0.0431119	0.0716595	-0.0704186	-0.0275038
0.0248626	0.0334161	-0.0229746	-0.0217609	-0.00279498
0.0107016	0.0225207	-0.0271383	-0.00964887	-0.00111155

Reaction 8

rxthermotab[pg3+de==pg2,{5,6,7,8,9},{0,.1,.25}]

5.53164	5.54601	5.66019	5.98488	6.13662
5.53714	5.5948	5.86614	6.10926	6.15453
5.54145	5.62832	5.9385	6.12755	6.1566
0.107375	0.106755	0.101949	0.0894334	0.0841235
0.107137	0.104674	0.0938216	0.0850572	0.0835179
0.106951	0.103268	0.0911224	0.0844318	0.0834482
0.00065925	0.0062929	0.0416282	0.0509505	0.00908086
0.00284928	0.0234401	0.0631637	0.0188138	0.00218857
0.00453314	0.0334474	0.0578654	0.0124135	0.00136563

Reaction 9

rxthermotab[pg2+de==pep+h2o,{5,6,7,8,9},{0,.1,.25}]

-0.828959	-0.989798	-2.0152	-3.7985	-4.37876
-0.891715	-1.47469	-3.25948	-4.28122	-4.44115
-0.939757	-1.76463	-3.59868	-4.34674	-4.44829
1.3971	1.49075	2.25448	4.62877	5.84954
1.43292	1.81281	3.7242	5.62385	5.99863
1.46096	2.03774	4.27029	5.77446	6.01593
-0.00760123	-0.0683949	-0.313273	-0.212886	-0.0319192
-0.0321025	-0.213665	-0.312677	-0.0680294	-0.00755601
-0.0501811	-0.274923	-0.257218	-0.0440393	-0.0047052

Reaction 10

rxthermotab[pep+adp+de==pyruvate+atp,{5,6,7,8,9},{0,.1,.25}]

-33.4613	-32.7661	-30.6224	-25.1722	-19.2526
-34.1841	-33.1449	-29.2617	-23.6955	-17.993
-34.4724	-33.1063	-28.8451	-23.2908	-17.5968
728017.	549974.	231625.	25700.8	2359.9
974465.	540772.	133784.	14166.3	1419.79
$1.09465\ 10^{6}$	630867.	113085.	12032.5	1210.04
0.150557	0.151752	0.700597	1.06297	1.01353
0.0719355	0.38045	0.915575	0.99698	0.999809
0.0826948	0.480624	0.927296	0.993667	0.999384

Net Reaction

**rxthermotab[glucose+2*pi+2*adp+2*nadox+de==2*pyruvate+2*atp+2*nadred+2*h2o,{5,6,7,8,9},
{0,.1,.25}]**

-26.9029	-48.5056	-67.9556	-80.9971	-92.2429
-38.2553	-59.9344	-77.5099	-90.3854	-101.978
-42.0903	-63.6158	-80.8106	-93.5651	-105.141
51659.7	$3.1461\ 10^{8}$	$8.0398\ 10^{11}$	$1.54878\ 10^{14}$	$1.44602\ 10^{16}$
$5.03504\ 10^{6}$	$3.16219\ 10^{10}$	$3.79375\ 10^{13}$	$6.83502\ 10^{15}$	$7.3388\ 10^{17}$
$2.36512\ 10^{7}$	$1.39618\ 10^{11}$	$1.43657\ 10^{14}$	$2.46488\ 10^{16}$	$2.62942\ 10^{18}$
-3.70662	-3.76388	-2.84378	-1.97548	-1.98561
-3.90829	-3.56983	-2.55969	-2.07742	-2.00807
-3.92322	-3.50274	-2.5161	-2.07045	-2.00732

4.9 Calculate the standard transformed reaction Gibbs energies, apparent equilibrium constants, and changes in the binding of hydrogen ions for the four reactions of gluconeogenesis that are different from those in glycolysis at 298.15 K, pHs 5, 6,

7, 8, and 9 and ionic strengths 0, 0.10, and 0.25 M. Also calculate the properties of the net of pyruvate carboxylase and phosphoenolpyruvate carboxykinase reactions and the net reaction of gluconeogenesis.

(BasicBiochemData2 must be loaded)

pyruvate carboxylase reaction

```
TableForm[Transpose[calctrGerx[pyruvate+co2tot+atp+de==oxaloacetate+adp+pi,{5,6,7,8,9},
{0,.1,.25}]],TableHeadings->{{"I = 0 M","I = 0.10 M","I = 0.25 M"},{"pH 5","pH 6","pH
7","pH 8","pH 9"}}]
```

	pH 5	pH 6	pH 7	pH 8	pH 9
I = 0 M	1.16519	-4.37511	-8.52558	-13.8718	-19.6143
I = 0.10 M	0.785196	-4.34584	-8.83247	-14.077	-19.4656
I = 0.25 M	0.754643	-4.31006	-8.8077	-14.0617	-19.3633

```
TableForm[Transpose[calckprime[pyruvate+co2tot+atp+de==oxaloacetate+adp+pi,{5,6,7,8,9},
{0,.1,.25}]],TableHeadings->{{"I = 0 M","I = 0.10 M","I = 0.25 M"},{"pH 5","pH 6","pH
7","pH 8","pH 9"}}]
```

	pH 5	pH 6	pH 7	pH 8	pH 9
I = 0 M	0.624984	5.84092	31.1607	269.294	2730.57
I = 0.10 M	0.728518	5.77236	35.2674	292.532	2571.58
I = 0.25 M	0.737553	5.68966	34.9167	290.728	2467.66

```
TableForm[(1/(8.31451*.29815*Log[10]))*D[(oxaloacetate+adp+pi-pyruvate-co2tot-
atp),pH]/.is->{0,.1,.25}/.pH->{5,6,7,8,9},TableHeadings->{{"I = 0 M","I = 0.10 M","I =
0.25 M"},{"pH 5","pH 6","pH 7","pH 8","pH 9"}}]
```

	pH 5	pH 6	pH 7	pH 8	pH 9
I = 0 M	-1.10545	-0.8173	-0.766313	-1.03032	-0.964759
I = 0.10 M	-0.980022	-0.801618	-0.843191	-0.96286	-0.885636
I = 0.25 M	-0.962049	-0.796006	-0.848177	-0.959722	-0.850021

phosphoenolpyruvate carboxykinase reaction

```
TableForm[Transpose[calctrGerx[oxaloacetate+atp+h2o+de==pep+adp+co2tot,{5,6,7,8,9},{0,.
1,.25}]],TableHeadings->{{"I = 0 M","I = 0.10 M","I = 0.25 M"},{"pH 5","pH 6","pH
7","pH 8","pH 9"}}]
```

	pH 5	pH 6	pH 7	pH 8	pH 9
I = 0 M	-3.00283	1.22754	1.54323	-3.45614	-9.42638
I = 0.10 M	0.103784	3.61904	1.59418	-3.70581	-9.63967
I = 0.25 M	1.15447	4.19971	1.61744	-3.72164	-9.74206

```
TableForm[Transpose[calckprime[oxaloacetate+atp+h2o+de==pep+adp+co2tot,{5,6,7,8,9},{0,.
1,.25}]],TableHeadings->{{"I = 0 M","I = 0.10 M","I = 0.25 M"},{"pH 5","pH 6","pH
7","pH 8","pH 9"}}]
```

	pH 5	pH 6	pH 7	pH 8	pH 9
I = 0 M	3.35791	0.609461	0.536586	4.03168	44.8148
I = 0.10 M	0.958999	0.23226	0.525671	4.45889	48.8413
I = 0.25 M	0.627693	0.183758	0.520761	4.48745	50.901

```
TableForm[(1/(8.31451*.29815*Log[10]))*D[(pep+adp+co2tot-oxaloacetate-atp-h2o),pH]/.is-
>{0,.1,.25}/.pH->{5,6,7,8,9},TableHeadings->{{"I = 0 M","I = 0.10 M","I = 0.25
M"},{"pH 5","pH 6","pH 7","pH 8","pH 9"}}]
```

	pH 5	pH 6	pH 7	pH 8	pH 9
I = 0 M	0.808208	0.547488	-0.512396	-1.04491	-1.05597
I = 0.10 M	0.862233	0.206083	-0.792542	-0.99541	-1.11014
I = 0.25 M	0.840963	0.0667518	-0.82107	-0.998722	-1.1457

Net of the preceding two reactions

```
TableForm[Transpose[calctrGerx[pyruvate+2*atp+h2o+de==pep+2*adp+pi,{5,6,7,8,9},{0,.1,.2
5}]],TableHeadings->{{"I = 0 M","I = 0.10 M","I = 0.25 M"},{"pH 5","pH 6","pH 7","pH
8","pH 9"}}]
```

	pH 5	pH 6	pH 7	pH 8	pH 9
I = 0 M	-1.83764	-3.14757	-6.98235	-17.328	-29.0407
I = 0.10 M	0.88898	-0.726795	-7.23829	-17.7828	-29.1052
I = 0.25 M	1.90911	-0.110354	-7.19026	-17.7833	-29.1054

```
TableForm[Transpose[calckprime[pyruvate+2*atp+h2o+de==pep+2*adp+pi,{5,6,7,8,9},{0,.1,.2
5}]],TableHeadings->{{"I = 0 M","I = 0.10 M","I = 0.25 M"},{"pH 5","pH 6","pH 7","pH
8","pH 9"}}]
```

	pH 5	pH 6	pH 7	pH 8	pH 9
I = 0 M	2.09864	3.55982	16.7204	1085.71	122370.
I = 0.10 M	0.698648	1.34069	18.539	1304.37	125599.
I = 0.25 M	0.462956	1.04552	18.1833	1304.63	125606.

```
TableForm[(1/(8.31451*.29815*Log[10]))*D[(pep+2*adp+pi-pyruvate-2*atp-h2o),pH]/.is-
>{0,.1,.25}/.pH->{5,6,7,8,9},TableHeadings->{{"I = 0 M","I = 0.10 M","I = 0.25
M"},{"pH 5","pH 6","pH 7","pH 8","pH 9"}}]
```

	pH 5	pH 6	pH 7	pH 8	pH 9
I = 0 M	-0.297246	-0.269812	-1.27871	-2.07523	-2.02073
I = 0.10 M	-0.11779	-0.595535	-1.63573	-1.95827	-1.99578
I = 0.25 M	-0.121086	-0.729254	-1.66925	-1.95844	-1.99572

Fructose 1,6-biphosphatase

```
TableForm[Transpose[calctrGerx[fructose16phos+h2o+de==fructose6phos+pi,{5,6,7,8,9},{0,.
1,.25}]],TableHeadings->{{"I = 0 M","I = 0.10 M","I = 0.25 M"},{"pH 5","pH 6","pH
7","pH 8","pH 9"}}]
```

	pH 5	pH 6	pH 7	pH 8	pH 9
I = 0 M	-22.7476	-22.1975	-20.1521	-18.6253	-18.3442
I = 0.10 M	-21.0371	-18.1741	-14.6161	-13.5633	-13.4289
I = 0.25 M	-20.2085	-16.2642	-12.7902	-11.9455	-11.8441

```
TableForm[Transpose[calckprime[fructose16phos+h2o+de==fructose6phos+pi,{5,6,7,8,9},{0,.
1,.25}]],TableHeadings->{{"I = 0 M","I = 0.10 M","I = 0.25 M"},{"pH 5","pH 6","pH
7","pH 8","pH 9"}}]
```

	pH 5	pH 6	pH 7	pH 8	pH 9
I = 0 M	9664.7	7741.4	3392.13	1832.26	1635.84
I = 0.10 M	4847.55	1527.39	363.586	237.781	225.229
I = 0.25 M	3470.27	706.875	174.073	123.807	118.847

```
TableForm[(1/(8.31451*.29815*Log[10]))*D[(fructose6phos+pi-fructose16phos-h2o),pH]/.is-
>{0,.1,.25}/.pH->{5,6,7,8,9},TableHeadings->{{"I = 0 M","I = 0.10 M","I = 0.25
M"},{"pH 5","pH 6","pH 7","pH 8","pH 9"}}]
```

	pH 5	pH 6	pH 7	pH 8	pH 9
I = 0 M	0.028674	0.211976	0.414941	0.11638	0.0136539
I = 0.10 M	0.215128	0.744345	0.384706	0.0586725	0.00618982
I = 0.25 M	0.371614	0.840829	0.324487	0.0445667	0.00463044

Glucose 6-phosphatasephosphatase

```
TableForm[Transpose[calctrGerx[glucose6phos+h2o+de==glucose+pi,{5,6,7,8,9},{0,.1,.25}]]
,TableHeadings->{{"I = 0 M","I = 0.10 M","I = 0.25 M"},{"pH 5","pH 6","pH 7","pH
8","pH 9"}}]
```

	pH 5	pH 6	pH 7	pH 8	pH 9
I = 0 M	-15.3426	-14.7682	-12.7076	-11.1844	-10.9041
I = 0.10 M	-15.2194	-14.0517	-11.8264	-10.9938	-10.8828
I = 0.25 M	-15.1497	-13.7484	-11.6155	-10.9608	-10.8793

```
TableForm[Transpose[calckprime[glucose6phos+h2o+de==glucose+pi,{5,6,7,8,9},{0,.1,.25}]]
,TableHeadings->{{"I = 0 M","I = 0.10 M","I = 0.25 M"},{"pH 5","pH 6","pH 7","pH
8","pH 9"}}]
```

	pH 5	pH 6	pH 7	pH 8	pH 9
I = 0 M	487.412	386.594	168.366	91.0791	81.3396
I = 0.10 M	463.784	289.565	118.	84.3364	80.6433
I = 0.25 M	450.911	256.21	108.375	83.2229	80.5302

```
TableForm[(1/(8.31451*.29815*Log[10]))*D[(glucose+pi-glucose6phos-h2o),pH]/.is-
>{0,.1,.25}/.pH->{5,6,7,8,9},TableHeadings->{{"I = 0 M","I = 0.10 M","I = 0.25
M"},{"pH 5","pH 6","pH 7","pH 8","pH 9"}}]
```

	pH 5	pH 6	pH 7	pH 8	pH 9
I = 0 M	0.0305457	0.218017	0.414317	0.116071	0.013619
I = 0.10 M	0.0764751	0.364857	0.291488	0.0482304	0.00513333
I = 0.25 M	0.101058	0.400861	0.242412	0.035721	0.00373907

Net reaction for gluconeogenesis

```
TableForm[Transpose[calctrGerx[2*pyruvate+6*atp+2*nadred+6*h2o+de==glucose+6*adp+6*pi+2
*nadox,{5,6,7,8,9},{0,.1,.25}]],TableHeadings->{{"I = 0 M","I = 0.10 M","I = 0.25
M"},{"pH 5","pH 6","pH 7","pH 8","pH 9"}}]
```

	pH 5	pH 6	pH 7	pH 8	pH 9
I = 0 M	-114.293	-95.1491	-82.4634	-89.0035	-100.93
I = 0.10 M	-94.9253	-75.5524	-68.4902	-75.5282	-86.4153
I = 0.25 M	-88.163	-69.2508	-63.3307	-70.7316	-81.6672

```
TableForm[Transpose[calckprime[2*pyruvate+6*atp+2*nadred+6*h2o+de==glucose+6*adp+6*pi+2
*nadox,{5,6,7,8,9},{0,.1,.25}]],TableHeadings->{{"I = 0 M","I = 0.10 M","I = 0.25
M"},{"pH 5","pH 6","pH 7","pH 8","pH 9"}}]
```

	pH 5	pH 6	pH 7	pH 8	pH 9
I = 0 M	1.05479×10^{20}	4.66988×10^{16}	2.79825×10^{14}	3.91429×10^{15}	4.80954×10^{17}
I = 0.10 M	4.26674×10^{16}	1.72242×10^{13}	9.97467×10^{11}	1.70563×10^{13}	1.37793×10^{15}
I = 0.25 M	2.78871×10^{15}	1.35564×10^{12}	1.24448×10^{11}	2.46364×10^{12}	2.02949×10^{14}

```
TableForm[(1/(8.31451*.29815*Log[10]))*D[(glucose+6*adp+6*pi+2*nadox-2*pyruvate-6*atp-2
*nadred-6*h2o),pH]/.is->{0,.1,.25}/.pH->{5,6,7,8,9},TableHeadings->{{"I = 0 M","I =
0.10 M","I = 0.25 M"},{"pH 5","pH 6","pH 7","pH 8","pH 9"}}]
```

	pH 5	pH 6	pH 7	pH 8	pH 9
I = 0 M	3.11986	3.29164	0.531329	-2.07356	-2.04318
I = 0.10 M	3.72487	2.70949	-0.320941	-1.76774	-1.9758
I = 0.25 M	3.76965	2.50822	-0.451703	-1.78866	-1.97804

4.10 Calculate the standard transformed reaction Gibbs energies, apparent equilibrium constants, and changes in the binding of hydrogen ions for the pyruvate dehydrogenase reaction and the nine reactions of the citric acid cycle at 298.15 K, pHs 5, 6, 7, 8, and 9 and ionic strengths 0, 0.10, and 0.25 M. Also calculate these properties for the net reaction of the citric acid cycle, the net reaction for pyruvate dehydrogenase plus the citric acid cycle, and the net reaction for glycolysis, pyruvate dehydrogenase, and the citric acid cycle.

(BasicBiochemData2 must be loaded)

Pyruvate dehydrogenase

```
TableForm[Transpose[calctrGerx[pyruvate+coA+nadox+h2o+de==co2tot+acetylcoA+nadred,{5,6,
7,8,9},{0,.1,.25}]],TableHeadings->{{"I = 0 M","I = 0.10 M","I = 0.25 M"},{"pH 5","pH
6","pH 7","pH 8","pH 9"}}]
```

	pH 5	pH 6	pH 7	pH 8	pH 9
I = 0 M	-23.6334	-24.4065	-27.5641	-32.0578	-34.6066
I = 0.10 M	-24.9212	-26.0596	-29.7622	-34.0761	-36.2027
I = 0.25 M	-25.3442	-26.6231	-30.4807	-34.7047	-36.7463

```
TableForm[Transpose[calckprime[pyruvate+coA+nadox+h2o+de==co2tot+acetylcoA+nadred,{5,6,
7,8,9},{0,.1,.25}]],TableHeadings->{{"I = 0 M","I = 0.10 M","I = 0.25 M"},{"pH 5","pH
6","pH 7","pH 8","pH 9"}}]
```

	pH 5	pH 6	pH 7	pH 8	pH 9
I = 0 M	13815.3	18871.6	67452.	413294.	1.15554×10^{6}
I = 0.10 M	23226.8	36763.7	163714.	932920.	2.1999×10^{6}
I = 0.25 M	27548.	46145.7	218750.	1.20217×10^{6}	2.73926×10^{6}

```
TableForm[(1/(8.31451*.29815*Log[10]))*D[(co2tot+acetylcoA+nadred-pyruvate-coA-nadox-
h2o),pH]/.is->{0,.1,.25}/.pH->{5,6,7,8,9},TableHeadings->{{"I = 0 M","I = 0.10 M","I =
0.25 M"},{"pH 5","pH 6","pH 7","pH 8","pH 9"}}]
```

	pH 5	pH 6	pH 7	pH 8	pH 9
I = 0 M	-0.0408172	-0.296603	-0.771728	-0.687435	-0.235697
I = 0.10 M	-0.0651484	-0.406673	-0.812946	-0.592257	-0.237883
I = 0.25 M	-0.0755408	-0.444665	-0.81951	-0.559928	-0.257158

Citrate synthase

```
TableForm[Transpose[calctrGerx[acetylcoA+oxaloacetate+h2o+de==citrate+coA,{5,6,7,8,9},{
0,.1,.25}]],TableHeadings->{{"I = 0 M","I = 0.10 M","I = 0.25 M"},{"pH 5","pH 6","pH
7","pH 8","pH 9"}}]
```

	pH 5	pH 6	pH 7	pH 8	pH 9
I = 0 M	-37.3535	-37.1217	-40.2877	-46.2697	-55.1332
I = 0.10 M	-36.9892	-38.7412	-43.6084	-50.3226	-59.8309
I = 0.25 M	-37.0791	-39.5563	-44.7706	-51.6711	-61.3642

```
TableForm[Transpose[calckprime[acetylcoA+oxaloacetate+h2o+de==citrate+coA,{5,6,7,8,9},{
0,.1,.25}]],TableHeadings->{{"I = 0 M","I = 0.10 M","I = 0.25 M"},{"pH 5","pH 6","pH
7","pH 8","pH 9"}}]
```

	pH 5	pH 6	pH 7	pH 8	pH 9
I = 0 M	3.49956×10^{6}	3.18713×10^{6}	1.14305×10^{7}	1.27658×10^{8}	4.55901×10^{9}
I = 0.10 M	3.02131×10^{6}	6.12526×10^{6}	4.36333×10^{7}	6.54759×10^{8}	3.03295×10^{10}
I = 0.25 M	3.13288×10^{6}	8.51005×10^{6}	6.97298×10^{7}	1.12808×10^{9}	5.62969×10^{10}

```
TableForm[(1/(8.31451*.29815*Log[10]))*D[(citrate+coA-acetylcoA-oxaloacetate-
h2o),pH]/.is->{0,.1,.25}/.pH->{5,6,7,8,9},TableHeadings->{{"I = 0 M","I = 0.10 M","I =
0.25 M"},{"pH 5","pH 6","pH 7","pH 8","pH 9"}}]
```

	pH 5	pH 6	pH 7	pH 8	pH 9
I = 0 M	0.32869	-0.242031	-0.839947	-1.27037	-1.80428
I = 0.10 M	0.0241225	-0.634751	-1.01063	-1.40059	-1.87189
I = 0.25 M	-0.0928804	-0.742845	-1.04053	-1.44166	-1.88881

Aconitase

```
TableForm[Transpose[calctrGerx[citrate+de==aconitatecis+h2o,{5,6,7,8,9},{0,.1,.25}]],Ta
bleHeadings->{{"I = 0 M","I = 0.10 M","I = 0.25 M"},{"pH 5","pH 6","pH 7","pH 8","pH
9"}}]
```

	pH 5	pH 6	pH 7	pH 8	pH 9
I = 0 M	17.5023	11.5531	8.92005	8.43053	8.37612
I = 0.10 M	13.4649	9.49365	8.50571	8.38388	8.37139
I = 0.25 M	12.3689	9.12025	8.45494	8.37861	8.37086

```
TableForm[Transpose[calckprime[citrate+de==aconitatecis+h2o,{5,6,7,8,9},{0,.1,.25}]],Ta
bleHeadings->{{"I = 0 M","I = 0.10 M","I = 0.25 M"},{"pH 5","pH 6","pH 7","pH 8","pH
9"}}]
```

	pH 5	pH 6	pH 7	pH 8	pH 9
I = 0 M	0.000858528	0.00946228	0.0273705	0.033346	0.034086
I = 0.10 M	0.00437581	0.0217167	0.0323499	0.0339794	0.034151
I = 0.25 M	0.0068089	0.025247	0.0330192	0.0340517	0.0341583

```
TableForm[(1/(8.31451*.29815*Log[10]))*D[(aconitatecis+h2o-citrate),pH]/.is-
>{0,.1,.25}/.pH->{5,6,7,8,9},TableHeadings->{{"I = 0 M","I = 0.10 M","I = 0.25
M"},{"pH 5","pH 6","pH 7","pH 8","pH 9"}}]
```

	pH 5	pH 6	pH 7	pH 8	pH 9
I = 0 M	-1.32911	-0.762126	-0.200124	-0.0241339	-0.00246441
I = 0.10 M	-1.02481	-0.372043	-0.0533865	-0.0055862	-0.000561228
I = 0.25 M	-0.907921	-0.265114	-0.0337371	-0.00346956	-0.000347946

Aconitase

```
TableForm[Transpose[calctrGerx[aconitatecis+h2o+de==citrateiso,{5,6,7,8,9},{0,.1,.25}]]
,TableHeadings->{{"I = 0 M","I = 0.10 M","I = 0.25 M"},{"pH 5","pH 6","pH 7","pH
8","pH 9"}}]
```

	pH 5	pH 6	pH 7	pH 8	pH 9
I = 0 M	-10.8055	-4.92275	-2.27779	-1.7815	-1.72622
I = 0.10 M	-6.81316	-2.85648	-1.85783	-1.73411	-1.72141
I = 0.25 M	-5.72527	-2.4798	-1.80629	-1.72875	-1.72088

```
TableForm[Transpose[calckprime[aconitatecis+h2o+de==citrateiso,{5,6,7,8,9},{0,.1,.25}]]
,TableHeadings->{{"I = 0 M","I = 0.10 M","I = 0.25 M"},{"pH 5","pH 6","pH 7","pH
8","pH 9"}}]
```

	pH 5	pH 6	pH 7	pH 8	pH 9
I = 0 M	78.1691	7.28489	2.5064	2.05165	2.0064
I = 0.10 M	15.6173	3.16542	2.11581	2.0128	2.00252
I = 0.25 M	10.0697	2.71919	2.07227	2.00846	2.00209

```
TableForm[(1/(8.31451*.29815*Log[10]))*D[(citrateiso-aconitatecis-h2o),pH]/.is-
>{0,.1,.25}/.pH->{5,6,7,8,9},TableHeadings->{{"I = 0 M","I = 0.10 M","I = 0.25
M"},{"pH 5","pH 6","pH 7","pH 8","pH 9"}}]
```

	pH 5	pH 6	pH 7	pH 8	pH 9
I = 0 M	1.30605	0.760857	0.202526	0.0245142	0.00250437
I = 0.10 M	1.0124	0.374671	0.0541865	0.00567633	0.00057035
I = 0.25 M	0.899474	0.267618	0.0342572	0.00352569	0.000353603

Isocitrate dehydrogenase

```
TableForm[Transpose[calctrGerx[citrateiso+nadox+h2o+de==ketoglutarate+co2tot+nadred,{5,
6,7,8,9},{0,.1,.25}]],TableHeadings->{{"I = 0 M","I = 0.10 M","I = 0.25 M"},{"pH
5","pH 6","pH 7","pH 8","pH 9"}}]
```

	pH 5	pH 6	pH 7	pH 8	pH 9
I = 0 M	8.34114	1.67592	-4.21769	-9.97115	-15.7852
I = 0.10 M	5.50847	0.398237	-4.45015	-10.0157	-15.9678
I = 0.25 M	4.78775	0.245626	-4.45696	-10.0274	-16.0697

```
TableForm[Transpose[calckprime[citrateiso+nadox+h2o+de==ketoglutarate+co2tot+nadred,{5,
6,7,8,9},{0,.1,.25}]],TableHeadings->{{"I = 0 M","I = 0.10 M","I = 0.25 M"},{"pH
5","pH 6","pH 7","pH 8","pH 9"}}]
```

	pH 5	pH 6	pH 7	pH 8	pH 9
I = 0 M	0.0345704	0.50862	5.48156	55.8289	582.674
I = 0.10 M	0.108383	0.851594	6.02043	56.8417	627.221
I = 0.25 M	0.144953	0.905667	6.03701	57.1107	653.532

```
TableForm[(1/(8.31451*.29815*Log[10]))*D[(ketoglutarate+co2tot+nadred-citrateiso-nadox-
h2o),pH]/.is->{0,.1,.25}/.pH->{5,6,7,8,9},TableHeadings->{{"I = 0 M","I = 0.10 M","I =
0.25 M"},{"pH 5","pH 6","pH 7","pH 8","pH 9"}}]
```

	pH 5	pH 6	pH 7	pH 8	pH 9
I = 0 M	-1.34729	-1.06162	-1.01433	-1.00645	-1.04494
I = 0.10 M	-1.07823	-0.788138	-0.931153	-1.00411	-1.1109
I = 0.25 M	-0.975816	-0.720242	-0.928031	-1.00858	-1.14667

ketoglutarate dehydrogenase

```
TableForm[Transpose[calctrGerx[ketoglutarate+nadox+coA+h2o+de==succinylcoA+co2tot+nadre
d,{5,6,7,8,9},{0,.1,.25}]],TableHeadings->{{"I = 0 M","I = 0.10 M","I = 0.25 M"},{"pH
5","pH 6","pH 7","pH 8","pH 9"}}]
```

	pH 5	pH 6	pH 7	pH 8	pH 9
I = 0 M	-32.2272	-32.6665	-35.7881	-40.2782	-42.8267
I = 0.10 M	-32.1519	-33.08	-36.7606	-41.0722	-43.1986
I = 0.25 M	-32.1464	-33.2446	-37.0834	-41.3056	-43.347

```
TableForm[Transpose[calckprime[ketoglutarate+nadox+coA+h2o+de==succinylcoA+co2tot+nadre
d,{5,6,7,8,9},{0,.1,.25}]],TableHeadings->{{"I = 0 M","I = 0.10 M","I = 0.25 M"},{"pH
5","pH 6","pH 7","pH 8","pH 9"}}]
```

	pH 5	pH 6	pH 7	pH 8	pH 9
I = 0 M	442519.	528319.	$1.86112 \ 10^{6}$	$1.13869 \ 10^{7}$	$3.18322 \ 10^{7}$
I = 0.10 M	429279.	624208.	$2.75508 \ 10^{6}$	$1.56857 \ 10^{7}$	$3.69848 \ 10^{7}$
I = 0.25 M	428331.	667072.	$3.1383 \ 10^{6}$	$1.72338 \ 10^{7}$	$3.92659 \ 10^{7}$

```
TableForm[(1/(8.31451*.29815*Log[10]))*D[(succinylcoA+co2tot+nadred-ketoglutarate-
nadox-coA-h2o),pH]/.is->{0,.1,.25}/.pH->{5,6,7,8,9},TableHeadings->{{"I = 0 M","I =
0.10 M","I = 0.25 M"},{"pH 5","pH 6","pH 7","pH 8","pH 9"}}]
```

	pH 5	pH 6	pH 7	pH 8	pH 9
I = 0 M	0.0991751	0.280585	-0.770103	-0.687272	-0.235681
I = 0.10 M	0.0252188	-0.396837	-0.811953	-0.592158	-0.237873
I = 0.25 M	0.00254914	-0.436266	-0.818663	-0.559843	-0.25715

Succinyl coA synthase

```
TableForm[Transpose[calctrGerx[succinylcoA+pi+adp+de==succinate+atp+coA,{5,6,7,8,9},{0,
.1,.25}]],TableHeadings->{{"I = 0 M","I = 0.10 M","I = 0.25 M"},{"pH 5","pH 6","pH
7","pH 8","pH 9"}}]
```

	pH 5	pH 6	pH 7	pH 8	pH 9
I = 0 M	11.3123	9.42916	6.0856	4.60067	1.48513
I = 0.10 M	8.64828	5.44782	2.53412	0.710067	-3.18751
I = 0.25 M	7.58697	4.09308	1.26214	-0.651704	-4.72216

```
TableForm[Transpose[calckprime[succinylcoA+pi+adp+de==succinate+atp+coA,{5,6,7,8,9},{0,
.1,.25}]],TableHeadings->{{"I = 0 M","I = 0.10 M","I = 0.25 M"},{"pH 5","pH 6","pH
7","pH 8","pH 9"}}]
```

	pH 5	pH 6	pH 7	pH 8	pH 9
I = 0 M	0.0104276	0.0222891	0.0858727	0.156315	0.549312
I = 0.10 M	0.0305419	0.111068	0.359786	0.750935	3.61763
I = 0.25 M	0.0468627	0.191835	0.601012	1.30069	6.71864

```
TableForm[(1/(8.31451*.29815*Log[10]))*D[(succinate+atp+coA-succinylcoA-pi-
adp),pH]/.is->{0,.1,.25}/.pH->{5,6,7,8,9},TableHeadings->{{"I = 0 M","I = 0.10 M","I =
0.25 M"},{"pH 5","pH 6","pH 7","pH 8","pH 9"}}]
```

	pH 5	pH 6	pH 7	pH 8	pH 9
I = 0 M	-0.0431903	-0.591126	-0.421843	-0.27808	-0.799126
I = 0.10 M	-0.347965	-0.659794	-0.328911	-0.443368	-0.876329
I = 0.25 M	-0.436744	-0.660859	-0.321521	-0.47926	-0.892712

Succinate dehydrogenase

```
TableForm[Transpose[calctrGerx[succinate+fadenzox+de==fumarate+fadenzred,{5,6,7,8,9},{0
,.1,.25}]],TableHeadings->{{"I = 0 M","I = 0.10 M","I = 0.25 M"},{"pH 5","pH 6","pH
7","pH 8","pH 9"}}]
```

	pH 5	pH 6	pH 7	pH 8	pH 9
I = 0 M	3.58609	0.778578	0.0656228	-0.0202505	-0.0290231
I = 0.10 M	2.15463	0.30828	0.00611379	-0.0263635	-0.0296361
I = 0.25 M	1.7504	0.220992	-0.00368738	-0.0273557	-0.0297354

```
TableForm[Transpose[calckprime[succinate+fadenzox+de==fumarate+fadenzred,{5,6,7,8,9},{0
,.1,.25}]],TableHeadings->{{"I = 0 M","I = 0.10 M","I = 0.25 M"},{"pH 5","pH 6","pH
7","pH 8","pH 9"}}]
```

	pH 5	pH 6	pH 7	pH 8	pH 9
I = 0 M	0.235369	0.730466	0.973876	1.0082	1.01178
I = 0.10 M	0.419303	0.883064	0.997537	1.01069	1.01203
I = 0.25 M	0.493565	0.914711	1.00149	1.0111	1.01207

```
TableForm[(1/(8.31451*.29815*Log[10]))*D[(fumarate+fadenzred-succinate-
fadenzox),pH]/.is->{0,.1,.25}/.pH->{5,6,7,8,9},TableHeadings->{{"I = 0 M","I = 0.10
M","I = 0.25 M"},{"pH 5","pH 6","pH 7","pH 8","pH 9"}}]
```

	pH 5	pH 6	pH 7	pH 8	pH 9
I = 0 M	-0.658653	-0.272415	-0.0377549	-0.00392427	-0.000393977
I = 0.10 M	-0.565671	-0.127042	-0.0144565	-0.0014658	-0.000146785
I = 0.25 M	-0.504786	-0.0961348	-0.0105565	-0.00106609	-0.000106714

Fumarase

```
TableForm[Transpose[calctrGerx[fumarate+h2o+de==malate,{5,6,7,8,9},{0,.1,.25}]],TableHe
adings->{{"I = 0 M","I = 0.10 M","I = 0.25 M"},{"pH 5","pH 6","pH 7","pH 8","pH 9"}}]
```

	pH 5	pH 6	pH 7	pH 8	pH 9
I = 0 M	-5.32376	-3.91639	-3.63473	-3.60351	-3.60035
I = 0.10 M	-4.53418	-3.72559	-3.61302	-3.60131	-3.60013
I = 0.25 M	-4.33506	-3.69229	-3.60948	-3.60095	-3.6001

```
TableForm[Transpose[calckprime[fumarate+h2o+de==malate,{5,6,7,8,9},{0,.1,.25}]],TableHe
adings->{{"I = 0 M","I = 0.10 M","I = 0.25 M"},{"pH 5","pH 6","pH 7","pH 8","pH 9"}}]
```

	pH 5	pH 6	pH 7	pH 8	pH 9
I = 0 M	8.56401	4.85419	4.33284	4.27862	4.27317
I = 0.10 M	6.22801	4.4946	4.29507	4.27482	4.27279
I = 0.25 M	5.74732	4.43464	4.28894	4.27421	4.27273

```
TableForm[(1/(8.31451*.29815*Log[10]))*D[(malate-fumarate-h2o),pH]/.is->{0,.1,.25}/.pH-
>{5,6,7,8,9},TableHeadings->{{"I = 0 M","I = 0.10 M","I = 0.25 M"},{"pH 5","pH 6","pH
7","pH 8","pH 9"}}]
```

	pH 5	pH 6	pH 7	pH 8	pH 9
I = 0 M	0.353078	0.115155	0.0138543	0.0014132	0.000141603
I = 0.10 M	0.271995	0.0486631	0.00523192	0.000527102	0.0000527496
I = 0.25 M	0.230733	0.0361476	0.00381207	0.00038328	0.0000383488

Malate dehydrogenase

```
TableForm[Transpose[calctrGerx[malate+nadox+de==oxaloacetate+nadred,{5,6,7,8,9},{0,.1,.
25}]],TableHeadings->{{"I = 0 M","I = 0.10 M","I = 0.25 M"},{"pH 5","pH 6","pH 7","pH
8","pH 9"}}]
```

	pH 5	pH 6	pH 7	pH 8	pH 9
I = 0 M	46.0468	38.1855	32.1083	26.3602	20.6481
I = 0.10 M	42.3126	35.4858	29.6322	23.9091	18.1995
I = 0.25 M	41.2328	34.6521	28.8372	23.1182	17.4091

```
TableForm[Transpose[calckprime[malate+nadox+de==oxaloacetate+nadred,{5,6,7,8,9},{0,.1,.
25}]],TableHeadings->{{"I = 0 M","I = 0.10 M","I = 0.25 M"},{"pH 5","pH 6","pH 7","pH
8","pH 9"}}]
```

	pH 5	pH 6	pH 7	pH 8	pH 9
I = 0 M	$8.57032 \cdot 10^{-9}$	$2.04279 \cdot 10^{-7}$	$2.3708 \cdot 10^{-6}$	0.0000240949	0.000241343
I = 0.10 M	$3.86543 \cdot 10^{-8}$	$6.07008 \cdot 10^{-7}$	$6.43723 \cdot 10^{-6}$	0.000064764	0.000648034
I = 0.25 M	$5.97546 \cdot 10^{-8}$	$8.49672 \cdot 10^{-7}$	$8.87102 \cdot 10^{-6}$	0.0000891027	0.000891421

```
TableForm[(1/(8.31451*.29815*Log[10]))*D[(oxaloacetate+nadred-malate-nadox),pH]/.is-
>{0,.1,.25}/.pH->{5,6,7,8,9},TableHeadings->{{"I = 0 M","I = 0.10 M","I = 0.25
M"},{"pH 5","pH 6","pH 7","pH 8","pH 9"}}]
```

	pH 5	pH 6	pH 7	pH 8	pH 9
I = 0 M	-1.64496	-1.15373	-1.01784	-1.00181	-1.00018
I = 0.10 M	-1.40356	-1.06337	-1.00672	-1.00068	-1.00007
I = 0.25 M	-1.3297	-1.04688	-1.00489	-1.00049	-1.00005

Net reaction for the citric acid cycle

```
TableForm[Transpose[calctrGerx[acetylcoA+3*nadox+fadenzox+adp+pi+4*h2o+de==2*co2tot+3*n
adred+fadenzred+atp+coA,{5,6,7,8,9},{0,.1,.25}]],TableHeadings->{{"I = 0 M","I = 0.10
M","I = 0.25 M"},{"pH 5","pH 6","pH 7","pH 8","pH 9"}}]
```

	pH 5	pH 6	pH 7	pH 8	pH 9
I = 0 M	1.07855	-17.0051	-39.0265	-62.5329	-88.5914
I = 0.10 M	-8.39957	-27.2695	-49.6119	-73.7693	-100.965
I = 0.25 M	-11.5591	-30.641	-53.1761	-77.5161	-105.074

```
TableForm[Transpose[calckprime[acetylcoA+3*nadox+fadenzox+adp+pi+4*h2o+de==2*co2tot+3*n
adred+fadenzred+atp+coA,{5,6,7,8,9},{0,.1,.25}]],TableHeadings->{{"I = 0 M","I = 0.10
M","I = 0.25 M"},{"pH 5","pH 6","pH 7","pH 8","pH 9"}}]
```

	pH 5	pH 6	pH 7	pH 8	pH 9
I = 0 M	0.647214	953.103	$6.87231 \ 10^{6}$	$9.02059 \ 10^{10}$	$3.31471 \ 10^{15}$
I = 0.10 M	29.6165	59893.3	$4.91554 \ 10^{8}$	$8.38962 \ 10^{12}$	$4.87771 \ 10^{17}$
I = 0.25 M	105.938	233370.	$2.07014 \ 10^{9}$	$3.80323 \ 10^{13}$	$2.55874 \ 10^{18}$

```
TableForm[(1/(8.31451*.29815*Log[10]))*D[(2*co2tot+3*nadred+fadenzred+atp+coA-
acetylcoA-3*nadox-fadenzox-adp-pi-4*h2o),pH]/.is->{0,.1,.25}/.pH-
>{5,6,7,8,9},TableHeadings->{{"I = 0 M","I = 0.10 M","I = 0.25 M"},{"pH 5","pH 6","pH
7","pH 8","pH 9"}}]
```

	pH 5	pH 6	pH 7	pH 8	pH 9
I = 0 M	-2.9362	-3.48762	-4.08556	-4.24612	-4.88442
I = 0.10 M	-3.08649	-3.61864	-4.0978	-4.44174	-5.09714
I = 0.25 M	-3.1151	-3.66458	-4.11986	-4.49046	-5.18546

Net reaction for the pyruvate dehydrogenase reaction plus the citric acid cycle (10 reactions)

```
TableForm[Transpose[calctrGerx[pyruvate+4*nadox+fadenzox+adp+pi+5*h2o+de==3*co2tot+4*na
dred+fadenzred+atp,{5,6,7,8,9},{0,.1,.25}]],TableHeadings->{{"I = 0 M","I = 0.10 M","I
= 0.25 M"},{"pH 5","pH 6","pH 7","pH 8","pH 9"}}]
```

	pH 5	pH 6	pH 7	pH 8	pH 9
I = 0 M	-22.5548	-41.4116	-66.5906	-94.5908	-123.198
I = 0.10 M	-33.3208	-53.3291	-79.3741	-107.845	-137.168
I = 0.25 M	-36.9033	-57.2641	-83.6568	-112.221	-141.82

```
TableForm[Transpose[calckprime[pyruvate+4*nadox+fadenzox+adp+pi+5*h2o+de==3*co2tot+4*na
dred+fadenzred+atp,{5,6,7,8,9},{0,.1,.25}]],TableHeadings->{{"I = 0 M","I = 0.10 M","I
= 0.25 M"},{"pH 5","pH 6","pH 7","pH 8","pH 9"}}]
```

	pH 5	pH 6	pH 7	pH 8	pH 9
I = 0 M	8941.45	1.79866×10^{7}	4.63551×10^{11}	3.72816×10^{16}	3.83029×10^{21}
I = 0.10 M	687894.	2.2019×10^{9}	8.04744×10^{13}	7.82684×10^{18}	1.07305×10^{24}
I = 0.25 M	2.91838×10^{6}	1.0769×10^{10}	4.52844×10^{14}	4.57215×10^{19}	7.00906×10^{24}

```
TableForm[(1/(8.31451*.29815*Log[10]))*D[(3*co2tot+4*nadred+fadenzred+atp-
pyruvate-4*nadox-fadenzox-adp-pi-5*h2o),pH]/.is->{0,.1,.25}/.pH-
>{5,6,7,8,9},TableHeadings->{{"I = 0 M","I = 0.10 M","I = 0.25 M"},{"pH 5","pH 6","pH
7","pH 8","pH 9"}}]
```

	pH 5	pH 6	pH 7	pH 8	pH 9
I = 0 M	-2.97701	-3.78422	-4.85728	-4.93355	-5.12012
I = 0.10 M	-3.15164	-4.02532	-4.91074	-5.034	-5.33503
I = 0.25 M	-3.19064	-4.10924	-4.93937	-5.05039	-5.44262

Net reaction for glycolysis, pyruvate dehydrogenase, and the citric acid cycle

```
TableForm[Transpose[calctrGerx[glucose+10*nadox+2*fadenzox+4*adp+4*pi+8*h2o+de==6*co2to
t+10*nadred+2*fadenzred+4*atp,{5,6,7,8,9},{0,.1,.25}]],TableHeadings->{{"I = 0 M","I =
0.10 M","I = 0.25 M"},{"pH 5","pH 6","pH 7","pH 8","pH 9"}}]
```

	pH 5	pH 6	pH 7	pH 8	pH 9
I = 0 M	-72.0125	-131.329	-201.137	-270.179	-338.639
I = 0.10 M	-104.897	-166.593	-236.258	-306.076	-376.313
I = 0.25 M	-115.897	-178.144	-248.124	-318.007	-388.781

```
TableForm[Transpose[calckprime[glucose+10*nadox+2*fadenzox+4*adp+4*pi+8*h2o+de==6*co2to
t+10*nadred+2*fadenzred+4*atp,{5,6,7,8,9},{0,.1,.25}]],TableHeadings->{{"I = 0 M","I =
0.10 M","I = 0.25 M"},{"pH 5","pH 6","pH 7","pH 8","pH 9"}}]
```

	pH 5	pH 6	pH 7	pH 8	pH 9
I = 0 M	4.13017×10^{12}	1.01782×10^{23}	1.72759×10^{35}	2.15268×10^{47}	2.12147×10^{59}
I = 0.10 M	2.38257×10^{18}	1.53315×10^{29}	2.45688×10^{41}	4.18709×10^{53}	8.4501×10^{65}
I = 0.25 M	2.01437×10^{20}	1.61917×10^{31}	2.94593×10^{43}	5.15271×10^{55}	1.29175×10^{68}

```
TableForm[(1/(8.31451*.29815*Log[10]))*D[(6*co2tot+10*nadred+2*fadenzred+4*atp-
glucose-10*nadox-2*fadenzox-4*adp-4*pi-8*h2o),pH]/.is->{0,.1,.25}/.pH-
>{5,6,7,8,9},TableHeadings->{{"I = 0 M","I = 0.10 M","I = 0.25 M"},{"pH 5","pH 6","pH
7","pH 8","pH 9"}}]
```

	pH 5	pH 6	pH 7	pH 8	pH 9
I = 0 M	-9.66065	-11.3323	-12.5583	-11.8426	-12.2258
I = 0.10 M	-10.2116	11.6205	-12.3812	-12.1454	-12.6781
I = 0.25 M	-10.3045	-11.7212	-12.3948	-12.1712	-12.8926

Chapter 5 Matrices in Chemical and Biochemical Thermodynamics

5.1 (a) Carry out the operations involved in equations 5.1-12 to 5.1-27 using Mathematica^R. (b) Carry out the operations involved in equations 5.1-22 to 5.1-26. (c) Calculate the amounts of the components C, H, and O for a system containing one mole of each of the five species. (d) Calculate the amounts of the components CO, H_2, and CH_4 for the system containing one mole of each of the five species.

5.2 (a) Construct the conservation matrix for the hydrolysis of ATP to ADP in terms of species. (b) Calculate a basis for the stoichiometric matrix from the conservation matrix and show that it is consistent with equations 5.1-28 to 5.1-31.

5.3 (a) Construct the conservation matrix $\mathbf{A}'$ for the hydrolysis of ATP to ADP in terms of reactants. (b) Calculate a basis for the stoichiometric matrix from the conservation matrix and show that it is consistent with ATP + H_2O = ADP + P_i.

5.4 The glutamate-ammonia ligase reaction is
glutamate + ATP + ammonia = glutamine + ADP + P_i
It can be considered to be the sum of two reactions. (a) Write the stoichiometric number matrix for this enzyme-catalyzed reaction and use NullSpace to obtain a basis for the conservation matrix.. (b) Write a conservation matrix that includes a constraint to couple the two subreactions, and row reduce it to show that it is equivalent to the stoichiometric number matrix obtained in (a).

5.5 Carry out the matrix multiplications in equation 5.4-4 for the three chemical reactions involved in the hydration of fumarate to malate in the pH range 5 to 9.

5.1 (a) Carry out the operations involved in equations 5.1-12 to 5.1-27 using Mathematica[R]. (b) Carry out the operations involved in equations 5.1-22 to 5.1-26. (c) Calculate the amounts of the components C, H, and O for a system containing one mole of each of the five species. (d) Calculate the amounts of the components CO, H_2, and CH_4 for the system containing one mole of each of the five species.

(a) Equations 5.1-12 to 5.1-20
The conservation matrix is given by

```
a={{1,0,1,0,1},{0,2,4,2,0},{1,0,0,1,2}};
```

```
TableForm[a]
```

```
1    0    1    0    1
0    2    4    2    0
1    0    0    1    2
```

The stoichiometric number matrix is given by

```
nu={{-1,-1},{-3,1},{1,0},{1,-1},{0,1}};
```

```
TableForm[nu]
```

```
-1    -1
-3     1
 1     0
 1    -1
 0     1
```

The matrix product is given by

```
TableForm[a.nu]
```

```
0    0
0    0
0    0
```

The canonical form of the conservation matrix is

```
TableForm[ared=RowReduce[a]]
```

```
1    0    0    1     2
0    1    0    3     2
0    0    1    -1    -1
```

The matrix product using the row reduced form of the conservation matrix isactivated by using a period.

```
TableForm[ared.nu]
```

```
0    0
0    0
0    0
```

Thus the row reduced form represents the conservation relations equally well.
Calculate a basis for the nullspace of the conservation matrices a.

```
TableForm[Transpose[NullSpace[a]]]
```

```
-2   -1
-2   -3
 1    1
 0    1
 1    0
```

These are the same two reactions as in equation 5.1-20.

(b) Equations 5.1-22 to 5.1-26.
Now we start with the transpose of the stoichiometric number matrix.

```
nutr=Transpose[nu]
```

```
{{-1, -3, 1, 1, 0}, {-1, 1, 0, -1, 1}}
```

```
TableForm[nutr.Transpose[a]]
```

```
0   0   0
0   0   0
```

Now calculate a basis for the transpose of the conservation matrix from the transposed stoichiometric number matrix.

```
TableForm[NullSpace[nutr]]
```

```
 3   -1   0   0   4
-1    1   0   2   0
 1    1   4   0   0
```

This looks different from equation 5.1-8, but yields 5.1-15 when RowReduction is used.

```
TableForm[RowReduce[NullSpace[nutr]]]
```

```
1   0   0    1    2
0   1   0    3    2
0   0   1   -1   -1
```

The rank of the conservation matrix is the number of components, which is 3.

```
Dimensions[a]
```

```
{3, 5}
```

The rank of the stoichiometric number matrix is equal to the number of independent reactions, which is 2.

```
Dimensions[nu]
```

```
{5, 2}
```

(c) Calculate the amounts of the components C, H, and O for a system containing one mole of each of the five species.
Equation 5.1-27 yields

```
a.{1,1,1,1,1}
```

```
{3, 8, 4}
```

Note that *Mathematica* makes no distinction between "row" and "column" vectors.

(d) Calculate the amounts of the components CO, H_2, and CH_4 for the system containing one mole of each of the five species.

```
ared.{1,1,1,1,1}
```

```
{4, 6, -1}
```

Note that the amount of a component can be negative.

5.2 (a) Construct the conservation matrix for the hydrolysis of ATP to ADP in terms of species. (b) Calculate a basis for the stoichiometric matrix from the conservation matrix and show that it is consistent with equations 5.1-28 to 5.1-31.

(a) Construct the conservation matrix for the hydrolysis of ATP to ADP in terms of species.

```
a={{10,0,0,0,10,10,10,0},{12,1,2,1,12,13,13,2},{13,0,1,4,10,13,10,4},{3,0,0,1,2,3,2,1}}
;
```

```
TableForm[a,TableHeadings->{{"C","H","O","P"},{"ATP4-","H+","H2O","HPO42-","ADP3-
","HATP3-","HADP2-","H2PO4-"}},TableSpacing->{1,1.5}]
```

	ATP4-	H+	H2O	HPO42-	ADP3-	HATP3-	HADP2-	H2PO4-
C	10	0	0	0	10	10	10	0
H	12	1	2	1	12	13	13	2
O	13	0	1	4	10	13	10	4
P	3	0	0	1	2	3	2	1

```
TableForm[RowReduce[a],TableHeadings->{{"ATP4-","H+","H2O","HPO42-"},{"ATP4-
","H+","H2O","HPO42-","ADP3-","HATP3-","HADP2-","H2PO4-"}},TableSpacing->{1,1.5}]
```

	ATP4-	H+	H2O	HPO42-	ADP3-	HATP3-	HADP2-	H2PO4-
ATP4-	1	0	0	0	1	1	1	0
H+	0	1	0	0	-1	1	0	1
H2O	0	0	1	0	1	0	1	0
HPO42-	0	0	0	1	-1	0	-1	1

(b) Calculate a basis for the stoichiometric matrix from the conservation matrix and show that it is consistent with equations 5.1-28 to 5.1-31.

```
TableForm[nu=NullSpace[a]]
```

0	-1	0	-1	0	0	0	1
-1	0	-1	1	0	0	1	0
-1	-1	0	0	0	1	0	0
-1	1	-1	1	1	0	0	0

The transposed stoichiometric matrix for reactions 5.1-28 to 5.1-31 is

```
nutrexpected={{-1,1,-1,1,1,0,0,0},{1,1,0,0,0,-1,0,0},{0,1,0,0,1,0,-1,0},{0,1,0,1,0,0,0,
-1}};
```

```
TableForm[nutrexpected,TableHeadings->{{"rx1","rx2","rx3","rx4"},{"ATP4-
","H+","H2O","HPO42-","ADP3-","HATP3-","HADP2-","H2PO4-"}}]
```

	ATP4-	H+	H2O	HPO42-	ADP3-	HATP3-	HADP2-	H2PO4-
rx1	-1	1	-1	1	1	0	0	0
rx2	1	1	0	0	0	-1	0	0
rx3	0	1	0	0	1	0	-1	0
rx4	0	1	0	1	0	0	0	-1

We can check whether NullSpace[a] and nuexpected are equivalent by row reducing each of them.

```
TableForm[RowReduce[NullSpace[a]]]
```

1	0	0	0	-1	-1	1	0
0	1	0	0	1	0	-1	0
0	0	1	0	0	1	-1	-1
0	0	0	1	-1	0	1	-1

```
TableForm[RowReduce[nutrexpected]]
```

1	0	0	0	-1	-1	1	0
0	1	0	0	1	0	-1	0
0	0	1	0	0	1	-1	-1
0	0	0	1	-1	0	1	-1

Since these last two matrices are identical, NullSpace[a] and nuexpected are equivalent.

5.3 (a) Construct the conservation matrix **A'** for the hydrolysis of ATP to ADP in terms of reactants. (b) Calculate a basis for the stoichiometric matrix from the conservation matrix and show that it is consistent with ATP + H_2O = ADP + P_i.

(a) Construct the conservation matrix for the hydrolysis of ATP to ADP in terms of reactants.

```
aa={{10,0,10,0},{13,1,10,4},{3,0,2,1}};
```

```
TableForm[aa,TableHeadings->{{"C","O","P"},{"ATP","H2O","ADP","Pi"}}]
```

	ATP	H2O	ADP	Pi
C	10	0	10	0
O	13	1	10	4
P	3	0	2	1

```
TableForm[RowReduce[aa],TableHeadings->{{"ATP","H2O","ADP"},{"ATP","H2O","ADP","Pi"}}]
```

	ATP	H2O	ADP	Pi
ATP	1	0	0	1
H2O	0	1	0	1
ADP	0	0	1	-1

(b) Calculate a basis for the stoichiometric number matrix from the conservation matrix.

```
TableForm[nu=NullSpace[aa],TableHeadings->{{"rx"},{"ATP","H2O","ADP","Pi"}}]
```

	ATP	H2O	ADP	Pi
rx	-1	-1	1	1

This is the expected stoichiometric number matrix.

5.4 The glutamate-ammonia ligase reaction is

glutamate + ATP + ammonia = glutamine + ADP + P$_i$

It can be considered to be the sum of two reactions. (a) Write the stoichiometric number matrix for this enzyme-catalyzed reaction and use NullSpace to obtain a basis for the conservation matrix.. (b) Write a conservation matrix that includes a constraint to couple the two subreactions, and row reduce it to show that it is equivalent to the stoichiometric number matrix obtained in (a).

(a) Calculate a basis for the conservation matrix from the stoichiometric number matrix and row reduce it.

> `TableForm[NullSpace[{{-1,-1,-1,1,1,1}}]]`

1	0	0	0	0	1
1	0	0	0	1	0
1	0	0	1	0	0
-1	0	1	0	0	0
-1	1	0	0	0	0

> `TableForm[RowReduce[NullSpace[{{-1,-1,-1,1,1,1}}]]]`

1	0	0	0	0	1
0	1	0	0	0	1
0	0	1	0	0	1
0	0	0	1	0	-1
0	0	0	0	1	-1

This shows that there are five components, in spite of the fact that there are just four elements (C, O, N, P).

(b) Write the apparent conservation matrix for the glutamate-ammonia ligase reaction and row reduce it. One way to couple the two subreactions is to require that every time an ATP molecule disappears, a glutamine molecule appears; this leads to the conservation equation

n(ATP) + n(glutamine) = const

> `a={{5,10,0,10,0,5},{4,13,0,10,4,3},{1,5,1,5,0,2},{0,3,0,2,1,0},{0,1,0,0,0,1}};`

> `TableForm[a,TableHeadings-`
> `>{{"C","O","N","P","con1"},{"Glutmate","ATP","Amm","ADP","Pi","Glutmine"}}]`

	Glutmate	ATP	Amm	ADP	Pi	Glutmine
C	5	10	0	10	0	5
O	4	13	0	10	4	3
N	1	5	1	5	0	2
P	0	3	0	2	1	0
con1	0	1	0	0	0	1

> `TableForm[RowReduce[a],TableHeadings-`
> `>{{"Glamate","ATP","Amm","ADP","Pi"},{"Glutmate","ATP","Amm","ADP","Pi","Glutmine"}}]`

	Glutmate	ATP	Amm	ADP	Pi	Glutmine
Glamate	1	0	0	0	0	1
ATP	0	1	0	0	0	1
Amm	0	0	1	0	0	1
ADP	0	0	0	1	0	-1
Pi	0	0	0	0	1	-1

Since this yields the same row reduced conservation matrix as (a), the stoichiometric number matrix and the conservation matrix are equivalent.

5.5 Carry out the matrix multiplications in equation 5.4-4 for the three chemical reactions involved in the hydration of fumarate to malate in the pH range 5 to 9.

First we construct the stoichiometric number matrix (eqation 5.4-9) and the vector of chemical potentials. Matrix 5.4-9 is given by

nu={{-1,1,0},{1,0,1},{0,-1,0},{0,0,-1},{0,1,1},{-1,0,0}};

TableForm[nu,TableHeadings->{{"F2-","M2-","HF-","HM-","H+","H2O"},{"rx5","rx6","rx7"}}]

	rx5	rx6	rx7
F2-	-1	1	0
M2-	1	0	1
HF-	0	-1	0
HM-	0	0	-1
H+	0	1	1
H2O	-1	0	0

mu={{muF,muM,muHF,muHM,muH,muH2O}};

Next we calclate the dot product of the chemical potential vector and the stoichiometric number matrix.

mu.nu

{{-muF - muH2O + muM, muF + muH - muHF, muH - muHM + muM}}

The column matrix of extents of reaction is given by

x={{x1},{x2},{x3}};

TableForm[x]

x1
x2
x3

The last term in equation 5.4-4 is given by

mu.nu.x

{{(-muF - muH2O + muM) x1 + (muF + muH - muHF) x2 + (muH - muHM + muM) x3}}

This corresponds with equation 5.4-10.

Chapter 6 Systems of Biochemical Reactions

6.1 Make the stoichiometric number matrix for reactions 6.5-17 to 6.5-19 and use it to print out these reactions.

6.2 Calculate the pathway from glucose to carbon dioxide and water in terms of (1) the net reaction for glycolysis, (2) the net reaction catalyzed by the pyruvate dehydrogenase complex, (3) the net reaction for the citric acid cycle, and (4) the net reaction for oxidative phosphorylation.

6.3 (a) Type in the stoichiometric number matrix for glycolysis at a specified pH that is given in Fig. 6.1. (b) Use nameMatrix to type out the 10 reactions. (c) Use equation 6.1-3 to calculate the net reaction for glycolysis. (d) Calculate a basis for the conservation matrix for glycolysis and row reduce it to determine the number of components.

6.4 The following reaction is carried out enzymatically at 298.15 K, pH 7, and 0.25 M ionic strength.
ATP + glucose = glucose6phos + ADP
The initial concentrations of ATP and glucose are 0.001 M. What is the equilibrium composition? Solve this problem using *Mathematica* in two ways.

6.5 Calculate the apparent equilibrium constant for the net reaction for the first five reactions of glycolyis that is, reactions 6.6-1 to 6.6-5) using equation 6.1-6. Check this by simply using the net reaction.

6.6 A liter of aqueous solution contains 0.01 mol $H_2 PO_4^-$, 0.01 mol HPO_4^{2-}, and 0.01 mol Mg^{2+}. Assuming that the ionic strength is zero, what is the equilibrium composition of the solution in terms of species at 298.15 K?

6.7 When 0.01 M glucose 6-phosphate is hydrolyzed to glucose and phosphate at 298.15 K in the pH range 9-10, only four species have to be considered.
$GlcP^{2-} + H_2 O = Glc + HPO_4^{2-}$
What is the equilibrium composition assuming the ionic strength is zero?

6.8 A liter of aqueous solution contains 0.02 mol phosphate and acid and NaCl are added to bring it to pH 7, and 0.25 M ionic strength. What is the equilibrium composition in terms of phosphate species?

6.1 Make the stoichiometric number matrix for reactions 6.5-17 to 6.5-19 and use it to print out these reactions.

```
nu={{-1,0,0},{-1,0,-1},{1,-1,0},{1,0,1},{0,1,-1},{0,0,1}};
```

```
TableForm[nu]
```

```
-1    0    0
-1    0    -1
1     -1   0
1     0    1
0     1    -1
0     0    1
```

```
names3={"Glc","ATP","G6P","ADP","F6P","F16BP"};
```

```
TableForm[nu,TableHeadings->{names3,{"rx 13","rx 14","rx 15"}}]
```

	rx 13	rx 14	rx 15
Glc	-1	0	0
ATP	-1	0	-1
G6P	1	-1	0
ADP	1	0	1
F6P	0	1	-1
F16BP	0	0	1

```
mkeqm[c_List,s_List]:=(*c_List is the list of stoichiometric numbers for a reaction.
s_List is a list of the names of species or reactants.  These names have to be put in
quotation marks.*)Map[Max[#,0]&,-c].s->Map[Max[#,0]&,c].s
```

```
nameMatrix[m_List,s_List]:=(*m_List is the transposed stoichiometric number matrix for
the system of reactions. s_List is a list of the names of species or reactants.  These
names have to be put in quotation marks.*)Map[mkeqm[#,s]&,m]
```

```
mkeqm[{-1,-1,1,1,0,0},names3]
```

```
ATP + Glc -> ADP + G6P
```

```
nameMatrix[Transpose[nu],names3]
```

```
{ATP + Glc -> ADP + G6P, G6P -> F6P, ATP + F6P -> ADP + F16BP}
```

6.2 Calculate the pathway from glucose to carbon dioxide and water in terms of (1) the net reaction for glycolysis, (2) the net reaction catalyzed by the pyruvate dehydrogenase complex, (3) the net reaction for the citric acid cycle, and (4) the net reaction for oxidative phosphorylation.

The stoichiometric number matrix for these four reactions is

```
nu4={{-1,0,0,0},{-2,0,-1,-3},{-2,0,-1,-3},{-2,-1,-4,1},{2,-1,0,0},{2,0,1,3},{2,1,4,-1},
{2,0,-2,4},{0,-1,1,0},{0,1,-1,0},{0,1,2,0},{0,0,0,-1/2}};
```

```
names={"Glc","Pi","ADP","NADox","Pyr","ATP","NADred","H2O","CoA","acetylCoA","CO2","O2"
};
```

```
TableForm[nu4,TableHeadings->{names,Automatic}]
```

	1	2	3	4
Glc	-1	0	0	0
Pi	-2	0	-1	-3
ADP	-2	0	-1	-3
NADox	-2	-1	-4	1
Pyr	2	-1	0	0
ATP	2	0	1	3
NADred	2	1	4	-1
H2O	2	0	-2	4
CoA	0	-1	1	0
acetylCoA	0	1	-1	0
CO2	0	1	2	0
O2	0	0	0	$-(\frac{1}{2})$

```
mkeqm[c_List,s_List]:=(*c_List is the list of stoichiometric numbers for a reaction.
s_List is a list of the names of species or reactants.  These names have to be put in
quotation marks.*)Map[Max[#,0]&,-c].s->Map[Max[#,0]&,c].s
```

```
nameMatrix[m_List,s_List]:=(*m_List is the transposed stoichiometric number matrix for
the system of reactions. s_List is a list of the names of species or reactants.  These
names have to be put in quotation marks.*)Map[mkeqm[#,s]&,m]
```

```
nameMatrix[Transpose[nu4],names]
```

{2 ADP + Glc + 2 NADox + 2 Pi -> 2 ATP + 2 H2O + 2 NADred + 2 Pyr,
  CoA + NADox + Pyr -> acetylCoA + CO2 + NADred,
  acetylCoA + ADP + 2 H2O + 4 NADox + Pi -> ATP + 2 CO2 + CoA + 4 NADred,
  3 ADP + NADred + $\frac{O2}{2}$ + 3 Pi -> 3 ATP + 4 H2O + NADox}

As indicated in equation 6.2-2 the stoichiometric numbers for the net reaction are given by

```
nunet={-1,-40,-40,0,0,40,0,46,0,0,6,-6}
```

{-1, -40, -40, 0, 0, 40, 0, 46, 0, 0, 6, -6}

The pathway can be calculated by use of

```
LinearSolve[nu4,nunet]
```

{1, 2, 2, 12}

Thus to oxidize a mole of glucose to carbon dioxide and water, reaction 1 has to occure once, reaction 2 has to occur twice, reaction 3 has to occur twice, and reaction 4 has to occur 12 times.

6.3 (a) Type in the stoichiometric number matrix for glycolysis at a specified pH that is given in Fig. 6.1. (b) Use nameMatrix to type out the 10 reactions. (c) Use equation 6.1-3 to calculate the net reaction for glycolysis. (d) Calculate a basis for the conservation matrix for glycolysis and row reduce it to determine the number of components.

(a) The stoichiometric number matrix for glycolysis

```
nu2={{-1,0,0,0,0,0,0,0,0,0},{-1,0,-1,0,0,0,1,0,0,1},{1,0,1,0,0,0,-1,0,0,-1},{0,0,0,0,0,
-1,0,0,0,0},{0,0,0,0,0,1,0,0,0,0},{0,0,0,0,0,-1,0,0,0,0},{1,-1,0,0,0,0,0,0,0,0},{0,1,-1
,0,0,0,0,0,0,0},{0,0,1,-1,0,0,0,0,0,0},{0,0,0,1,-1,0,0,0,0,0},{0,0,0,0,0,1,-1,0,0,0},{0
,0,0,0,0,0,1,-1,0,0},{0,0,0,0,0,0,0,1,-1,0},{0,0,0,0,0,0,0,0,1,-1},{0,0,0,1,1,-1,0,0,0,
0},{0,0,0,0,0,0,0,0,0,1}};
```

```
names2={"Glc","ATP","ADP","NADox","NADred","Pi","G6P","F6P","FBP","DHAP","13BPG","3PG",
"2PG","PEP","GAP","Pyr"};
```

```
TableForm[nu2,TableHeadings->{names2,{1,2,3,4,5,6,7,8,9,10}}]
```

	1	2	3	4	5	6	7	8	9	10
Glc	-1	0	0	0	0	0	0	0	0	0
ATP	-1	0	-1	0	0	0	1	0	0	1
ADP	1	0	1	0	0	0	-1	0	0	-1
NADox	0	0	0	0	0	-1	0	0	0	0
NADred	0	0	0	0	0	1	0	0	0	0
Pi	0	0	0	0	0	-1	0	0	0	0
G6P	1	-1	0	0	0	0	0	0	0	0
F6P	0	1	-1	0	0	0	0	0	0	0
FBP	0	0	1	-1	0	0	0	0	0	0
DHAP	0	0	0	1	-1	0	0	0	0	0
13BPG	0	0	0	0	0	1	-1	0	0	0
3PG	0	0	0	0	0	0	1	-1	0	0
2PG	0	0	0	0	0	0	0	1	-1	0
PEP	0	0	0	0	0	0	0	0	1	-1
GAP	0	0	0	1	1	-1	0	0	0	0
Pyr	0	0	0	0	0	0	0	0	0	1

This is Fig. 6.1.

(b) Use mathematica to type out the reactins of glycolysis

```
mkeqm[c_List,s_List]:=Map[Max[#,0]&,-c].s->Map[Max[#,0]&,c].s
```

```
nameMatrix[m_List,s_List]:=Map[mkeqm[#,s]&,m]
```

```
nameMatrix[Transpose[nu2],names2]
```

```
{ATP + Glc -> ADP + G6P, G6P -> F6P, ATP + F6P -> ADP + FBP, FBP -> DHAP + GAP,
    DHAP -> GAP, GAP + NADox + Pi -> 13BPG + NADred, 13BPG + ADP -> 3PG + ATP,
    3PG -> 2PG, 2PG -> PEP, ADP + PEP -> ATP + Pyr}
```

(c) Equation 6.1-3 ($v.s = nunet$) is given by

```
nu2.{1,1,1,1,1,2,2,2,2,2}
```

{-1, 2, -2, -2, 2, -2, 0, 0, 0, 0, 0, 0, 0, 0, 0, 2}

mkeqm[nu2.{1,1,1,1,1,2,2,2,2,2},names2]

2 ADP + Glc + 2 NADox + 2 Pi -> 2 ATP + 2 NADred + 2 Pyr

(d) Use *Mathematica* to make Fig. 6.2.

TableForm[Transpose[RowReduce[NullSpace[Transpose[nu2]]]],TableHeadings->{names2,{"Glc","ATP","ADP","NADox","NADred","Pi"}}]

	Glc	ATP	ADP	NADox	NADred	Pi
Glc	1	0	0	0	0	0
ATP	0	1	0	0	0	0
ADP	0	0	1	0	0	0
NADox	0	0	0	1	0	0
NADred	0	0	0	0	1	0
Pi	0	0	0	0	0	1
G6P	1	1	-1	0	0	0
F6P	1	1	-1	0	0	0
FBP	1	2	-2	0	0	0
DHAP	$\frac{1}{2}$	1	-1	0	0	0
13BPG	$\frac{1}{2}$	1	-1	1	-1	1
3PG	$\frac{1}{2}$	0	0	1	-1	1
2PG	$\frac{1}{2}$	0	0	1	-1	1
PEP	$\frac{1}{2}$	0	0	1	-1	1
GAP	$\frac{1}{2}$	1	-1	0	0	0
Pyr	$\frac{1}{2}$	-1	1	1	-1	1

This is Fig. 6.2. When the reactants are listed in this order, the components are Glc, ATP, ADP, NADox, NADred, and Pi. This means that when the concentrations of the last five are specified, the equilibrium compositions of the remaining reactants can be calculated without using equcalcrx.

6.4 The following reaction is carried out enzymatically at 298.15 K, pH 7, and 0.25 M ionic strength.
ATP + glucose = glucose6phos + ADP
The initial concentrations of ATP and glucose are 0.001 M. What is the equilibrium composition? Solve this problem using *Mathematica* in two ways.

(BasicBiochemData2 has to be loaded)

Use equcalcrx: First calculate the apparent equilibrium constant.

```
kprime=calckprime[atp+glucose+de==glucose6phos+adp,7,.25]

18973.7


equcalcrx[nt_,lnkr_,no_]:=Module[{as,lnk},
(*nt=transposed stoichiometric number matrix
lnkr=ln of equilibrium constants of rxs (vector)
no=initial composition vector*)
(*Setup*)
lnk=LinearSolve[nt,lnkr];
as=NullSpace[nt];
equcalcc[as,lnk,no]
]


equcalcrx[{{-1,-1,1,1}},{Log[1.90*10^4]},{.001,.001,0,0}]

             -6             -6
{7.20251 10  , 7.20251 10  , 0.000992797, 0.000992797}
```

These are the equilibrium concentrations of ATP, glucose, glucose6phos, and ATP.
This problem can also be solved using NSolve. Since there are four unknown concentrations, there have to be four independent relations between these four unknowns. There is one equilibrium constant expression and three conservations equations, which we take here to be the conservation of adenosine, glucose entity, and phosphorus.

```
NSolve[{cglucose6phos*cadp/(catp*cglucose)==1.90*10^4,catp+cadp==0.001,cglucose+cglucos
e6phos==.001,catp+cglucose6phos==.001},{catp,cglucose,cglucose6phos,cadp}]

                         -6                    -6
{{cglucose -> -7.30778 10  , catp -> -7.30778 10  , cadp -> 0.00100731,

                                                      -6                    -6
   cglucose6phos -> 0.00100731}, {cglucose -> 7.20251 10  , catp -> 7.20251 10  ,

   cadp -> 0.000992797, cglucose6phos -> 0.000992797}}
```

The first solution is impossible because concentratins have to be positive. The second solution agrees with that obtained using equcalcrx.

6.5 Calculate the apparent equilibrium constant for the net reaction for the first five reactions of glycolyis that is, reactions 6.6-1 to 6.6-5) using equation 6.1-6. Check this by simply using the net reaction.

(BasicBiochemData has to be loaded)

The stoichiometric number matrix is

```
nu={{-1,0,0,0,0},{-1,0,-1,0,0},{1,-1,0,0,0},{1,0,1,0,0},{0,1,-1,0,0},{0,0,1,-1,0},{0,0,
0,1,1},{0,0,0,1,-1}};

TableForm[nu,TableHeadings-
>{{"glc","atp","g6p","adp","f6p","f16bip","dihydroxyacetonephos","glyceraldehydephos"},
Automatic}]
```

	1	2	3	4	5
glc	-1	0	0	0	0
atp	-1	0	-1	0	0
g6p	1	-1	0	0	0
adp	1	0	1	0	0
f6p	0	1	-1	0	0
f16bip	0	0	1	-1	0
dihydroxyacetonephos	0	0	0	1	1
glyceraldehydephos	0	0	0	1	-1

The vector of standard transformed Gibbs energies of formation at pH 7 and is -0.25 M is

vectorG={glucose,atp,glucose6phos,adp,fructose6phos,fructose16phos,dihydroxyacetonephos ,glyceraldehydephos}/.pH->7/.is->.25

{-426.708, -2292.5, -1318.92, -1424.7, -1315.74, -2206.78, -1095.7, -1088.04}

The pathway vector is

vectors={1,1,1,1,1};

vectorG.nu.vectors

-29.105

apparentK=Exp[29.105/(8.31451*.29815)]

125587.

The net reaction is
glc + 2ATP = 2dihydroxyacetonephos + 2ADP

calckprime[glucose+2*atp+de==2*dihydroxyacetonephos+2*adp,7,.25]

125587.

6.6 A liter of aqueous solution contains 0.01 mol $H_2 PO_4^-$, 0.01 mol HPO_4^{2-}, and 0.01 mol Mg^{2+}. Assuming that the ionic strength is zero, what is the equilibrium composition of the solution in terms of species at 298.15 K?

(BasicBiochemData has to be loaded)

This problem canbe solved in two different ways. Using equcalcc, the conservation matrix with H, Mg, and P as components and H^+, Mg^{2+}, $H_2 PO_4^-$, HPO_4^{2-}, and $MgHPO_4$ as species is given by

aa={{1,0,2,1,1},{0,1,0,0,1},{0,0,1,1,1}};

{{1, 0, 2, 1, 1}, {0, 1, 0, 0, 1}, {0, 0, 1, 1, 1}}

TableForm[aa,TableHeadings->{{"H","Mg","P"},{"H","Mg","H2PO4-","HPO42-","MgHPO4"}}]

	H	Mg	H2PO4-	HPO42-	MgHPO4
H	1	0	2	1	1
Mg	0	1	0	0	1
P	0	0	1	1	1

The standard Gibbs energies of formation in kJ mol^{-1} are used to construct lnk.

```
lnka=-(1/(8.31451*.29815))*{0,-455.3,-1137.3,-1096.1,-1566.87};
```

```
{0, 183.665, 458.779, 442.159, 632.065}
```

The initial composition matrix is given by

```
noa={0,.01,.01,.01,0};
```

```
{0, 0.01, 0.01, 0.01, 0}
```

The amounts of the species at equilibrium (in the order used above) are given by

```
equcalcc[as_,lnk_,no_]:=Module[{l,x,b,ac,m,n,e,k},
(*  as=conservation matrix
lnk=-(1/RT)(Gibbs energy of formation vector at T)
no=initial composition vector *)
(*Setup*)
{m,n}=Dimensions[as];
b=as.no;
ac=as;
(*Initialize*)
l=LinearSolve[ as.Transpose[as],-as.(lnk+Log[n]) ];
(*Solve*)
Do[ e=b-ac.(x=E^(lnk+l.as) );
If[(10^-10)>Max[ Abs[e] ], Break[] ];
l=l+LinearSolve[ac.Transpose[as*Table[x,{m}]],e],
{k,100}];
If[ k=100,Return["Algorithm Failed"] ];
Return[x]
]
```

```
equcalcc[aa,lnka,noa]
```

```
{1.70726 10
          -7
            , 0.00354638, 0.00999983, 0.00354655,
  0.00645362}
```

```
TableForm[{equcalcc[aa,lnka,noa]},TableHeadings->{{"c/M"},{"H","Mg","H2PO4-","HPO42-
","MgHPO4"}}]
```

	H	Mg	H2PO4-	HPO42-	MgHPO4
c/M	1.70726 10^{-7}	0.00354638	0.00999983	0.00354655	0.00645362

This problem can also be solved using the stoichiometric number matrix for the two reactions and equcalcrx

$H_2 PO_4^- = H^+ + HPO_4^{2-}$ $K_1 = 6.05499*10^\wedge-8$
$MgHPO_4^- = Mg^{2+} + HPO_4^{2-}$ $K_2 = 1.9489*10^\wedge-3$

The transformed stoichiometric number matrix is given by

```
tnua={{1,0,-1,1,0},{0,1,0,1,-1}};
```

```
{{1, 0, -1, 1, 0}, {0, 1, 0, 1, -1}}
```

```
TableForm[tnua]
```

```
1    0    -1    1    0
0    1     0    1   -1
```

```
lnkra=Log[{6.05499*10^-8,1.9489*10^-3}]
```

```
{-16.6198, -6.24049}
```

The initial amount vector is unchanged, and so the equilibrium composition is given by

```
equcalcrx[tnua,lnkra,noa]
```

$\{1.70726\ 10^{-7}$, 0.00354638, 0.00999983, 0.00354656, $0.00645362\}$

```
TableForm[{equcalcrx[tnua,lnkra,noa]},TableHeadings->{{"c/M"},{"H","Mg","H2PO4-
","HPO42-","MgHPO4"}}]
```

	H	Mg	H2PO4-	HPO42-	MgHPO4
c/M	$1.70726\ 10^{-7}$	0.00354638	0.00999983	0.00354656	0.00645362

6.7 When 0.01 M glucose 6-phosphate is hydrolyzed to glucose and phosphate at 298.15 K in the pH range 9-10, only four species have to be considered.

$GlcP^{2-} + H_2O = Glc + HPO_4^{2-}$

What is the equilibrium composition assuming the ionic strength is zero?

This calculation using equcalcc involves the problem that $\Delta_f G°$ (H_2O) is used in the calculation of the equilibrium constant K, but the expression for the equilibrium constant does not involve the concentration of H_2O. Thus in effect oxygen atoms are not conserved, because in dilute aqueous solutions they are drawn for th essentially infinite reservoir of the solvent. Therefore, the further transformed Gibbs energy G' has to be used. The conservation matrix with C and P as components and $GlcP^{2-}$, Glc, and HPO_4^{2-} as species is given by

```
ab={{6,6,0},{1,0,1}}
```

```
{{6, 6, 0}, {1, 0, 1}}
```

```
TableForm[ab]
```

```
6  6  0
1  0  1
```

The H row and charge row are redundant, and so they are omitted. The transformed Gibbs energy has to be used because the "concentration" of water is held constant. These transformed Gibbs energies of formation are given by

```
tgeGlcP=-1763.94-9*(-237.19)
```

```
370.77
```

```
tgeGlc=-915.9-6*(-237.19)
```

```
507.24
```

```
tgeHPO4=-1096.1-4*(-237.19)
```

```
-147.34
```

where the standard Gibbs energy of formaton of H_2O is -237.19 kJ mol^{-1}.

```
lnkb=(-1/(8.31451*.29815))*{tgeGlcP,tgeGlc,tgeHPO4}
```

{-149.566, -204.617, 59.4359}

nob={0.1,0,0}

{0.1, 0, 0}

equcalcc[ab,lnkb,nob]

{0.000124334, 0.0998757, 0.0998757}

This problem can also be solved using the stoichiometric number matrix. Water is omitted from the stoichiometric number matrix, and so the transposed stoichiometric number matrix is given by

tnub={{-1,1,1}};

TableForm[tnub]

-1   1   1

The transformed Gibbs energy of reaction is given by

tgeGlc+tgeHPO4-tgeGlcP

-10.87

lnkrb={(-1/(8.31451*.29815))*(-10.87)}

{4.38488}

equcalcrx[tnub,lnkrb,nob]

{0.000124334, 0.0998757, 0.0998757}

Thus the concentration of $G6P^{2-}$ is 1.24×10^{-4}.

6.8 A liter of aqueous solution contains 0.02 mol phosphate and acid andNaCl are added to bring it to pH 7, and 0.25 M ionic strength. What is the equilibrium composition in terms of phosphate species?

(BasicBiochemData2 has to be loaded)

Since H and Mg are not conserved, the conservation matrix is

as={{1,1,1}}

{{1, 1, 1}}

where the species are $H_2PO_4^-$, HPO_4^{2-}, and $MgHPO_4$. The transformed Gibbs energies of these three species at 25 °C, pH 7, pMg 3, and I = 0.25 M have been calculated by Alberty and Goldberg (1992), and they can be used to calculate

lnk=-(1/(8.31451*.29815))*{-1056.58,-1058.57,-1050.44}

{426.217, 427.02, 423.74}

The equilibrium concentrations of the three phosphate species are given by

```
equcalcc[as,lnk,{0.01,0.01,0}]
```

{0.00603194, 0.0134613, 0.000506736}

Alternatively, we can use the transposed stoichiometric number matrix

```
trnu={{-1,1,0},{0,1,-1}}
```

{{-1, 1, 0}, {0, 1, -1}}

where the reactions are

$H_2PO_4^- = HPO_4^{2-}$ $K_1 = \exp[(1058.57 - 1056.58)/RT]$

$MgHPO_4 = HPO_4^{2-}$ $K_2 = \exp[(1058.57-1050.44)/RT]$

Thus lnkr is given by

```
lnkr={(1058.57 - 1056.58)/(8.31451*.29815),(1058.57-1050.44)/(8.31451*.29815)}
```

{0.802752, 3.27959}

The equilibrium concentrations of the three species are the same as obtained using the conservation matrix.

```
equcalcrx[trnu,lnkr,{0.01,0.01,0}]
```

{0.00603194, 0.0134613, 0.000506736}

Chapter 7 Thermodynamics of Binding of Ligands by Proteins

7.1 Use the equilibrium constants in equations 7.1-3 to 7.1-6 to calculate the further transformed Gibbs energies of formation of the forms of the tetramer of hemoglobin and of the pseudoisomer group at $[O_2] = 5 \times 10^{-6}$, 10^{-5}, and 2×10^{-5} M. (a) Make a table with the last three columns like Table 7.1. (b) Calculate the equilibrium mole fractions of forms of the tetramer at 21.4 °C, 1 bar, pH 7.4, $[Cl^-] = 0.2$ M, and 0.2 M ionic strength and make a table like Table 7.2.

7.2 Use the equilibrium constants in equations 7.3-2 and 7.3-3 to calculate the further transformed Gibbs energies of formation of the forms of the dimer of hemoglobin and of the pseudoisomer group at $[O_2] = 5 \times 10^{-6}$, 10^{-5}, and 2×10^{-5} M. Make a table with the last three columns and first four rows of Table 7.3.

7.3 Calculate the fractional saturation Y_T of the tetramer of human hemoglobin with molecular oxygen using the equilibrium constants determined by Mills, Johnson, and Akers (1976) at 21.5 °C, 1 bar, pH 7.4, $[Cl^-] = 0.2$ M and 0.2 M ionic strength. Make the calculation with the Adair equation and also by using the binding polymomial Y_T.

7.4 Calculate the fractional saturation Y_D of the dimer of human hemoglobin with molecular oxygen using the equilibrium constants determined by Mills, Johnson, and Akers (1976) at 21.5 °C, 1 bar, pH 7.4, $[Cl^-] = 0.2$ M and 0.2 M ionic strength. Make the calculation with the Adair equation and also by using the binding polymomial P_D.

7.5 (a) Calculate the oxygen binding curve Y for human hemoglobin when the dimer and tetramer are in equilibrium using the equilibrium constants determined by Mills, Johnson, and Akers (1976) at 21.5 °C, 1 bar, pH 7.4, $[Cl^-] = 0.2$ M and 0.2 M ionic strength. This plot depends on the heme concentration. The currently accessible range of heme concentrations is about 0.04 μM to 5 mM. Plot logK" versus $[O_2]$. (b) Use the equation for Y to extrapolate to values of Y_T at high heme concentrations. (c) Use the equation for Y to extrapolate to values of Y_D at low heme concentrations.

7.1 Use the equilibrium constants in equations 7.1-3 to 7.1-6 to calculate the further transformed Gibbs energies of formation of the forms of the tetramer of hemoglobin and of the pseudoisomer group at $[O_2] = 5 \times 10^{-6}$, 10^{-5}, and 2×10^{-5} M. (a) Make a table with the last three columns like Table 7.1. (b) Calculate the equilibrium mole fractions of forms of the tetramer at 21.4 °C, 1 bar, pH 7.4, $[Cl^-] = 0.2$ M, and 0.2 M ionic strength and make a table like Table 7.2.

(a) Calculate further transformed Gibbs energies of forms of the tretramer and the pseudoisomer group Ttot.

```
calctgfT[keq_,conco2_]:=Module[{b,m,tgf,ge,nO2,gtrans,giso,output},(*This program
returns the standard further transformed Gibbs energies of formation of the forms of
the tetramer of hemoglobin and the standard further transformed Gibbs energy of the
pseudoisomer group at a specified molar concentration of molecular oxygen and
specified equilibrium constants.  keq_ is a vector of equilibrium constants.  conco2
is the molar concentration of molecular oxygen.*)
b=-1*(8.31451*10^-3)*294.65*Log[keq]+16.1;
m={{1,0,0,0},{-1,1,0,0},{0,-1,1,0},{0,0,-1,1}};
ge=LinearSolve[m,b];
tgf={Flatten[{0,ge}]};
nO2={{0,1,2,3,4}};
gtrans=Transpose[tgf]-Transpose[nO2]*(16.1+(8.31451*10^-3)*294.65*Log[conco2]);
giso=-1*(8.31451*10^-3)*294.65*Log[Apply[Plus,Exp[-gtrans/((8.31451*10^-3)*294.65)]]];
output={gtrans,giso};
Return[output]]


keq={43970.,12210.,404900.,664400.};


col1=Flatten[calctgfT[keq,5*10^-6]];


col2=Flatten[calctgfT[keq,10^-5]];


col3=Flatten[calctgfT[keq,2*10^-5]];


TableForm[Transpose[{col1,col2,col3}],TableHeadings-
>{{"T","T(O2)","T(O2)2","T(O2)3","T(O2)4","ΔfG''o(TotT)"},{"5x10^-6 M","10^-5
M","2x10^-5 M"}}]
```

T	5x10^-6 M	10^-5 M	2x10^-5 M
	0	0	0
T(O2)	3.71109	2.01297	0.314846
T(O2)2	10.5611	7.16484	3.76859
T(O2)3	8.83313	3.73877	-1.3556
T(O2)4	5.89189	-0.900589	-7.69307
ΔfG''o(TotT)	-0.736508	-2.8149	-8.06907

(b) Calculate the equilibrium mole fractions of the various foms of the tetramer.

```
col11=Flatten[Exp[(Table[Take[col1,-1],{5}]-Take[col1,5])/(8.31451*.29465)]];

col12=Flatten[Exp[(Table[Take[col2,-1],{5}]-Take[col2,5])/(8.31451*.29465)]];

col13=Flatten[Exp[(Table[Take[col3,-1],{5}]-Take[col3,5])/(8.31451*.29465)]];

TableForm[Transpose[{col11,col12,col13}],TableHeadings-
>{{"T","T(O2)","T(O2)2","T(O2)3","T(O2)4"},{"5x10^-6 M","10^-5 M","2x10^-5 M"}}]
```

	5x10^-6 M	10^-5 M	2x10^-5 M
T	0.740351	0.316953	0.0371174
T(O2)	0.162766	0.139364	0.032641
T(O2)2	0.00993687	0.0170164	0.00797093
T(O2)3	0.0201172	0.0688993	0.0645486
T(O2)4	0.0668293	0.457767	0.857722

7.2 Use the equilibrium constants in equations 7.3-2 and 7.3-3 to calculate the further transformed Gibbs energies of formation of the forms of the dimer of hemoglobin and of the pseudoisomer group at $[O_2] = 5 \times 10^{-6}$, 10^{-5}, and 2×10^{-5} M. Make a table with the last three columns and first four rows of Table 7.3.

```
calctgfD[keq_,conco2_]:=Module[{b,b1,b2,m,tgf,ge,nO2,gtrans,giso,output},(*This
program returns the standard frurther transformed Gibbs energies of formation of the
forms of the dimer of hemoglobin and the standard further transformed Gibbs energy of
the pseudoisomer group at a specified molar concentration of molecular oxygen and
specified equilibrium constants.  keq_ is a vector of equilibrium constants.  conco2
is the molar concentration of molecular oxygen.*)
b1=-1*(8.31451*10^-3)*294.65*Log[keq[[1]]]+16.1+30.0832;
b2=-1*(8.31451*10^-3)*294.65*Log[keq[[2]]]+16.1;
b={b1,b2};
m={{1,0},{-1,1}};
ge=LinearSolve[m,b];
tgf={Flatten[{30.0832,ge}]};
nO2={{0,1,2}};
gtrans=Transpose[tgf]-Transpose[nO2]*(16.1+(8.31451*10^-3)*294.65*Log[conco2]);
giso=-1*(8.31451*10^-3)*294.65*Log[Apply[Plus,Exp[-gtrans/((8.31451*10^-3)*294.65)]]];
output={gtrans,giso};
Return[output]]


keq={3.253*10^6,8.155*10^5};

col21=Flatten[calctgfD[keq,5*10^-6]];

col22=Flatten[calctgfD[keq,10^-5]];

col23=Flatten[calctgfD[keq,2*10^-5]];
```

Table 7.3 Standard further transformed Gibbs energies of formation of dimer at 21.5 C, pH 7.4, $[Cl^-] = 0.2$ M and 0.2 M ionic strength

```
TableForm[Transpose[{col21,col22,col23}],TableHeadings-
>{{"D","D(O2)","D(O2)2","ΔfG''o(TotD)"},{"5x10^-6 M","10^-5 M","2x10^-5 M"}}]
```

	5x10^-6 M	10^-5 M	2x10^-5 M
D	30.0832	30.0832	30.0832
D(O2)	23.2505	21.5524	19.8542
D(O2)2	19.8072	16.411	13.0147
ΔfG''o(TotD)	19.2404	16.1194	12.8668

7.3 Calculate the fractional saturation Y_T of the tetramer of human hemoglobin with molecular oxygen using the equilibrium constants determined by Mills, Johnson, and Akers (1976) at 21.5 °C, 1 bar, pH 7.4, $[Cl^-] = 0.2$ M and 0.2 M ionic strength. Make the calculation with the Adair equation and also by using the binding polymomial Y_T.

```
kT1=4.397*10^4;
```

```
kT2=1.221*10^4;

kT3=4.049*10^5;

kT4=6.644*10^5;

yT=(kT1*o2+2*kT1*kT2*o2^2+3*kT1*kT2*kT3*o2^3+4*kT1*kT2*kT3*kT4*o2^4)/(4*(1+kT1*o2+kT1*k
T2*o2^2+kT1*kT2*kT3*o2^3+kT1*kT2*kT3*kT4*o2^4));

Plot[yT,{o2,0,2*10^-5},PlotRange->{0,1},AxesLabel->{"[\!\(O\_2\)]","\!\(Y\_T\)"}];
```

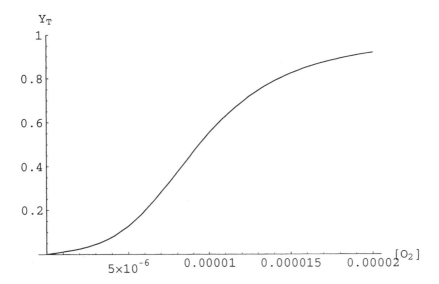

```
pT=1+kT1*o2+kT1*kT2*o2^2+kT1*kT2*kT3*o2^3+kT1*kT2*kT3*kT4*o2^4;

Plot[Evaluate[o2*D[Log[pT]/4,o2]],{o2,10^-8,2*10^-5},PlotRange->{0,1},AxesLabel-
>{"[\!\(O\_2\)]","\!\(Y\_T\)"}];
```

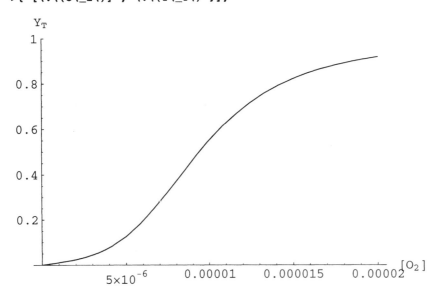

7.4 Calculate the fractional saturation Y_D of the dimer of human hemoglobin with molecular oxygen using the equilibrium constants determined by Mills, Johnson, and Akers (1976) at 21.5 °C, 1 bar, pH 7.4, $[Cl^-] = 0.2$ M and 0.2 M ionic strength. Make the calculation with the Adair equation and also by using the binding polynomial P_D.

```
kD1=3.253*10^6;

kD2=8.155*10^5;

yD=(kD1*o2+2*kD1*kD2*o2^2)/(2*(1+kD1*o2+kD1*kD2*o2^2));

Plot[yD,{o2,0,5*10^-6},PlotRange->{0,1},AxesLabel->{"\!\(O\_2\)]","\!\(Y\_D\)"}];
```

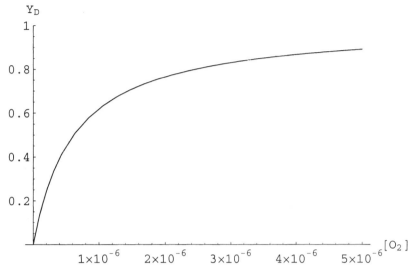

```
pD=1+kD1*o2+kD1*kD2*o2^2
```

$$1 + 3.253 \ 10^6 \ o2 + 2.65282 \ 10^{12} \ o2^2$$

```
Plot[Evaluate[o2*D[Log[pD]/2,o2]],{o2,0,5*10^-6},PlotRange->{0,1},AxesLabel-
>{"[\!\(O\_2\)]","\!\(Y\_D\)"}];
```

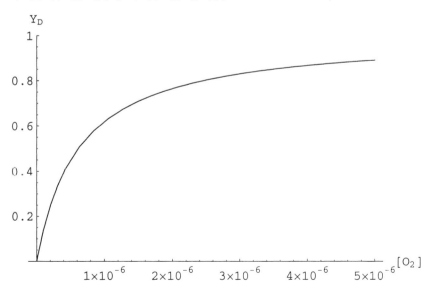

7.5 (a) Calculate the oxygen binding curve Y for human hemoglobin when the dimer and tetramer are in equilibrium using the equilibrium constants determined by Mills, Johnson, and Akers (1976) at 21.5 °C, 1 bar, pH 7.4, $[Cl^-] = 0.2$ M and 0.2 M ionic strength. This plot depends on the heme concentration. The currently accessible range of heme concentrations is about 0.04 μM to 5 mM. Plot logK" versus $[O_2]$. (b) Use the equation for Y to extrapolate to values of Y_T at high heme concentrations. (c) Use the equation for Y to extrapolate to values of Y_D at low heme concentrations.

(a) The values of Y_T and Y_D are calclted as follows:

```
kT1=4.397*10^4;

kT2=1.221*10^4;

kT3=4.049*10^5;

kT4=6.644*10^5;

yT=(kT1*o2+2*kT1*kT2*o2^2+3*kT1*kT2*kT3*o2^3+4*kT1*kT2*kT3*kT4*o2^4)/(4*(1+kT1*o2+kT1*kT2*o2^2+kT1*kT2*kT3*o2^3+kT1*kT2*kT3*kT4*o2^4));

kD1=3.253*10^6;

kD2=8.155*10^5;

yD=(kD1*o2+2*kD1*kD2*o2^2)/(2*(1+kD1*o2+kD1*kD2*o2^2));
```

The apparent association constant K"(2D=T) for human hemoglobin is given by

```
k=(4.633*10^10)*(1+kT1*o2+kT1*kT2*o2^2+kT1*kT2*kT3*o2^3+kT1*kT2*kT3*kT4*o2^4)/((1+kD1*o2+kD1*kD2*o2^2)^2);

Plot[Log[10,k],{o2,0,2*10^-5},AxesLabel->{"[\!\(O\_2\)]","logK''"}];
```

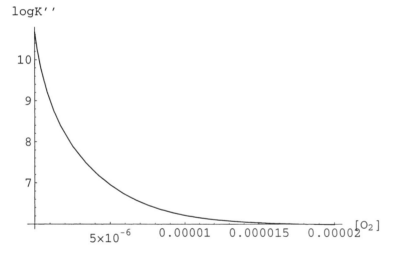

The dependence of Y on [heme] is given by

```
y=yT+2*(yD-yT)/(1+(1+4*k*heme)^.5);
```

The fractional saturation Y at the highest possible hemoglobin concentration is given by:

```
yhigh=y/.heme->5*10^-3;

plot1=Plot[yhigh,{o2,0,2*10^-5},PlotRange->{0,1},AxesLabel->{"[\!\(O\_2\)]","Y"}];
```

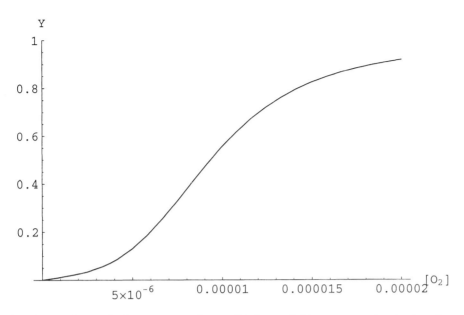

The fractional saturation at the lowest currently possible hemoglobin concentration is given by:

```
ylow=y/.heme->4*10^-8;

plot2=Plot[ylow,{o2,0,2*10^-5},PlotRange->{0,1},AxesLabel->{"[\!\(O\_2\)]","Y"}];
```

The fractional saturation at a heme concentration 1/10th of the lowest currently possible hemoglobin concentration is given by:

```
ylower=y/.heme->4*10^-9;

plot3=Plot[ylower,{o2,0,2*10^-5},PlotRange->{0,1},AxesLabel->{"[\!\(O\_2\)]","Y"}];
```

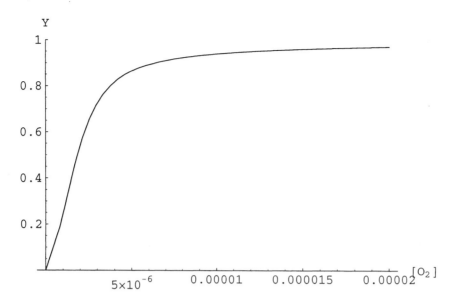

Show[{plot1,plot2,plot3}];

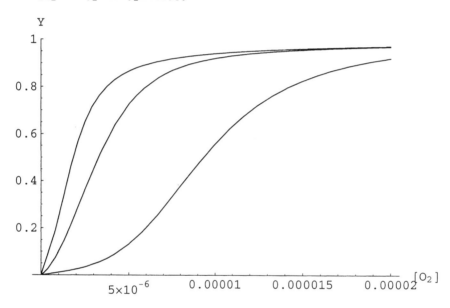

(b) Plot Y versus $[heme]^{-.5}$ at high [heme] to obtain Y_T. Y is plotted versus x, where x = $[heme]^{-.5}$ or [heme] = $1/x^2$.

```
Clear[x]

yfnx=yT+2*(yD-yT)/(1+(1+4*k/(x^2))^.5);

Plot[{yfnx/.o2->10^-4,yfnx/.o2->2*10^-5,yfnx/.o2->10^-5,yfnx/.o2->5*10^-6,yfnx/.o2-
>10^-6},{x,0,2000},PlotRange->{0,1},AxesLabel->{"\!\([heme]\^-.5\)","Y"}];
```

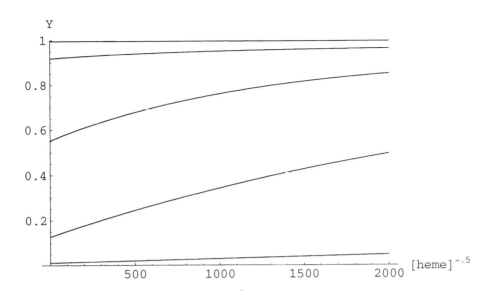

This is Fig. 1 in R. A. Alberty, Biophys. Chem. 63, 189-132 (1997)..

(c) Plot Y versus [heme] a very low concentrations of heme to obtain Y_D by extrapolation.

```
Plot[{y/.o2->10^-4,y/.o2->2*10^-5,y/.o2->10^-5,y/.o2->5*10^-6,y/.o2->10^-6,y/.o2-
>10^-7},{heme,0,10^-7},PlotRange->{0,1},AxesLabel->{"[heme]","Y"}];
```

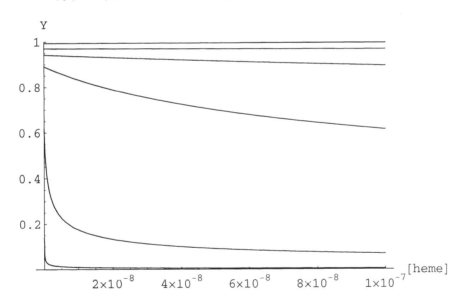

This is Fig. 3 in R. A. Alberty, Biophys. Chem. 63, 189-132 (1997)..

Chapter 8 Phase Equilibrium in Aqueous Solutions

8.1 (a) Calculate the the standard transformed Gibbs eneries for the reaction CO2tot = CO2(g) + H2O at 283.15 K, 298.15 K, and 313.15 K at pHs 5, 6, 7, 8, and 9 at ionic strength 0.25 M. (b) Plot the Henry's law constants versus pH from pH 2 to pH 12.

8.2 (a) Calculate the standard transformed Gibbs energies for the reaction CO2tot = CO2(g) + H2O at 298.15 K and pHs 5, 6, 7, 8, and 9 for ionic strengths of 0, 0.10, and 0.25 M. (b) Plot the Henry's law constants versus pH from pH 3 to pH 9. (c) Plot $\log K_H$ as a function of pH at three ionic strengths.

8.1 (a) Calculate the the standard transformed Gibbs eneries for the Reaction CO2tot = CO2(g) + H2O at 283.15 K, 298.15 K, and 313.15 K at pHs 5, 6, 7, 8, and 9 at ionic strength 0.25 M. (b) Plot the Henry's law constants versus pH from pH 2 to pH 12.

(a) Calculation of the standard transformed Gibbs energy of reaction:.

The following program can be used to calculate the function of temperature, pH, and ionic strength that represents the standard transformed Gobbs energy of formation of a reactant.

```
calcdGHT[speciesmat_] :=
Module[{dGzero, dGzeroT,dHzero,zi, nH, gibbscoeff,pHterm,
isterm,gpfnsp,dGfn,dHfn},(*This program produces the function of T (in Kelvin), pH and
ionic strength (is) that gives the standard transformed Gibbs energy of formation of a
reactant (sum of species) and the standard transformed enthalpy.  The input speciesmat
is a matrix that gives the standard Gibbs energy of formation at 298.15 K, the
standard enthalpy of formation at 298.15 K, the electric charge, and the number of
hydrogen atoms in each species. There is a row in the matrix for each species of the
reactant. gpfnsp is a list of the functions for the transformed Gibbs energies of the
species.  The output is in the form {dGfn,dHfn}, and energies are expressed in kJ
mol^-1.  The values of the standard transformed Gibbs energy of formation and the
standard transformed enthalpy of formation can be calculated at any temperature in the
range 273.15 K to 313.15 K, any pH in the range 5 to 9, and any ionic strength in the
range 0 to 0.35 m by use of the assignment operator(/.).*)
{dGzero,dHzero,zi,nH}=Transpose[speciesmat];
gibbscoeff=9.20483*10^-3*t-1.284668*10^-5*t^2+4.95199*10^-8*t^3;
dGzeroT=dGzero*t/298.15+dHzero*(1-t/298.15);
pHterm = nH*8.31451*(t/1000)*Log[10^-pH];
istermG = gibbscoeff*((zi^2) - nH)*(is^.5)/(1 + 1.6*is^.5);
gpfnsp=dGzeroT - pHterm - istermG;
dGfn=-8.31451*(t/1000)*Log[Apply[Plus,Exp[-1*gpfnsp/(8.31451*(t/1000))]]];
dHfn=-t^2*D[dGfn/t,t];
{dGfn,dHfn}]
```

The program calctrGerx is readily modified to take a list of temperatures.

```
calctrGerxT[eq_,pHlist_,islist_,tlist_]:=Module[{energy},(*Calculates the standard
transformed Gibbs energy of reaction in kJ mol^-1 at specified pHs, ionic strengths,
and temperatures for a biochemical equation typed in the form atpt+h2ot+de==adpt+pit,
where the functions include temperature (in K).  The names of the reactants call the
appropriate functions of pH, ionic strength, and temperature. pHlist, islist, and
tlist can be lists. This program can also be used to calculate the standard
transformed enthalpy of reaction.*)
energy=Solve[eq,de];
energy[[1,1,2]]/.pH->pHlist/.is->islist/.t->tlist]
```

```
co2gt=calcdGHT[co2gsp][[1]]
```

$$-0.00831451\ t\ \text{Log}[E^{(-120.272\ (-393.5\ (1\ -\ 0.00335402\ t)\ -\ 1.32269\ t))/t}]$$

```
co2tott=calcdGHT[co2totsp][[1]]
```

$$-0.00831451\ t\ \text{Log}[\text{Power}[E,\ (-120.272\ (-677.14\ (1\ -\ 0.00335402\ t)\ -\ 1.77028\ t\ -$$
$$(4\ is^{0.5}\ (0.00920483\ t\ -\ 0.0000128467\ t^2\ +\ 4.95199\ 10^{-8}\ t^3))/(1\ +\ 1.6\ is^{0.5})))$$
$$\text{Power}[E,\ (-120.272\ (-699.63\ (1\ -\ 0.00335402\ t)\ -\ 2.08992\ t\ +$$
$$(2\ is^{0.5}\ (0.00920483\ t\ -\ 0.0000128467\ t^2\ +\ 4.95199\ 10^{-8}\ t^3))/(1\ +\ 1.6\ is^{0.5})\ -$$
$$E^{(-120.272\ (-691.99\ (1\ -\ 0.00335402\ t)\ -\ 1.96804\ t\ -\ 0.00831451\ t\ \text{Log}[10^{-pH}]))/t}]$$

```
h2ot=calcdGHT[h2osp][[1]]
```

-0.00831451 t Log[Power[E, (-120.272 (-285.83 (1 - 0.00335402 t) - 0.795539 t +

$(2 \text{ is}^{0.5} \ (0.00920483 \text{ t} - 0.0000128467 \text{ t}^2 + 4.95199 \ 10^{-8} \text{ t}^3))/(1 + 1.6 \text{ is}^{0.5}) -$

```
tab1=calctrGerxT[co2tott+de==co2gt+h2ot,{5,6,7,8,9},.25,{283.15,298.15,313.15}]
```

{{-6.8368, -8.24314, -9.64371}, {-5.7574, -6.94646, -8.1138}, {-2.12304, -2.91737, -3.69
   {3.03905, 2.57555, 2.12561}, {8.67202, 8.60993, 8.59301}}

```
TableForm[Transpose[tab1],TableHeadings->{{"283.15 K","298.15 K","313.15 K"},{"pH
5","pH 6","pH 7","pH 8","pH 9"}}]
```

	pH 5	pH 6	pH 7	pH 8	pH 9
283.15 K	-6.8368	-5.7574	-2.12304	3.03905	8.67202
298.15 K	-8.24314	-6.94646	-2.91737	2.57555	8.60993
313.15 K	-9.64371	-8.1138	-3.6971	2.12561	8.59301

The standard transformed Gibbs energy of reaction becomes more negative at lower pH and higher temperatures.

(b) Plot the Henry's law constants at the three temperatures versus pH.

```
plot283=Plot[Exp[-
calctrGerxT[co2tott+de==co2gt+h2ot,pH,.25,283.15]/(8.31451*.28315)],{pH,2,12},DisplayFu
nction->Identity];

plot298=Plot[Exp[-
calctrGerxT[co2tott+de==co2gt+h2ot,pH,.25,298.15]/(8.31451*.29815)],{pH,2,12},DisplayFu
nction->Identity];

plot313=Plot[Exp[-
calctrGerxT[co2tott+de==co2gt+h2ot,pH,.25,313.15]/(8.31451*.31315)],{pH,2,12},DisplayFu
nction->Identity];

Show[plot283,plot298,plot313,AxesLabel->{"pH","\!\(K\_H\)"},DisplayFunction-
>$DisplayFunction];
```

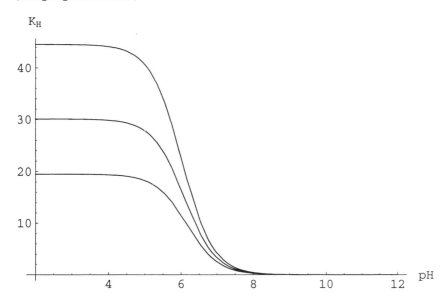

This is the equilibrium pressure in bars for an ideal 1 M solution of carbondioxide as a function of temperature. The equilibrium pressure increases with the temperature. This is Fig. 8.1.

8.2 (a) Calculate the standard transformed Gibbs energies for the reaction CO2tot = CO2(g) + H2O at 298.15 K and pHs 5, 6, 7, 8, and 9 for ionic strengths of 0, 0.10, and 0.25 M. (b) Plot the Henry's law constants versus pH from pH 3 to pH 9. (c) Plot $\log K_H$ as a function of pH at three ionic strengths.

(BasicBiochemData2 has to be loaded)

(a) Calculation of the standard transformed Gibbs energy of reaction:.

```
calctrGerx[eq_,pHlist_,islist_]:=Module[{energy},(*Calculates the standard transformed
Gibbs energy of reaction in kJ mol^-1 at specified pHs and ionic strengths for a
biochemical equation typed in the form atp+h2o+de==adp+pi.  The names of the reactants
call the appropriate functions of pH and ionic strength. pHlist and islist can be
lists. This program can be used to calculate the standard transformed enthalpy of
reaction by appending an h to the name of each reactant.*)
energy=Solve[eq,de];
energy[[1,1,2]]/.pH->pHlist/.is->islist]

tab1=calctrGerx[co2tot+de==co2g+h2o,{5,6,7,8,9},{0,.1,.25}];

TableForm[Transpose[tab1],TableHeadings->{{"I=0","I=0.1","I=0.25"},{"pH 5","pH 6","pH
7","pH 8","pH 9"}}]
```

	pH 5	pH 6	pH 7	pH 8	pH 9
I=0	-8.33561	-7.55317	-4.30451	0.952641	6.71138
I=0.1	-8.27119	-7.11762	-3.26789	2.17395	8.11334
I=0.25	-8.24314	-6.94648	-2.9174	2.57552	8.60988

The standard transformed Gibbs energy of reaction becomes more negative at lower pH and lower ionic strengths.

(b) Plot the Henry's law constants at the three ionic strengths versus pH.

```
Plot[Evaluate[Exp[-
calctrGerx[co2tot+de==co2g+h2o,pH,{0,.1,.25}]/(8.31451*.29815)],{pH,3,9},AxesLabel-
>{"pH","\!\(K\_H\)"}]];
```

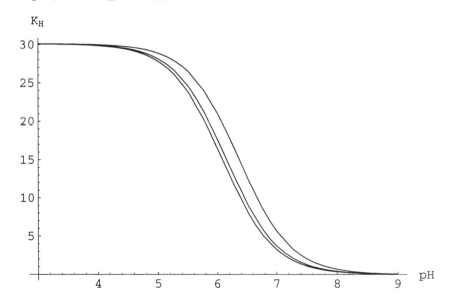

The upper curve is for zero ionic strength. This is Fig. 8.2.

(c) Plot $logK_H$ as a function of pH at three ionic strengths.

```
Plot[Evaluate[-
calctrGerx[co2tot+de==co2g+h2o,pH,{0,.1,.25}]/(8.31451*.29815*Log[10])],{pH,3,11},AxesO
rigin->{3,-5},AxesLabel->{"pH","\!\(K\_H\)"}];
```

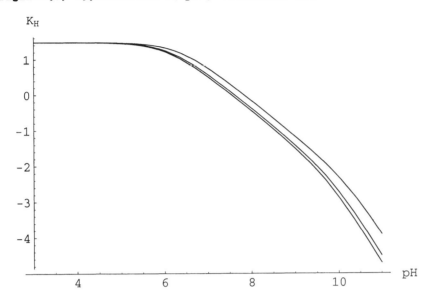

Note that the greatest effects of raising the ionic strength are at the highest pHs.

Chapter 9 Redox Reactions

9.1 Calculate the standard apparent reduction potentials for 298.15 K at pHs 5, 6, 7, 8, and 9 and ionic strengths 0, 0.10, and 0.25 M for the following biochemical half reactions and plot their pH dependencies at 0.25 M ionic strength.

(1) $O_2(aq) + 4e^- = 2H_2O$

(2) $O_2(g) + 4e^- = 2H_2O$

(3) $O_2(aq) + 2e^- = 2H_2O_2$

(4) $O_2(g) + 2e^- = 2H_2O_2$

(5) cytochrome $c_{ox} + e^- =$ cytochrome c_{red}

(6) pyruvate $+ 2e^- =$ lactate

(7) acetaldehyde $+ 2e^- =$ ethanol

(8) $FMN_{ox} + 2e^- = FMN_{red}$

(9) retinal $+ 2e^- =$ retinol

(10) acetone $+ 2e^- =$ 2-propanol

(11) $NAD_{ox} + 2e^- = NAD_{red}$

(12) $NADP_{ox} + 2e^- = NADP_{red}$

(13) $ferredoxin_{ox} + e^- = ferredoxin_{red}$

(14) acetylcoA $+ 2e^- =$ coA $+$ acetaldehyde

(15) $2e^- = H_2(g)$

(16) $2e^- = H_2(aq)$

9.2 Calculate the apparent equilibrium constants K' at 298.15 K and pHs 5, 6, 7, 8, and 9 for the following four reactions and plot the base 10 logarithm of K' versus pH:

(1) $N_2(g) + 8$ ferredoxin$_{red} = 2$ ammonia $+ H_2(g) + 8$ ferredoxin$_{ox}$

(2) $N_2(aq) + 8$ ferredoxin$_{red} = 2$ ammonia $+ H_2(aq) + 8$ ferredoxin$_{ox}$

(3) 2 ferredoxin$_{red} = H_2(g) + 2$ ferredoxin$_{ox}$

(4) 2 ferredoxin$_{red} = H_2(aq) + 2$ ferredoxin$_{ox}$

Note the ionic strength is not mentioned because the reference reactions and the dissociation of NH_4^+ do not depend on the ionic strength. Note that the apparent equilibrium constaants of the first two reactions change by a factor of 10 when the pH is changed 0.10 in the neutral region.

9.3 Calculate the standard apparent reduction potentials of the following half reactions involving carbon dioxide at 298.15 K, pHs 5, 6, 7, 8, and 9 and ionic strengths of 0, 0.10, and 0.25 M and plot them versus pH at ionic strength 0.25 M.

(1) CO_2 tot $+ 2e^- =$ formate $+ H_2O$

(2) $CO_2(g) + 2e^- =$ formate

(3) CO_2 tot $+$ pyruvate $+ 2e^- =$ malate $+ H_2O$

(4) $CO_2(g) +$ pyruvate $+ 2e^- =$ malate

(5) CO_2 tot $+$ acetylCoA $+ 2e^- =$ pyruvate $+$ CoA $+ H_2O$

9.4 Calculate the change in binding of hydrogen ions in the following 10 biochemical half reactions at 298.15 K, pHs 5, 6, 7, 8, and 9 and ionic strengths of 0, 0.10, and 0.25 M. Plot the change in binding at ionic strength 0.25 M versus pH.

(1) CO_2 tot $+ 2e^- =$ formate $+ H_2O$

(2) CO_2 tot $+$ acetyl-CoA $+ 2e^- =$ pyruvate $+$ CoA $+ H_2O$

(3) ketoglutarate $+$ ammonia $+ 2e^- =$ glutamate $+ H_2O$

(4) pyruvate $+$ ammonia $+ 2e^- =$ alanine $+ H_2O$

(5) pyruvate $+ CO_2$ tot $+ 2e^- =$ malate $+ H_2O$

(6) cystine $+ 2e^- = 2$ cysteine $+ H_2$

(7) citrate $+$ CoA $+ 2e^- =$ malate $+$ acteyl-CoA $+ H_2O$

(8) $O_2(g) + 4e^- = 2\,H_2O$

(9) $NAD_{ox} + 2e^- = NAD_{red}$

(10) $N_2(g) + 8e^- = 2$ ammonia $+ H_2(g)$

9.5 Plot the change in binding of hydrogen ions versus pH at ionic strength 0.25 M for the following five biochemical reactions:

(1) NAD_{ox} + formate + H_2O = NAD_{red} + CO_2 tot

(2) NAD_{ox} + malate + H_2O = NAD_{red} + CO_2 tot + pyruvate

(3) NAD_{ox} + ethanol = NAD_{red} + acetaldehyde

(4) NAD_{ox} + alanine + H_2O = NAD_{red} + pyruvate + ammonia

(5) NAD_{ox} + malate + acetylcoA + H_2O = NAD_{red} + citrate + coA

9.1 Calculate the standard apparent reduction potentials for 298.15 K at pHs 5, 6, 7, 8, and 9 and ionic strengths 0, 0.10, and 0.25 M for the following biochemical half reactions and plot their pH dependencies at 0.25 M ionic strength.

(1) $O_2(aq) + 4e^- = 2H_2O$

(2) $O_2(g) + 4e^- = 2H_2O$

(3) $O_2(aq) + 2e^- = 2H_2O_2$

(4) $O_2(g) + 2e^- = 2H_2O_2$

(5) cytochrome $c_{ox} + e^- =$ cytochrome c_{red}

(6) pyruvate $+ 2e^- =$ lactate

(7) acetaldehyde $+ 2e^- =$ ethanol

(8) $FMN_{ox} + 2e^- = FMN_{red}$

(9) retinal $+ 2e^- =$ retinol

(10) acetone $+ 2e^- =$ 2-propanol

(11) $NAD_{ox} + 2e^- = NAD_{red}$

(12) $NADP_{ox} + 2e^- = NADP_{red}$

(13) $ferredoxin_{ox} + e^- = ferredoxin_{red}$

(14) acetylcoA $+ 2e^- =$ coA + acetaldehyde

(15) $2e^- = H_2(g)$

(16) $2e^- = H_2(aq)$

(BasicBiochemData has to be loaded)

```
calcappredpot[eq_, nu_, pHlist_, islist_] :=
 Module[{energy}, (*Calculates the standard apparent reduction potential of
      a half reaction at specified pHs and ionic
      strengths for a biochemical equation typed in the form
      nadox+de==nadred. The names of the reactants call the corresponding
      functions of pH and ionic strength. nu is the number of
      electrons involved. pHlist and islist can be lists.*)
  energy = Solve[eq, de];
  (-1 * energy[[1, 1, 2]] / (nu * 96.485)) /. pH → pHlist /. is → islist
  ]
```

```
TableForm[Transpose[calcappredpot[nadox + de == nadred, 2, {5, 6, 7, 8, 9}, {0, .1, .25}]],
  TableHeadings → {{"I=0 M", "I=0.10 M", "I=0.25 M"},
    {"pH 5", "pH 6", "pH 7", "pH 8", "pH 9"}}, TableSpacing → {1, 1}]
```

	pH 5	pH 6	pH 7	pH 8	pH 9
I=0 M	-0.265275	-0.294855	-0.324435	-0.354015	-0.383595
I=0.10 M	-0.258932	-0.288512	-0.318092	-0.347672	-0.377252
I=0.25 M	-0.256884	-0.286464	-0.316044	-0.345624	-0.375204

```
plot11 = Plot[calcappredpot[nadox + de == nadred, 2, pH, .25],
   {pH, 5, 9}, AxesLabel → {"pH", "E'°/V"}];
```

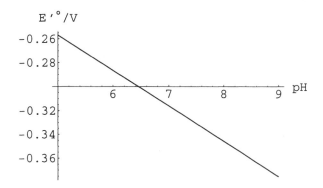

```
TableForm[Transpose[calcappredpot[nadpox + de == nadpred, 2, {5, 6, 7, 8, 9}, {0, .1, .25}]],
  TableHeadings → {{"I=0 M", "I=0.10 M", "I=0.25 M"},
    {"pH 5", "pH 6", "pH 7", "pH 8", "pH 9"}}, TableSpacing → {1, 1}]
```

	pH 5	pH 6	pH 7	pH 8	pH 9
I=0 M	-0.282584	-0.312164	-0.341744	-0.371324	-0.400904
I=0.10 M	-0.263553	-0.293133	-0.322713	-0.352293	-0.381873
I=0.25 M	-0.257409	-0.286989	-0.316569	-0.346149	-0.375729

```
plot12 = Plot[calcappredpot[nadpox + de == nadpred, 2, pH, .25],
  {pH, 5, 9}, AxesLabel → {"pH", "E'°/V"}];
```

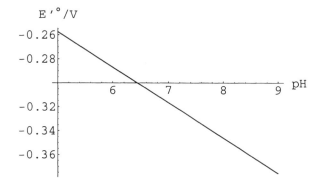

```
TableForm[Transpose[calcappredpot[o2g + de == 2 * h2o, 4, {5, 6, 7, 8, 9}, {0, .1, .25}]],
  TableHeadings → {{"I=0 M", "I=0.10 M", "I=0.25 M"},
    {"pH 5", "pH 6", "pH 7", "pH 8", "pH 9"}}, TableSpacing → {1, 1}]
```

	pH 5	pH 6	pH 7	pH 8	pH 9
I=0 M	0.933355	0.874195	0.815036	0.755876	0.696716
I=0.10 M	0.927012	0.867852	0.808692	0.749532	0.690372
I=0.25 M	0.924964	0.865804	0.806644	0.747484	0.688324

```
plot2 =
  Plot[calcappredpot[o2g + de == 2 * h2o, 4, pH, .25], {pH, 5, 9}, AxesLabel → {"pH", "E'°/V"}];
```

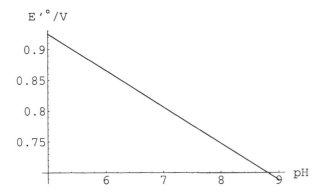

```
TableForm[Transpose[calcappredpot[o2aq + de == 2 * h2o, 4, {5, 6, 7, 8, 9}, {0, .1, .25}]],
  TableHeadings → {{"I=0 M", "I=0.10 M", "I=0.25 M"},
    {"pH 5", "pH 6", "pH 7", "pH 8", "pH 9"}}, TableSpacing → {1, 1}]
```

	pH 5	pH 6	pH 7	pH 8	pH 9
I=0 M	0.975849	0.916689	0.857529	0.798369	0.739209
I=0.10 M	0.969505	0.910345	0.851186	0.792026	0.732866
I=0.25 M	0.967457	0.908297	0.849138	0.789978	0.730818

```
plot1 =
  Plot[calcappredpot[o2aq + de == 2 * h2o, 4, pH, .25], {pH, 5, 9}, AxesLabel → {"pH", "E'°/V"}];
```

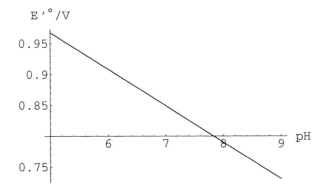

```
TableForm[Transpose[calcappredpot[o2g + de == h2o2aq, 2, {5, 6, 7, 8, 9}, {0, .1, .25}]],
  TableHeadings → {{"I=0 M", "I=0.10 M", "I=0.25 M"},
    {"pH 5", "pH 6", "pH 7", "pH 8", "pH 9"}}, TableSpacing → {1, 1}]
```

	pH 5	pH 6	pH 7	pH 8	pH 9
I=0 M	0.398764	0.339605	0.280445	0.221285	0.162125
I=0.10 M	0.392421	0.333261	0.274101	0.214941	0.155781
I=0.25 M	0.390373	0.331213	0.272053	0.212893	0.153733

```
plot4 =
  Plot[calcappredpot[o2g + de == h2o2aq, 2, pH, .25], {pH, 5, 9}, AxesLabel → {"pH", "E'°/V"}];
```

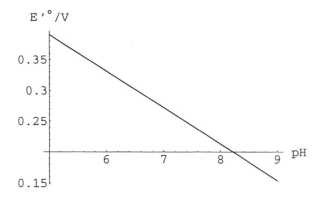

```
TableForm[Transpose[calcappredpot[o2aq+de == h2o2aq, 2, {5, 6, 7, 8, 9}, {0, .1, .25}]],
  TableHeadings → {{"I=0 M", "I=0.10 M", "I=0.25 M"},
    {"pH 5", "pH 6", "pH 7", "pH 8", "pH 9"}}, TableSpacing → {1, 1}]
```

	pH 5	pH 6	pH 7	pH 8	pH 9
I=0 M	0.483752	0.424592	0.365432	0.306272	0.247112
I=0.10 M	0.477408	0.418248	0.359088	0.299928	0.240769
I=0.25 M	0.47536	0.4162	0.35704	0.29788	0.238721

```
plot3 =
  Plot[calcappredpot[o2aq+de == h2o2aq, 2, pH, .25], {pH, 5, 9}, AxesLabel → {"pH", "E'°/V"}]
```

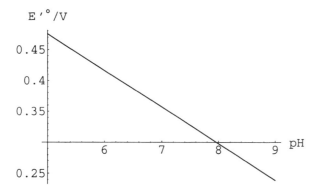

Calcappredpot fails when the oxidized and reduced forms contain the same number of hydrogen atoms. In this case, there is no pH dependence, and the standard apparent reduction potential can be calculated by using

```
(-1/96.485)*(cytochromecred-cytochromecox/.is->{0,.1,.25})
```

```
{0.254029, 0.222311, 0.212071}
```

```
plot5=Plot[.2121,{pH,5,9},PlotRange->{0,1}];
```

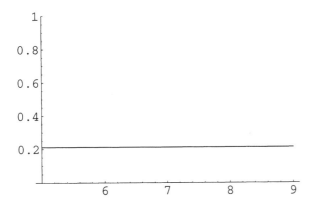

```
TableForm[
  Transpose[calcappredpot[acetaldehyde + de == ethanol, 2, {5, 6, 7, 8, 9}, {0, .1, .25}]],
  TableHeadings → {{"I=0 M", "I=0.10 M", "I=0.25 M"},
    {"pH 5", "pH 6", "pH 7", "pH 8", "pH 9"}}, TableSpacing → {1, 1}]
```

	pH 5	pH 6	pH 7	pH 8	pH 9
I=0 M	-0.0748325	-0.133992	-0.193152	-0.252312	-0.311472
I=0.10 M	-0.0811761	-0.140336	-0.199496	-0.258656	-0.317816
I=0.25 M	-0.0832242	-0.142384	-0.201544	-0.260704	-0.319864

```
plot7 = Plot[calcappredpot[acetaldehyde + de == ethanol, 2, pH, .25],
  {pH, 5, 9}, AxesLabel → {"pH", "E'°/V"}];
```

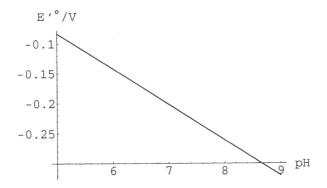

```
TableForm[Transpose[calcappredpot[fmnox + de == fmnred, 2, {5, 6, 7, 8, 9}, {0, .1, .25}]],
  TableHeadings → {{"I=0 M", "I=0.10 M", "I=0.25 M"},
    {"pH 5", "pH 6", "pH 7", "pH 8", "pH 9"}}, TableSpacing → {1, 1}]
```

	pH 5	pH 6	pH 7	pH 8	pH 9
I=0 M	-0.0943174	-0.153477	-0.212637	-0.271797	-0.330957
I=0.10 M	-0.100661	-0.159821	-0.218981	-0.278141	-0.337301
I=0.25 M	-0.102709	-0.161869	-0.221029	-0.280189	-0.339349

```
plot8 = Plot[calcappredpot[fmnox + de == fmnred, 2, pH, .25],
  {pH, 5, 9}, AxesLabel → {"pH", "E'°/V"}];
```

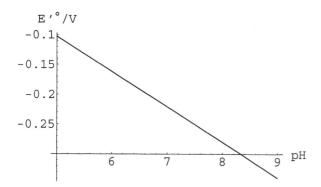

```
TableForm[
 Transpose[calcappredpot[pyruvate + de == lactate, 2, {5, 6, 7, 8, 9}, {0, .1, .25}]],
 TableHeadings → {{"I=0 M", "I=0.10 M", "I=0.25 M"},
  {"pH 5", "pH 6", "pH 7", "pH 8", "pH 9"}}, TableSpacing → {1, 1}]
```

	pH 5	pH 6	pH 7	pH 8	pH 9
I=0 M	-0.0654528	-0.124613	-0.183773	-0.242932	-0.302092
I=0.10 M	-0.0717964	-0.130956	-0.190116	-0.249276	-0.308436
I=0.25 M	-0.0738445	-0.133004	-0.192164	-0.251324	-0.310484

```
plot6 = Plot[calcappredpot[pyruvate + de == lactate, 2, pH, .25],
 {pH, 5, 9}, AxesLabel → {"pH", "E'°/V"}];
```

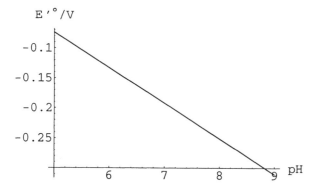

```
TableForm[Transpose[calcappredpot[de == h2g, 2, {5, 6, 7, 8, 9}, {0, .1, .25}]],
 TableHeadings → {{"I=0 M", "I=0.10 M", "I=0.25 M"},
  {"pH 5", "pH 6", "pH 7", "pH 8", "pH 9"}}, TableSpacing → {1, 1}]
```

	pH 5	pH 6	pH 7	pH 8	pH 9
I=0 M	-0.295799	-0.354959	-0.414119	-0.473279	-0.532439
I=0.10 M	-0.302143	-0.361303	-0.420463	-0.479623	-0.538783
I=0.25 M	-0.304191	-0.363351	-0.422511	-0.481671	-0.540831

```
plot15 = Plot[calcappredpot[de == h2g, 2, pH, .25], {pH, 5, 9}, AxesLabel → {"pH", "E'°/V"}];
```

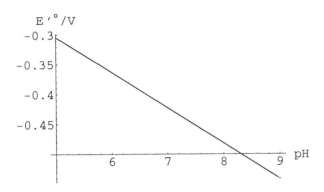

```
TableForm[Transpose[calcappredpot[de == h2aq, 2, {5, 6, 7, 8, 9}, {0, .1, .25}]]],
  TableHeadings → {{"I=0 M", "I=0.10 M", "I=0.25 M"},
    {"pH 5", "pH 6", "pH 7", "pH 8", "pH 9"}}, TableSpacing → {1, 1}]
```

	pH 5	pH 6	pH 7	pH 8	pH 9
I=0 M	-0.387005	-0.446165	-0.505325	-0.564485	-0.623645
I=0.10 M	-0.393349	-0.452509	-0.511669	-0.570829	-0.629989
I=0.25 M	-0.395397	-0.454557	-0.513717	-0.572877	-0.632037

```
plot16 = Plot[calcappredpot[de == h2aq, 2, pH, .25], {pH, 5, 9}, AxesLabel → {"pH", "E'°/V"}];
```

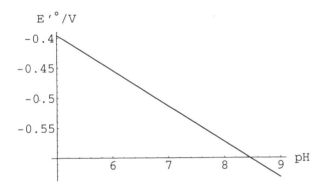

Since neither ferredoxinox or ferredoxinred contains hydrogen atoms, the pH-dependent apparent reduction potential is calculated using

```
(-1/96.485)*(ferredoxinred-ferredoxinox/.is->{0,.1,.25})
```

```
{-0.394569, -0.400913, -0.402961}
```

```
plot13=Plot[-.4030,{pH,5,9},PlotRange->{-1,1}];
```

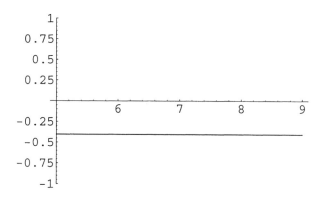

```
TableForm[Transpose[calcappredpot[retinal + de == retinol, 2, {5, 6, 7, 8, 9}, {0, .1, .25}]],
  TableHeadings → {{"I=0 M", "I=0.10 M", "I=0.25 M"},
   {"pH 5", "pH 6", "pH 7", "pH 8", "pH 9"}}, TableSpacing → {1, 1}]
```

	pH 5	pH 6	pH 7	pH 8	pH 9
I=0 M	-0.151166	-0.210325	-0.269485	-0.328645	-0.387805
I=0.10 M	-0.157509	-0.216669	-0.275829	-0.334989	-0.394149
I=0.25 M	-0.159557	-0.218717	-0.277877	-0.337037	-0.396197

```
plot9 = Plot[calcappredpot[retinal + de == retinol, 2, pH, .25],
  {pH, 5, 9}, AxesLabel → {"pH", "E'°/V"}];
```

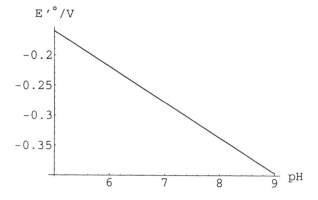

```
TableForm[
  Transpose[calcappredpot[acetone + de == propanol2, 2, {5, 6, 7, 8, 9}, {0, .1, .25}]],
  TableHeadings → {{"I=0 M", "I=0.10 M", "I=0.25 M"},
   {"pH 5", "pH 6", "pH 7", "pH 8", "pH 9"}}, TableSpacing → {1, 1}]
```

	pH 5	pH 6	pH 7	pH 8	pH 9
I=0 M	-0.163499	-0.222659	-0.281819	-0.340979	-0.400139
I-0.10 M	-0.169843	-0.229003	0.288162	0.347322	0.406482
I=0.25 M	-0.171891	-0.231051	-0.290211	-0.34937	-0.40853

```
plot10 = Plot[calcappredpot[acetone + de == propanol2, 2, pH, .25],
  {pH, 5, 9}, AxesLabel → {"pH", "E'°/V"}];
```

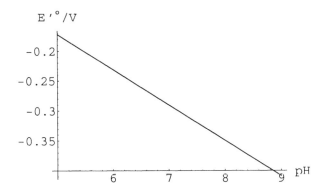

```
TableForm[Transpose[
    calcappredpot[acetylcoA + de == coA + acetaldehyde, 2, {5, 6, 7, 8, 9}, {0, .1, .25}]],
  TableHeadings → {{"I=0 M", "I=0.10 M", "I=0.25 M"},
    {"pH 5", "pH 6", "pH 7", "pH 8", "pH 9"}}, TableSpacing → {1, 1}]
```

	pH 5	pH 6	pH 7	pH 8	pH 9
I=0 M	-0.262266	-0.321377	-0.380065	-0.435269	-0.477795
I=0.10 M	-0.268606	-0.327687	-0.386084	-0.439399	-0.4788
I=0.25 M	-0.270652	-0.32972	-0.387991	-0.440576	-0.479044

```
plot14 = Plot[calcappredpot[acetylcoA + de == coA + acetaldehyde, 2, pH, .25],
  {pH, 5, 9}, AxesLabel → {"pH", "E'°/V"}];
```

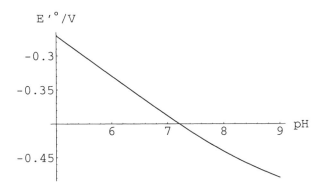

```
Show[{plot1, plot2, plot3, plot4, plot5, plot6, plot7,
    plot8, plot9, plot10, plot11, plot12, plot13, plot14, plot15, plot16},
  AxesOrigin -> {5, -.8}, AspectRatio → 1.5, PlotRange → {-.8, 1}];
```

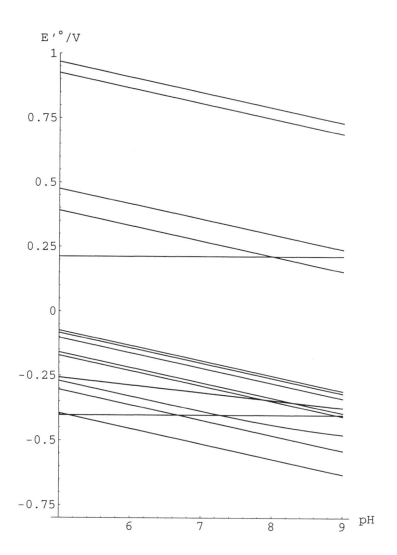

9.2 Calculate the apparent equilibrium constants K' at 298.15 K and pHs 5, 6, 7, 8, and 9 for the following four reactions and plot the base 10 logarithm of K' versus pH:

(1) N_2 (g) + 8 ferredoxin$_{red}$ = 2 ammonia + H_2 (g) + 8 ferredoxin$_{ox}$

(2) N_2 (aq) + 8 ferredoxin$_{red}$ = 2 ammonia + H_2 (aq) + 8 ferredoxin$_{ox}$

(3) 2 ferredoxin$_{red}$ = H_2 (g) + 2 ferredoxin$_{ox}$

(4) 2 ferredoxin$_{red}$ = H_2 (aq) + 2 ferredoxin$_{ox}$

Note the ionic strength is not mentioned because the reference reactions and the dissociation of NH_4^+ do not depend on the ionic strength. Note that the apparent equilibrium constants of the first two reactions change by a factor of 10 when the pH is changed 0.10 in the neutral region.

(BasicBiochemData2 has to be loaded)

These calculations and plots can be made by using the program calckprime.

```
kprime1=calckprime[n2g+8*ferredoxinred+de==2*ammonia+h2g+8*ferredoxinox,pH,.25];

kprime1/.pH->{5,6,7,8,9}
```

$\{1.39705\ 10^{31},\ 1.39846\ 10^{21},\ 1.41258\ 10^{11},\ 15.5771,\ 3.39921\ 10^{-9}\}$

Note that the apparent equilibrium constants decrease by a factor of 10^{10} per pH unit and becomes non-spontaneous just above pH 8

```
Plot[Log[10,kprime1],{pH,5,9},AxesLabel->{"pH","log K'"}];
```

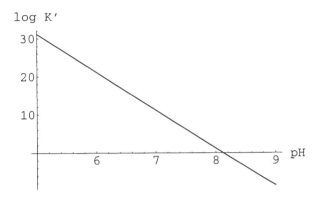

```
kprime2=calckprime[n2aq+8*ferredoxinred+de==2*ammonia+h2aq+8*ferredoxinox,pH,.25];
```

```
kprime2/.pH->{5,6,7,8,9}
```

$\{2.17732\ 10^{31},\ 2.17951\ 10^{21},\ 2.20152\ 10^{11},\ 24.2771,\ 5.29772\ 10^{-9}\}$

```
Plot[Log[10,kprime2],{pH,5,9},AxesLabel->{"pH","log K'"}];
```

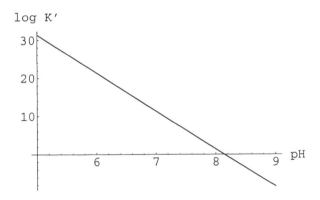

```
kprime3=calckprime[2*ferredoxinred+de==h2g+2*ferredoxinox,pH,.25];
```

```
kprime3/.pH->{5,6,7,8,9}
```

$\{2183.11,\ 21.8311,\ 0.218311,\ 0.00218311,\ 0.0000218311\}$

```
Plot[Log[10,kprime3],{pH,5,9},AxesLabel->{"pH","log K'"}];
```

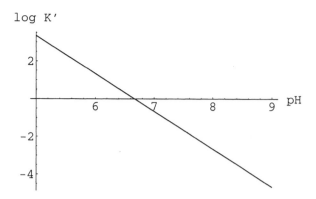

```
kprime4=calckprime[2*ferredoxinred+de==h2aq+2*ferredoxinox,pH,.25];
```

```
kprime4/.pH->{5,6,7,8,9}
```

$\{1.8018, 0.018018, 0.00018018, 1.8018\ 10^{-6}, 1.8018\ 10^{-8}\}$

```
Plot[Log[10,kprime4],{pH,5,9},AxesLabel->{"pH","log K'"}];
```

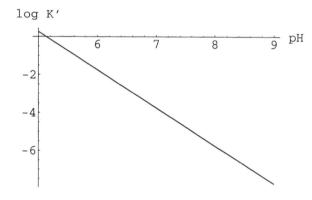

9.3 Calculate the standard apparent reduction potentials of the following half reactions involving carbon dioxide at 298.15 K, pHs 5, 6, 7, 8, and 9 and ionic strengths of 0, 0.10, and 0.25 M and plot them versus pH at ionic strength 0.25 M.

(1) CO_2 tot + $2e^-$ = formate + H_2O

(2) $CO_2(g)$ + $2e^-$ = formate

(3) CO_2 tot + pyruvate + $2e^-$ = malate + H_2O

(4) $CO_2(g)$ + pyruvate + $2e^-$ = malate

(5) CO_2 tot + acetylCoA + $2e^-$ = pyruvate + CoA + H_2O

(BasicBiochemData2 has to be loaded)

```
calcappredpot[eq_, nu_, pHlist_, islist_] := Module[{energy},
   (*Calculates the standard apparent reduction potential of a half reaction at
      specified pHs and ionic strengths for a biochemical equation typed in the form
      nadox+de==nadred.  The names of the reactants call the corresponding
      functions of pH and ionic strength. nu is the number of
      electrons involved. pHlist and islist can be lists.*)
   energy = Solve[eq, de];
   (-1*energy[[1, 1, 2]] / (nu*96.485)) /. pH → pHlist /. is → islist
   ]
```

```
TableForm[
 Transpose[calcappredpot[co2tot + de == formate + h2o, 2, {5, 6, 7, 8, 9}, {0, .1, .25}]],
 TableHeadings → {{"I=0 M", "I=0.10 M", "I=0.25 M"},
   {"pH 5", "pH 6", "pH 7", "pH 8", "pH 9"}}, TableSpacing → {1, 1}]
```

	pH 5	pH 6	pH 7	pH 8	pH 9
I=0 M	-0.329401	-0.363036	-0.409451	-0.466274	-0.525697
I=0.10 M	-0.329735	-0.365293	-0.414823	-0.472603	-0.532962
I=0.25 M	-0.329881	-0.36618	-0.416639	-0.474684	-0.535535

```
plot1 = Plot[calcappredpot[co2tot + de == formate + h2o, 2, pH, .25],
  {pH, 5, 9}, AxesLabel → {"pH", "E'°/V"}];
```

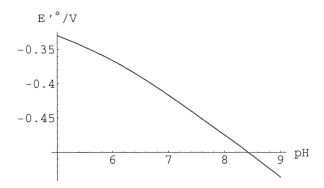

The chemical reaction corresponding with reaction 2 is not affected by ionic strength because $z_i^2 = N_H$, and so the table of apparent reduction potentials can be made more simply;

```
calcappredpot[co2g+de==formate,2,{5,6,7,8,9},.25]
```

{-0.372598, -0.402178, -0.431758, -0.461338, -0.490918}

```
plot2 = Plot[calcappredpot[co2g + de == formate, 2, pH, .25],
  {pH, 5, 9}, AxesLabel → {"pH", "E'°/V"}];
```

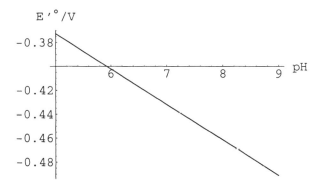

```
TableForm[Transpose[
   calcappredpot[pyruvate + co2tot + de == malate + h2o, 2, {5, 6, 7, 8, 9}, {0, .1, .25}]],
  TableHeadings → {{"I=0 M", "I=0.10 M", "I=0.25 M"},
   {"pH 5", "pH 6", "pH 7", "pH 8", "pH 9"}}, TableSpacing → {1, 1}]
```

	pH 5	pH 6	pH 7	pH 8	pH 9
I=0 M	-0.215617	-0.26041	-0.308738	-0.365769	-0.425213
I=0.10 M	-0.216271	-0.257627	-0.307911	-0.365769	-0.426136
I=0.25 M	-0.215868	-0.25669	-0.307703	-0.365804	-0.426661

```
plot3 = Plot[calcappredpot[pyruvate + co2tot + de == malate + h2o, 2, pH, .25],
  {pH, 5, 9}, AxesLabel → {"pH", "E'°/V"}];
```

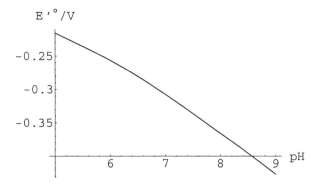

```
TableForm[
  Transpose[calcappredpot[pyruvate + co2g + de == malate, 2, {5, 6, 7, 8, 9}, {0, .1, .25}]],
  TableHeadings → {{"I=0 M", "I=0.10 M", "I=0.25 M"},
    {"pH 5", "pH 6", "pH 7", "pH 8", "pH 9"}}, TableSpacing → {1, 1}]
```

	pH 5	pH 6	pH 7	pH 8	pH 9
I=0 M	-0.258813	-0.299552	-0.331045	-0.360832	-0.390433
I=0.10 M	-0.259134	-0.294511	-0.324846	-0.354503	-0.384091
I=0.25 M	-0.258585	-0.292687	-0.322821	-0.352458	-0.382043

```
plot4 = Plot[calcappredpot[pyruvate + co2g + de == malate, 2, pH, .25],
  {pH, 5, 9}, AxesLabel → {"pH", "E'°/V"}];
```

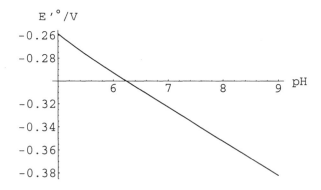

```
TableForm[Transpose[calcappredpot[
    acetylcoA + co2tot + de == coA + pyruvate + h2o, 2, {5, 6, 7, 8, 9}, {0, .1, .25}]],
  TableHeadings → {{"I=0 M", "I=0.10 M", "I=0.25 M"},
    {"pH 5", "pH 6", "pH 7", "pH 8", "pH 9"}}, TableSpacing → {1, 1}]
```

	pH 5	pH 6	pH 7	pH 8	pH 9
I=0 M	-0.387747	-0.421334	-0.467277	-0.520144	-0.562932
I=0.10 M	-0.388078	-0.423557	-0.472324	-0.524259	-0.56486
I=0.25 M	-0.388221	-0.424428	-0.473999	-0.525469	-0.565628

```
plot5 = Plot[calcappredpot[acetylcoA + co2tot + de == coA + pyruvate + h2o, 2, pH, .25],
  {pH, 5, 9}, AxesLabel → {"pH", "E'°/V"}];
```

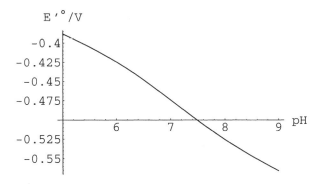

```
Show[{plot1, plot2, plot3, plot4, plot5}, AxesOrigin → {5, -.7}, PlotRange → {-.7, 0}];
```

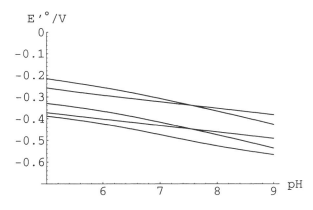

9.4 Calculate the change in binding of hydrogen ions in the following 10 biochemical half reactions at 298.15 K, pHs 5, 6, 7, 8, and 9 and ionic strengths of 0, 0.10, and 0.25 M. Plot the change in binding at ionic strength 0.25 M versus pH.

(1) CO_2 tot + 2e$^-$ = formate + H_2O

(2) CO_2 tot + acetyl-CoA + 2e$^-$ = pyruvate + CoA + H_2O

(3) ketoglutarate + ammonia +2e$^-$ = glutamate + H_2O

(4) pyruvate + ammonia +2e$^-$ = alanine + H_2O

(5) pyruvate + CO_2 tot + 2e$^-$ = malate + H_2O

(6) cystine + 2e$^-$ = 2 cysteine + H_2

(7) citrate + CoA + 2e$^-$ = malate +acteyl-CoA + H_2O

(8) O_2 (g) + 4e$^-$ = 2 H_2O

(9) NAD_{ox} + 2e$^-$ = NAD_{red}

(10) N_2 (g) + 8e$^-$ = 2 ammonia + H_2 (g)

(BasicBiochemData2 has to be loaded)

```
calcappredpot[eq_, nu_, pHlist_, islist_] :=
 Module[{energy}, (*Calculates the standard apparent
       reduction potential of a half reaction at specified pHs and ionic
       strengths for a biochemical half reaction typed in the form
       nadox+de==nadred.
       The names of the reactants call the
       corresponding functions of pH and ionic strength. nu is the
       number of electrons involved. pHlist and islist can be lists.*)
    energy = Solve[eq, de];
    (-1 * energy[[1, 1, 2]] / (nu * 96.485)) /. pH → pHlist /. is → islist
 ]
```

```
plot5 = Plot[Evaluate[- (2 * 96.485 / (8.31451 * .29815 * Log[10])) *
      D[calcappredpot[pyruvate + co2tot + de == malate + h2o, 2, pH, .25], pH]],
   {pH, 5, 9}, PlotRange → {0, 3}, AxesOrigin → {5, 0}, AxesLabel -> {"pH", "ΔN_H"}];
```

This result is reasonable because the chemical reference reaction shows that 2 hydrogen ions are bound. This is affected in the neighborhood of pH 6.2 by the binding of a hydrogen ion by bicarbonate and malate. As pH 9 is approached, the dissociation of the hydrogen ion from bicarbonate begins to have an effect.

```
plot8 = Plot[Evaluate[- (4 * 96.485 / (8.31451 * .29815 * Log[10])) *
      D[calcappredpot[o2g + de == 2 * h2o, 4, pH, .25], pH]], {pH, 5, 9},
   PlotRange -> {0, 5}, AxesOrigin -> {5, 0}, AxesLabel -> {"pH", "ΔN_H"}];
```

```
plot9 = Plot[Evaluate[- (2 * 96.485 / (8.31451 * .29815 * Log[10])) *
      D[calcappredpot[nadox + de == nadred, 2, pH, .25], pH]], {pH, 5, 9},
   PlotRange -> {0, 5}, AxesOrigin -> {5, 0}, AxesLabel -> {"pH", "ΔN_H"}];
```

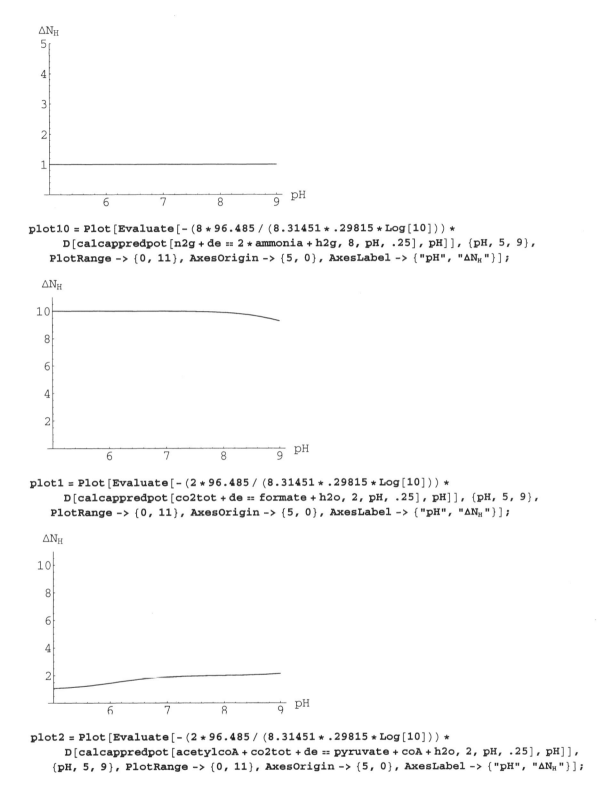

```
plot10 = Plot[Evaluate[-(8 * 96.485 / (8.31451 * .29815 * Log[10])) *
    D[calcappredpot[n2g + de == 2 * ammonia + h2g, 8, pH, .25], pH]], {pH, 5, 9},
  PlotRange -> {0, 11}, AxesOrigin -> {5, 0}, AxesLabel -> {"pH", "ΔN_H"}];
```

```
plot1 = Plot[Evaluate[-(2 * 96.485 / (8.31451 * .29815 * Log[10])) *
    D[calcappredpot[co2tot + de == formate + h2o, 2, pH, .25], pH]], {pH, 5, 9},
  PlotRange -> {0, 11}, AxesOrigin -> {5, 0}, AxesLabel -> {"pH", "ΔN_H"}];
```

```
plot2 = Plot[Evaluate[-(2 * 96.485 / (8.31451 * .29815 * Log[10])) *
    D[calcappredpot[acetylcoA + co2tot + de == pyruvate + coA + h2o, 2, pH, .25], pH]],
    {pH, 5, 9}, PlotRange -> {0, 11}, AxesOrigin -> {5, 0}, AxesLabel -> {"pH", "ΔN_H"}];
```

```
plot3 = Plot[Evaluate[-(2*96.485/(8.31451*.29815*Log[10]))*
      D[calcappredpot[oxoglutarate2 + ammonia + de == glutamate + h2o, 2, pH, .25], pH]],
   {pH, 5, 9}, PlotRange -> {0, 11}, AxesOrigin -> {5, 0}, AxesLabel -> {"pH", "ΔN_H"}];
```

```
plot4 = Plot[Evaluate[-(2*96.485/(8.31451*.29815*Log[10]))*
      D[calcappredpot[pyruvate + ammonia + de == alanine + h2o, 2, pH, .25], pH]], {pH, 5, 9},
   PlotRange -> {0, 11}, AxesOrigin -> {5, 0}, AxesLabel -> {"pH", "ΔN_H"}];
```

```
plot6 = Plot[Evaluate[-(2*96.485/(8.31451*.29815*Log[10]))*
      D[calcappredpot[cystineL + de == 2*cysteineL + h2o, 2, pH, .25], pH]], {pH, 5, 9},
   PlotRange -> {0, 11}, AxesOrigin -> {5, 0}, AxesLabel -> {"pH", "ΔN_H"}];
```

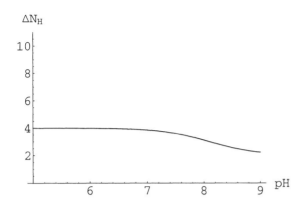

```
plot7 = Plot[Evaluate[-(2 * 96.485 / (8.31451 * .29815 * Log[10])) *
    D[calcappredpot[citrate + coA + de == malate + acetylcoA + h2o, 2, pH, .25], pH]],
  {pH, 5, 9}, PlotRange -> {0, 11}, AxesOrigin -> {5, 0}, AxesLabel -> {"pH", "ΔNₕ"}];
```

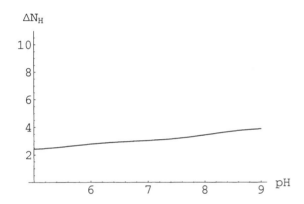

```
Show[{plot5, plot6, plot7, plot9, plot10}, PlotRange → {0, 11}, AspectRatio → 1.5];
```

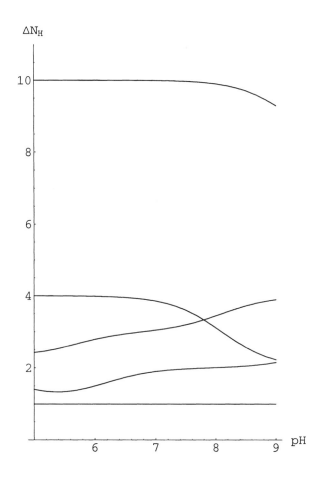

starting at top
10
6
7
5
9

9.5 Plot the change in binding of hydrogen ions versus pH at ionic strength 0.25 M for the following five biochemical reactions:

(1) NAD_{ox} + formate + H_2O = NAD_{red} + CO_2 tot

(2) NAD_{ox} + malate + H_2O = NAD_{red} + CO_2 tot + pyruvate

(3) NAD_{ox} + ethanol = NAD_{red} + acetaldehyde

(4) NAD_{ox} + alanine + H_2O = NAD_{red} + pyruvate + ammonia

(5) NAD_{ox} + malate + acetylcoA + H_2O = NAD_{red} + citrate + coA

(BasicBiochemData2 has to be loaded)

In[12]:= `calctrGerx[eq_, pHlist_, islist_] :=`
`    Module[{energy}, (*Calculates the standard transformed Gibbs`
`        energy of reaction in kJ mol^-1 at specified pHs and ionic`
`        strengths for a biochemical reaction typed in the form atp+h2o+de==`
`    adp+pi.  The names of reactants call the appropriate functions of`
`        pH and ionic strength.  pHlist and is list can be lists.*)`
`    energy = Solve[eq, de];`
`    energy[[1, 1, 2]] /. pH → pHlist /. is → islist]`

In[13]:= `plot1=Plot[Evaluate[(1/(8.31451*.29815*Log[10]))*D[calctrGerx[nadox+ethanol+de==nadred`
`    +acetaldehyde,pH,.25],pH]],{pH,5,9},AxesOrigin->{5,0},\!\(AxesLabel -> {"\<pH\>",`
`    \*"\"\<\!\(ΔN\_H\)\>\""}\)];`

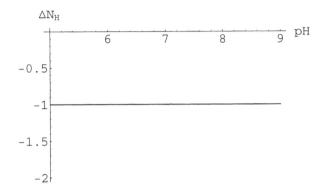

In[14]:= `plot2=Plot[Evaluate[(1/(8.31451*.29815*Log[10]))*D[calctrGerx[nadox+malate+acetylcoA+`
`    h2o+de==nadred+citrate+coA,pH,.25],pH]],{pH,5,9},AxesOrigin->{5,0},\!\(AxesLabel ->`
`    {"\<pH\>", \*"\"\<\!\(ΔN\_H\)\>\""}\),PlotRange->{0,-2.5}];`

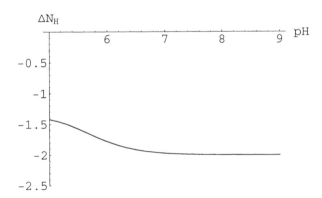

In[15]:= `plot3=Plot[Evaluate[(1/(8.31451*.29815*Log[10]))*D[calctrGerx[nadox+alanine+h2o+de==na`
`    dred+pyruvate+ammonia,pH,.25],pH]],{pH,5,9},AxesOrigin->{5,0},\!\(AxesLabel ->`
`    {"\<pH\>", \*"\"\<\!\(ΔN\_H\)\>\""}\),PlotRange->{0,-2.5}];`

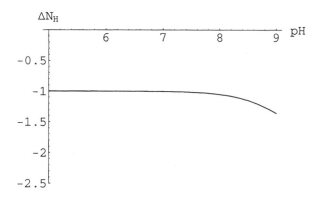

In[16]:= **plot4=Plot[Evaluate[(1/(8.31451*.29815*Log[10]))*D[calctrGerx[nadox+formate+h2o+de==na dred+co2tot,pH,.25],pH]],{pH,5,9},AxesOrigin->{5,0},\!\(AxesLabel -> {"\<pH\>", *"\"\<\!\(ΔN_H\)\>\""}\),PlotRange->{0,-2.5}];**

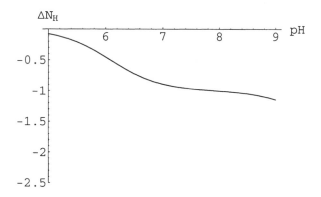

In[17]:= **plot5=Plot[Evaluate[(1/(8.31451*.29815*Log[10]))*D[calctrGerx[nadox+malate+h2o+de==nad red+pyruvate+co2tot,pH,.25],pH]],{pH,5,9},AxesOrigin->{5,-2.5},\!\(AxesLabel -> {"\<pH\>", *"\"\<\!\(ΔN_H\)\>\""}\),PlotRange->{0,-2.5}];**

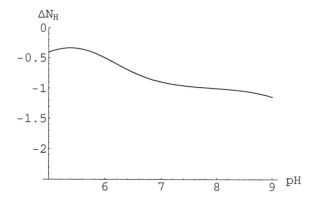

In[18]:= **Show[{plot1,plot2,plot3,plot4,plot5},AxesOrigin->{5,-2.5},PlotRange->{0,-2.5},AspectRatio->1.5];**

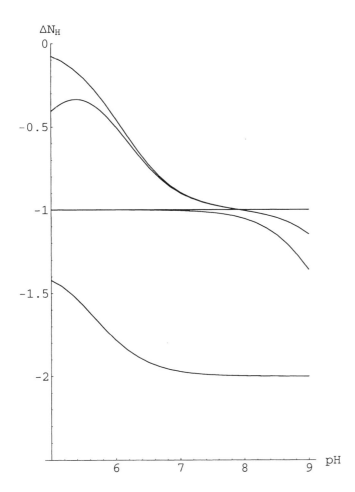

Chapter 11 Use of Semigrand Partition Functions

11.1 (a) Write out the semigrand partition function Γ' for a system containing a weak acid and its basic form at a specified pH. (b) Write out the equation for $\Delta_f G'^\circ$ of acetate at zero ionic strength. (c) Evaluate this function for $\Delta_f G'^\circ$ at pH 5 and zero ionic strength. Use calcdGmat to confirm this value. (d) Calculate the semigrand partition function Γ' and use it to calculate $\Delta_f G'^\circ$ of acetate at pH 5.

11.2 (a) Calculate the further transformed Gibbs energies of formation of G6P, F6P, and F16BP at pH 7, I = 0, [ATP] $= 10^{-4}$ M, and [ADP] $= 10^{-2}$ M at 298.15 K. (b) Calculate the further transformed Gibbs energy of formation of the pseudoisomer group. (c) Calculate the equilibrium mole fractions for G6P, F6P, and F16BP. (d) Repeat this calculation at [ATP] $= 10^{-2}$ M, and [ADP] $= 10^{-2}$ M at 298.15 K.

11.3 (a) Print out the equation for the standard transformed Gibbs energy of formation of the pseudoisomer group consisting of G6P, F6P, and F16BP symbolically. Assume that the system contains one mole of the pseudoisomer group. (b) Print out the equation for the corresponding semigrand partition function $\Gamma = \exp[-G''/RT]$. (c) Calculate the amount of ATP bound by the pseudoisomer group at [ATP] = 0.0001 M and [ADP] = 0.01 M and then at [ATP] = 0.01 M and [ADP] = 0.01 M. (d) Make the same calculation for the binding of ADP.

11.1 (a) Write out the semigrand partition function Γ ' for a system containing a weak acid and its basic form at a specified pH. (b) Write out the equation for $\Delta_f G'^\circ$ of acetate at zero ionic strength. (c) Evaluate this function for $\Delta_f G'^\circ$ at pH 5 and zero ionic strength. Use calcdGmat to confirm this value. (d) Calculate the semigrand partition function Γ ' and use it to calculate $\Delta_f G'^\circ$ of acetate at pH 5.

(a) The standard transformed Gibbs energy for one mole of the pseudoisomer group at zero ionic strength is given by

```
g'=-r*t*Log[Exp[-μ1/(r*t)]*Exp[-NH1*ln10*pH]+Exp[-μ2/(r*t)]*Exp[-NH2*ln10*pH]]
```

$$-(r\ t\ \text{Log}[E^{-(\ln 10\ \text{NH1}\ \text{pH})\ -\ \mu 1/(r\ t)} + E^{-(\ln 10\ \text{NH2}\ \text{pH})\ -\ \mu 2/(r\ t)}])$$

The ln10 has been put in in this way so that the partition function would print in a simple way.

Since G ' =-RTlnΓ ', Γ ' =exp(-G '/RT).

```
Γ' = Exp[-(1/(r*t))*g']
```

$$E^{-(\ln 10\ \text{NH1}\ \text{pH})\ -\ \mu 1/(r\ t)} + E^{-(\ln 10\ \text{NH2}\ \text{pH})\ -\ \mu 2/(r\ t)}$$

This is the semigrand partition function for the system. It was calculated from $\Delta_f G'^\circ$, but it could have been written down directly.

(b) Write the equation for $\Delta_f G'^\circ$ of acetate at a specified pH.

For the anion

```
-369.31+3*r*t*Log[10]*pH
```

-369.31 + 3 pH r t Log[10]

For the acid

```
-396.45+4*r*t*Log[10]*pH
```

-396.45 + 4 pH r t Log[10]

The standard transformed Gibbs energy of the pseudoisomer group is given by

```
gacetate=-r*t*Log[Exp[-(-369.31+3*r*t*Log[10]*pH)/(r*t)]+Exp[-
(-396.45+4*r*t*Log[10]*pH)/(r*t)]]
```

$$-(r\ t\ \text{Log}[E^{-((-369.31\ +\ 3\ \text{pH}\ r\ t\ \text{Log}[10])/(r\ t))} + E^{-((-396.45\ +\ 4\ \text{pH}\ r\ t\ \text{Log}[10])/(r\ t))}])$$

```
gacetate/.r->8.31451/.t->.29815/.pH->5
```

-284.805

This value can be confirmed by use of calcdGmat.

```
calcdGmat[acetatesp]/.pH->5/.is->0
```

-284.805

(d) Use Γ' to calculate $\Delta_f G'^\circ$ at pH 5. The partition function that is evident in the calculation of gacetate can be printed in traditional form.

TraditionalForm[Exp[-(-369.31+3*r*t*Log[10]*pH)/(r*t)]+Exp[-(-396.45+4*r*t*Log[10]*pH)/(r*t)]]

$$e^{-\frac{3\,\mathrm{pH}\,r\,t\log(10)-369.31}{r\,t}} + e^{-\frac{4\,\mathrm{pH}\,r\,t\log(10)-396.45}{r\,t}}$$

The numerical value of the partition function for acetate at pH 5 is given by

(Exp[-(-369.31+3*r*t*Log[10]*pH)/(r*t)]+Exp[-(-396.45+4*r*t*Log[10]*pH)/(r*t)])/.r->8.31451/.t->.29815/.pH->5

`7.85988 10`^{49}

Since $G' = -RT\ln\Gamma'$,

-8.31451*.29815*Log[Exp[-(-369.31+3*r*t*Log[10]pH)/(r*t)]+Exp[-(-396.45+4*r*t*Log[10]pH)/(r*t)]]/.r->8.31451/.t->.29815/.pH->5

-284.805

Thus the semigrand partition function contains all the information on the thermodynamics of the system containing acetate at a specified pH.

11.2 (a) Calculate the further transformed Gibbs energies of formation of G6P, F6P, and F16BP at pH 7, I = 0, [ATP] = 10^{-4} M, and [ADP] = 10^{-2} M at 298.15 K. (b) Calculate the further transformed Gibbs energy of formation of the pseudoisomer group. (c) Calculate the equilibrium mole fractions for G6P, F6P, and F16BP. (d) Repeat this calculation at [ATP] = 10^{-2} M, and [ADP] = 10^{-2} M at 298.15 K.

(BasicBiochemData2 has to be loaded)

(a) Calculate the standard transformed Gibbs energies of formation of the reactants at pH 7 and zero ionic strength

gpatp=atp/.pH->7/.is->0

-2292.61

gpadp=adp/.pH->7/.is->0

-1428.93

gpglucose6phos=glucose6phos/.pH->7/.is->0

-1325.

gpfructose6phos=fructose6phos/.pH->7/.is->0

-1321.71

gpfructose16phos=fructose16phos/.pH->7/.is->0

-2202.84

Now calculate the further transformed Gibbs energies of formation of these reactants at the specified concentrations of ATP and ADP

```
gppglucose6phos=gpglucose6phos-
(gpatp+8.31451*.29815*Log[.0001])+(gpadp+8.31451*.29815*Log[.01])
```

-449.906

```
gppfructose6phos=gpfructose6phos-
(gpatp+8.31451*.29815*Log[.0001])+(gpadp+8.31451*.29815*Log[.01])
```

-446.611

```
gppfructose16phos=gpfructose16phos-2*(gpatp+8.31451*.29815*Log[.0001])+2*(gpadp+8.31451
*.29815*Log[.01])
```

-452.647

(b) Calculate the further transformed Gibbs energy of formation of the speudoisomer group

```
calciso[transG_] :=
  Module[{},(*This program produces the function of pH and ionic strength that gives
the standard transformed Gibbs energy of formation of a
pseudoisomer group at 298.15 K.  The input is a list of the names of the functions for
the pseudoisomers in the groups.  Energies are expressed in kJ mol^-1.*)
  -8.31451*.29815*
Log[Apply[Plus, Exp[-1*transG/(8.31451*.29815)]]]]]

gppiso=calciso[{gppglucose6phos,gppfructose6phos,gppfructose16phos}]
```

-453.514

(c) Calculate the equilibrium mole fractions

```
ri=Exp[(gppiso-{gppglucose6phos,gppfructose6phos,gppfructose16phos})/(8.31451*.29815)]
```

{0.233262, 0.061748, 0.70499}

(d) Now increase the ATP steady state concentration by a factor of 100

```
gppglucose6phos=gpglucose6phos-
(gpatp+8.31451*.29815*Log[.01])+(gpadp+8.31451*.29815*Log[.01])
```

-461.322

```
gppfructose6phos=gpfructose6phos-
(gpatp+8.31451*.29815*Log[.01])+(gpadp+8.31451*.29815*Log[.01])
```

-458.027

```
gppfructose16phos=gpfructose16phos-2*(gpatp+8.31451*.29815*Log[.01])+2*(gpadp+8.31451*.
29815*Log[.01])
```

-475.48

```
gppiso=calciso[{gppglucose6phos,gppfructose6phos,gppfructose16phos}]
```

-475.49

```
ri=Exp[(gppiso-{gppglucose6phos,gppfructose6phos,gppfructose16phos})/(8.31451*.29815)]
```

{0.00329493, 0.00087222, 0.995833}

A higher concentration of ATP causes more F16BP to exist at equilibrium.

11.3 (a) Print out the equation for the standard further transformed Gibbs energy of formation of the pseudoisomer group consisting of G6P, F6P, and F16BP symbolically. Assume that the system contains one mole of the pseudoisomer group. (b) Print out the equation for the corresponding semigrand partition function $\Gamma'' = \exp[-G''/RT]$. (c) Calculate the amount of ATP bound by the pseudoisomer group at [ATP] = 0.0001 M and [ADP] = 0.01 M and then at [ATP] = 0.01 M and [ADP] = 0.01 M. (d) Make the same calculation for the binding of ADP.

This should be run when Problem 11.2 is still in the workspace.

(a) Print out the equation for the standard further transformed Gibbs energy of formation of the C_6 pseudoisomer group

```
-8.31451*.29815*Log[Exp[-(gpglucose6phos-
(gpatp+8.31451*.29815*Log[ATP])+(gpadp+8.31451*.29815*Log[ADP]))/(8.31451*.29815)]+Exp[
-(gpfructose6phos-
(gpatp+8.31451*.29815*Log[ATP])+(gpadp+8.31451*.29815*Log[ADP]))/(8.31451*.29815)]+Exp[
-
(gpfructose16phos-2*(gpatp+8.31451*.29815*Log[ATP])+2*(gpadp+8.31451*.29815*Log[ADP]))/
(8.31451*.29815)]]
```

$$-2.47897 \, Log[E^{-0.403393 \, (-461.322 \, + \, 2.47897 \, Log[ADP] \, - \, 2.47897 \, Log[ATP])} \, + $$
$$E^{-0.403393 \, (-458.027 \, + \, 2.47897 \, Log[ADP] \, - \, 2.47897 \, Log[ATP])} \, + $$
$$Power[E, \, -0.403393 \, (-2202.84 \, + \, 2 \, (-1428.93 \, + \, 2.47897 \, Log[ADP]) \, - $$
$$2 \, (-2292.61 \, + \, 2.47897 \, Log[ATP]))]]$$

```
TraditionalForm[-8.31451*.29815*Log[Exp[-(gpglucose6phos-
(gpatp+8.31451*.29815*Log[ATP])+(gpadp+8.31451*.29815*Log[ADP]))/(8.31451*.29815)]+Exp[
-(gpfructose6phos-
(gpatp+8.31451*.29815*Log[ATP])+(gpadp+8.31451*.29815*Log[ADP]))/(8.31451*.29815)]+Exp[
-
(gpfructose16phos-2*(gpatp+8.31451*.29815*Log[ATP])+2*(gpadp+8.31451*.29815*Log[ADP]))/
(8.31451*.29815)]]]
```

$$-2.47897 \log(e^{-0.403393 \, (2.47897 \log(ADP) - 2.47897 \log(ATP) - 461.322)} \, + $$
$$e^{-0.403393 \, (2.47897 \log(ADP) - 2.47897 \log(ATP) - 458.027)} \, + $$
$$e^{-0.403393 \, (2 \, (2.47897 \log(ADP) - 1428.93) - 2 \, (2.47897 \log(ATP) - 2292.61) - 2202.84)})$$

Note that the species concentrations of ATP and ADP are in the exponents

(b) Print out the semigrand partition function

```
gamma=Exp[Log[Exp[-(gpglucose6phos-
(gpatp+8.31451*.29815*Log[ATP])+(gpadp+8.31451*.29815*Log[ADP]))/(8.31451*.29815)]+Exp[
-(gpfructose6phos-
(gpatp+8.31451*.29815*Log[ATP])+(gpadp+8.31451*.29815*Log[ADP]))/(8.31451*.29815)]+Exp[
-
(gpfructose16phos-2*(gpatp+8.31451*.29815*Log[ATP])+2*(gpadp+8.31451*.29815*Log[ADP]))/
(8.31451*.29815)]]]
```

$$E^{-0.403393\ (-461.322\ +\ 2.47897\ \text{Log[ADP]}\ -\ 2.47897\ \text{Log[ATP]})}\ +$$

$$E^{-0.403393\ (-458.027\ +\ 2.47897\ \text{Log[ADP]}\ -\ 2.47897\ \text{Log[ATP]})}\ +$$

```
   Power[E, -0.403393 (-2202.84 + 2 (-1428.93 + 2.47897 Log[ADP]) -
       2 (-2292.61 + 2.47897 Log[ATP]))]]
```

TraditionalForm[gamma]

$$e^{-0.403393\,(2.47897\,\log(\text{ADP})-2.47897\,\log(\text{ATP})-461.322)}\ +$$

$$e^{-0.403393\,(2.47897\,\log(\text{ADP})-2.47897\,\log(\text{ATP})-458.027)}\ +$$

$$e^{-0.403393\,(2\,(2.47897\,\log(\text{ADP})-1428.93)-2\,(2.47897\,\log(\text{ATP})-2292.61)-2202.84)}$$

Note that the first Exp and Log in gamma cancel so that the same result is obtained with

```
gamma2=Exp[-(gpglucose6phos-
(gpatp+8.31451*.29815*Log[ATP])+(gpadp+8.31451*.29815*Log[ADP]))/(8.31451*.29815)]+Exp[
-(gpfructose6phos-
(gpatp+8.31451*.29815*Log[ATP])+(gpadp+8.31451*.29815*Log[ADP]))/(8.31451*.29815)]+Exp[
-
(gpfructose16phos-2*(gpatp+8.31451*.29815*Log[ATP])+2*(gpadp+8.31451*.29815*Log[ADP]))/
(8.31451*.29815)]
```

$$E^{-0.403393\ (-461.322\ +\ 2.47897\ \text{Log[ADP]}\ -\ 2.47897\ \text{Log[ATP]})}\ +$$

$$E^{-0.403393\ (-458.027\ +\ 2.47897\ \text{Log[ADP]}\ -\ 2.47897\ \text{Log[ATP]})}\ +$$

```
   Power[E, -0.403393 (-2202.84 + 2 (-1428.93 + 2.47897 Log[ADP]) -
       2 (-2292.61 + 2.47897 Log[ATP]))]]
```

TraditionalForm[gamma2]

$$e^{-0.403393\,(2.47897\,\log(\text{ADP})-2.47897\,\log(\text{ATP})-461.322)}\ +$$

$$e^{-0.403393\,(2.47897\,\log(\text{ADP})-2.47897\,\log(\text{ATP})-458.027)}\ +$$

$$e^{-0.403393\,(2\,(2.47897\,\log(\text{ADP})-1428.93)-2\,(2.47897\,\log(\text{ATP})-2292.61)-2202.84)}$$

(c) The fundamental equation for G" shows that the amount of ATP bound by the pseudoisomer group is given by
$$nc(\text{ATP}) = \partial\ln\Gamma/\partial\ln[\text{ATP}]$$
at constant [ADP]. There is a corresponding equation for ADP.

boundATP=D[Log[gamma2],Log[ATP]]

$$\left(\frac{1.}{E^{0.403393\ (-461.322\ +\ 2.47897\ \text{Log[ADP]}\ -\ 2.47897\ \text{Log[ATP]})}}\ +\right.$$

$$\frac{1.}{E^{0.403393\ (-458.027\ +\ 2.47897\ \text{Log[ADP]}\ -\ 2.47897\ \text{Log[ATP]})}}\ +$$

```
   2. / Power[E, 0.403393 (-2202.84 + 2 (-1428.93 + 2.47897 Log[ADP]) -
       2 (-2292.61 + 2.47897 Log[ATP]))]) /
```

$$\left(E^{-0.403393\ (-461.322\ +\ 2.47897\ \text{Log[ADP]}\ -\ 2.47897\ \text{Log[ATP]})}\ +\right.$$

$$E^{-0.403393\ (-458.027\ +\ 2.47897\ \text{Log[ADP]}\ -\ 2.47897\ \text{Log[ATP]})}\ +$$

```
   Power[E, -0.403393 (-2202.84 + 2 (-1428.93 + 2.47897 Log[ADP]) -
       2 (-2292.61 + 2.47897 Log[ATP]))])
```

TraditionalForm[boundATP]

$$(1. e^{-0.403393 \,(2.47897 \log(ADP) - 2.47897 \log(ATP) - 461.322)} +$$
$$1. e^{-0.403393 \,(2.47897 \log(ADP) - 2.47897 \log(ATP) - 458.027)} +$$
$$2. e^{-0.403393 \,(2\,(2.47897 \log(ADP) - 1428.93) - 2\,(2.47897 \log(ATP) - 2292.61) - 2202.84)}) /$$
$$(e^{-0.403393 \,(2.47897 \log(ADP) - 2.47897 \log(ATP) - 461.322)} +$$
$$e^{-0.403393 \,(2.47897 \log(ADP) - 2.47897 \log(ATP) - 458.027)} +$$
$$e^{-0.403393 \,(2\,(2.47897 \log(ADP) - 1428.93) - 2\,(2.47897 \log(ATP) - 2292.61) - 2202.84)})$$

boundATP/.ATP->.0001/.ADP->.01

1.70499

boundATP/.ATP->.01/.ADP->.01

1.99583

As the concentration of ATP is increased, F16BP will predominate more and more and this average number bound will approach 2.

(d)

boundADP=D[Log[gamma2],Log[ADP]]

```
(        -1.                                                -
 ---------------------------------------------------------
   0.403393 (-461.322 + 2.47897 Log[ADP] - 2.47897 Log[ATP])
  E

             1.                                             -
 ---------------------------------------------------------
   0.403393 (-458.027 + 2.47897 Log[ADP] - 2.47897 Log[ATP])
  E

  2. / Power[E, 0.403393 (-2202.84 + 2 (-1428.93 + 2.47897 Log[ADP]) -
       2 (-2292.61 + 2.47897 Log[ATP]))]) /
   -0.403393 (-461.322 + 2.47897 Log[ADP] - 2.47897 Log[ATP])
  (E                                                          +

   -0.403393 (-458.027 + 2.47897 Log[ADP] - 2.47897 Log[ATP])
  E                                                           +

  Power[E, -0.403393 (-2202.84 + 2 (-1428.93 + 2.47897 Log[ADP]) -
       2 (-2292.61 + 2.47897 Log[ATP]))])
```

TraditionalForm[boundADP]

$$(-1. e^{-0.403393 \,(2.47897 \log(ADP) - 2.47897 \log(ATP) - 461.322)} -$$
$$1. e^{-0.403393 \,(2.47897 \log(ADP) - 2.47897 \log(ATP) - 458.027)} -$$
$$2. e^{-0.403393 \,(2\,(2.47897 \log(ADP) - 1428.93) - 2\,(2.47897 \log(ATP) - 2292.61) - 2202.84)}) /$$
$$(e^{-0.403393 \,(2.47897 \log(ADP) - 2.47897 \log(ATP) - 461.322)} +$$
$$e^{-0.403393 \,(2.47897 \log(ADP) - 2.47897 \log(ATP) - 458.027)} +$$
$$e^{-0.403393 \,(2\,(2.47897 \log(ADP) - 1428.93) - 2\,(2.47897 \log(ATP) - 2292.61) - 2202.84)})$$

boundADP/.ATP->.0001/.ADP->.01

-1.70499

```
boundADP/.ATP->.01/.ADP->.01
```

-1.99583

As the concentration of ATP is increased, F16BP will predominate more and more and this average number bound will approach -2. Note the binding of ADP is the same as the binding of ATP except for sign.

Index of Mathematica Programs

(Problem numbers are given, and Pkg indicates that the program is in the package BasicBiochemData2.)

Index